An Introduction to Distance Measurement in Astronomy

An Introduction to Distance Measurement in Astronomy

RICHARD DE GRIJS | 何锐思

Kavli Institute for Astronomy and Astrophysics,
Peking University, China
(中国北京大学科维理天文与天体物理研究所)

WILEY
A John Wiley & Sons, Ltd., Publication

This edition first published 2011

Registered office
John Wiley & Sons Ltd, The Atrium, Southern Gate, Chichester, West Sussex, PO19 8SQ, United Kingdom

For details of our global editorial offices, for customer services and for information about how to apply for permission to reuse the copyright material in this book please see our website at www.wiley.com.

Library of Congress Cataloging-in-Publication Data

De Grijs, Richard, 1969-
An introduction to distance measurement in astronomy / Richard de Grijs.
p. cm.
Includes bibliographical references and index.
ISBN 978-0-470-51179-4 (cloth) – ISBN 978-0-470-51180-0 (paper) – ISBN 978-1-119-97818-3 (ePDF) – ISBN 978-1-119-97817-6 (ebook) – ISBN 978-1-119-97980-7 (ePub) – ISBN 978-1-119-97981-4 (eMobi)
1. Cosmological distances–Measurement. 2. Astronomy–Methodology. I. Title.
QB991.C66D34 2011
522.87–dc23

2011014927

A catalogue record for this book is available from the British Library.

Print ISBN: 9780470511794 (H/B) 9780470511800 (P/B)
ePDF ISBN: 9781119978183
oBook ISBN: 9781119978176
ePub ISBN: 9781119979807
eMobi ISBN: 9781119979814
Set in 10/12 pt Times-Roman by Thomson Digital, Noida, India

Printed and bound in Singapore by Markono Print Media Pte. Ltd.

To 那洁 (Jie)

for her unconditional love and support throughout the years

Contents

Preface

Knowing the distance of an astrophysical object is key to understanding it: without an accurate distance, we do not know how bright it is, how large it is, or even (for great distances) when it existed. But astronomical distance measurements are difficult. Distances to stars were first measured in 1838 by Bessel, Struve and Henderson, and accurate distances to other galaxies – even the nearest – date only from the 1950s. This is not really surprising, since the only information we have about any object beyond our solar system is its position (perhaps as a function of time), its brightness (as a function of wavelength and time) and perhaps its radial velocity or chemical composition. Yet, from this unpromising starting point, modern astronomers have developed methods of measuring distances which can take us from the nearest star to the most distant galaxy, using techniques that vary from the mundane (the astronomical equivalent of the surveyor's theodolite) to the exotic (the bending of light in general relativity, wiggles in the spectrum of the cosmic microwave background). Nevertheless, the most accurate optical and near-infrared (near-IR) methods of distance determination, from the solar neighbourhood to the highest redshifts, in use today rely heavily on having access to accurate spectroscopy, supplemented by astrometric measurements in the Milky Way and slightly beyond.

In 1997, the *Hipparcos* space mission provided (for the first time) a significant number of absolute trigonometric parallaxes at milli-arcsecond-level precision across the whole sky, which had a major impact on all fields of astrophysics. In addition, during the past 10 years, the use of ground-based 8–10 m-class optical and near-IR telescopes (including the Keck Observatory, the Very Large Telescope, the twin Gemini telescopes and the Japanese Subaru telescope) and space observatories (such as the *Hubble Space Telescope*, the *Spitzer Space Telescope*, the *Chandra X-ray Observatory* and the European *XMM–Newton* satellite) have provided an unprecedented wealth of accurate photometric and spectroscopic data for stars and galaxies in the local Universe. Radio observations, particularly with the Very Large Baseline Array and the Japanese *VERA* (VLBI Exploration of Radio Astrometry, where VLBI stands for Very Long Baseline Interferometry) array, have achieved 10 micro-arsecond astrometric accuracy. Moreover, stellar models and numerical simulations are providing accurate predictions of a broad range of physical phenomena, which can now – in principle – be tested using accurate spectroscopic and astrometric observations (including measurements of e.g. spectral line ratios and shapes, spectral slopes, radial velocities and velocity dispersions). However, at present, comparisons of theory and observations are mainly hampered by precision (or lack thereof) in distance measurements/estimates.

This is a very exciting time for numerous fields relying on astronomical distance determinations. VLBI sensitivity is being expanded, allowing (for example) direct measurement of distances throughout the Milky Way and even to Local Group galaxies. The field will likely make a major push forward into the era of *Gaia*, optical interferometer and Extremely Large Telescope-driven science, which (for example) will allow us to determine Coma cluster distances without having to rely on secondary distance indicators, thus finally making the leap to accurate distance measurements well beyond the Local Group of galaxies.

In this book, we combine various aspects of distance determinations and, most importantly, the underlying physics enabling this (without being restrictive in areas where statistical and observational approaches are more relevant), from the solar neighbourhood to the edge of the Universe, exploring on the way the various methods employed to define the milestones along the road. We will emphasize recent advances made to further our physical insights. We aim to provide a snapshot of the field of distance measurement, offering not only up-to-date results and a cutting-edge account of recent progress but also full discussion of the pitfalls encountered and the uncertainties which remain. We aim to provide a roadmap for future efforts in this field, both theoretically and observationally. This book is aimed at senior undergraduate and postgraduate students, as well as researchers in the various fields touched upon by the plethora of techniques covered here. For that reason, we have tried to both explain basic physical concepts which may not necessarily be intuitively obvious and provide extensive referencing to the primary literature for follow-up reading and research.

Although our focus is on techniques of distance determination, this is intimately linked to many other aspects of astrophysics and cosmology. On our journey from the solar neighbourhood to the edge of the Universe, we shall encounter stars of all types, alone, in pairs and in clusters, their life cycles, and their explosive ends: binary stars, in particular, play an important role both in this context, e.g. in pinning down accurate distances to the Pleiades open cluster and Local Group galaxies, and in future ground- and space-based surveys (including *Gaia*, RAVE: the Radial Velocity Experiment, and others); the stellar content, dynamics and evolution of galaxies and groups of galaxies; the gravitational bending of starlight; and the expansion, geometry and history of the Universe. As a result, this book offers not only a comprehensive study of distance measurement but also a tour of many recent and exciting advances in astrophysics.

It has taken significant time and effort to collect and shape the contents of this book. Along the way, numerous people generously assisted or gave their time, answering my questions, providing me with feedback on earlier drafts of (parts of) chapters, keeping my imagination in check, and helping me put my thoughts (and the book's outline) in order. I would specifically like to express my gratitude to (in alphabetical order) Giuseppe Bono, Susan Cartwright, 范祖辉 (Zuhui Fan), Stefan Gillessen, Stephen Justham, Michael Merrifield, Brent Miszalski, Göran Östlin, Mike Reid, Stephen Smartt, Nial Tanvir, Max Tegmark, Floor van Leeuwen and 徐仁新 (Renxin Xu), as well as to my publishing contacts at Wiley, particularly Andy Slade, Jenny Cossham, John Peacock, Sarah Tilley and Janine Maer, for believing despite all odds that this project would eventually materialize. Finally, I acknowledge partial funding from the National Natural Science Foundation of China through grants 11043006 and 11073001.

Richard de Grijs
Beijing 北京,
February 2011

1

The Importance of Astrophysical Distance Measurements

When we try to pick out anything by itself, we find it hitched to everything else in the Universe.
– John Muir (1838–1914), American naturalist and explorer

Each problem that I solved became a rule, which served afterwards to solve other problems.
– René Descartes (1596–1650), French philosopher

Accurate distance measurements are of prime importance for our understanding of the fundamental properties of both the Universe as a whole and the large variety of astrophysical objects contained within it. But astronomical distance measurement is a challenging task: the first distance to another star was measured as recently as 1838, and accurate distances to other galaxies – even the nearest – date only to the 1950s, despite evidence of the existence of 'spiral nebulae' as early as Lord Rosse's observations in the mid-nineteenth century. This is not surprising, since the only information we have about any object beyond our solar system includes its position (perhaps as a function of time), its brightness (as a function of wavelength and time) and possibly its radial velocity and chemical composition.

While we can determine highly accurate distances to objects in our solar system using active radar measurements, once we leave the Sun's immediate environment, most distance measurements depend on inferred physical properties and are, therefore, fundamentally uncertain. Yet at the same time, accurate distance measurements on scales of galaxies and beyond are crucial to get a handle on even the most basic questions related to the age and size of the Universe as a whole as well as its future evolution. The primary approach to obtaining distance measurements at increasingly greater distances is by means of the so-called *distance ladder*, where – in its most simplistic form – each rung is calibrated using the rung immediately below it. It is, therefore, of paramount importance to reduce the

---

An Introduction to Distance Measurement in Astronomy, First Edition. Richard de Grijs.

statistical uncertainties inherent to measuring distances to even the nearest star clusters in our Milky Way, because these objects are the key benchmarks for calibrating the cosmic distance scale locally. In this book, we take the reader on a journey from the solar neighbourhood to the edge of the Universe, *en passant* discussing the range of applicable distance measurement methods at each stage. Modern astronomers have developed methods of measuring distances which vary from the mundane (the astronomical equivalent of the surveyor's theodolite) to the exotic, such as the bending of light in **general relativity**[1] or using wiggles in the spectrum of the **cosmic microwave background** (CMB).

Not only do we provide an up-to-date account of the progress made in a large number of subfields in astrophysics, in turn leading to improved distance estimates, but we also focus in particular on the physics underlying the sometimes surprising notion that all of these methods work remarkably well and give reasonably consistent results. In addition, we point out the pitfalls one encounters in all of these areas, and particularly emphasize the state of the art in each field: we discuss the impact of the remaining uncertainties on a complete understanding of the properties of the Universe at large.

Before embarking on providing detailed accounts of the variety of distance measurement methods in use, here we will first provide overviews of some of the wide-ranging issues that require accurate determinations of distances, with appropriate forward referencing to the relevant chapters in this book. We start by discussing the distance to the Galactic Centre (Section 1.1). We then proceed to discuss the long-standing, although largely historical controversy surrounding the distance to the Large Magellanic Cloud (LMC) (Section 1.2). Finally, in Section 1.3 we go beyond the nearest extragalactic yardsticks and offer our views on the state of the art in determining the 3D structure of large galaxy clusters and large-scale structure, at increasing **redshifts**.

1.1 The Distance to the Galactic Centre

The Galactic Centre hosts a dense, luminous star cluster with the compact, nonthermal radio source **Sagittarius** (Sgr) **A*** at its core. The position of the latter object coincides with the Galaxy's kinematic centre. It is most likely a massive black hole with a mass of $M_{\rm BH} \sim 4.4 \times 10^6$ $M_\odot$ (see the review of Genzel *et al.* 2010), which is – within the uncertainties – at rest with respect to the stellar motions in this region. The exact distance from the Sun to the Galactic Centre, R_0, serves as a benchmark for a variety of methods used for distance determination, both inside and beyond the Milky Way. Many parameters of Galactic objects, such as their distances, masses and luminosities, and even the Milky Way's mass and luminosity as a whole, are directly related to R_0. Most luminosity and many mass estimates scale as the square of the distance to a given object, while masses based on total densities or orbit modelling scale as distance cubed. This dependence sometimes involves adoption of a rotation model of the Milky Way, for which we also need to know the Sun's circular velocity with high accuracy. As the best estimate of R_0 is refined, so are the estimated distances, masses and luminosities of numerous Galactic and extragalactic objects, as well as our best estimates of the rate of Galactic rotation and size of the Milky Way. Conversely, if we could achieve a highly accurate *direct* distance determination to the Galactic Centre, this would allow reliable recalibration of the zero points of a range of secondary distance

[1] Terms and concepts which appear in the Glossary are rendered in boldface font at first occurrence in the text.

calibrators, including **Cepheid, RR Lyrae** and **Mira variable stars** (Sections 3.5.2, 3.5.5 and 3.5.3, respectively), thus reinforcing the validity of the extragalactic distance scale (cf. Olling 2007). In turn, this would enable better estimates of globular cluster (GC) ages, the **Hubble constant** – which relates a galaxy's recessional velocity to its distance, in the absence of '**peculiar motions**' (see Section 5.1) – and the age of the Universe, and place tighter constraints on a range of cosmological scenarios (cf. Reid *et al.* 2009b).

1.1.1 Early Determinations of R_0

The American astronomer, Harlow Shapley (1918a,b), armed with observations of GCs taken with the Mount Wilson 60-inch telescope (California, USA) since 1914, used the light curves of Cepheid variables and, hence, their **period–luminosity relation** to draw a map of the distribution of 69 GCs with respect to the Sun's position and the plane of the Milky Way (see Figure 1.1). He eventually extended this to include all 93 Galactic GCs known at the time. He concluded that the Sun was not located in or near the Galactic Centre – as previously deduced from star counts that were, in fact, heavily affected by interstellar **extinction** in the Galactic plane (e.g. Herschel 1785; Kapteyn 1922) – but at Galactic longitude $\ell \simeq 325°$ (in the direction of the constellation Sagittarius), at a distance of $\sim$13 – 25 kpc, i.e. significantly greater than the current best estimate of $8.28 \pm 0.15 \pm 0.29$ kpc, where the two errors represent the statistical and systematic uncertainties (Genzel *et al.* 2010; see also Reid 1993; Eisenhauer *et al.* 2003; Horrobin *et al.* 2004; Ghez *et al.* 2008;

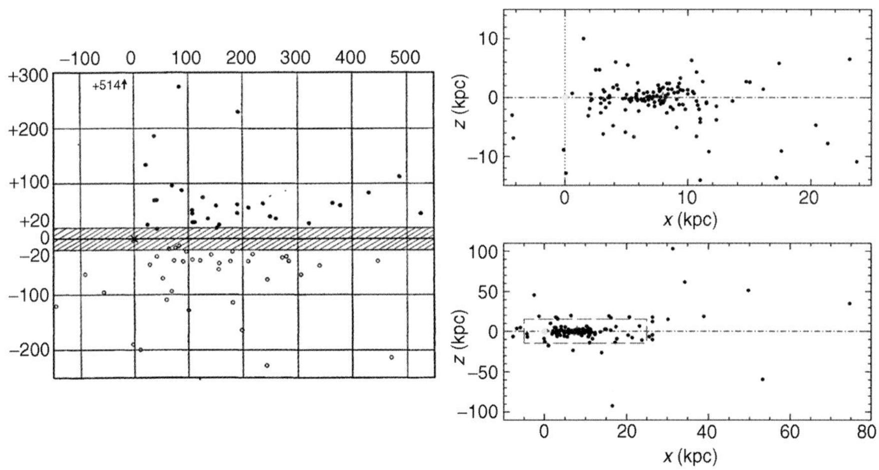

Figure 1.1 *(Left) Projection of the positions of globular clusters perpendicularly to the Galactic midplane (Shapley 1918a,b). Cross: position of the Sun. The unit of distance is 100* **parsec** *(pc). The position of the GC NGC 4147 is indicated by the arrow (outside the figure boundaries).* (Reprinted from H. Shapley and M. J. Reid, *Astrophysical Journal*, **48**, Studies based on the colors and magnitudes in stellar clusters. VII. The distances, distribution in space, and dimensions of 69 globular clusters, p. 154–181, Copyright 1918, with permission of the AAS.) *(Right) Up-to-date distribution of Galactic GCs (data collected by Harris 1996; 2010 edition). The Sun is located at the origin (indicated in yellow) and distances are based on RR Lyrae period–luminosity calibration (Section 3.5.5). The Galactic Centre is indicated by a red star. The blue dashed box in the bottom panel represents the area shown in the top panel.*

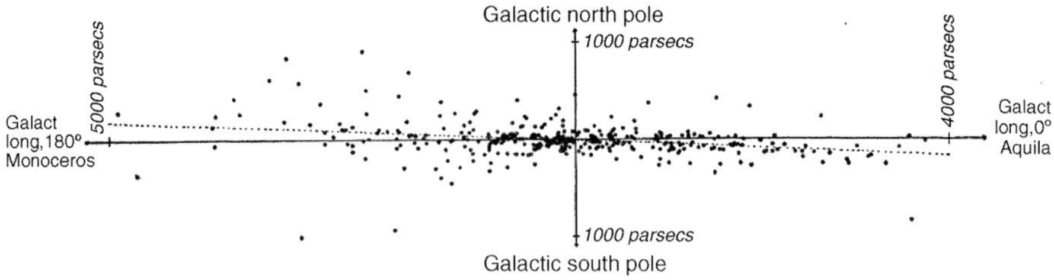

Figure 1.2 *Projection of Galactic open clusters on the same plane as in Figure 1.1 (Trumpler 1930). Dotted line: plane of symmetry of the open clusters. The Sun is located slightly to the left of the vertical axis, in the midst of a subset of open clusters shown as open circles (clusters within 1 kpc of the Sun).* (Reprinted from R. J. Trumpler, *Lick Observatory Bulletin*, **XIV**, Preliminary results on the distances, dimensions, and space distribution of open star clusters, p. 154–188, Copyright 1930, with permission of UC Regents/Lick Observatory.)

Gillessen *et al.* 2009a; Majaess *et al.* 2009). He also found that the distribution of GCs above and below the Galactic plane was approximately symmetrical, with no clusters seen closer than 1300 pc from the plane.

Although Shapley's lower limit of $R_0 \sim 13$ kpc is within a factor of 2 of the currently accepted value, his method of distance determination was affected by a number of partially compensating systematic errors. His Cepheid period–luminosity calibration was too faint by $\sim$1 mag, while he used 'Population II' Cepheids (**W Virginis stars**; see Section 3.5.4) instead of the type I Cepheids he thought he had observed. The former are generally some 2 mag fainter than the latter, leading to a distance scale that was $\approx$1 mag too bright and a distance overestimate by a factor of $\sim$1.6 (cf. Reid 1993).

By taking advantage of radial velocity and proper motion measurements tracing the differential rotation of stars in the solar neighbourhood – in the sense that stars closer to the Galactic Centre travelled faster than their counterparts at greater distances – the Dutch astronomer Jan Hendrik Oort (1927) established the centre of rotation about the Milky Way to within 2° of Shapley's estimate, at $\ell = 323°$. Note that this was at a much smaller distance, approximately 5.9 kpc, than Shapley's estimate. He adopted Lindblad's (1927) Galactic rotation model and assumed a circular velocity at the solar circle of $v_c = 272$ km s^{-1}, which is now known to be considerably greater than International Astronomical Union recommendation of $v_c = 220$ km s^{-1}.[2]

The discrepancy between Shapley's and Oort's distance estimates to the Galactic Centre was predominantly caused by interstellar extinction, which was largely unknown at the time until Robert J. Trumpler's discovery of the effects of interstellar dust grains in the 1930s.[3] In 1929, Trumpler, a Swiss–American astronomer based at Lick Observatory (California, USA), tried to use open star clusters to repeat what Shapley had done with the Milky Way's GC population. He knew that open clusters tended to lie in the disc of the Galaxy and reasoned that this was a reasonable way to clarify the disc's shape (see Figures 1.2 and 1.3).

[2] This would also have contributed to a smaller estimated distance to the Galactic Centre than that resulting from adoption of $R_0 = 8.0$–8.5 kpc, which is currently generally adopted and is in line with the International Astronomical Union recommendation of $R_0 = 8.0$ kpc. Note that Reid *et al.* (2009a) recently obtained a revised best-fitting Galactic rotation velocity of $v_c = 254 \pm 16$ km s^{-1} based on a Galaxy-wide survey of masing sources in high-mass star-forming regions.

[3] Although Trumpler is often credited with this discovery, its effects were first reported by Friedrich Georg Wilhelm von Struve in 1847.

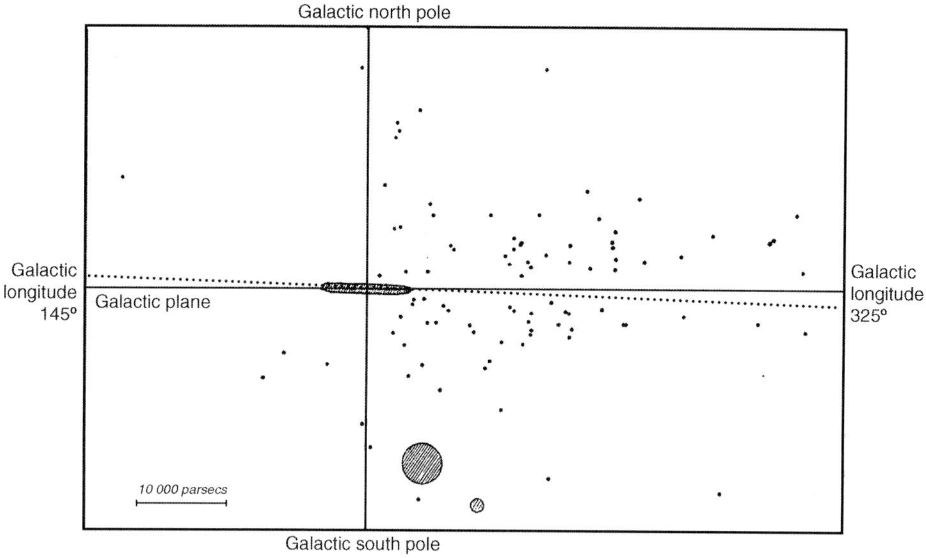

Figure 1.3 *Trumpler's (1930) view of the distribution of open and globular clusters in the Milky Way. Solid dots: GCs. Shaded area: open clusters (see Figure 1.2). Shaded circles: Magellanic Clouds.* (Reprinted from R. J. Trumpler, *Lick Observatory Bulletin*, **XIV**, Preliminary results on the distances, dimensions, and space distribution of open star clusters, p. 154–188, Copyright 1930, with permission of UC Regents/Lick Observatory.)

Determining the distances to his sample of open clusters was key. Trumpler devised two ways to achieve this. First, he used a version of the **main-sequence fitting technique** (see Section 3.2.1) to estimate distances, in essence relying on the unproven principle of 'faintness equals farness', which was unproven in the sense that the idea had not been shown to work reliably. In an alternative technique, he deduced that if all open clusters had approximately the same physical, linear size, then the more distant ones would have smaller angular sizes, a 'smallness equals farness' argument. When he compared the results of both methods, he found that the main-sequence fitting technique gave systematically larger distances.

Unknowingly, Trumpler had stumbled on the evidence that the space between the stars is not entirely empty. Before Trumpler, it was known that there were obvious dark clouds in the sky which blocked the light from behind, such as the Coalsack Nebula. However, Trumpler showed that such effects were not confined to distinct clouds, but to a general 'fogginess' of space (see also Section 6.1.1). Its effect on the main-sequence fitting technique is to add a term, A, to the **distance modulus** equation to make the shift of the apparent magnitudes of Trumpler's clusters larger than they would have been in the absence of absorption:

$$m_V - M_V = 5\log(d/\mathrm{pc}) - 5 + A_V, \tag{1.1}$$

where m_V and M_V are the apparent and absolute magnitudes in the optical V filter, and d is the distance sought. Meanwhile, Trumpler's distance to the Galactic Centre, properly corrected for the effects of extinction, was actually very close to the present-day value. Note that because interstellar dust is most concentrated in the Galactic plane, Shapley's experiment with GCs was not affected, at least not significantly, by the interference of absorption and scattering by dust. However, there is a clear lack of objects in his work in the direction of the Galactic midplane, where dust blocked his view and created the so-called 'zone of avoidance' for good GC targets.

Interestingly, Shapley commented that '... within 2000 parsecs of that plane there are only five [GCs], four of which are among the clusters nearest the sun.' Discussing the frequency distribution of his observed GCs as a function of distance from the Galactic plane, he notes that '[t]he completion of that curve, in a form naturally to be expected for the frequency of objects concentrated toward the Galaxy, would require at least 50 globular clusters within 1500 parsecs of the plane; there is, however, only one, Messier 22, ...' and '[h]ence we conclude that this great mid-galactic region, which is particularly rich in all types of stars, planetary nebulae, and open clusters, is unquestioningly a region unoccupied by globular clusters.'

1.1.2 Modern Results

Since the presence of interstellar dust severely hampers our view of the Galactic Centre, longer-wavelength (IR and radio) observations (see e.g. Figure 1.4) have been employed extensively to arrive at more accurate Galactic Centre distance estimates. Reid (1993) and Genzel *et al.* (2010) provide extensive reviews of the range of methods used, as well as their accuracy at the time of these publications. In this section, we focus on the *primary, direct* methods of distance determination to the Galactic Centre (the reader is referred to Reid 1993, Genzel *et al.* 2010, as well as the relevant chapters elsewhere in this book for alternative methods) and summarize the current state of the art in this field.

Following Reid (1993), we distinguish the variety of methods used to determine R_0 into primary, secondary and indirect measurements. Primary measurements determine R_0

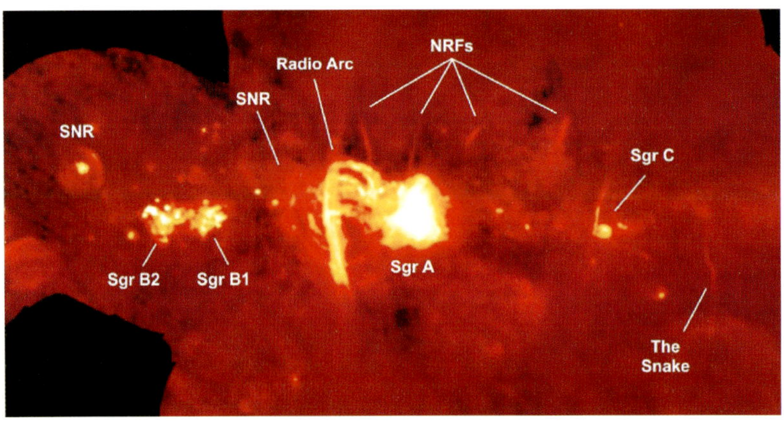

Figure 1.4 Combined radio image, covering a range of radio wavelengths, of the Galactic Centre region based on observations obtained with the Very Large Array and the Green Bank Telescope. The horizontal and vertical coordinates represent Galactic longitude and latitude, respectively. The linear filaments near the top are nonthermal radio filaments (NRFs). SNR: supernova remnant. Sgr: Sagittarius. (Reprinted from Yusef-Zadeh *et al.*, NRAO Press Release (Online), Origin of enigmatic Galactic Center filaments revealed, Copyright 2004, with permission of NRAO/AUI/NSF.)

directly, without having to rely on **standard candle** calibration (secondary methods) or a Galactic rotation model (one type of indirect method). The former include using proper motions and **trigonometric parallax** measurements of **masing** interstellar molecules (see also Section 3.7.4),[4] OH/IR stars – late-type stars that exhibit 1612 MHz OH maser emission following far-IR 'pumping' of the population levels – near the Galactic Centre, direct Keplerian orbit measurements and statistical estimates based on assuming equality of radial and tangential (i.e. isotropic) velocity dispersions of the Galactic Centre star cluster. Secondary measurements include Shapley's method of using the centroid of the GC distribution – which is, in essence, based on adoption of a suitable period–luminosity relation for variable stars and assumes that the GC population is symmetrically distributed with respect to the Galactic Centre (see Reid 1993 and Figure 1.1 for a more recent update) – and of other, presumably symmetrically distributed, bright objects, and calibration based on RR Lyrae and Mira period–luminosity relations (cf. Section 3.5). In addition to these methods, indirect methods rely on either Galactic rotation models, the **Eddington luminosity** of X-ray sources (e.g. Reid 1993 and references therein) or the planetary nebula luminosity function (e.g. Dopita *et al.* 1992; see Section 4.4), among other endeavours (see also Vanhollebeke *et al.* 2009, their Table 1).

1.1.2.1 Maser-Based Geometric Distances

The molecular material associated with massive stars at the time of starbirth is closely traced by water vapour. Population inversion of the H_2O energy levels – which refers to a configuration with higher occupancy of excited than the lower energy states – by ionization caused by collisional pumping of the rotational energy levels and other shock-related physical processes (e.g. Elitzur 1992; Lo 2005) by the intense radiation from these massive stars and subsequent coherent de-excitation causes 22 GHz masing 'spots' to appear in the dust-rich envelopes of asymptotic giant branch stars, with sizes of $\sim 10^{11}$ m and brightness temperatures[5] as high as 10^{12}–10^{15} K (cf. Reid 1993). These small sizes and high brightness temperatures render these objects ideal tracers for proper-motion measurements using **Very Long Baseline Interferometry** (VLBI) techniques (see also Section 3.7.4) because of the associated micro-arcsecond (μas) astrometric accuracy over fields of view of a few arcseconds in diameter.

Early geometric distance estimates to the Galactic Centre were based on the '**expanding cluster parallax**' method (equivalent to the **moving groups method**; see Section 2.1.3) applied to two H_2O masers in the dominant, high-mass star-forming region near the Galactic Centre, Sgr B2, resulting in distances of 7.1 and between 6 and 7 kpc (Reid 1993) for Sgr B2 North[6] and Middle, respectively, with a combined statistical and systematic uncertainty of ± 1.5 kpc (1σ). The accuracy attainable for distance determinations to H_2O maser sources is limited by (i) the motions of the individual spots (which exhibit random motions in all spatial coordinates of ≈ 15 km s^{-1}, compared to typical measurement uncertainties of a few km s^{-1}); (ii) their distribution around the exciting star: a nonuniform distribution, as observed for the Galactic Centre maser source Sgr B2 (North)

---

[4] 'Masers' (microwave amplification by stimulated emission of radiation) are the microwave analogues of optical and IR lasers.

[5] The brightness temperature is the temperature of a blackbody in thermal equilibrium with its surroundings which resembles the observed intensity distribution of a 'grey-body' object at a given frequency.

[6] Sgr B2 (North) is almost certainly located within 0.3 kpc of the Galactic Centre (Reid *et al.* 1988; Reid 1993; Snyder *et al.* 1994; Belloche *et al.* 2008). Reid *et al.* (2009b) derive a Galactocentric distance of 0.13 ± 0.06 kpc by adopting $R_0 \approx 8.0$ kpc and $v_{LSR,Sgr\,B2} \approx 62$ km s^{-1} for the velocity of Sgr B2 with respect to the local standard of rest and a low-eccentricity orbit.

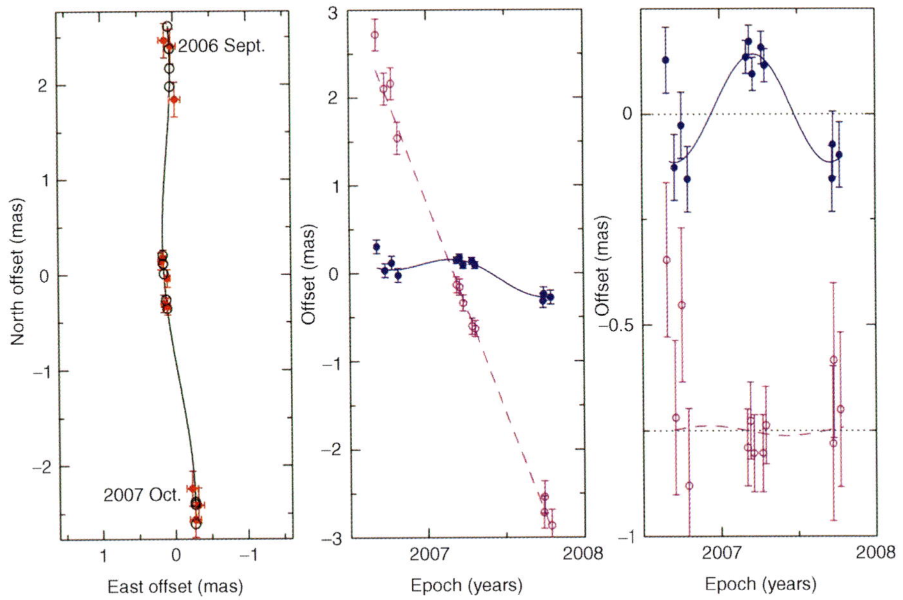

Figure 1.5 H_2O maser parallax and proper-motion data and fits for Sgr B2N (Reid et al. 2009b). (Left) Positions on the sky (red circles). The expected positions from the parallax and proper motion fit are indicated (black circles and solid line, respectively). (Middle) East (filled blue circles and solid line) and North (open magenta circles and dashed line) position offsets and best-fitting parallax and proper motions as a function of time. (Right) Same as the middle panel, except the best-fitting proper motion has been removed, allowing the effects of only the parallax to be seen. (Reprinted from M. J. Reid et al., *Astrophysical Journal*, **705**, A trigonometric parallax of Sgr B2, p. 1548–1553, Copyright 2009, with permission of the AAS.)

(Reid *et al.* 1988), combined with the requirement to determine the line-of-sight distance from the central star for each spot, results in correlations between the maser source's expansion speed and its distance (cf. Reid 1993); and (iii) tropospheric signal propagation delays after calibration (cf. Reid *et al.* 2009b).

Reid *et al.* (2009b) recently provided the first trigonometric parallax measurement for the Galactic Centre (see Figure 1.5), using H_2O maser astrometry with the Very Long Baseline Array (VLBA), the US VLBI network. Their measured parallax for Sgr B2 (North) and Sgr B2 (Middle) is 0.128 ± 0.015 and 0.130 ± 0.012 milli-arcseconds (mas), respectively, leading to a combined parallax for the Sgr B2 region of 0.129 ± 0.012 mas and, thus, $R_0 = 7.8^{+0.8}_{-0.7}$ kpc. Correcting for the small offset between Sgr B2 and Sgr A* (Sgr B2 is thought to be closer to the Sun than Sgr A*), they find $R_0 = 7.9^{+0.8}_{-0.7}$ kpc. Their associated measurement uncertainty, of order 10% for the first year's data, will further reduce with an increasing time baseline, as $\sigma_{R_0} \propto 1/\sqrt{N}$ for N similar yearly observations.

OH/IR stars, of which many are found close to the Galactic Centre, can also potentially be used for determination of R_0. To do so would require direct measurements of both the angular diameter of the OH maser shell[7] using radio interferometry and the light travel time

[7] SiO masers typically occur at radii of ~8 a.u. (astronomical units) for Mira variables, which corresponds to ~1 mas at $R_0 \approx 8$ kpc.

across the shell, based on the time lag between red- and blueshifted emission from the shell's far and near sides, respectively (cf. Schultz *et al.* 1978; Jewell *et al.* 1980). However, VLBI observations have shown that the angular sizes of OH/IR shells near the Galactic Centre are strongly affected by scattering off electrons in the interstellar medium (van Langevelde and Diamond 1991; Frail *et al.* 1994; Lazio *et al.* 1999), hence preventing measurements of intrinsic shell sizes and, thus, a direct determination of R_0. The scattering scales with wavelength as λ^2 (Lo *et al.* 1981), so that only the highest-frequency (> 20 GHz) masers are potentially suitable for precision astrometry.

OH/IR stars often also host 22 GHz H_2O and/or 43 GHz ($J = 1 - 0$) silicon oxide (SiO) masers in their circumstellar shells (e.g. Habing 1996 and references therein). Although H_2O masers are highly variable, SiO masers are more stable. The latter can, therefore, potentially be used for astrometry in the Galactic Centre region (e.g. Menten *et al.* 1997; Sjouwerman *et al.* 1998, 2002; Reid *et al.* 2003). However, relatively few SiO masers are known to be associated with OH/IR stars near the Galactic Centre (Lindqvist *et al.* 1991; Sjouwerman *et al.* 2002), which has triggered searches for 43 GHz masers in other types of mid-IR sources with colours typical of circumstellar envelopes and in blind surveys of the Galactic Centre (see Sjouwerman *et al.* 2004 for a review). SiO masers at 43 GHz or even at 86 GHz ($J = 2 - 1$) should be readily observable in the much more numerous Mira, long-period and semi-regular variables, as well as red supergiant stars (e.g. Habing 1996; Messineo *et al.* 2002, 2004; Sjouwerman *et al.* 2004; and references therein). This potentially offers an independent confirmation of distances determined based on period–luminosity analysis. The latter are also affected by numerous systematic uncertainties, such as an ambiguous **extinction law**, a bias for smaller values of R_0 because of preferential sampling of variable stars towards the near side of the bulge owing to extinction, and an uncertainty in characterizing how a mean distance to the group of variable stars relates to R_0 (cf. Gould *et al.* 2001; Udalski 2003; Ruffle *et al.* 2004; Kunder and Chaboyer 2008; Majaess 2010; see also Chapter 6).

Current best estimates of R_0 using secondary distance indicators include 8.24 ± 0.08 (statistical) $\pm$ 0.42 (systematic) kpc based on Mira variables (Matsunaga *et al.* 2009; but see Groenewegen and Blommaert 2005: $R_0 = 8.8 \pm 0.4$ kpc), 7.7 ± 0.4 kpc for RR Lyrae based on statistical-parallax solutions (Dambis 2010) versus 8.1 ± 0.6 kpc using RR Lyrae observed as part of the Optical Gravitational Lensing Experiment (Majaess 2010), 7.9 ± 0.3 kpc based on δ **Scuti** stars (McNamara *et al.* 2000; see Section 3.5.6) and 7.8 ± 0.6 kpc using Cepheids (Majaess *et al.* 2009). Vanhollebeke *et al.* (2009) considered both the full 3D stellar population mixture in the Galactic bulge, including its **metallicity** distribution, and the red clump stars alone, and concluded that $R_0 = 8.7^{+0.57}_{-0.42}$ kpc. Their large distance disagrees, however, with the recent Babusiaux and Gilmore (2005) and Nishiyama *et al.* (2006) distance determinations based on near-IR data for the red clump (cf. Section 3.2.2).

1.1.2.2 Orbital Modelling

Careful analysis of the stellar motions in the inner regions of the Milky Way can potentially result in a distance estimate to the Galactic Centre with significantly reduced uncertainties. Genzel *et al.* (2000) derived a primary distance (statistical parallax; see Section 2.1.3) of 8.0 ± 0.9 kpc (1σ) based on a statistical comparison of proper motions and line-of-sight velocities of stars in the central 0.5 pc, updated to $R_0 = 7.2 \pm 0.9$ kpc by Eisenhauer *et al.* (2003) and subsequently to $R_0 = 8.07 \pm 0.32 \pm 0.12$ kpc (statistical and systematic uncertainties, respectively) by Trippe *et al.* (2008). Diffraction-limited near-IR observations

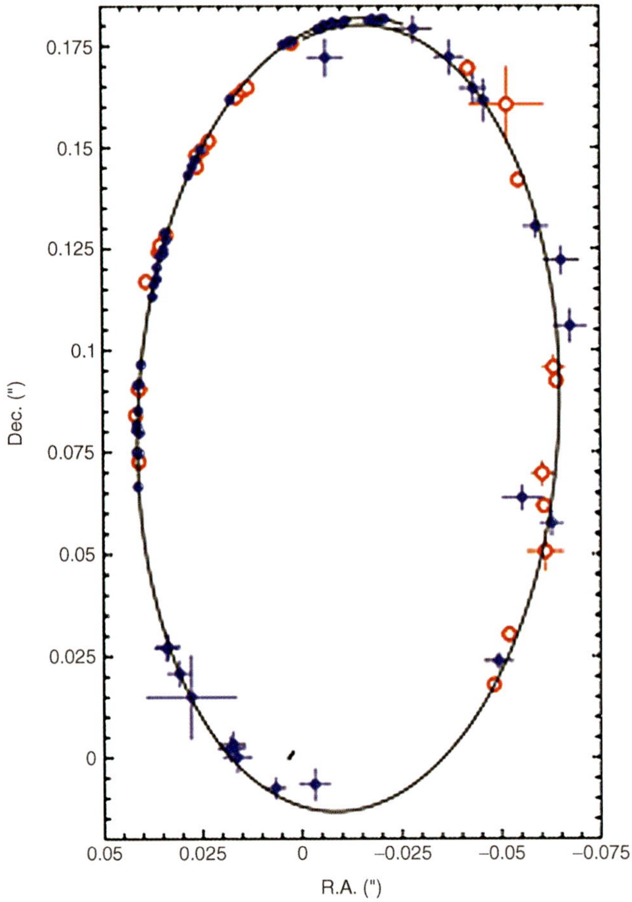

Figure 1.6 *Orbital fit for the Galactic Centre star S2 (Gillessen et al. 2009b). Blue: New Technology Telescope/Very Large Telescope (European Southern Observatory) measurements. Red: Keck telescope measurements. Black line: Keplerian fit. R.A., Dec.: right ascension, declination.* (Reprinted from S. Gillessen, *et al.*, *Astrophysical Journal*, **707**, The orbit of the star S2 around Sgr A* from Very Large Telescope and Keck data, L114–L117, Copyright 2009, with permission of the AAS and S. Gillessen.)

of the Galactic Centre reveal ≈100 S stars[8] within 1″ of Sgr A*. The positions of the brightest sources can be measured to astrometric accuracies of 200–300 μas (limited by crowding effects) using K-band adaptive-optics observations (Ghez *et al.* 2008; Fritz *et al.* 2010), while radial velocities with a precision of $\sim$15 km s^{-1} for the brightest early-type stars are typical (based on adaptive optics-assisted integral-field spectroscopy), decreasing to $\sim$50–100 km s^{-1} for fainter objects. The combined data set has enabled determination of

[8] Eckart and Genzel (1996) identified the remarkably fast-moving stars in the 'Sgr A* cluster' which were known at that time by 'S' followed by a number. The number of S stars has since grown to more than 200 (Ghez *et al.* 2008; Gillessen *et al.* 2009b; but note that both groups use different nomenclature).

the orbits of some 30 stars (Eisenhauer *et al.* 2003, 2005; Ghez *et al.* 2005, 2008; Gillessen *et al.* 2009b).

S2, the brightest of the S stars in the Galactic Centre, has completed a full revolution around Sgr A* since high-resolution astrometric observations first became possible in 1992. It has an orbital period of 15.9 years and traces a highly elliptical Keplerian orbit with an orbital semi-major axis of 125 mas (see Figure 1.6; Schödel *et al.* 2002; Ghez *et al.* 2008; Gillessen *et al.* 2009a,b). Using a '**dynamical parallax**' approach (see Section 2.2) allows estimation of R_0 (cf. Salim and Gould 1999), of which the accuracy is currently limited by systematic uncertainties: $R_0 = 8.28 \pm 0.15$ (statistical) $\pm$ 0.30 (systematic) kpc (Gillessen *et al.* 2009a). This, in turn, constrains the black hole mass contained within its orbit to $M_{BH} =$ [4.30 $\pm$ 0.06 (statistical) $\pm$ 0.35 (due to uncertainties in R_0)] $\times$ 10^6 $M_\odot$. A similar black hole mass of $M_{BH} = (4.5 \pm 0.4) \times 10^6$ $M_\odot$ was derived independently by Ghez *et al.* (2008), for $R_0 = 8.4 \pm 0.4$ kpc (see also the review of Genzel *et al.* 2010).

In principle, to fully solve the equations governing two masses orbiting each other requires determination of six phase-space coordinates for each mass, as well as the two masses (e.g. Salim and Gould 1999). However, given the observational and systematic uncertainties, the mass of the star ($m_{S2}/m_{\mathrm{Sgr\ A^*}} \sim 5 \times 10^{-6}$) and the three velocity components of Sgr A* can be neglected without accuracy penalties (Eisenhauer *et al.* 2003), provided that it is at rest with respect to the stellar cluster at the Galactic Centre. In addition, after subtraction of the motions of the Earth and the Sun around the Galactic Centre, the proper motion of Sgr A* is -7.2 ± 8.5 and -0.4 ± 0.9 km s^{-1} in the plane of the Milky Way in the direction of the rotation and towards the Galactic pole, respectively (Reid and Brunthaler 2004, with updates by Reid *et al.* 2009a; see also Backer and Sramek 1999; Reid *et al.* 1999, 2003), so that the velocity of Sgr A* is <1% of that of S2. Thus, the current best estimate of the distance to the Galactic Centre has an associated combined uncertainty of $\pm$0.34 kpc.

1.2 The Distance to the Large Magellanic Cloud

The Magellanic Clouds, and in particular the Large Magellanic Cloud, represent the first rung on the extragalactic distance ladder. The galaxy hosts statistically large samples of potential 'standard candles' (objects with the same absolute magnitude), including many types of variable stars. They are all conveniently located at roughly the same distance – although for detailed distance calibration the LMC's line-of-sight depth and 3D morphology must also be taken into account – and relatively unaffected by foreground extinction. The LMC's unique location allows us to compare and, thus, cross-correlate and calibrate a variety of largely independent distance indicators, which can, in turn, be applied to more distant targets. The distance to the LMC has played an important role in constraining the value of the Hubble constant, H_0, the single most important parameter for determining the age and size of the Universe and (with CMB fluctuations) the amount of dark matter and the **equation of state** of **dark energy**, i.e. the ratio of the dark energy's pressure and density. The *Hubble Space Telescope (HST)* **Key Project** estimated $H_0 = 72 \pm 3$ (statistical) $\pm$ 7 (systematic) km s^{-1} Mpc^{-1} (Freedman *et al.* 2001). Most notably, the ~10% systematic uncertainty is predominantly driven by the uncertainty in the assumed distance to the LMC (Reid *et al.* 2009b). Closer to home, proper-motion measurements of objects in the LMC are

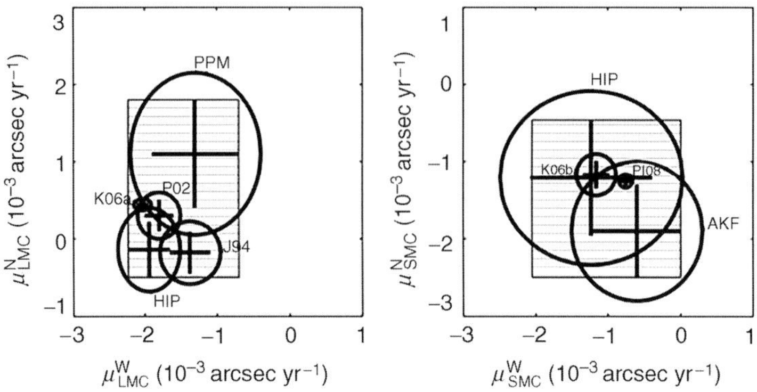

Figure 1.7 *Two-dimensional projections of the proper motions (North, West) for both the LMC and the SMC (Růžička et al. 2009). K06a,b: Kallivayalil et al. (2006a,b). PI08: Piatek et al. (2008). J94: Jones et al. (1994). PPM: Kroupa et al. (1994). HIP: Kroupa and Bastian (1997). P02: Pedreros et al. (2002). AKF: combination of Freire et al. (2003) and Anderson and King (2004a,b). The ellipses show the 68.3% confidence regions.* (Reprinted from A. Růžička et al., *Astrophysical Journal*, **691**, Spatial motion of the Magellanic Clouds: tidal models ruled out?, p. 1807–1815, Copyright 2009, with permission of the AAS and A. Růžička.)

now coming within reach (cf. Gardiner and Noguchi 1996; Kallivayalil *et al.* 2006a,b; Piatek *et al.* 2008; Costa *et al.* 2009), with major progress in this area expected from precision astrometric measurements in the *Gaia* era (see Section 2.1.2). A reliable distance estimate to the LMC is of crucial importance to assess future evolution scenarios of the Milky Way–LMC–Small Magellanic Cloud (SMC) system in the context of the Local Group of galaxies (e.g. Kallivayalil *et al.* 2006a,b, 2009; Besla *et al.* 2007; Bekki 2008; Růžička *et al.* 2009; and references therein; see also Figure 1.7).

It has become common practice to quote the distance to the LMC as a reddening-corrected distance modulus, $(m - M)_0$. Most modern determinations cluster around $(m - M)_0 = 18.5$ mag (e.g. Schaefer 2008; Szewczyk *et al.* 2008; and references therein). This was the value eventually settled on by the *HST* Key Project (Freedman *et al.* 2001; see also Section 4.1), $(m - M)_0 = 18.50 \pm 0.10$ mag ($50.1^{+1.4}_{-1.2}$ kpc; cf. Freedman and Madore 1991), and is currently considered the consensus distance modulus. Freedman *et al.* (2001) used a revised calibration of the Cepheid period–luminosity relation based on the maser-based distance to NGC 4258 (see Section 3.7.4) as well as several secondary distance measurement techniques – including Cepheids, RR Lyrae, Mira and eclipsing variables, the tip of the red giant branch (TRGB) as a standard candle, calibration of the red clump and **supernova** (SN) 1987A **light echoes** (see, respectively, Sections 3.5.2, 3.5.5, 3.5.3, 3.7.3, 3.3.1, 3.2.2 and 3.7.2) over the range from approximately 60 to 400 Mpc – to estimate the distance to the LMC. Many articles have focussed on obtaining a reliable distance to the LMC (see, for recent compilations, Westerlund 1997; Cole 1998; Gibson 2000; Freedman *et al.* 2001; Benedict *et al.* 2002; Clementini *et al.* 2003; Tammann *et al.* 2003; Walker 2003; Alves 2004; Schaefer 2008) using a range of methods, each of which has, in turn, been calibrated based on numerous independent techniques. For instance, calibration of the most often used Cepheid period–luminosity relation is commonly achieved using the surface brightness/**Baade–Wesselink method**, main-sequence fitting based on Galactic

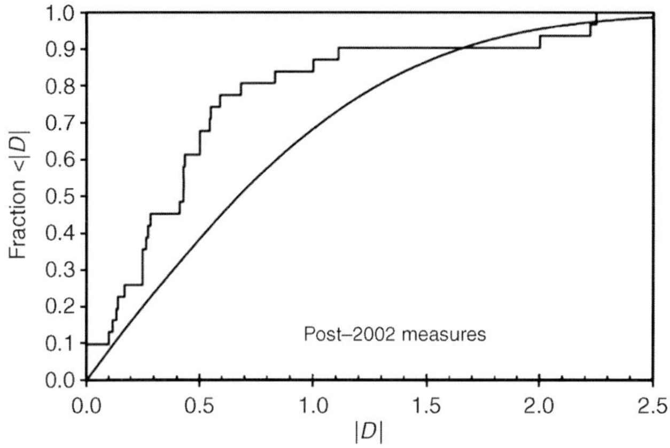

Figure 1.8 *Cumulative distributions of |D| from observations and for Gaussian errors (Schaefer 2008), where $D = (\mu - 18.50)/\sigma$, μ is the distance modulus and σ the observational uncertainty. The Kolmogorov–Smirnov test is a comparison between the cumulative distributions of |D| from observations (stepped curve) and the model (smooth curve). If the published values of the LMC distance modulus are unbiased and have correctly reported error bars, the two curves should lie relatively close together. If all but a few of the 31 post-2002 values included are too tightly clustered about the HST Key Project value of $\mu = 18.50$ mag, the observed curve should step high above the model curve. The maximum deviation between the two curves is 0.33 at |D| = 0.59, which is very unlikely if the published data report unbiased values with correct error bars.* (Reprinted from B. E. Schaefer, *Astronomical Journal*, **135**, A problem with the clustering of recent measures of the distance to the Large Magellanic Cloud, p. 112–119, Copyright 2008, with permission of the AAS and B. E. Schaefer.)

open cluster and/or GC colour–magnitude diagrams, nonlinear pulsation modelling, and *Hipparcos* and *HST* parallaxes (see, respectively, Sections 3.5.1, 3.2.1, 3.5.5 and 2.1.2).

Although the published LMC distance moduli before Freedman *et al.*'s 2001 article covered the range from ~18.1 to 18.8 mag, corresponding to distances from 42 to 58 kpc,[9] with much smaller individual error bars than the overall spread of the values, the wide scatter suddenly disappeared after the results of the *HST* Key Project were published, with a 'true' distance modulus of 18.50 ± 0.02 mag implied by the 14 measurements published between 2001 and 2004 (Alves 2004; see also Schaefer 2008). Schaefer (2008) notes that this situation, in which most methods were originally dominated by large, mostly unrecognized systematic errors, which then essentially disappeared overnight, is disturbing. (The same is not seen for the smaller number of distance determinations to the SMC, which might imply that the LMC effect is caused by 'sociological' or 'bandwagon' behaviour, also known as 'publication bias'. The SMC was not included in the *HST* Key Project.) He argues that all 14 $(m - M)_0$ values published between 2001 and 2004 are *too* consistent with the *HST* Key Project's result: $(m - M)_0 = 18.50$ mag falls within the 1σ uncertainty for all 14 determinations, corresponding to an improbably low χ^2 statistical probability of 0.0022 (Schaefer 2008; see Figure 1.8).

[9] Historically, there has been vigorous debate supporting a 'long' versus 'short' distance modulus: see Fouqué *et al.* (2007) and Sandage *et al.* (2009) for recent results in favour of the 'short' distance scale, often resulting from application of statistical parallaxes and the Baade–Wesselink method (see also de Vaucouleurs 1993a,b; Gratton *et al.* 1997; Clementini *et al.* 2003). However, this has largely disappeared after publication of Freedman *et al.* (2001).

In fact, Schaefer (2008) further extended his analysis of LMC distance moduli published after 2002 – including a total of 31 independent measurements, without substantial overlap of targets or correlations between publications – and concluded that there is a clear statistical overabundance of determinations that agree with the *HST* Key Project to much greater accuracy than the quoted error bars: a Kolmogorov–Smirnov test (a nonparametric test that allows statistical comparison of two one-dimensional distributions; Press *et al.* 1992) proves that the distribution of the published distance moduli deviates from the expected Gaussian profile at the $>3\sigma$ level. This calls into serious doubt the reliability of LMC distance moduli determined since 2002, because there are only two ways in which such a statistical condition can be met, either by artificially adjusting or selecting the published values to be near $(m - M)_0 = 18.50$ mag or by systematically overestimating the error bars (which is unlikely; Schaefer 2008). Clearly, this is a very unfortunate situation, given that the distance to the LMC is a crucial step towards the calibration of extragalactic distances! To remedy this situation, a comprehensive and independent recalibration, including realistic error bars, of the current best data set of reliable distance indicators seems unavoidable. Alternatively, new maser- or **eclipsing binary**-based *direct* methods of distance determination may provide an independent means of calibrating the first rung of the extragalactic distance ladder (cf. Herrnstein *et al.* 1999; Macri *et al.* 2006; Di Benedetto 2008; Pietrzyński *et al.* 2009; see also Sections 3.7.4, and 1.3 and 3.7.3, respectively).

An interesting alternative is offered by the coming online of large-scale near-IR survey capabilities with access to the Magellanic Clouds, which will essentially eliminate the effects of reddening and provide an independent and highly reliable calibration approach (e.g. Nemec *et al.* 1994; Bono 2003; Szewczyk *et al.* 2008; and references therein). Although efforts are continuing to further refine the LMC distance based on near-IR observations (see Table 1.1 for an update since Schaefer 2008), large-scale surveys such as the *VISTA* near-IR

Table 1.1 *Published LMC distance determinations since Schaefer (2008)*

Date	Article	$(m - M)_0$ (mag)	(Opt./NIR) Method
Aug. 2007	van Leeuwen *et al.*	18.39 ± 0.05	(Opt.) Cepheids
Jan. 2008	Clement *et al.*	18.49 ± 0.11	(Opt.) RR Lyrae
Mar. 2008	Sollima *et al.*	18.56 ± 0.13	(NIR) RR Lyrae[a]
Apr. 2008	Catelan and Cortés	18.44 ± 0.11	(Opt.) RR Lyrae
Jun. 2008	Ngeow and Kanbur	18.48 ± 0.03	(NIR) Cepheids[b]
		18.49 ± 0.04	(NIR) Cepheids[b]
Jul. 2008	Szewczyk *et al.*	18.58 ± 0.03 (stat.) ± 0.11 (syst.)	(NIR) RR Lyrae
Nov. 2008	Di Benedetto	18.559 ± 0.003 (stat.) ± 0.026 (syst.)	(Opt.) Cepheids
May 2009	Pietrzyński *et al.*	18.50 ± 0.55	Eclipsing binary
Jun. 2009	Dambis	18.27 ± 0.08	(NIR) RR Lyrae
Jul. 2009	Koerwer	18.54 ± 0.06	(NIR) Red clump
Aug. 2009	Borissova *et al.*	18.53 ± 0.13	(NIR) RR Lyrae
		18.46 ± 0.07	(NIR) Red clump
Aug. 2009	Matsunaga *et al.*	18.46 ± 0.10	(NIR) Cepheids
Jun. 2010	Reid and Parker	18.46 ± 0.2	Planetary nebulae

[a] To Reticulum.
[b] Using a linear and a broken period–luminosity relation, respectively.

YJK_S survey of the Magellanic system (Cioni *et al.* 2008, 2011) hold the promise of finally reducing the systematic uncertainties and settling the distance to the LMC conclusively, with remaining uncertainties in the distance modulus of $\ll 0.1$ mag.

1.3 Benchmarks Beyond the Magellanic Clouds: the 3D Universe on Large(r) Scales

Beyond the Magellanic Clouds, the next logical object for distance benchmarking is the Andromeda galaxy (M31), the other large spiral galaxy – in addition to the Milky Way – in the Local Group[10] (see also Brunthaler *et al.* 2005 for a case in favour of M33 as distance anchor, although see footnote 11). Once its distance is known to sufficient accuracy, all of its various stellar populations are available as potential standard candles. M31 is a potentially crucial rung on the extragalactic distance ladder (Clementini *et al.* 2001; Vilardell *et al.* 2006, 2010). First, its distance is sufficiently large, 744 ± 33 kpc or $(m - M)_0 = 24.36 \pm 0.08$ mag (Vilardell *et al.* 2010: direct estimate based on two eclipsing binary systems), that poorly constrained geometry effects do not cause additional significant systematic uncertainties, as for the Magellanic Clouds. Second, individual stars suitable for calibration of extragalactic distances (Cepheid or RR Lyrae variables, eclipsing binaries, **novae** and SNe, as well as GCs, for instance) can be observed fairly easily and are affected by only moderate extinction and reddening, with a **colour excess** $E(B - V) \equiv A_B - A_V = 0.16 \pm 0.01$ mag (Massey *et al.* 1995). Finally, as a mid-type spiral galaxy, it has a chemical composition and morphology similar to that of the Milky Way and other galaxies commonly used for distance determination (e.g. Freedman *et al.* 2001) and it can be used for absolute local calibration of the **Tully–Fisher relation**, one of the commonly used distance indicators to more distant spiral galaxies (see Section 4.5).

The compilation of published distance estimates of Vilardell *et al.* (2006) shows that most methods return best estimates between $(m - M)_0 = 24.0$ and 24.5 mag, with the majority of recent measurements tending towards the greater-distance end of this range. For instance, Holland (1998), Stanek and Garnavich (1998), Durrell *et al.* (2001), Joshi *et al.* (2003, 2010), Brown *et al.* (2004), McConnachie *et al.* (2005), Clementini *et al.* (2009) and Sarajedini *et al.* (2009) all reported $(m - M)_0 \in [24.46, 24.52]$ mag – based on analysis of tracers as diverse as the red giant branch (Section 3.3, particularly Section 3.3.2), red clump (Section 3.2.2), Cepheids (Section 3.5.2), RR Lyrae (Section 3.5.5) and the TRGB (Section 3.3.1) – with uncertainties of generally $\Delta(m - M)_0 < 0.10$ mag, although the type I and II Cepheid-based distances reported by Vilardell *et al.* (2007) and Majaess *et al.* (2010) are somewhat shorter. This situation is reminiscent of that of the LMC, in the sense that the distribution is narrower than the expected Gaussian profile. Hence, exercise of caution is needed. The direct, eclipsing binary-based distance determinations to M31 (Ribas *et al.* 2005; Vilardell *et al.* 2010; see also Bonanos *et al.* 2003) agree very well with independent Cepheid distances (e.g. Vilardell *et al.* 2007; see also Vilardell *et al.* 2006, their Table 1). In turn, these are based on either an eclipsing binary calibration of the LMC

[10] The Local Group is a loose galaxy association with a core radius of order 1 Mpc centred on the Milky Way–M31 **barycentre**. Its member galaxies are characterized by velocities which are close to the velocity–position relations satisfied by most known Local Group members.

distance (e.g. Fitzpatrick *et al.* 2003) or a maser-based distance determination to NGC 4258 (Macri *et al.* 2006).[11] The latter object has been suggested as an alternative yet highly robust benchmark for anchoring of the local distance calibration (Riess *et al.* 2009a). Prospects of 1% level *direct, geometric* distance determinations to M31 and M33 look promising, e.g. by employing time-delayed, dust-scattered X-ray haloes (Section 7.3; see for application to M31, Draine and Bond 2004) or – with significantly improved astrometric precision (see Section 2.1.2) – using novel '**rotational parallax**' measurements (Peterson and Shao 1997; Olling and Peterson 2000; Brunthaler *et al.* 2005; Olling 2007) combined with galactic velocity fields (see also Gould 2000) and, thus, improved distance precision (cf. Shaya and Olling 2009). The latter are the extragalactic equivalent to the 'orbital parallax' method for resolved binary systems, where radial velocities and proper motions of visual binaries are combined to derive the orbital parameters as well as the distance (e.g. Armstrong *et al.* 1992; Davis *et al.* 2005).

The importance of accurate distance determinations cannot be overstated. Despite significant efforts and worldwide coordination, even for the nearest objects the field is not free from controversy. For instance, the long-standing distance determination to the Orion Nebula – in particular to the high-mass star-forming Becklin–Neugebauer/Kleinmann–Low (KL) region – of 480 ± 80 pc (Genzel *et al.* 1981), which was based on VLBI observations of 22 GHz H_2O maser features, was recently significantly revised to $d = 389^{+24}_{-21}$ pc (see Figure 1.9; Sandstrom *et al.* 2007; and review of previous determinations therein). Sandstrom *et al.* (2007) employed 15 GHz VLBA radio-continuum observations, which yielded the parallax (see also Bertout *et al.* 1999 for *Hipparcos*-based results) and proper motion of the flaring, nonthermal radio star GMR A in the Orion Nebula Cluster (see also Hirota *et al.* 2007; Jeffries 2007; Kraus *et al.* 2007). Similarly, Menten *et al.* (2007) used VLBA radio-continuum observations at 8.4 GHz to determine a trigonometric parallax of several member stars of the Orion Nebula Cluster which exhibited nonthermal radio emission. They concluded that $d = 414 \pm 7$ pc, in agreement with the results of both Kraus *et al.* (2007) and Kim *et al.* (2008). The latter were based on orbital solution modelling of the θ^1 Orionis C close binary system and parallactic SiO maser observations of the Orion-KL region using VLBI, respectively. These more modern parallactic determinations are fully model independent, as opposed to Genzel *et al.*'s measurement, which required assumptions about the distribution of the masers and application of an expanding, thick-shell model. A 10% shorter distance than previously adopted results in 10% lower masses, 20% fainter luminosities and 20–30% younger ages for the stars in the Orion Nebula region.

Accurate distances are clearly also important to determine the physical properties of the stars, star clusters, peculiar objects (such as ultraluminous X-ray sources) and gas clouds in galaxies beyond the Local Group, and to assess their structure and internal as well as external dynamics (for the latter, see e.g. de Grijs and Robertson 2006). The recent controversy surrounding the Antennae interacting galaxies (NGC 4038/4039) provides an illustrative example. New *HST* observations of the TRGB in this system seemed to imply a significantly shorter distance (from $d \sim 20$ to 13.3 ± 1.0 Mpc; Saviane *et al.* 2008) than previously adopted based on a careful assessment of the system's

[11] For the third largest Local Group galaxy M33, the Triangulum galaxy, a systematic discrepancy remains between the eclipsing binary distance of Bonanos *et al.* (2006), the Cepheid determination of Freedman *et al.* (2001) and the maser-based distance of Brunthaler *et al.* (2005), in the sense that the eclipsing binary determination places the galaxy 0.3 mag more distant than the other methods.

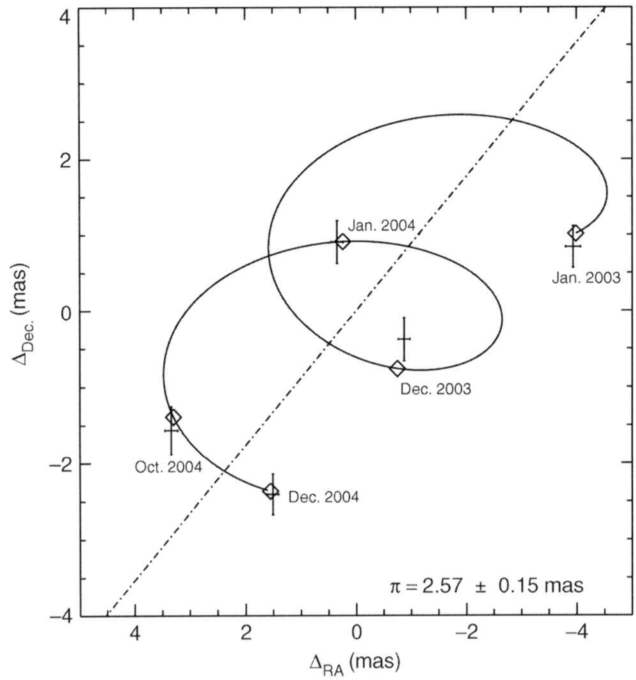

Figure 1.9 *Measured positions (diamonds) of the flaring, nonthermal radio star GMR A in the Orion Nebula Cluster with the best-fitting parallax and proper motion (Sandstrom et al. 2007). The dot-dashed line is the proper motion of the source, with the parallactic motion subtracted. The trigonometric parallax (π) corresponds to a distance of 389^{+24}_{-21} pc.* (Reprinted from K. M. Sandstrom *et al.*, *Astrophysical Journal*, **667**, A parallactic distance of 389 parsecs to the Orion Nebula Cluster from Very Long Baseline Array observations, p. 1161–1169, Copyright 2007, with permission of the AAS and K. M. Sandstrom.)

recession velocity, adoption of a reasonable value for H_0 and including proper corrections for the local **Hubble flow** (see Section 5.1). However, Schweizer *et al.* (2008) pointed out that not only would the Antennae system's size, mass and luminosity – as well as the equivalent properties of the galaxies' constituents – reduce, but its heliocentric velocity would also deviate by close to 3σ from the best large-scale flow model were this shorter distance adopted (see Figure 1.10). Using observations of the Type Ia SN 2007sr, an excellent standard candle[12] (see Section 5.2.1), they derive an independent distance estimate to the interacting system of $d = 22.3 \pm 2.8$ kpc. They suggest that Saviane *et al.*'s (2008) shorter distance determination may have had its origin in a misidentification of the TRGB. Schweizer *et al.* report a preliminary $d_{\rm TRGB} = 20.0 \pm 1.6$ Mpc, based on a reanalysis of the same *HST* data.

Beyond the nearest, well-resolved galaxies, the tool of choice for distance determinations has traditionally been the use of galaxies' recessional velocities and, hence, redshifts.

[12] If Saviane *et al.*'s (2008) distance were correct, SN 2007sr's peak luminosity would differ by $\sim 7\sigma$ (or 1.1 mag) from the mean peak luminosity of SNe Ia.

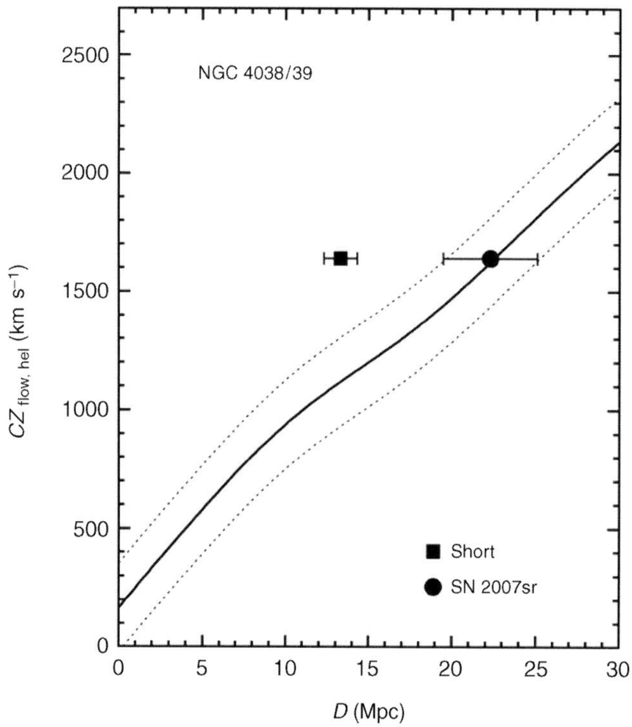

Figure 1.10 *Comparison of distances (D) measured for NGC 4038/39 (points) with a relevant large-scale flow model (solid line). cz_{hel}: heliocentric recession velocity. Plotted at the new distance based on Type Ia SN 2007sr (filled dot), the recession velocity of NGC 4038/39 falls well within $1\sigma_{th}$ (dotted lines) of the large-scale flow, where σ_{th} is the cosmic random radial velocity. In contrast, when plotted at the short distance based on the TRGB (square, Saviane et al. 2008), the recession velocity of NGC 4038/39 lies 522 km s^{-1} or 2.8σ above the flow (Schweizer et al. 2008).* (Reprinted from F. Schweizer *et al.*, *Astronomical Journal*, **136**, A new distance to the Antennae galaxies (NGC 4038/39) based on the Type Ia supernova 2007sr, p. 1482–1489, Copyright 2008, with permission of the AAS and F. Schweizer.)

Although this technique works reasonably well in the smooth Hubble flow (see Section 5.1.2), the mutual attractive forces of galaxies within the gravitational potential wells of large galaxy clusters cause significant 'peculiar motions' and, hence, distort the distance–redshift relationship. Observationally, this results in an elongation of the positions, in redshift space, of galaxy cluster members along the line of sight, which is commonly referred to as the **'Finger-of-God' effect** (see also Section 5.1.2).

In recent years, distance measurements to significant numbers of nearby cluster galaxies have become available, thus allowing studies of the true 3D distributions of the Virgo and Fornax clusters. This has led to the realization that the Virgo cluster is, in fact, highly substructured (e.g. West and Blakeslee 2000; Solanes *et al.* 2002; Jerjen 2003; Mei *et al.* 2005, 2007). Using the technique of **surface brightness fluctuations** (SBFs; see Section 4.2), West and Blakeslee (2000) and Jerjen *et al.* (2004) revealed strong 3D substructure and bimodality along the line of sight in the cluster's northern regions. The latter authors concluded that the northern subcluster consists of two dynamically distinct systems, with a small group around M86 falling into the M87 subcluster from behind.

Based on their detailed analysis of the SBFs in the statistically homogeneous and significant samples of Virgo and Fornax cluster galaxies, Mei *et al.* (2005, 2007) and Blakeslee *et al.* (2009, 2010) recently provided new and much improved insights into the 3D distribution of these clusters' member galaxies. Blakeslee *et al.* (2009) find a very tight correlation between the mean z_{850} magnitudes and $(g_{475} - z_{850})$ colours (where the subscripts in the filter names denote their central wavelengths in nanometres) of early-type galaxies in the Fornax cluster ($d = 20.0 \pm 0.2 \pm 1.4$ Mpc, where the errors are statistical and systematic, respectively), which allows these authors to obtain a first estimate of the 'cosmic' scatter in the relation, i.e. the scatter caused by the cluster's depth, $\sigma_{\rm cos} \approx 0.06 \pm 0.01$ mag, assuming a 20% depth uncertainty. This estimated scatter is approximately 40% smaller than that for the Virgo cluster (Mei *et al.* 2005, 2007; after correction of the latter depth estimates by a factor $1/\sqrt{2}$, which was omitted by these authors; see also Tonry *et al.* 2000), which implies that the former is more compact along the line of sight and exhibits less substructure (see e.g. Dunn and Jerjen 2006 and references therein). Blakeslee *et al.* (2009) derive a true linear root-mean-square depth for the bright ($B_T \leq 15.5$ mag), early-type galaxies in the Fornax cluster of $\sigma_d = 0.49^{+0.11}_{-0.15}$ Mpc, implying a back-to-front $\pm 2\sigma_d$ distance depth of $2.0^{+0.4}_{-0.6}$ Mpc (see Figure 1.11). This is comparable to the earlier, independent depth estimate of

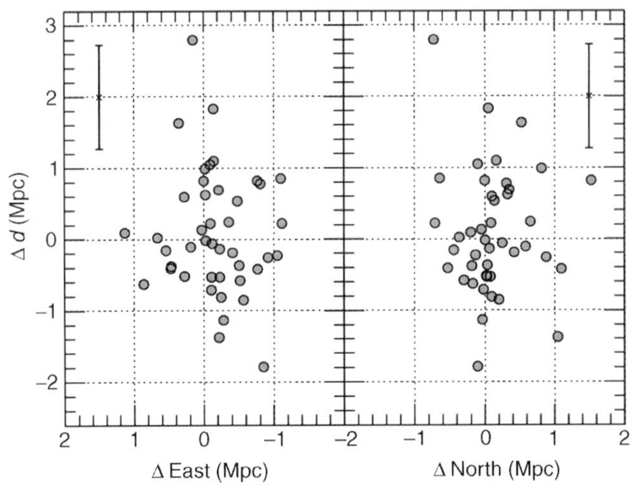

Figure 1.11 *Galaxy distance in the Fornax cluster, with respect to the mean distance of 20 Mpc, versus physical offset in Mpc east–west (left panel) and north–south (right panel) with respect to the central, giant elliptical galaxy NGC 1399 (Blakeslee et al. 2009). The median error in Δd is shown in both panels. There is a bias towards the cluster appearing elongated along the line of sight caused by distance errors and because galaxies more than about 1.5 Mpc from the cluster mean would not be included in the underlying catalogue if the offset were in the plane of the sky rather than along the line of sight.* (Reprinted from J. Blakeslee et al., *Astrophysical Journal*, **694**, The ACS Fornax Cluster Survey. V. Measurement and recalibration of surface brightness fluctuations and a precise value of the Fornax–Virgo relative distance, p. 556–572, Copyright 2009, with permission of the AAS and J. Blakeslee.)

the Fornax cluster core by Dunn and Jerjen (2006), $\sigma_{\rm int} = 0.74^{+0.52}_{-0.74}$ Mpc, which in turn is similar to the projected cluster size on the sky.

The prospects are promising for application of SBF approaches, based on high-resolution observations with the *HST* or large ground-based observatories, to galaxies in the Coma cluster, at $d \sim 100$ Mpc (cf. Liu and Graham 2001). However, the choice of suitable distance tracers on these scales is limited. Here, well into the smooth Hubble flow, we must predominantly rely on bright standard candles for reliable distances, including Type Ia SNe (see Section 5.2.1). Although we do not yet fully understand the physics governing SNe Ia explosions, the relationship between their absolute magnitude at peak brightness, their colour and their rate of decline is among the tightest empirical tools available for distance determinations at moderate redshifts. It allowed Riess *et al.* (1998) and Perlmutter *et al.* (1999) to conclude independently that SNe Ia at $z \sim 0.5$ appear to be approximately 10% fainter than their local counterparts. On the assumption that they were dealing with the same type of objects, they postulated that this implied that the expansion rate of the Universe is accelerating (see Figure 5.3; see, for reviews, Filippenko 2005; Frieman *et al.* 2008). From a physical perspective, this implies that the Universe must be subject to a large negative pressure, which has since been associated with Einstein's **cosmological constant** Λ and a vacuum energy denoted by Ω_Λ, which has been coined 'dark energy'.

Beyond the redshifts currently accessible with best-established, fairly 'local' distance tracers, cosmologists are particularly interested in reducing the uncertainties in and precisely establishing the main cosmological parameters that determine the evolution of the Universe on the largest scales. The latter include, of course, the Hubble constant, but also the matter–energy density ($\Omega_{\rm M}$), the curvature of the Universe – represented by the constant k, where $k < 0$, $k = 0$ and $k > 0$ corresponds to an open, flat and closed Universe – the dark energy's equation-of-state parameter (w) and σ_8, which measures the amplitude of the linear power spectrum on the scale of $8h^{-1}$ Mpc, where h denotes the value of the Hubble constant in units of 100 km s^{-1} Mpc^{-1}. In the context of this chapter and with the aim of achieving improved distance determinations, of these the current value of the Hubble constant is of greatest relevance (for a discussion of the prevailing cosmological model, see Section 5.1.3).

Significant strides have been made towards the goal of establishing the value of H_0 with an accuracy of better than 10%. Locally, two independent teams have endeavoured to achieve this aim using a variety of well-established distance anchors (Freedman *et al.* 2001; Sandage *et al.* 2006), although the resulting values of H_0 differ systematically: the *HST* Key Project (Freedman *et al.* 2001) derived a statistically weighted value of $H_0 = 72 \pm 8$ km s^{-1} Mpc^{-1}, while Sandage *et al.* (2006) found $H_0 = 62.3 \pm 1.3$ (statistical) $\pm$ 5.0 (systematic) km s^{-1} Mpc^{-1}, an unsatisfactory outcome that remains an issue of contention (but see Riess *et al.* 2009a,b; see also Chapter 6).

Much progress has been achieved in the last few years. In particular, the 7-year results of the *Wilkinson Microwave Anisotropy Probe (WMAP)* (e.g. Komatsu *et al.* 2009, 2011 and references therein) have reduced the uncertainties in the large-scale Hubble constant to unprecedentedly low levels, $H_0 = 70.4 \pm 2.5$ km s^{-1} Mpc^{-1}. However, *WMAP* results rely on many different '**priors**' (pre-imposed constraints) and do not allow an independent determination of the Hubble constant. This is because the value of H_0 is degenerate with the total curvature of the Universe. For instance, decreasing H_0 by 20 km s^{-1} Mpc^{-1} reduces the total matter–energy density in the Universe by 0.1 in terms of $\Omega_{\rm M}$. In particular, the *WMAP*-supported value of H_0 relies on the assumption of a flat geometry. When that

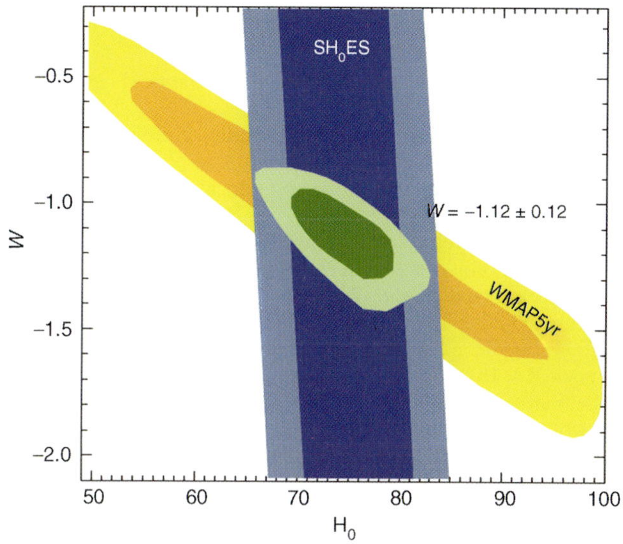

Figure 1.12 *Confidence regions in the plane of H_0 and the dark energy's equation of state, w (Riess et al. 2009a). The localization of the third acoustic peak in the WMAP 5-year data (Komatsu et al. 2009) produces a confidence region which is narrow but highly degenerate in this space. The best constraints on the Hubble constant are $H_0 = 74.2 \pm 3.6$ km s^{-1} Mpc^{-1}, while $w = 1.12 \pm 0.12$ for a constant equation of state. This result is comparable in precision to determinations of w from baryon acoustic oscillations and high-redshift SNe Ia, but is independent of both. The inner and outer regions represent 68 and 95% confidence, respectively. SH$_0$ES: supernovae and H_0 for the equation of state.* (Reprinted from A. Riess et al., *Astrophysical Journal*, **699**, A redetermination of the Hubble constant with the Hubble Space Telescope from a differential distance ladder, p. 539–563, Copyright 2009, with permission of the AAS and A. Riess.)

constraint is relaxed, the fitted value moves to $H_0 = 53^{+15}_{-13}$ km s^{-1} Mpc^{-1}: the central value has changed considerably, and the precision is much reduced.

Because of the dominant degeneracies precluding the direct and unambiguous determination of the value of the Hubble constant, cosmologists must rely on combining constraints resulting from many different approaches and tracers (see Section 5.3.4). For instance, the 5-year *WMAP* observations in combination with both SNe Ia and constraints from **baryon acoustic oscillations** (see Section 5.3.3) result in $H_0 = 70.5 \pm 1.3$ km s^{-1} Mpc^{-1} (Komatsu *et al.* 2009), while combining observational redshift distributions of galaxies with constraints on the baryon and CDM densities from *WMAP-5* and SNe Ia, assuming essentially a flat geometry, yields $H_0 = 68 \pm 2$ km s^{-1} Mpc^{-1} (Freedman and Madore 2010). Figure 1.12 shows an example of such an approach, in which the value of H_0 is constrained based on a combination of *WMAP-5* data and observations of SNe Ia. Although the Hubble constant is currently known to better than 5%, provided that all priors and assumptions on a flat Universe hold, further improvements are urgently required to better constrain the nature of the elusive dark energy (cf. Riess *et al.* 2009a; see also Section 5.3.4).

Bibliography

Alves DR 2004 A review of the distance and structure of the Large Magellanic Cloud. *New Astron. Rev.* **48**, 659–665.

Anderson J and King IR 2004a Multi-filter PSFs and distortion corrections for the HRC. *ACS Instrument Science Report* **2004-15**, 51 pp.

Anderson J and King IR 2004b Erratum: The rotation of the globular cluster 47 Tucanae in the plane of the sky. *Astron. J.* **128**, 950–950.

Armstrong JT, Mozurkewich D, Vivekanand M, Simon RS, Denison CS, Johnston KJ, Pan X-P, Shao M and Colavita MM 1992 The orbit of α Equulei measured with long-baseline optical interferometry: component masses, spectral types, and evolutionary state. *Astron. J.* **104**, 241–252.

Babusiaux C and Gilmore G 2005 The structure of the Galactic bar. *Mon. Not. R. Astron. Soc.* **358**, 1309–1319.

Backer DC and Sramek RA 1999 Proper motion of the compact, nonthermal radio source in the Galactic Center, Sagittarius A*. *Astrophys. J.* **524**, 805–815.

Bekki K 2008 A possible common halo of the Magellanic Clouds. *Astrophys. J.* **684**, L87–L90.

Belloche A, Menten KM, Comito C, Müller HSP, Schilke P, Ott J, Thorwirth S and Hieret C 2008 Detection of amino acetonitrile in Sgr B2(N). *Astron. Astrophys.* **482**, 179–196.

Benedict GF, McArthur BE, Fredrick LW, Harrison TE, Lee J, Slesnick CL, Rhee J, Patterson RJ, Nelan E, Jefferys WH, van Altena W, Shelus PJ, Franz OG, Wasserman LH, Hemenway PD, Duncombe RL, Story D, Whipple AL and Bradley AJ 2002 Astrometry with the Hubble Space Telescope: a parallax of the fundamental distance calibrator RR Lyrae. *Astron. J.* **123**, 473–484.

Bertout C, Robichon N and Arenou F 1999 Revisiting Hipparcos data for pre-main sequence stars. *Astron. Astrophys.* **352**, 574–586.

Besla G, Kallivayalil N, Hernquist L, Robertson B, Cox TJ, van der Marel RP and Alcock C 2007 Are the Magellanic Clouds on their first passage about the Milky Way? *Astrophys. J.* **668**, 949–967.

Blakeslee JP, Jordán A, Mei S, Côté P, Ferrarese L, Infante L, Peng EW, Tonry JL and West MJ 2009 The ACS Fornax cluster survey. V. Measurement and recalibration of surface brightness fluctuations and a precise value of the Fornax–Virgo relative distance. *Astrophys. J.* **694**, 556–572.

Bonanos AZ, Stanek KZ, Sasselov DD, Mochejska BJ, Macri LM and Kałużny J 2003 DIRECT distances to nearby galaxies using detached eclipsing binaries and Cepheids. IX. Variables in the field M31Y discovered with image subtraction. *Astron. J.* **126**, 175–186.

Bonanos AZ, Stanek KZ, Kudritzki R-P, Macri LM, Sasselov DD, Kałużny J, Stetson PB, Bersier D, Bresolin F, Matheson T, Mochejska BJ, Przybilla N, Szentgyorgyi AH, Tonry J and Torres G 2006 The first DIRECT distance determination to a detached eclipsing binary in M33. *Astrophys. J.* **652**, 313–322.

Bono G 2003 RR Lyrae distance scale: theory and observations. In *Proc. Stellar Candles for the Extragalactic Distance Scale* (eds Alloin D and Gieren W), *Lect. Notes Phys.* **635**, 85–104.

Borissova J, Rejkuba M, Minniti D, Catelan M and Ivanov VD 2009 Properties of RR Lyrae stars in the inner regions of the Large Magellanic Cloud. III. Near-infrared study. *Astron. Astrophys.* **502**, 505–514.

Brown TM, Ferguson HC, Smith E, Kimble RA, Sweigart AV, Renzini A and Rich RM 2004 RR Lyrae stars in the Andromeda halo from deep imaging with the Advanced Camera for Surveys. *Astron. J.* **127**, 2738–2752.

Brunthaler A, Reid MJ, Falcke H, Greenhill LJ and Henkel C 2005 The geometric distance and proper motion of the Triangulum galaxy (M33). *Science* **307**, 1440–1443.

Catelan M and Cortés C 2008 Evidence for an overluminosity of the variable star RR Lyrae, and a revised distance to the LMC. *Astrophys. J.* **676**, L135–L138.

Cioni M-RL, Bekki K, Clementini G, de Blok WJG, Emerson JP, Evans CJ, de Grijs R, Gibson BK, Girardi L, Groenewegen MAT, Ivanov VD, Leisy P, Marconi M, Mastropietro C, Moore B, Naylor T, Oliveira JM, Ripepi V, van Loon JT, Wilkinson MI and Wood PR 2008 The Magellanic Clouds as a template for the study of stellar populations and galaxy interactions. *Publ. Astron. Soc. Aust.* **25**, 121–128.

Cioni M-RL, Clementini G, Girardi L, Guandalini R, Gullieuszik M, Miszalski B, Moretti M-I, Ripepi V, Rubele S, Bagheri G, Bekki K, Cross N, de Blok WJG, de Grijs R, Emerson JP, Evans CJ, Gonzales-Solares E, Groenewegen MAT, Irwin M, Ivanov VD, Kerber L, Lewis J, Marconi M, Marquette J-B, Mastropietro C, Moore B, Napiwotzki R, Naylor T, Oliveira JM, Read M, Sutorius E, van Loon JT, Wilkinson MI and Wood PR 2011 The VMC Survey. I. Strategy and first data. *Astron. Astrophys.* **527**, A116.

Clement CM, Xu X and Muzzin AV 2008 The distance of the first overtone RR Lyrae variables in the MACHO Large Magellanic Cloud database: a new method to correct for the effects of crowding. *Astron. J.* **135**, 83–91.

Clementini G, Federici L Corsi C, Cacciari C, Bellazzini M and Smith HA 2001 RR Lyrae variables in the globular clusters of M31: a first detection of likely candidates. *Astrophys. J.* **559**, L109–L112.

Clementini G, Gratton R, Bragaglia A, Carretta E, Di Fabrizio L and Maio M 2003 Distance to the Large Magellanic Cloud: the RR Lyrae stars. *Astron. J.* **125**, 1309–1329.

Clementini G, Contreras R, Federici L, Cacciari C, Merighi R, Smith HA, Catelan M, Fusi Pecci F, Marconi M, Kinemuchi K and Pritzl BJ 2009 The variable star population of the globular cluster B514 in the Andromeda galaxy. *Astrophys. J.* **704**, L103–L107.

Cole AA 1998 Age, metallicity, and the distance to the Magellanic Clouds from red clump stars. *Astrophys. J.* **500**, L137–L140.

Costa E, Méndez RA, Pedreros MH, Moyano M, Gallart C, Noël N, Baume G and Carraro G 2009 The proper motion of the Magellanic Clouds. I. First results and description of the program. *Astron. J.* **137**, 4339–4360.

Dambis AK 2009 The kinematics and zero-point of the log $P - \langle M_K \rangle$ relation for Galactic RR Lyrae variables via statistical parallax. *Mon. Not. R. Astron. Soc.* **396**, 553–569.

Dambis AK 2010 Estimating the kinematic parameters and the distance-scale zero point for the thin-disk, thick-disk, and halo population tracers via 3D velocity data. In *Proc. Variable Stars, the Galactic Halo and Galaxy Formation* (eds Sterken C, Samus N and Szabados L), pp. 177–180. Moscow, Sternberg Astronomical Institute.

Davis J, Mendez A, Seneta EB, Tango WJ, Booth AJ, O'Byrne JW, Thorvaldson ED, Ausseloos M, Aerts C and Uytterhoeven K 2005 Orbital parameters, masses and distance to β Centauri determined with the Sydney University Stellar Interferometer and high-resolution spectroscopy. *Mon. Not. R. Astron. Soc.* **356**, 1362–1370.

de Grijs R and Robertson ARI 2006 Arp 116: interacting system or chance alignment? *Astron. Astrophys.* **460**, 493–498.

Descartes R 1637 *Discourse on the Method of Rightly Conducting One's Reason and of Seeking Truth in the Sciences* (French original: *Discours de la méthode pour bien conduire sa raison, et chercher la vérité dans les sciences*). Leiden, Ian Maire Publ.

de Vaucouleurs G 1993a Tests of the long and short extragalactic distance scales. *Publ. Nat'l Acad. Sci. USA* **90**, 4811–4813.

de Vaucouleurs G 1993b The extragalactic distance scale. VIII. A comparison of distance scales. *Astrophys. J.* **415**, 10–32.

Di Benedetto GP 2008 The Cepheid distance to the Large Magellanic Cloud and NGC 4258 by the surface brightness technique and improved calibration of the cosmic distance scale. *Mon. Not. R. Astron. Soc.* **390**, 1762–1776.

Dopita MA 1992 A theoretical calibration of the planetary nebular cosmic distance scale. *Astrophys. J.* **389**, 27–38.

Draine BT and Bond NA 2004 Direct extragalactic distance determination using X-ray scattering. *Astrophys. J.* **617**, 987–1003.

Dunn LP and Jerjen H 2006 First results from SAPAC: toward a three-dimensional picture of the Fornax cluster core. *Astron. J.* **132**, 1384–1395.

Durrell PR, Harris WE and Pritchet CJ 2001 Photometry and the metallicity distribution of the outer halo of M31. *Astron. J.* **121**, 2557–2571.

Eckart A and Genzel R 1996 Observations of stellar proper motions near the Galactic Centre. *Nature* **383**, 415–417.

Eisenhauer F, Schödel R, Genzel R, Ott T, Tecza M, Abuter R, Eckart A and Alexander T 2003 A geometric determination of the distance to the Galactic Center. *Astrophys. J.* **597**, L121–L124.

Eisenhauer F, Genzel R, Alexander T, Abuter R, Paumard T, Ott T, Gilbert A, Gillessen S, Horrobin M, Trippe S, Bonnet H, Dumas C, Hubin N, Kaufer A, Kissler-Patig M, Monnet G, Ströbele S, Szeifert T, Eckart A, Schödel R and Zucker S 2005 SINFONI in the Galactic Center: young stars and infrared flares in the central light-month. *Astrophys. J.* **628**, 246–259.

Elitzur M 1992 Astronomical masers. *Annu. Rev. Astron. Astrophys.* **30**, 75–112.

Filippenko AV 2005 Type Ia supernovae and cosmology. In *White Dwarfs: Cosmological and Galactic Probes* (eds Sion E, Vennes S and Shipman H), *Astrophys. Space Sci. Libr.* **332**, 97–133.

Fitzpatrick EL, Ribas I, Guinan EF, Maloney FP and Claret A 2003 Fundamental properties and distances of Large Magellanic Cloud eclipsing binaries. IV. HV 5936. *Astrophys. J.* **587**, 685–700.

Fouqué P, Arriagada P, Storm J, Barnes TG, Nardetto N, Mérand A, Kervella P, Gieren W, Bersier D, Benedict GF and McArthur BE 2007 A new calibration of Galactic Cepheid period–luminosity relations from B to K bands, and a comparison to LMC relations. *Astron. Astrophys.* **476**, 73–81.

Frail DA, Diamond PJ, Cordes JM and van Langevelde HJ 1994 Anisotropic scattering of OH/IR stars toward the Galactic Center. *Astrophys. J.* **427**, L43–L46.

Freedman WL and Madore BF 1991 Metallicity effects on the Cepheid distance scale. In *The Magellanic Clouds* (eds Haynes R and Milne D), *Proc. Int'l Astron. Union Symp.* **148**, 471–472.

Freedman WL and Madore BF 2010 The Hubble constant. *Annu. Rev. Astron. Astrophys.* **48**, 673–710.

Freedman WL, Madore BF, Gibson BK, Ferrarese L, Kelson DD, Sakai S, Mould JR, Kennicutt Jr RC, Ford HC, Graham JA, Huchra JP, Hughes SMG, Illingworth GD, Macri LM and Stetson PB 2001 Final results from the Hubble Space Telescope Key Project to measure the Hubble constant. *Astrophys. J.* **553**, 47–72.

Freire PC, Camilo F, Kramer M, Lorimer DR, Lyne AG, Manchester RN and D'Amico N 2003 Further results from the timing of the millisecond pulsars in 47 Tucanae. *Mon. Not. R. Astron. Soc.* **340**, 1359–1374.

Frieman JA, Turner MS and Huterer D 2008 Dark energy and the accelerating Universe. *Annu. Rev. Astron. Astrophys.* **46**, 385–432.

Fritz T, Gillessen S, Trippe S, Ott T, Bartko H, Pfuhl O, Dodds-Eden K, Davies R, Eisenhauer F and Genzel R 2010 What is limiting near-infrared astrometry in the Galactic Centre? *Mon. Not. R. Astron. Soc.* **401**, 1177–1188.

Gardiner LT and Noguchi M 1996 N-body simulations of the Small Magellanic Cloud and the Magellanic Stream. *Mon. Not. R. Astron. Soc.* **278**, 191–208.

Genzel R, Reid MJ, Moran JM and Downes D 1981 Proper motions and distances of H_2O maser sources. I. The outflow in Orion-KL. *Astrophys. J.* **244**, 884–902.

Genzel R, Pichon C, Eckart A, Gerhard OE and Ott T 2000 Stellar dynamics in the Galactic Centre: proper motions and anisotropy. *Mon. Not. R. Astron. Soc.* **317**, 348–374.

Genzel R, Eisenhauer F and Gillessen S 2010 The Galactic Center massive black hole and nuclear star cluster. *Rev. Mod. Phys.* **82**, 3121–3195.

Ghez AM, Salim S, Hornstein SD, Tanner A, Lu JR, Morris M, Becklin EE and Duchêne G 2005 Stellar orbits around the Galactic Center black hole. *Astrophys. J.* **620**, 744–757.

Ghez AM, Salim S, Weinberg NN, Lu JR, Do T, Dunn JK, Matthews K, Morris MR, Yelda S, Becklin EE, Kremenek T, Milosavljevic M and Naiman J 2008 Measuring distance and properties of the Milky Way's central supermassive black hole with stellar orbits. *Astrophys. J.* **689**, 1044–1062.

Gibson BK 2000 The distance to the Large Magellanic Cloud. *Mem. Soc. Astron. It.* **71**, 693–700.

Gillessen S, Eisenhauer F, Trippe S, Alexander T, Genzel R, Martins F and Ott T 2009a Monitoring stellar orbits around the massive black hole in the Galactic Center. *Astrophys. J.* **692**, 1075–1109.

Gillessen S, Eisenhauer F, Fritz TK, Bartko H, Dodds-Eden K, Pfuhl O, Ott T and Genzel R 2009b The orbit of the star S2 around Sgr A* from Very Large Telescope and Keck data. *Astrophys. J.* **707**, L114–L117.

Gould A 2000 A new kinematic distance estimator to the Large Magellanic Cloud. *Astrophys. J.* **528**, 156–160.

Gould A, Stutz A and Frogel JA 2001 A method to measure the ratio of total to selective extinction toward Baade's Window. *Astrophys. J.* **547**, 590–593.

Gratton RG, Fusi Pecci F, Carretta E, Clementini G, Corsi CE and Lattanzi M 1997 Ages of globular clusters from Hipparcos parallaxes of local subdwarfs. *Astrophys. J.* **491**, 749–771.

Groenewegen MAT and Blommaert JADL 2005 Mira variables in the OGLE bulge fields. *Astron. Astrophys.* **443**, 143–156.

Habing HJ 1996 Circumstellar envelopes and asymptotic giant branch stars. *Astron. Astrophys. Rev.* **7**, 97–207.

Harris WE 1996 A catalog of parameters for globular clusters in the Milky Way. *Astron. J.* **112**, 1487–1488. December 2010 edition, http://physwww.mcmaster.ca/%7Eharris/mwgc.dat, accessed 24 December 2010 (arXiv:1012.3224v1).

Herrnstein JR, Moran JM, Greenhill LJ, Diamond PJ, Inoue M, Nakai N, Miyoshi M, Henkel C and Riess A 1999 A geometric distance to the galaxy NGC4258 from orbital motions in a nuclear gas disk. *Nature* **400**, 539–541.

Herschel W 1785 On the construction of the heavens. *Phil. Trans. R. Soc. London* **75**, 213–266. (Reprinted in: Dreyer JLE 1912 *The Scientific Papers of Sir William Herschel* **I**, 223–259, London, UK.)

Hirota T, Bushimata T, Choi YK, Honma M, Imai H, Iwadate K, Jike T, Kameno S, Kameya O, Kamohara R, Kan-Ya Y, Kawaguchi N, Kijima M, Kim MK, Kobayashi H, Kuji S, Kurayama T, Manabe S, Maruyama K, Matsui M, Matsumoto N, Miyaji T, Nagayama T, Nakagawa A, Nakamura K, Oh CS, Omodaka T, Oyama T, Sakai S, Sasao T, Sato K, Sato M, Shibata KM, Shintani M, Tamura Y, Tsushima M and Yamashita K 2007 Distance to Orion KL measured with VERA. *Publ. Astron. Soc. Jpn* **59**, 897–903.

Holland S 1998 The distance to the M31 globular cluster system. *Astron. J.* **115**, 1916–1920.

Horrobin M, Eisenhauer F, Tecza M, Thatte N, Genzel R, Abuter R, Iserlohe C, Schreiber J, Schegerer A, Lutz D, Ott T and Schödel I 2004 The Galactic Center. *Astron. Nachr.* **325**, 120–123.

Jeffries RD 2007 The distance to the Orion Nebula Cluster. *Mon. Not. R. Astron. Soc.* **376**, 1109–1119.

Jerjen H 2003 Surface brightness fluctuation distances for dwarf elliptical galaxies in the Fornax cluster. *Astron. Astrophys.* **398**, 63–79.

Jerjen H, Binggeli B and Barazza FD 2004 Distances, metallicities, and ages of dwarf elliptical galaxies in the Virgo cluster from surface brightness fluctuations. *Astron. J.* **127**, 771.

Jewell PR, Webber JC and Snyder LE 1980 The linear shell diameter of IRC+10011. *Astrophys. J.* **242**, L29–L31.

Jones BF, Klemola AR and Lin DNC 1994 Proper motion of the Large Magellanic Cloud and the mass of the Galaxy. I. Observational results. *Astron. J.* **107**, 1333–1337.

Joshi YC, Pandey AK, Narasimha D, Sagar R and Giraud-Héraud Y 2003 Identification of 13 Cepheids and 333 other variables in M31. *Astron. Astrophys.* **402**, 113–123.

Joshi YC, Narasimha D, Pandey AK and Sagar R 2010 Nainital Microlensing Survey – detection of short period Cepheids in the disk of M31. *Astron. Astrophys.* **512**, A66.

Kallivayalil N, van der Marel RP, Alcock C, Axelrod T, Cook KH, Drake AJ and Geha M 2006a The proper motion of the Large Magellanic Cloud using HST. *Astrophys. J.* **638**, 772–785.

Kallivayalil N, van der Marel RP and Alcock C 2006b Is the SMC bound to the LMC? The Hubble Space Telescope proper motion of the SMC. *Astrophys. J.* **652**, 1213–1229.

Kallivayalil N, Besla G, Sanderson R and Alcock C 2009 Revisiting the role of M31 in the dynamical history of the Magellanic Clouds. *Astrophys. J.* **700**, 924–930.

Kapteyn JC 1922 First attempt at a theory of the arrangement and motion of the sidereal system. *Astrophys. J.* **55**, 302–328.

Kim MK, Hirota T, Honma M, Kobayashi H, Bushimata T, Choi YK, Imai H, Iwadate K, Jike T, Kameno S, Kameya O, Kamohara R, Kan-Ya Y, Kawaguchi N, Kuji S, Kurayama T, Manabe S, Matsui M, Matsumoto N, Miyaji T, Nagayama T, Nakagawa A, Oh CS, Omodaka T, Oyama T, Sakai S, Sasao T, Sato K, Sato M, Shibata KM, Tamura Y and Yamashita K 2008 SiO maser observations toward Orion-KL with VERA. *Publ. Astron. Soc. Jpn* **60**, 991–999.

Koerwer JF 2009 Large Magellanic Cloud distance and structure from near-infrared red clump observations. *Astron. J.* **138**, 1–6.

Komatsu E, Dunkley J, Nolta MR, Bennett CL, Gold B, Hinshaw G, Jarosik N, Larson D, Limon M, Page L, Spergel DN, Halpern M, Hill RS, Kogut A, Meyer SS, Tucker GS, Weiland JL, Wollack E and Wright EL 2009 Five-year Wilkinson Microwave Anisotropy Probe observations: cosmological interpretation. *Astrophys. J. Suppl. Ser.* **180**, 330–376.

Komatsu E, Smith KM, Dunkley J, Bennett CL, Gold B, Hinshaw G, Jarosik N, Larson D, Nolta MR, Page L, Spergel DN, Halpern M, Hill RS, Kogut A, Limon M, Meyer SS, Odegard N, Tucker GS, Weiland JL, Wollack E and Wright EL 2011 Seven-year Wilkinson Microwave Anisotropy Probe (WMAP) observations: cosmological interpretation. *Astrophys. J. Suppl. Ser.* **192**, 18.

Kraus S, Balega YY, Berger J-P, Hofmann K-H, Millan-Gabet R, Monnier JD, Ohnaka K, Pedretti E, Preibisch T, Schertl D, Schloerb FP, Traub WA and Weigelt G 2007 Visual/infrared interferometry of Orion Trapezium stars: preliminary dynamical orbit and aperture synthesis imaging of the θ^1 Orionis C system. *Astron. Astrophys.* **466**, 649–659.

Kroupa P and Bastian U 1997 The Hipparcos proper motion of the Magellanic Clouds. *New Astron.* **2**, 77–90

Kroupa P, Röser S and Bastian U 1994 On the motion of the Magellanic Clouds. *Mon. Not. R. Astron. Soc.* **266**, 412–420.

Kunder A and Chaboyer B 2008 Metallicity analysis of MACHO Galactic bulge RR0 Lyrae stars from their light curves. *Astron. J.* **136**, 2441–2452.

Lazio TJW, Anantharamaiah KR, Goss WM, Kassim NE and Cordes JM 1999 G359.87+0.18, an FR II radio galaxy 15′ from Sagittarius A*: implications for the scattering region in the Galactic Center. *Astrophys. J.* **515**, 196–205.

Lindblad B 1927 On the nature of the spiral nebulae. *Mon. Not. R. Astron. Soc.* **87**, 420–426.

Lindqvist M, Winnberg A, Johansson LEB and Ukita N 1991 SiO maser emission from OH/IR stars close to the Galactic Centre. *Astron. Astrophys.* **250**, 431–436.

Liu MC and Graham JR 2001 Infrared surface brightness fluctuations of the Coma elliptical galaxy NGC 4874 and the value of the Hubble constant. *Astrophys. J.* **557**, L31–L34.

Lo KY 2005 Mega-masers and galaxies. *Annu. Rev. Astron. Astrophys.* **43**, 625–676.

Lo KY, Cohen MH, Readhead ASC and Backer DC 1981 Multiwavelength VLBI observations of the Galactic Center. *Astrophys. J.* **249**, 504–512.

Macri LM, Stanek KZ, Bersier D, Greenhill LJ and Reid MJ 2006 A new Cepheid distance to the maser-host galaxy NGC 4258 and its implications for the Hubble constant. *Astrophys. J.* **652**, 1133–1149.

Majaess DJ 2010 Concerning the distance to the center of the Milky Way and its structure. *Acta Astron.* **60**, 55–74.

Majaess DJ, Turner DG and Lane DJ 2009 Characteristics of the Galaxy according to Cepheids. *Mon. Not. R. Astron. Soc.* **398**, 263–270.

Majaess D, Turner D and Lane D 2010 Type II Cepheids as extragalactic distance candles. *Acta Astron.* **59**, 403–418.

Massey P, Armandroff TE, Pyke R, Patel K and Wilson CD 1995 Hot, luminous stars in selected regions of NGC 6822, M31, and M33. *Astron. J.* **110**, 2715–2738.

Matsunaga N, Kawadu T, Nishiyama S, Nagayama T, Hatano H, Tamura M, Glass IS and Nagata T 2009 A near-infrared survey of Miras and the distance to the Galactic Centre. *Mon. Not. R. Astron. Soc.* **399**, 1709–1729.

McConnachie AW, Irwin MJ, Ferguson AMN, Ibata RA, Lewis GF and Tanvir N 2005 Distances and metallicities for 17 Local Group galaxies. *Mon. Not. R. Astron. Soc.* **356**, 979–997.

McNamara DH, Madsen JB, Barnes J and Ericksen BF 2000 The distance to the Galactic Center. *Publ. Astron. Soc. Pac.* **112**, 202–216.

Mei S, Blakeslee JP, Tonry JL, Jordán A, Peng EW, Côté P, Ferrarese L, West MJ, Merritt D and Milosavljević M 2005 The Advanced Camera for Surveys Virgo Cluster Survey. V. Surface brightness fluctuation calibration for giant and dwarf early-type galaxies. *Astrophys. J.* **625**, 121–129.

Mei S, Blakeslee JP, Côté P, Tonry JL, West MJ, Ferrarese L, Jordán A, Peng EW, Anthony A and Merritt D 2007 The ACS Virgo Cluster Survey. XIII. SBF distance catalog and the three-dimensional structure of the Virgo cluster. *Astrophys. J.* **655**, 144–162.

Menten KM, Reid MJ, Eckart A and Genzel R 1997 The position of Sagittarius A*: accurate alignment of the radio and infrared reference frames at the Galactic Center. *Astrophys. J.* **475**, L111–L114.

Menten KM, Reid MJ, Forbrich J and Brunthaler A 2007 The distance to the Orion Nebula. *Astron. Astrophys.* **474**, 515–520.

Messineo M, Habing HJ, Sjouwerman LO, Omont A and Menten KM 2002 86 GHz SiO maser survey of late-type stars in the inner Galaxy. I. Observational data. *Astron. Astrophys.* **393**, 115–128.
Messineo M, Habing HJ, Menten KM, Omont A and Sjouwerman LO 2004 86 GHz SiO maser survey of late-type stars in the inner Galaxy. II. Infrared photometry of the SiO target stars. *Astron. Astrophys.* **418**, 103–116.
Muir J 1911 *My First Summer in the Sierra*. Houghton Mifflin Co.
Nemec JM, Nemec AFL and Lutz TE 1994 Period–luminosity–metallicity relations, pulsation modes, absolute magnitudes, and distances for Population 2 variable stars. *Astron. J.* **108**, 222–246.
Ngeow C and Kanbur SM 2008 Large Magellanic Cloud distance from Cepheid variables using least squares solutions. In *Proc. Galaxies in the Local Volume* (eds Koribalski BS and Jerjen H), *Astrophys. Space Sci. Proc.*, pp. 317–318.
Nishiyama S, Nagata T, Sato S, Kato D, Nagayama T, Kusakabe N, Matsunaga N, Naoi T, Sugitani K and Tamura M 2006 The distance to the Galactic Center derived from infrared photometry of bulge red clump stars. *Astrophys. J.* **647**, 1093–1098.
Olling RP 2007 Accurate extragalactic distances and dark energy: anchoring the distance scale with rotational parallaxes. *Mon. Not. R. Astron. Soc.* **378**, 1385–1399.
Olling RP and Peterson DM 2000 One percent distances to Local Group galaxies via rotational parallaxes. *Bull. Am. Astron. Soc.* **32**, 1576–1576.
Oort JH 1927 Observational evidence confirming Lindblad's hypothesis of a rotation of the galactic system. *Bull. Astron. Inst. Neth.* **3**, 275–282.
Pedreros MH, Anguita C and Maza J 2002 Proper motion of the Large Magellanic Cloud using QSOs as an inertial reference system: the Q0459−6427 field. *Astron. J.* **123**, 1971–1977.
Perlmutter S, Aldering G, Goldhaber G, Knop RA, Nugent P, Castro PG, Deustua S, Fabbro S, Goobar A, Groom DE, Hook IM, Kim AG, Kim MY, Lee JC, Nunes NJ, Pain R, Pennypacker CR, Quimby R, Lidman C, Ellis RS, Irwin M, McMahon RG, Ruiz-Lapuente P, Walton N, Schaefer B, Boyle BJ, Filippenko AV, Matheson T, Fruchter AS, Panagia N, Newberg HJM, Couch WJ and the Supernova Cosmology Project 1999 Measurements of Ω and Λ from 42 high-redshift supernovae. *Astrophys. J.* **517**, 565–586.
Peterson D and Shao M 1997 The scientific basis for the Space Interferometry Mission. In *Proc. ESA Symp. Hipparcos – Venice '97* (ed. Perryman MAC), *ESA Spec. Publ.* **402**, pp. 749–754.
Piatek S, Pryor C and Olszewski EW 2008 Proper motions of the Large Magellanic Cloud and Small Magellanic Cloud: re-analysis of Hubble Space Telescope data. *Astron. J.* **135**, 1024–1038.
Pietrzyński G, Thompson IB, Graczyk D, Gieren W, Udalski A, Szewczyk O, Minniti D, Kołaczkowski Z, Bresolin F and Kudritzki R-P 2009 The Araucaria project. Determination of the Large Magellanic Cloud distance from late-type eclipsing binary systems. I. OGLE-051019.64−685812.3. *Astrophys. J.* **697**, 862–866.
Press WH, Flannery BP, Teukolsky SA and Vetterling WT 1992 Kolmogorov–Smirnov test. In *Numerical Recipes in Fortran 77: The Art of Scientific Computing*, 2nd edn, pp. 617–622. Cambridge University Press.
Reid MJ 1993 The distance to the center of the Galaxy. *Annu. Rev. Astron. Astrophys.* **31**, 345–372.
Reid MJ and Brunthaler A 2004 The proper motion of Sagittarius A*. II. The mass of Sagittarius A*. *Astrophys. J.* **616**, 872–884.
Reid MJ, Schneps MH, Moran JM, Gwinn CR, Genzel R, Downes D and Rönnäng B 1988 The distance to the center of the Galaxy: H_2O maser proper motions in Sagittarius B2(N). *Astrophys. J.* **330**, 809–816.
Reid MJ, Readhead ACS, Vermeulen RC and Treuhaft RN 1999 The proper motion of Sagittarius A*. I. First VLBA results. *Astrophys. J.* **524**, 816–823.
Reid MJ, Menten KM, Genzel R, Ott T, Schödel R and Eckart A 2003 The position of Sagittarius A*. II. Accurate positions and proper motions of stellar SiO masers near the Galactic Center. *Astrophys. J.* **587**, 208–220.
Reid MJ, Menten KM, Zheng XW, Brunthaler A, Moscadelli L, Xu Y, Zhang B, Sato M, Honma M, Hirota T, Hachisuka K, Choi YK, Moellenbrock GA and Bartkiewicz A 2009a Trigonometric parallaxes of massive star-forming regions. VI. Galactic structure, fundamental parameters, and noncircular motions. *Astrophys. J.* **700**, 137–148.

Reid MJ, Menten KM, Zheng XW, Brunthaler A and Xu Y 2009b A trigonometric parallax of Sgr B2. *Astrophys. J.* **705**, 1548–1553.
Reid WA and Parker QA 2010 A new population of planetary nebulae discovered in the Large Magellanic Cloud. III. The luminosity function. *Mon. Not. R. Astron. Soc.* **405**, 1349–1374.
Ribas I, Jordi C, Vilardell F, Fitzpatrick EL, Hilditch RW and Guinan EF 2005 First determination of the distance and fundamental properties of an eclipsing binary in the Andromeda galaxy. *Astrophys. J.* **635**, L37–L40.
Riess AG, Filippenko AV, Challis P, Clocchiatti A, Diercks A, Garnavich PM, Gilliland RL, Hogan CJ, Jha S, Kirshner RP, Leibundgut B, Phillips MM, Reiss D, Schmidt BP, Schommer RA, Smith RC, Spyromilio J, Stubbs C, Suntzeff NB and Tonry J 1998 Observational evidence from supernovae for an accelerating Universe and a cosmological constant. *Astron. J.* **116**, 1009–1038.
Riess AG, Macri L, Casertano S, Sosey M, Lampeitl H, Ferguson HC, Filippenko AV, Jha SW, Li W, Chornock R and Sarkar D 2009a A redetermination of the Hubble constant with the Hubble Space Telescope from a differential distance ladder. *Astrophys. J.* **699**, 539–563.
Riess AG, Macri L, Li W, Lampeitl H, Casertano S, Ferguson HC, Filippenko AV, Jha SW, Chornock R, Greenhill L, Mutchler M, Ganeshalingham M and Hicken M 2009b Cepheid calibrations of modern Type Ia supernovae: implications for the Hubble constant. *Astrophys. J. Suppl. Ser.* **183**, 109–141.
Ruffle PME, Zijlstra AA, Walsh JR, Gray MD, Gesicki K, Minniti D and Comeron F 2004 Angular diameters, fluxes and extinction of compact planetary nebulae: further evidence for steeper extinction towards the bulge. *Mon. Not. R. Astron. Soc.* **353**, 796–812.
Růžička A, Theis C and Palouš J 2009 Spatial motion of the Magellanic Clouds: tidal models ruled out? *Astrophys. J.* **691**, 1807–1815.
Salim S and Gould A 1999 Sagittarius A* 'visual binaries': a direct measurement of the Galactocentric distance. *Astrophys. J.* **523**, 633–641.
Sandage A, Tammann GA, Saha A, Reindl B, Macchetto FD and Panagia N 2006 The Hubble constant: a summary of the Hubble Space Telescope program for the luminosity calibration of Type Ia supernovae by means of Cepheids. *Astrophys. J.* **653**, 843–860.
Sandage A, Tammann GA and Reindl B 2009 New period–luminosity and period–color relations of classical Cepheids. III. Cepheids in SMC. *Astron. Astrophys.* **493**, 471–479.
Sandstrom KM, Peek JEG, Bower GC, Bolatto AD and Plambeck RL 2007 A parallactic distance of 389^{+24}_{-21} parsecs to the Orion Nebula Cluster from Very Long Baseline Array observations. *Astrophys. J.* **667**, 1161–1169.
Sarajedini A, Mancone CL, Lauer TR, Dressler A, Freedman W, Trager SC, Grillmair C and Mighell KJ 2009 RR Lyrae variables in two fields in the spheroid of M31. *Astron. J.* **138**, 184–195.
Saviane I, Momany Y, da Costa GS, Rich RM and Hibbard JE 2008 A new red giant-based distance modulus of 13.3 Mpc to the Antennae galaxies and its consequences. *Astrophys. J.* **678**, 179–186.
Schaefer BE 2008 A problem with the clustering of recent measures of the distance to the Large Magellanic Cloud. *Astron. J.* **135**, 112–119.
Schödel R, Ott T, Genzel R, Hofmann R, Lehnert M, Eckart A, Mouawad N, Alexander T, Reid MJ, Lenzen R, Hartung M, Lacombe F, Rouan D, Gendron E, Rousset G, Lagrange A-M, Brandner W, Ageorges N, Lidman C, Moorwood AFM, Spyromilio J, Hubin N and Menten KM 2002 A star in a 15.2-year orbit around the supermassive black hole at the centre of the Milky Way. *Nature* **419**, 694–696.
Schultz GV, Sherwood WA and Winnberg A 1978 Radial diameters of type II OH/IR sources. *Astron. Astrophys.* **63**, L5–L7.
Schweizer F, Burns CR, Madore BF, Mager VA, Phillips MM, Freedman WL, Boldt L, Contreras C, Folatelli G, González S, Hamuy M, Krzeminski W, Morrell NI, Persson SE, Roth MR and Stritzinger MD 2008 A new distance to the Antennae galaxies (NGC 4038/39) based on the type Ia supernova 2007sr. *Astron. J.* **136**, 1482–1489.
Shapley H 1918a Studies based on the colors and magnitudes in stellar clusters. Seventh paper: The distances, distribution in space, and dimensions of 69 globular clusters. *Contrib. Mount Wilson Obs.* **152**, 1–28.

Shapley H 1918b Studies based on the colors and magnitudes in stellar clusters. VII. The distances, distribution in space, and dimensions of 69 globular clusters. *Astrophys. J.* **48**, 154–181.
Shaya EJ and Olling R 2009 Rotational parallax: a SIM science study. *Bull. Am. Astron. Soc.* **41**, 356–356.
Sjouwerman LO, van Langevelde HJ and Diamond PJ 1998 Stellar positions from SiO masers in the Galactic Center. *Astron. Astrophys.* **339**, 897–903.
Sjouwerman LO, Lindqvist M, van Langevelde HJ and Diamond PJ 2002 H_2O and SiO maser emission in Galactic Center OH/IR stars. *Astron. Astrophys.* **391**, 967–978.
Sjouwerman LO, Messineo M and Habing HJ 2004 43 GHz SiO masers and astrometry with VERA in the Galactic Center. *Publ. Astron. Soc. Jpn* **56**, 45–50.
Snyder LE, Kuan Y-J and Miao Y 1994 Where is the heavy molecule Heimat in Sgr B2? In *Proc. The Structure and Content of Molecular Clouds* (eds Wilson TJ and Johnston KJ), *Lect. Notes Phys.* **439**, 187–198.
Solanes JM, Sanchis T, Salvador-Solé E, Giovanelli R and Haynes MP 2002 The three-dimensional structure of the Virgo cluster region from Tully–Fisher and HI data. *Astron. J.* **124**, 2440–2452.
Sollima A, Cacciari C, Arkharov AAH, Larionov VM, Gorshanov DL, Efimova NV and Piersimoni A 2008 The infrared *JHK* light curves of RR Lyr. *Mon. Not. R. Astron. Soc.* **384**, 1583–1587.
Stanek KZ and Garnavich PM 1998 Distance to M31 with the Hubble Space Telescope and Hipparcos red clump stars. *Astrophys. J.* **503**, L131–L134.
Szewczyk O, Pietrzyński G, Gieren W, Storm J, Walker A, Rizzi L, Kinemuchi K, Bresolin F, Kudritzki R-P and Dall'Ora M 2008 The Araucaria project. The distance of the Large Magellanic Cloud from near-infrared photometry of RR Lyrae variables. *Astron. J.* **136**, 272–279.
Tammann GA, Sandage A and Reindl B 2003 New period–luminosity and period–color relations of classical Cepheids. I. Cepheids in the Galaxy. *Astron. Astrophys.* **404**, 423–448.
Tonry JL, Blakeslee JP, Ajhar EA and Dressler A 2000 The surface brightness fluctuation survey of galaxy distances. II. Local and large-scale flows. *Astrophys. J.* **530**, 625–651.
Trippe S, Gillessen S, Gerhard OE, Bartko H, Fritz TK, Maness HL, Eisenhauer F, Martins F, Ott T, Dodds-Eden K and Genzel R 2008 Kinematics of the old stellar population at the Galactic Centre. *Astron. Astrophys.* **492**, 419–439.
Trumpler RJ 1930 Preliminary results on the distances, dimensions and space distribution of open star clusters. *Lick Obs. Bull.* **XIV**, 154–188.
Udalski A 2003 The Optical Gravitational Lensing Experiment: is interstellar extinction toward the Galactic Center anomalous? *Astrophys. J.* **590**, 284–290.
Vanhollebeke E, Groenewegen MAT and Girardi L 2009 Stellar populations in the Galactic bulge. Modelling the Galactic bulge with TRILEGAL. *Astron. Astrophys.* **498**, 95–107.
van Langevelde HJ and Diamond PJ 1991 Interstellar scattering of OH/IR stars at the Galactic Centre. *Mon. Not. R. Astron. Soc.* **249**, 7P–10P.
van Leeuwen F, Feast MW, Whitelock PA and Laney CD 2007 Cepheid parallaxes and the Hubble constant. *Mon. Not. R. Astron. Soc.* **379**, 723–737.
Vilardell F, Ribas I and Jordi C 2006 Eclipsing binaries suitable for distance determination in the Andromeda galaxy. *Astron. Astrophys.* **459**, 321–331.
Vilardell, F, Jordi C and Ribas I 2007 A comprehensive study of Cepheid variables in the Andromeda galaxy. Period distribution, blending, and distance determination. *Astron. Astrophys.* **473**, 847–855.
Vilardell F, Ribas I, Jordi C, Fitzpatrick EL and Guinan EF 2010 The distance to the Andromeda galaxy from eclipsing binaries. *Astron. Astrophys.* **509**, A70.
von Struve FGW 1847 *Etudes d'astronomie stellaire: sur la voie lactée et sur la distance des etoiles fixes.* **IV**. St. Petersburg, Tip. Acad. Imper.
Walker AR 2003 Distances to Local Group galaxies. In *Proc. Stellar Candles for the Extragalactic Distance Scale* (eds Alloin D and Gieren W), *Lect. Notes Phys.* **635**, 265–279.
West MJ and Blakeslee JP 2000 The principal axis of the Virgo cluster. *Astrophys. J.* **543**, L27–L30.
Westerlund BE 1997 The Magellanic Clouds. *Cambridge Astrophys. Ser.* **29**.
Yusef-Zadeh F, Hewitt JW and Cotton W 2004 A 20 centimeter survey of the Galactic Center region. I. Detection of numerous linear filaments. *Astrophys. J. Suppl. Ser.* **155**, 421–550.

2

The Solar Neighbourhood

... there were three things of which I diligently sought the reasons why they were so, and not otherwise: the number, size, and motion of the spheres.

–Johannes Kepler (1571–1630), German astronomer

Direct observations of stellar tracers in the immediate solar neighbourhood offer the distinct advantage of accurate and unambiguous geometric distance determinations, which in turn allow accurate determinations of luminosities (equivalent to the total energy emitted), masses, sizes and other physical quantities of cosmic tracers. Beyond radar measurements and orbit modelling in the solar system, these methods represent the first fundamental rung on the cosmic distance ladder. Because distance determinations to objects at much greater distances are fundamentally constrained by the accuracy of these local measurements, it is imperative that we understand and minimize both the systematic and random uncertainties involved.

2.1 Geometric Parallax Measurements

2.1.1 Trigonometric Parallax

As the Earth orbits the Sun, the nearest stars appear to trace equivalent elliptical orbits against the background stars, which appear to remain in fixed positions because of their much greater distances: see Figure 2.1. The size of the apparent ellipse (the trigonometric parallax) depends on the distance to the star of interest: the distance is inversely proportional to the parallax angle, $d = 1/\varpi$ (where d is expressed in parsecs and ϖ in arcseconds), obtained from astrometric measurements usually taken 6 months apart. Distances derived from parallax measurements are expressed in units of parsecs: an annual parallax angle of one arcsecond ($1''$) corresponds to a distance of 1 pc (3.086×10^{16} m or 3.26 lightyears). The measurement precision decreases from the ecliptic pole to the equator owing to

An Introduction to Distance Measurement in Astronomy, First Edition. Richard de Grijs.

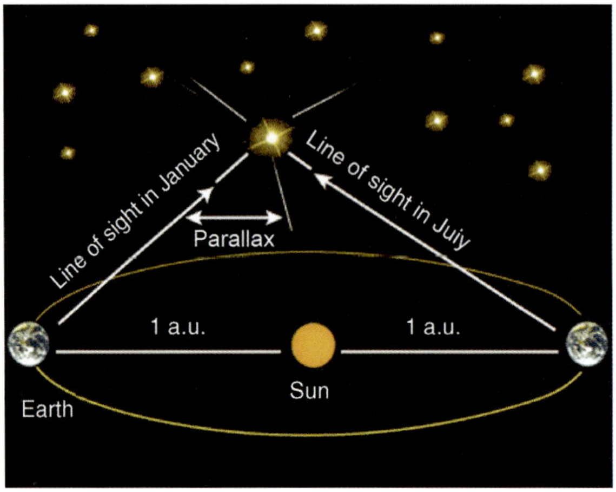

Figure 2.1 *Measuring the trigonometric parallax of nearby stars against the 'fixed' background stars. a.u.: astronomical unit.* (Reprinted from ESA (online) Science and Technology: Stellar distances – stellar parallax, Copyright 2009, with permission of ESA.)

projection of the parallax ellipse on the plane of the sky, which is referred to as the 'parallax factor'. It is a function of ecliptic latitude and the difference in longitude between the directions towards the Sun and the relevant star. Note that when using the extremes of the projected orbit to benefit from the maximum achievable baseline of 2 astronomical units (a.u., the average distance from the Earth to the Sun), the angle measured is twice the parallax angle.

This technique was first applied in 1838 by the German mathematician and astronomer Friedrich Bessel, who measured a distance of 10.4 lightyears to the star 61 Cygni (9.6% too small). Before Bessel's publication, the Baltic–German astronomer Friedrich Georg Wilhelm von Struve had announced a parallax measurement for Vega (α Lyrae) of $\varpi = 0.125''$ (see Berry 1899), which was surprisingly close to the currently accepted value of $0.129''$. However, he later almost doubled his determination and, thus, cast significant doubt on his result, so that Bessel is now generally credited with the first published parallax measurement. The Scottish astronomer Thomas Henderson published a distance of 3.25 lightyears (33.7% too small) to α Centauri based on his own parallax measurements, but was relegated to second place because of his doubts regarding the accuracy of his instruments.

The only star with a parallax greater than $1''$ is the Sun. The next nearest star, Proxima Centauri, has an annual trigonometric parallax of $0.77''$, corresponding to a distance of 1.3 pc. Within 10 lightyears ($\sim$3 pc) of the Sun, there are only 11 known stars (see Table 2.1). In practice, it is feasible to measure parallax angles down to $\sim 0.01''$ using ground-based observations under nominal seeing conditions of $1''$ (requiring measurements of stellar centroids to this accuracy), which implies that this method is applicable to distances of $\ll$100 pc. A realistic distance limit for a parallax uncertainty of $0.01''$ is 20 pc. This limits the applicability to only a few hundred stars, while distance uncertainties increase rapidly beyond a few parsecs.

Note that while the basic underlying principle is straightforward, obtaining true distances (i.e. absolute parallaxes) and true absolute magnitudes from measured trigonometric

Table 2.1 *Stars within 3 pc of the Sun*

Object	ϖ ('')	d (pc)	m_V (mag)	L_V ($L_{\odot,V}$)
Proxima Centauri	0.77	1.3	11.05	6×10^{-5}
α Cen A	0.75	1.3	−0.01	1.6
α Cen B	0.75	1.3	1.33	0.45
Barnard's star	0.545	1.8	9.45	4.5×10^{-5}
Wolf 359	0.421	2.4	13.5	2×10^{-5}
BD +36°2147	0.397	2.5	12.52	5.5×10^{-3}
Luyten 726-8A	0.387	2.6	12.52	6×10^{-5}
Luyten 726-8B	0.387	2.6	13.02	4×10^{-5}
Sirius A	0.377	2.6	−1.46	23.5
Sirius B	0.377	2.6	8.3	3×10^{-3}
Ross 154	0.345	2.9	10.45	4.8×10^{-4}

parallaxes is complicated (Smith and Eichhorn 1996; Brown *et al.* 1997; Kovalevsky 1998; Arenou and Luri 1999) because of the nonlinear relationships involved (e.g. Butkevich *et al.* 2005). In all ground-based parallax measurements, calibration to absolute parallax remains the largest contribution to the error budget and is often underestimated. We return to this issue in Chapter 6 in the context of the **Lutz–Kelker bias**. In addition, when using parallaxes for distance determinations, nonlinear uncertainty progression becomes important: the first- and second-order uncertainties in distance caused by uncertainties in the parallax angles, for small errors, are given by

$$\delta d = \frac{\delta\varpi}{\varpi^2}\left(-1 + \frac{\delta\varpi}{\varpi}\right). \tag{2.1}$$

However, distances are more often expressed as distance moduli,

$$m - M = 5\log d(\text{kpc}) + 10 = 10 - 5\log\varpi(\text{mas}), \tag{2.2}$$

where m and M are the apparent and absolute magnitude of the object of interest, respectively. Errors in parallax propagate into distance uncertainties as

$$\delta(m - M) = -\frac{5}{\ln 10}\left[\frac{\delta\varpi}{\varpi} - 0.5\left(\frac{\delta\varpi}{\varpi}\right)^2\right]. \tag{2.3}$$

As a rule of thumb, the quadratic terms become sufficiently important to significantly affect one's interpretation for formal errors exceeding 10%.

2.1.2 Astrometric Advances: Space-Based Missions and Interferometry

2.1.2.1 Hipparcos

Achieving significant gains in astrometric precision with respect to ground-based measurements requires dedicated space-based observatories covering large angular swathes of the sky – based on the so-called 'connectivity requirement' that allows precision calibration of the parallax zero point – to derive accurate **absolute parallaxes**. The European

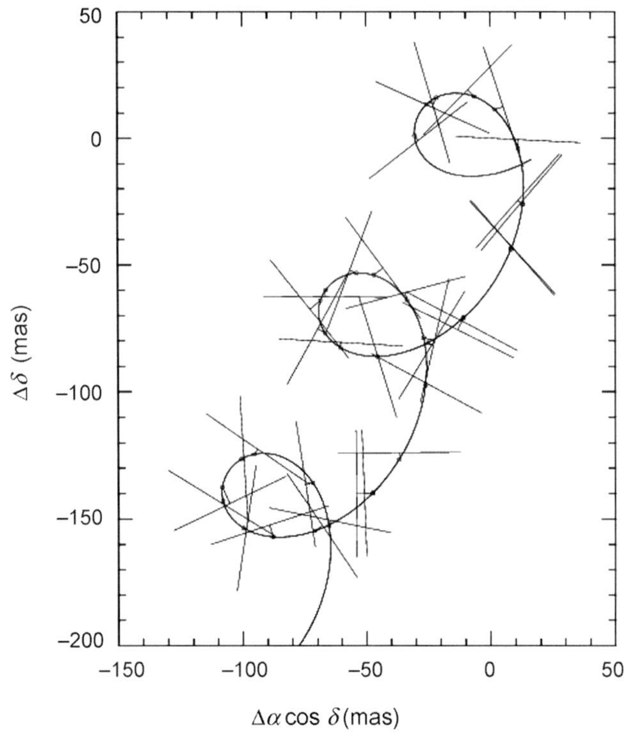

Figure 2.2 *Path on the sky of one object in the Hipparcos Catalogue over a period of 3 years. Each straight line indicates the observed position of the star at a particular epoch. Because the measurement is one dimensional, the precise location along this line is undetermined. The curve is the modelled stellar path fitted to all measurements. The inferred position at each epoch is indicated by a dot, and the residual by a short line joining the dot to the corresponding position line. The amplitude of the oscillatory motion gives the star's parallax, with the linear component representing the star's proper motion.* (Courtesy of Michael Perryman/Wikimedia Commons, licensed under the Creative Commons Attribution–Share Alike 3.0 Unported license.)

Space Agency's pioneering *Hipparcos*[1] *Space Astrometry Mission* (1989–1993) measured absolute parallaxes of 118 218 stars (ESA 1997; Perryman *et al.* 1997) with a median astrometric precision of approximately 1 milli-arcsec (mas; ∼10% accuracy, despite a serious problem with the satellite's orbit) out to some 100 pc using its main photomultiplier-based instrument (see Figure 2.2). *Hipparcos* measured large angles using a telescope with two apertures aimed at directions separated by 58 degrees. The two fields were joined into one optical path towards the focal-plane assembly using a beam combiner. By scanning the sky in great circles, it was possible to calibrate the angle between the two fields of view to high accuracy. An auxiliary star mapper, eventually resulting in the Tycho Catalogue (Høg *et al.* 2000a,b), observed 1 058 332 stars with a median precision of 25 mas (standard 1σ error)

[1] High Precision Parallax Collecting Satellite.

for all stars brighter than $V_T = 11.5$ mag and 7 mas for $V_T < 9$ mag. The majority of stars contained in the *Hipparcos* Catalogue are also contained in the Tycho Catalogue.

The Tycho-2 Catalogue includes B and V equivalent magnitudes of 99% of all stars brighter than $V_T = 11$ mag and extends down to $V_T \simeq 12.5$ mag (2 539 913 entries). The internal positional precision of the Tycho-2 Catalogue is better than ~10, 20, 50 and 70 mas for $V_T \leq 9$, 10, 11 and 11.5 mag, respectively, and decreases to ~110 mas for $V_T = 12.5$ mag. (For stars fainter than $V_T \simeq 10$ mag, the UCAC1 – US Naval Observatory CCD Astrograph Catalog – offers better internal precision, at approximately 16–17 mas down to $V_T \simeq 12.5$ mag; Zacharias *et al.* 2000.[2]) Proper motions in the Tycho-2 Catalogue were derived by combining the Tycho positional measurements with those obtained from the Carte du Ciel and Astrographic Catalogue photographic programmes, which started at the end of the nineteenth century.

The combined data set resulting from the *Hipparcos* mission represented a significant leap in our knowledge of the distances and space motions of stars in the solar neighbourhood. Its main scientific results range from the provision of an accurate positional reference frame and derivation of tight constraints on stellar evolutionary models for a wide variety of stellar types to substantial improvements in our understanding of the structure and dynamics of the Milky Way (see, for a comprehensive review, Perryman 2009).

The *Hipparcos* internal astrometric reference frame defines and uses the International Celestial Reference System (ICRS),[3] adopted by the International Astronomical Union (IAU). The system is based on coordinates of reference extragalactic radio sources, the International Celestial Reference Frame (ICRF), which in turn is based on Very Long Baseline Interferometry (VLBI) estimates. The origin of the ICRS is the solar system's barycentre – obtained through appropriate modelling of VLBI observations in the framework of Einstein's theory of general relativity – which is known to a precision of ±20 mas. The system's pole is in the direction defined by the conventional IAU models for precession (Lieske *et al.* 1977) and nutation (Seidelmann 1982), while the zero point of right ascension was defined by fixing the right ascension of the radio-loud quasar 3C 273B to the Hazard *et al.* (1971) FK5 (Fricke *et al.* 1988) value transferred to the J2000.0 epoch (Arias *et al.* 1995), to a precision of ±20 mas.

Nevertheless, the precision of the original *Hipparcos* Catalogue (ESA 1997) attracted significant (although not always well-founded) criticism related to the presence of systematic errors at the submilli-arcsecond level. Most of the problems originally identified in the initial release of the Catalogue have since been corrected through publication of a full rereduction of the data (van Leeuwen 2007a,b; see also, for an historical overview, van Leeuwen 2010). For conversion of the original 1D stellar transit-time measurements (referred to as the 'abscissae') to positional arcs on the sky, a high-accuracy reconstruction of, in particular, the along-scan attitude of the satellite was required (e.g. Fantino 2000; Makarov 2002; van Leeuwen 2005). Since the original data release, a new attitude-reconstruction programme was developed, based on the dynamics of a free-moving rigid body in space subjected to external and internal torques, instead of a purely mathematical model as used for the

[2] The UCAC3 is the current most up-to-date all-sky astrometric catalogue. It contains more than 100 million objects (mostly stars) and includes the *Hipparcos* and Tycho-2 Catalogues (Zacharias *et al.* 2010). The magnitude range from $R = 8$ to 16 is covered with a positional precision of 15–100 mas, depending on magnitude. Positional errors are approximately 15–20 mas for stars in the $R = 10$–14 mag range.

[3] http://www.iers.org/IERS/EN/Science/ICRS/ICRS.html.

original data release, the 'fully dynamic attitude model' (van Leeuwen and Fantino 2003). This new model reconstructs the underlying torques rather than the resulting directional variations. The rereduction of the original *Hipparcos* observations includes corrections for approximately 1600 'events' (perturbations of the satellite's motion) that occurred over the satellite's active lifetime and caused deterioration of the resulting measurement accuracy, such as scan-phase jumps caused by discrete thermal adjustments of the spacecraft and hits by dust particles. As a result of the improvements in attitude reconstruction, the distance over which *Hipparcos* can measure significant parallaxes has been extended by a factor of 3–4 (e.g. van Leeuwen 2007b).

2.1.2.2 Gaia

As follow-up to *Hipparcos* and using a very similar observational all-sky survey approach (cf. Høg 2008; Lindegren 2010), the European Space Agency's Cornerstone mission *Gaia*[4] (launch currently foreseen for Spring 2013 into a Lissajous orbit near the second Lagrangian point L2, 1.5 million km further from the Sun than Earth) will measure parallaxes to an accuracy of 10^{-5} arcsec, corresponding to distances accurate to 10% at $d \sim 10$ kpc. The instrument is equipped with a focal-plane mosaic of 106 CCDs covering a 40×40 arcmin^2 field of view in each viewing direction, separated by 106.5 degrees, allowing distance determinations to more than 200 million stars in the Milky Way (see Figure 2.3). This will enable construction of the most accurate 3D model of the Milky Way yet. Its data reduction, the Astrometric Global Iterative Solution, will largely follow the rereduction approach adopted for the *Hipparcos* observations (cf. Lindegren 2005; O'Mullane *et al.* 2007; van Leeuwen 2007b).

The mission's primary goal (see Figure 2.4) is to study Galactic structure and evolution by correlating the spatial distributions and kinematics of stars with their astrophysical properties (e.g. Lindegren 2010). Determining number densities and space motions for large, volume-complete samples of stars allows to trace the Galactic gravitational potential – and hence the mass distribution – in greater detail than ever before. The determination of very accurate (<1%) parallax distances to stellar samples of unprecedented size will provide stringent tests of stellar structure models and drive improvements of theoretical models of stellar atmospheres, interiors and evolution. *Gaia* will also play a pivotal role in the fields of exoplanetary system detection and characterization, and solar system physics, e.g. through detection of many tens of thousands of new minor planets, and even new trans-Neptunian objects, including Plutinos, may be discovered. Estimates suggest that *Gaia* will detect some 15 000 planets beyond our solar system by looking for tiny movements in stellar positions caused by the minute gravitational pulls of planets on the stars. The mission will also test Einstein's theory of general relativity by checking the perturbing effect of the Sun's gravity on starlight to approximately 2×10^{-6} and, thus, directly observe the structure of **spacetime**.

Its strength lies in its astrometric accuracy, combined with the expected limiting magnitude of $V \simeq 20$ mag. The former is expected to reach 7–25 micro-arcsecond (μas) for $V \leq 15$ mag at the mission's midpoint in early 2015, and a few hundred μas at $V \simeq 20$ mag. These astrometric precisions translate into distance uncertainties of 10% or better out to 10 kpc.

[4] Formerly known as the Global Astrometric Interferometer for Astrophysics, although the spacecraft is no longer an interferometric mission.

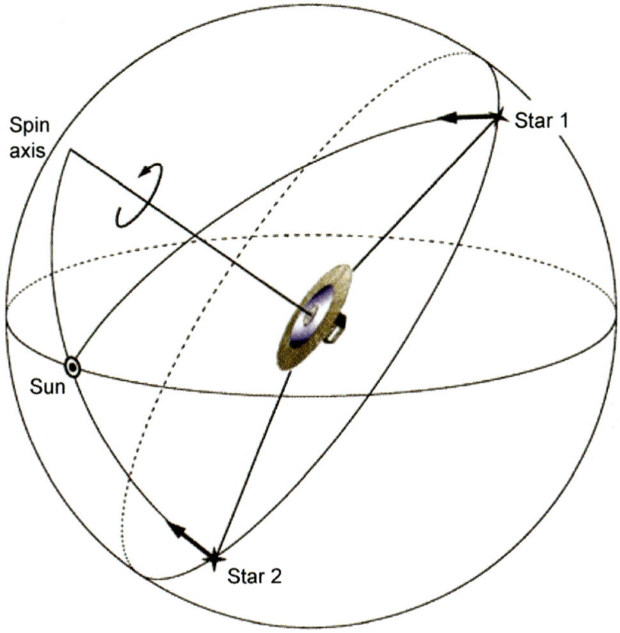

Figure 2.3 *Principle of absolute parallax measurement (Lindegren 2010). The parallactic displacement of a star is directed along the great-circle arc from the star to the solar system's barycentre. The observed angle between the stars, therefore, depends on the parallax of star 1, but is independent of the parallax of star 2. At some other time during the mission, the same pair may be observed with the Sun between the spin axis and star 1, in which case the observed angle will depend only on the parallax of star 2. In general, the observed angle will depend on both parallaxes, but their different projection factors will allow disentangling of the absolute values.* (Reprinted from D. Dravins, *et al.*, ESA Symposium Proceedings, Hipparcos – Venice '97 (ed. Perryman MAC), *ESA Special Publications*, **402**, Astrometric radial velocities from Hipparcos, p. 733–738, Copyright 1997, with permission of ESA and L. Lindegren.)

2.1.2.3 Interferometry

Interferometry is a proven technique for parallax determinations at radio wavelengths, which is particularly useful in regions where optical observations are obscured by dust (e.g. in the Galactic plane and in regions of active star formation), as well as in the optical regime.[5] To achieve the highest attainable angular resolution, on the order of milli-arcseconds, however, it is imperative to obtain multi-epoch observations using the longest possible baseline (see Figure 2.5). Although radio interferometry is a well-established technique, because of the much longer wavelengths of radio waves compared to the optical regime, similar angular resolution is attained only if the corresponding baselines are significantly larger. For instance, the UK's upgraded Multi-Element Radio-Linked Interferometer Network (e-MERLIN) has a maximum baseline of 217 km, resulting in an angular resolution of

[5] The Hubble Space Telescope's Fine Guidance Sensors led the way in space-based precision astrometry (e.g. Benedict *et al.* 2002, 2007; and references therein).

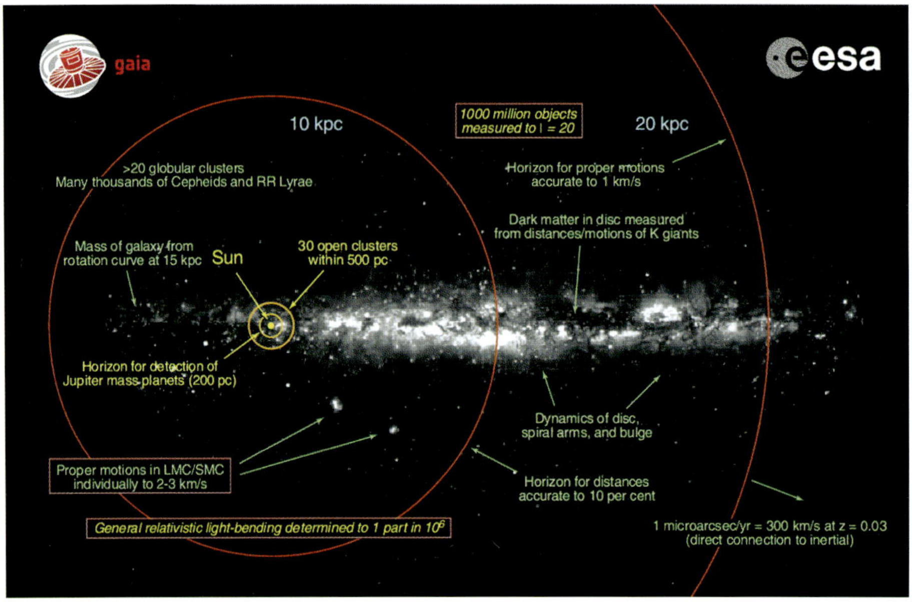

Figure 2.4 *Schematic diagram showing the distances to which Gaia will contribute to our knowledge of the Milky Way.* (Courtesy of the European Space Agency/C. Carreau.)

40 mas at a frequency of 5 GHz. This is comparable to that of the 2.4 m diameter *Hubble Space Telescope* at optical wavelengths.

In this context, significant progress has been made in recent years thanks to VLBI observations (Lestrade *et al.* 1999; Chatterjee *et al.* 2004; Loinard *et al.* 2005, 2007, 2008; Xu *et al.* 2006; Honma *et al.* 2007; Menten *et al.* 2007; Sandstrom *et al.* 2007; Torres *et al.* 2007, 2009; Kobayashi *et al.* 2008; Reid *et al.* 2009; Dzib *et al.* 2010). Various VLBI set-ups are operational across the globe. In addition to e-MERLIN, they include the US Very Long Baseline Array (VLBA), the European VLBI Network (EVN) and the Low Frequency Array (LOFAR), the Japanese VLBI Exploration of Radio Astrometry (VERA) project and the Chinese and Korean VLBI Networks (CVN and KVN), while the next-generation **Square Kilometre Array** (SKA) is under development and establishment of an East Asian VLBI network is being considered.

However, VLBI is only sensitive to high surface brightness emission and can only detect objects dominated by nonthermal processes, including masers (Section 3.7.4; Hirota *et al.* 2007; Kim *et al.* 2008) and magnetically active young stars in star-forming regions (Dzib *et al.* 2010). Its accuracy depends on access to regular and careful geodetic observations of the individual antennas to tie them to a general positional reference frame with respect to the Earth's centre of mass (e.g. Sovers *et al.* 1998; Mantovani and Kus 2004; Petrov *et al.* 2009). This is hard to achieve using ground-based stations, so that current efforts are looking at space-based radio interferometry missions, such as the Japanese VLBI Space Observatory Programme (VSOP and VSOP-2; e.g. Murata *et al.* 2008), enabling longer-baseline space–ground interferometry.

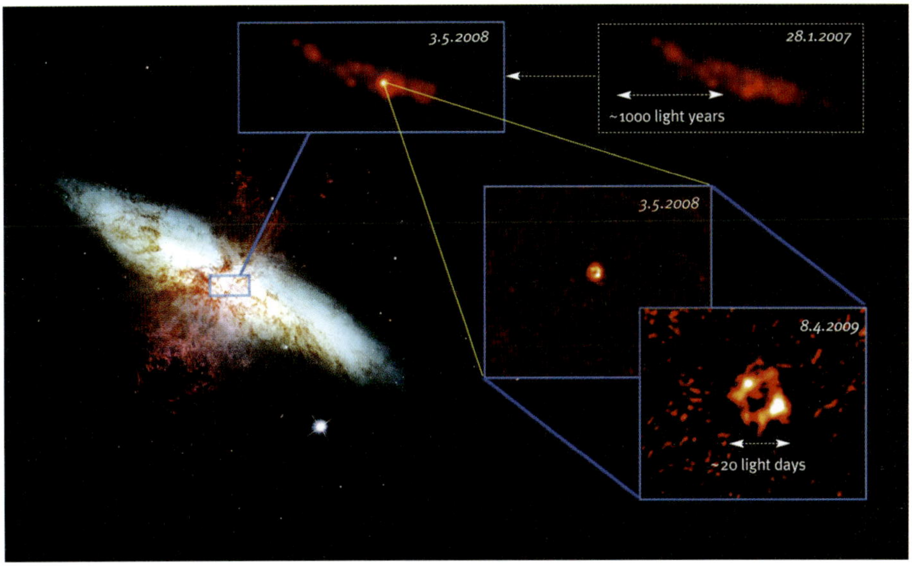

Figure 2.5 *Centre of the galaxy Messier 82. (Left) Hubble Space Telescope (HST) image showing the body of the galaxy in blue and hydrogen gas breaking out from the central starburst in red. The VLA image (top left) clearly shows the supernova (SN) 2008iz (Brunthaler et al. 2009). The higher-resolution VLBI images (bottom right) zoom in onto an expanding shell at a scale of a few lightdays.* (Graphics: Milde Science Communication. *HST* image: NASA/ESA and the Hubble Heritage Team. Radio images: A. Brunthaler, Max-Planck-Institut für Radioastronomie, Germany.)

2.1.3 Secular and Statistical Parallaxes: Moving Groups Method

We can also exploit the Sun's motion with respect to the local field stars in the direction of Vega through the Milky Way, $\sim$19.7 km s^{-1} ($\sim$4.14 a.u. yr^{-1}) and 229 $\pm$ 12 km s^{-1} (Dambis 2009) with respect to the disc and halo stellar populations, respectively. In essence, this technique extends the baseline of parallax measurements by the distance travelled by the Sun in the time interval considered. However, while the Sun is moving, all other nearby stars are presumably also moving along their own orbits, so that this method fails to improve the distance accuracies derived for individual stars. In fact, it is important to note that we also assume that both the Sun and the nearby stars of which we use parallax measurements for distance determination are not moving with transverse velocities with respect to each other. In practice, stars with significant peculiar ('proper') motions require at least three epochs of observation for accurate separation of proper motion and parallax effects (binary systems introduce further complications).[6] Nevertheless, we can still use the Sun's motion to extend the baseline for parallax measurements to determine the *average* distance to a

[6] For stars in the solar neighbourhood, proper motions are usually larger than their parallaxes. For instance, Barnard's star has a parallax of 0.55$''$ and the largest known proper motion, 10.55$''$ yr^{-1}.

group of stars for which the average motion with respect to that of the Sun approximates zero.

This statistical approach is referred to as the **secular parallax**. Its use has recently seen renewed interest in resolving the discrepancy between the Cepheid and RR Lyrae distance estimates to the Large Magellanic Cloud (see Section 1.2). Popowski and Gould (1998a,b) carefully explored the systematic uncertainties affecting the secular and *classical statistical parallax* methods (jointly referred to as the **statistical parallax**; e.g. Murray 1983; Hawley *et al.* 1986; Strugnell *et al.* 1986; Layden *et al.* 1996; Luri *et al.* 1996; Bochanski *et al.* 2011; see, for a review, Dambis 2009). Secular parallax works by requiring that the three first moments of the bulk velocity distribution are equal, while the classical statistical parallax approach forces equality of the six second moments, which determine the symmetric velocity covariance matrix. The Dutch astronomer Jacobus C. Kapteyn was one of the first to perfect and apply the classical statistical parallax method to the Hyades moving group, a physical association of a significant number of 100–200 stars moving along similar paths. Because of our distant vantage point, the Hyades stars appear to be moving towards a point of convergence (see Figure 2.6). Identification of that convergence point provides sufficient detail to translate each star's apparent motion into a real space velocity. The distance to a star in the cluster can be deduced by comparing its actual speed with how fast it appears to be moving. Kapteyn validated his results by comparing the distances he obtained using this *moving groups method* with those derived from trigonometric parallaxes for stars in some of the moving groups such as the Hyades and Praesepe clusters.

In pioneering work, Pavlovskaya (1953) first applied the statistical parallax technique to estimate the mean absolute magnitude of RR Lyrae stars. It has since been used, in

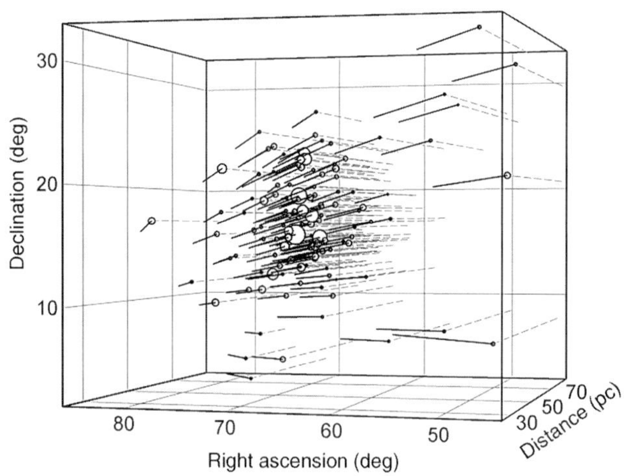

Figure 2.6 *Principle of determining astrometric radial velocities (Dravins et al. 1997). Stars in a moving cluster share the same mean velocity vector. Parallaxes give the distance, while proper motion vectors show the fractional change with time of the cluster's angular size. The latter equals the time derivative of distance, or the radial velocity. This plot shows positions for single stars in the Hyades, together with their measured distances and proper motions, shown over 100 000 years.*

increasingly sophisticated form, on ever growing samples of stars (see, for a review, Dambis 2009). Starting with the seminal study of Murray (1983), rigorous modern approaches (e.g. Strugnell *et al.* 1986; Layden *et al.* 1996; Dambis 2009; and references therein) apply maximum-likelihood analyses to simultaneously determine the three first and six second moments (the second moments define the shape of the velocity ellipsoid, often approximated by a 3D normal or Gaussian distribution), as well as a distance-dependent scaling factor. The latter converts the observed transverse, proper motions to velocities in the same reference frame as the radial velocity measurements. Further complications are introduced by the choice of reference frame: the resulting heliocentric velocity of a given star in the sample is set equal to the sum of the bulk velocity of the stellar sample considered at the corresponding location, with respect to either the sample's centroid or the 'local standard of rest'. The latter also requires assumptions on the position-dependent velocity implied by the adopted Galactic rotation curve model, and as regards the star's peculiar velocity, whose distribution has zero mean. We must further assume that the stellar peculiar velocities are independent of each other (see, for a detailed description and mathematical basis, Dambis 2009).

To account for e.g. a possible multicomponent structure in the RR Lyrae stellar population considered, Luri *et al.* (1996, 1998) proposed and applied a further generalization of the statistical parallax method (see Dambis 2009 for the mathematical basis and a recent application). This extension first diagonalizes the velocity ellipsoid matrices of all subpopulations – thus reducing the six second moments to three for a given population – and leaves the fractions of stars belonging to each of the subsamples as free parameters. For a two-population mixture, this results in 14 free parameters: three components of the relative bulk motion of the objects in both subpopulations with respect to the local rest frame, three diagonal components of the velocity ellipsoids of both subsamples, the distance scale correction factor – which is identical for both subsamples – and the fraction of stars belonging to the first subsample. Most current studies of RR Lyrae samples use the statistical parallax technique to constrain the mean V-band absolute magnitude or set the zero point of the metallicity/iron-abundance ([Fe/H])–mean absolute V-band magnitude ($\langle M_V \rangle$) relation (see Section 3.5.5). See also Dambis (2009) for arguments in favour of using the near-infrared (near-IR) log P (period)–$\langle M_V \rangle$ relation instead. This thus implies that the method has evolved to become a fundamental lower rung of the cosmic distance ladder.

Popowski and Gould (1998a) and Dambis (2009) lucidly review the underlying principles of the approach. The main problem associated with long time baselines is the intrinsic velocity dispersion (σ) of the stellar population used. They point out that, even for perfect measurements, the precision of distance determination is inversely proportional to the Mach number, $\kappa = W/\sigma$ (of order unity), where W represents the value of a given sample's bulk motion. The uncertainty can be reduced by averaging over N stars (i.e. by increasing the sample size), to reach a precision $\propto N^{-1/2}\kappa^{-1}$. In addition, the velocity dispersion can be used as input to the classical statistical parallax method by requiring that the dispersions in radial velocity and proper motion of one's sample stars are the same. This is, of course, an assumption which may affect subsequent luminosity calibration, but it appears reasonable for roughly spherical cluster-like stellar samples. The resulting precision is $\propto N^{-1/2}$, while the combined precision of both determinations is $\propto [N(1 + g\kappa^2)]^{-1/2}$, where $g \simeq \frac{1}{6}$ is a geometrical factor (Popowski and Gould 1998a). This method is suitable for use beyond those distances where we can still achieve precision astrometry (it is readily applicable to $d \sim 500$ pc), although one should bear in mind that the results may be affected by significant **Malmquist (1920) bias** (cf. Section 6.1.3; e.g. Smith 1987a,b; Ratnatunga and Upgren 1997; Popowski and Gould 1998a).

Moreover, additional uncertainties in the resulting distance determinations may be introduced by the choice of stellar population considered. The ideal stellar population is composed of standard candles drawn from a single velocity distribution, i.e. dynamically homogeneous but of unconstrained shape (cf. Popowski and Gould 1998a). Although these requirements are unlikely met in reality, suitable stellar samples can be selected from the stars in the solar neighbourhood.

Dambis (2009) applied the detailed methodology to a state-of-the-art sample of Galactic thick-disc and halo RR Lyrae stars in the near-IR K_s band. He derives a correction to the zero point, and hence an updated near-IR period–luminosity (PL) relation compared to that originally advocated by Jones *et al.* (1992),

$$\langle M_{K_s} \rangle = -2.33 \log P_F - 0.818 \pm 0.081, \tag{2.4}$$

where P_F refers to the fundamental period of his RR Lyrae sample stars.

Based on this recalibration of the near-IR RR Lyrae PL relation, the resulting distance to the Galactic Centre is $R_0 = 7.58 \pm 0.40$ kpc. This represents slight downscaling with respect to previous determinations based on near-IR observations of RR Lyrae stars, but is consistent with both Feast *et al.*'s (2008) distance determination to the Galactic Centre, $R_0 = 7.64 \pm 0.21$ kpc, based on the PL relations of RR Lyrae and type II Cepheids (see Section 3.5.4) and the dynamical estimate by Ghez *et al.* (2008) of $R_0 = 8.0 \pm 0.6$ kpc, using a fully unconstrained Keplerian orbit solution for the short-period star S2 orbiting the Milky Way's central black hole (see Section 1.2.2). In addition, the Dambis (2009) near-IR PL relation leads to a distance modulus to the Large Magellanic Cloud of $(m - M)_0 \equiv 5 \log(d/\text{pc}) - 5 = 18.27 \pm 0.08$ mag, again consistent with Feast *et al.*'s (2008) estimate, $(m - M)_0 = 18.37 \pm 0.09$ mag, although the resulting Hubble constant ($H_0 = 80 \pm 9$ km s^{-1} Mpc^{-1}) is only marginally consistent with currently accepted results (see Section 5.1).

2.2 Dynamical Parallax

Geometric distances to visual binary systems can be determined accurately based on careful observations of the orbits of the individual binary components (see Figure 2.7). To do so, we require good estimates of the masses of both components, ideally based on a *Hipparcos* parallax and an astrometric orbit, or an astrometric orbit and a radial velocity curve, since these are the only systems that can provide direct mass measurements (but see below and Section 3.7.3 for methods based on observations of eclipsing binary systems). We also need to know the orbital size and its period (P). Based on observations of the orbital semi-major axis, a, and the apparent **bolometric** luminosities of the stars in the binary system, we can then use the mass–luminosity (ML) relation (in the absence of direct orbital measurements; see Section 2.2.1) and Newton's physical generalization of Kepler's third law:

$$P^2 = \frac{4\pi^2}{G(M_{C1} + M_{C2})} a^3 \tag{2.5}$$

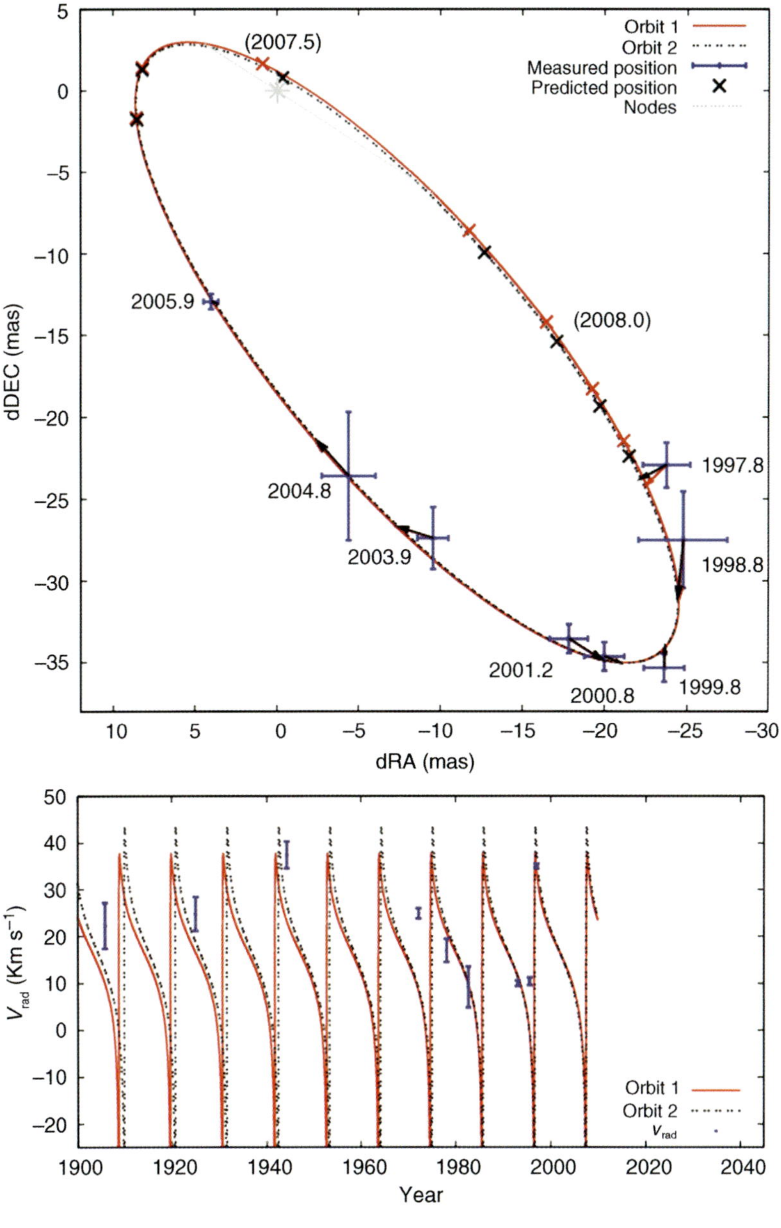

Figure 2.7 *(Top) Two best-fitting apparent orbits of the θ^1 Ori C binary system (solid and dashed lines). Each position measurement is connected to its predicted position. The line of nodes and the predicted positions for the interval 2007.0–2008.5 (increments of 0.25 yr) are also shown. North is up and east to the left. (Bottom) Radial velocity variations of both components, assuming a stellar mass ratio of 0.47 (Kraus et al. 2007).* (Reprinted from S. Kraus et al., *Astronomy and Astrophysics*, **466**, Visual/infrared interferometry of Orion Trapezium stars: preliminary dynamical orbit and aperture synthesis imaging of the θ^1 Orionis C system, p. 649–659, Copyright 2007, with permission of ESO.)

(where M_{C1} and M_{C2} are the individual stellar component masses and G is the usual Newtonian gravitational constant, all expressed in SI units) to obtain a first estimate of the distance to the binary system's centre of mass.

This process requires an iterative approach. From initial estimates of the binary components' stellar masses, we obtain an estimate of the linear separation between both stars. Combined with their angular separation, a preliminary estimate of the distance to the system is obtained, which – combined with an appropriate bolometric correction (see Buzzoni *et al.* 2010 for a recent review of the pitfalls associated with the latter) – can be used to convert the apparent stellar magnitudes into bolometric luminosities. Based on the ML relation,

$$\phi(M_i) = -\xi(m)\frac{\mathrm{d}m}{\mathrm{d}M_i} \tag{2.6}$$

(where m, M_i and $\xi(m)$ are the stellar mass and absolute magnitude in a particular passband, i, and the present-day stellar mass function, MF, respectively), we can now derive better estimates of the individual stellar masses, which can then be fed back into the first step of the iterative process.

In practice, we can also relate the system's orbital parameters through Kepler's third law directly to the dynamical parallax angle, ϖ_{dyn},

$$\varpi_{\mathrm{dyn}} = \frac{a}{P^{2/3}(M_{C1} + M_{C2})^{1/3}}, \tag{2.7}$$

where a and ϖ_{dyn} are expressed in units of mas, P in years and the component masses in $M_\odot$. To separate the system mass and parallax without having access to spectroscopic orbital parameters (e.g. radial velocities), the **Baize and Romaní (1946) algorithm** can be applied to main-sequence stars (cf. Heintz 1978, p. 62; Reed 1984). This method assumes that the component masses follow a generic ML relation, which can be combined with a bolometric correction and extinction-corrected magnitudes to solve for the system mass ($M_{\mathrm{sys}} = M_{C1} + M_{C2}$) and the dynamical parallax simultaneously (cf. Angelov 1993).

The method's accuracy depends on the uncertainties in the orbital parameters, the bolometric and extinction corrections[7] and the choice of a particular nonlinear ML relation (Angelov 1993), i.e. $\log m - M_V$. In turn, the ML relation depends on the total stellar mass, internal chemical composition and the mechanisms of energy release and transfer, which we will discuss in more detail below.

With the advent of optical and near-IR interferometers, we can now obtain much more accurate orbital parameters for smaller-separation visual and spectroscopic binary stars (see Figure 2.7), and – combined with spectroscopic measurements – better distance determinations (e.g. Hummel *et al.* 1998; Zwahlen *et al.* 2004; Cunha *et al.* 2007; see Boden and Quirrenbach 2008 for a review). This technique, which involves combining the light from multiple independent telescopes to form and measure interference fringes (variations in measured power as a function of optical path-length differences between telescopes), was first envisioned by Michelson in 1920 (cf. Michelson and Pease 1921). Anderson (1920) and Merrill (1922) used Michelson's interferometer at Mt Wilson Observatory and published

[7] This may not be straightforward if the luminosity of the system is dominated by the primary component and the secondary's spectral type is unknown. However, in most cases, and particularly if the spectral types of both stars are similar, neglecting bolometric corrections does not introduce significant errors (cf. Reed 1984).

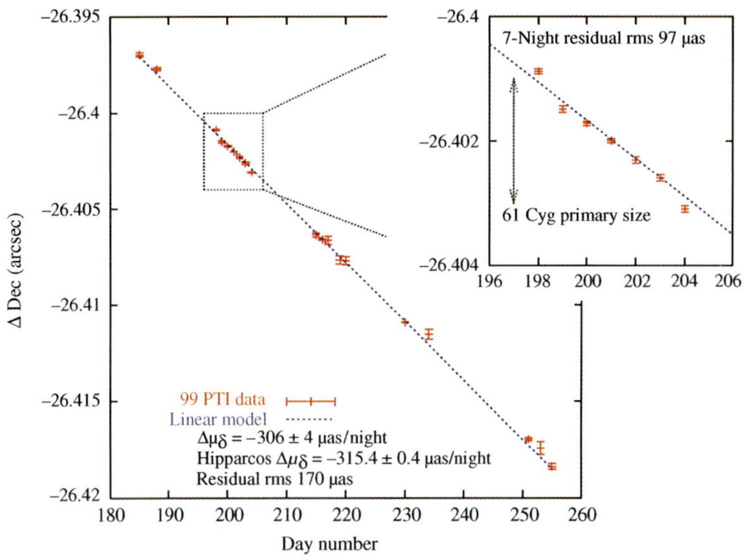

Figure 2.8 *Dual-beam, narrow-angle astrometry demonstration (Boden and Quirrenbach 2008). Roughly 3 months of relative astrometry data on the 30″ visual binary 61 Cygni are shown projected onto the principal declination measurement axis as a function of time. The apparent night-to-night relative motion owing to the estimated 678-year orbital period is clear. The apparent relative declination motion of 300 μas day^{-1} is very close to that measured by Hipparcos. The difference between the two estimates is consistent with acceleration from the binary orbit. The root-mean-square (rms) scatter of these data around the best-fitting linear motion model is 170 μas; data from a particularly stable 7-night run (inset) show an rms scatter of approximately 100 μas.* (Reprinted from A. Boden and A. Quirrenbach, International Astronomical Union Symposium Proceedings, **248**, Astrometry with ground-based interferometers, p. 36–43, Copyright 2008, with permission of the IAU.)

the first estimates of Capella's visual orbit. With current, state-of-the-art optical/near-IR interferometers, orbital positions can be derived with relative astrometric precisions on the order of 15 μas over a 0.5 arcsec field (e.g. Boden and Quirrenbach 2008). However, the drawback of this technique is that it is limited to arcsecond-scale binary systems, because the fringes from the individual components need to be covered within a single coherence length, i.e. the propagation distance over which the emission retains a certain degree of coherence or correlation. However, Shao and Colavita (1992), as well as Gorshanov *et al.* (2006), proposed and demonstrated a technique known as 'dual-beam, narrow-angle astrometry', taking advantage of significant correlation in the atmospheric phase noise over an isoplanatic angle of several tens of arcseconds[8] (see also Figure 2.8 and Boden and Quirrenbach 2008 for a review).

[8] The isoplanatic angle is the angular separation at which the atmospheric perturbations applied to the light from two astronomical objects become uncorrelated.

2.2.1 Mass–Luminosity Relations

Empirical ML relations based on observations of resolved binary systems are hard to come by, and have so far been obtained only for solar metallicity binaries (e.g. Popper 1980; Andersen 1991; Henry and McCarthy 1993, hereafter HM93; Kroupa *et al.* 1993, hereafter KTG93). The ML relation is, however, a strong function of stellar metallicity and one needs to include corrections for hidden companion stars to avoid introducing a systematic bias in the derived stellar MF (e.g. KTG93; Kroupa 2000). Although the ML relation is relatively well established for stars more massive than $\sim 0.8\ M_\odot$, our rather limited understanding of the lower-mass, more metal-poor stars, especially of the boundary conditions between the stellar interior and their atmospheres, has until recently severely limited the applicability of reliable ML relations at the low-mass end.

As shown by Equation (2.6), it is in fact the *slope* of the ML relation at a given absolute magnitude which determines the corresponding mass. This is, therefore, quite model dependent and has been addressed in detail by D'Antona and Mazzitelli (1983), Kroupa *et al.* (1990, 1993), Elson *et al.* (1995) and Kroupa and Tout (1997). The slope of the ML relation varies significantly with absolute magnitude, or mass. As shown by Kroupa *et al.* (1990, 1993) for solar metallicity stars with masses $m \leq 1\ M_\odot$, it has a local maximum at $M_V \approx 7$ mag and reaches a minimum at $M_V \approx 11.5$ mag (see also Kroupa 2000). This pronounced minimum corresponds to a maximum in the present-day luminosity function (LF), while the local maximum at $M_V \approx 7$ mag corresponds to the **Wielen dip** in the present-day LF of nearby stars, a plateau – or shallower slope – in the stellar LF between approximately $M_V = 6$ and 9 mag (e.g. Kroupa *et al.* 1990; D'Antona and Mazzitelli 1996; and references therein). The local maximum in the derivative of the ML relation at $M_V \approx 7$ mag ($m \approx 0.7\ M_\odot$) is caused by the increased importance of the H^- opacity in low-mass stars with decreasing mass (KTG93; Kroupa and Tout 1997). The ML relation steepens near $M_V = 10$ mag ($m \sim 0.4 - 0.5\ M_\odot$) because of the increased importance of H_2 formation in the outer shells of main-sequence stars, which in turn leads to core contraction (e.g. Chabrier and Baraffe 1997; Baraffe *et al.* 1998; Kroupa 2000).

Given the nonlinear shape of the ML relation and the small slope at the low-mass end, any attempt to model the ML relation by either a polynomial fit or a power-law dependence will yield intrinsically unreliable MFs (cf. Elson *et al.* 1995; Chabrier and Méra 1997), in particular in the low-mass regime. This model dependence is clearly illustrated by Ferraro *et al.* (1997) and de Grijs *et al.* (2002), who compared the MFs for the globular cluster (GC) NGC 6752 and a sample of Large Magellanic Cloud clusters, respectively, derived from a variety of ML relations available in the literature.

The exact shape of the ML relation is sensitive to metallicity. Metallicity changes affect the stellar spectral energy distribution and, therefore, the absolute magnitude in a given optical passband (cf. Brewer *et al.* 1993). In fact, it has been argued (cf. Baraffe *et al.* 1998) that, although the *V*-band ML relation is strongly metallicity dependent, the *K*-band ML relation is only a very weak function of metal abundance, yielding similar *K*-band fluxes for global metallicities $[M/H] = -0.5$ and 0.0 dex. Although the ML relation is currently relatively well determined for solar metallicity stars with $m \geq 0.8\ M_\odot$, at low metallicities the relation remains very uncertain. This is partially owing to the lack of an empirical comparison, and to our still relatively poor understanding of the physical properties of these stars, although major efforts are currently underway to alleviate the latter problem. As long as we consider only unevolved main-sequence stars, it is generally assumed that age effects are negligible, so that they can be ignored for practical purposes (cf. Brewer *et al.* 1993; Ferraro *et al.* 1997).

Since no empirical ML relations are available for low-mass, low-metallicity main-sequence stars, a test of the goodness of ML relations in this regime must, therefore, bear on a comparison of different models. Several recent studies have adopted this approach (e.g. Alexander *et al.* 1997; Ferraro *et al.* 1997; Kroupa and Tout 1997; Piotto *et al.* 1997; Saviane *et al.* 1998; de Grijs *et al.* 2002).

For solar metallicity stars in the mass range $0.1 < m \leq 1$ $M_\odot$, Leggett *et al.* (1996) and Kroupa and Tout (1997) concluded that, although *all* models considered provided reasonable fits to the empirical ML relation, the Baraffe *et al.* (1995) theoretical ML relations provided the best overall agreement with all recent observational constraints. On the other hand, Bedin *et al.* (2001) show that these are poor at low metallicity. Note that the Baraffe *et al.* (1995) models were based on *grey* model atmospheres (i.e. characterized by wavelength-independent absorption).

Both Piotto *et al.* (1997) and Saviane *et al.* (1998), from a comparison of largely the same theoretical ML relations available in the literature with observational data for the low-metallicity Galactic GCs NGC 6397 ([Fe/H] $\simeq -1.9$ dex) and NGC 1851 ([Fe/H] $\simeq -1.3$ dex), respectively, concluded that the Alexander *et al.* (1997) theoretical ML relations for the appropriate metallicity provided the best match for masses $m \leq 0.6$–0.8 $M_\odot$. Similar conclusions were drawn by Piotto *et al.* (1997) for three other Galactic GCs, M15, M30 and M92. Alexander *et al.* (1997) themselves found good to excellent overall agreement between their models and those of the Lyon group, in particular the updated Chabrier *et al.* (1996) ones, which employ the most recent nongrey model atmospheres. The theoretical ML relation for solar abundance by Chabrier *et al.* (1996) closely follows the most recent semi-empirical ML relation compiled by Kroupa (KTG93; Kroupa and Tout 1997).

Figure 2.9a shows the available empirical data, on which these comparisons are based for solar metallicity stellar populations. The solid bullets represent the HM93 sample; the open circles the higher-mass Andersen (1991) binary stars. In Figure 2.9b, we show the $m \leq 2$ $M_\odot$ subsample. Overplotted are the best-fitting relation of HM93 (solid line), the fit to the semi-empirical ML relation (dotted line) of KTG93 and Kroupa and Tout (1997), and the theoretical ML relation of Chabrier *et al.* (1996) (dashed line) for $0.075 \leq m \leq 0.6$ $M_\odot$. The figure shows that the observational data allow for significant local differences in the slope of the solar metallicity ML relation; these uncertainties propagate through the derivative of the relation when converting LFs to MFs.

The theoretical ML relation for solar abundance by Chabrier *et al.* (1996) closely follows the semi-empirical ML relation compiled by Kroupa (KTG93, Kroupa and Tout 1997). In Figure 2.9c, we compare the current theoretical ML relations for subsolar metallicity: the solid lines represent the Alexander *et al.* (1997) ML relations for (top to bottom) [M/H] $\simeq -1.3$, -1.5 and -2.0 dex; for reasons of clarity, we show only the [M/H] = -1.5 dex ML relation of Baraffe *et al.* (1997), but the spread caused by metallicity differences is similar to that exhibited by the Alexander *et al.* (1997) relations. The most significant differences between both sets of models are seen at masses $m \geq 0.4$ $M_\odot$. This is likely owing to the slightly different treatment of the stellar atmospheres and radiative opacities. Finally, for comparison we also show the solar metallicity theoretical ML relation of Chabrier *et al.* (1996) and Baraffe *et al.* (1997).

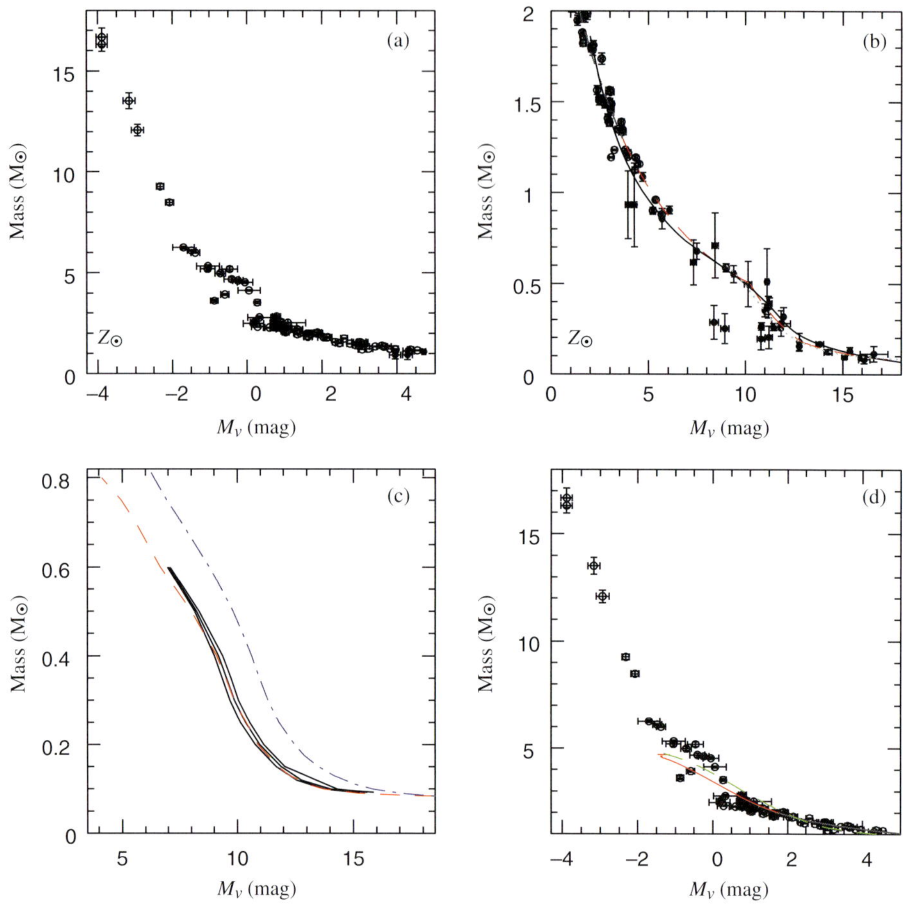

Figure 2.9 *Empirical and theoretical ML relations (de Grijs et al. 2002). (a) Solid bullets: HM93; open circles: Andersen (1991). (b) Solid line: HM93 fit; dotted line: KTG93 and Kroupa and Tout (1997) semi-empirical ML relation; dashed line: Chabrier et al. (1996) theoretical ML relation for $m \geq 0.6\ M_\odot$, based on a third-order polynomial fit. (c) Theoretical ML relations for subsolar abundances: Alexander et al. (1997) (solid lines, for [M/H] = −1.3, −1.5, −2.0 dex (top to bottom)), and Baraffe et al. (1997) (dashed line, [M/H] = −1.5 dex). For comparison, the solar-abundance ML relation of Baraffe et al. (1997) is also shown (dash-dotted line). (d) Observational data for $m \geq 1.0\ M_\odot$ stars (Andersen 1991), and – for $m \geq 5\ M_\odot$ – theoretical models by Girardi et al. (2000) for solar abundance (solid line) and [M/H] = −1.3 dex (dashed line).*

Figure 2.10 shows the derivatives of the ML relations as a function of absolute visual magnitude. From Figure 2.10a, it is immediately clear that the empirical fit to the HM93 ML relation inherently leads to unreliable luminosity-to-mass conversions because of the two sharp discontinuities in the slope. Figure 2.10b shows the metallicity dependence of the slope of the ML relation; the solid lines represent the Alexander *et al.* (1997) ML relations

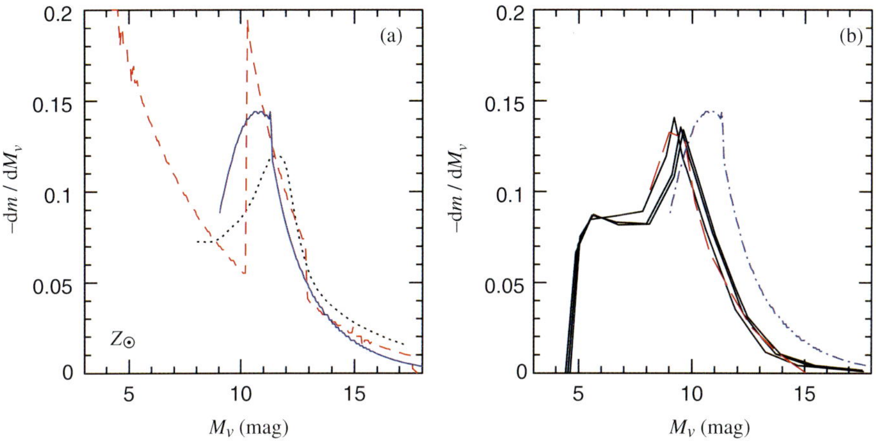

Figure 2.10 *Slope of the ML relation (de Grijs et al. 2002). (a) Solar metallicity ML relations: HM93 (dashed line, empirical), KTG93 and Kroupa and Tout (1997) (dotted line, semi-empirical), and Chabrier et al. (1996) and Baraffe et al. (1997) (solid line, theoretical). (b) Theoretical ML relations for subsolar abundances: Alexander et al. (1997) (solid lines, for [M/H] = −1.3, −1.5, −2.0 dex, peaking from right to left), and Baraffe et al. (1997) (dashed line, [M/H] = −1.5 dex). For comparison, the solar-abundance ML relation of Chabrier et al. (1996) and Baraffe et al. (1997) is also shown (dash-dotted line).*

with [M/H] = −1.3, −1.5 and −2.0 dex, peaking from right to left. The Baraffe *et al.* (1997) models (cf. the dashed line, for [M/H] = −1.5 dex) closely follow those of Alexander *et al.* (1997). For comparison, we have also included the solar-abundance model of Chabrier *et al.* (1996) and Baraffe *et al.* (1997), as in panel (a).

The main uncertainties for the luminosity evolution of stars with masses $m \geq 0.8\ M_\odot$ are in the treatment of the degree of mass loss and convective-core overshooting. Girardi *et al.* (2000) computed a grid of stellar evolutionary models for stars in the mass range $0.15 \leq m \leq 7\ M_\odot$ for metallicities between $\frac{1}{50}$ and 1.5× solar, using updated input physics, as well as moderate core overshooting.

de Grijs *et al.* (2002) compared their stellar mass estimates based on HM93, KTG93, Baraffe *et al.* (1998) and Girardi *et al.* (2000) with Tout *et al.*'s (1996; TPEH96) mathematical ML relation parameterization, for both solar and subsolar ([Fe/H] = −0.5 dex) metallicities and ages of ∼10 and 25 Myr. Significant differences are appreciated among the individual models, in particular between TPEH96 and the high-mass end ($\log m/M_\odot \geq 0.3$) of KTG93 and between TPEH96 and the models of BCAH98, which show systematic deviations from the one-to-one relation. It is, therefore, not unlikely that the differences among the models dominate the uncertainties.

Nevertheless, and despite the remaining uncertainties, a satisfactory first-order comparison between stellar masses obtained by application of a suitable ML relation and those resulting from using Kepler's third law should suffice to derive stellar distances based on the dynamical parallax method to within 5–10%.

2.3 Spectroscopic and Photometric Parallaxes

Most stars are too far away to measure their distances directly using the trigonometric parallax method, but by employing spectroscopy and photometry, we can determine an approximate distance. Stellar spectral classification, based on measurements of their spectral line widths, becomes a powerful tool in combination with their position in the **Hertzsprung–Russell diagram**. The spectral type (OBAFGKM) and luminosity class of a given star can be derived from its spectrum, assuming that nearby and distant stars are, in essence, similar (cf. Adams and Kohlschütter 1914). The behaviour of luminosity classes Ia (luminous supergiants) to V (main-sequence stars) is, to first order, reasonably well understood as a function of spectral type or, equivalently, photometric colour (see Figure 2.11). Luminosity classes, in the Yerkes spectral classification (or MKK classes, from the authors' initials: Morgan, Keenan and Kellman 1943; updated, revised and reviewed by Johnson and Morgan 1953 and Morgan and Keenan 1973 so that they are now known as MK classes), express the widths and intensities of certain spectral absorption lines that are sensitive to temperature and surface gravity – the full grid is defined by an array of carefully chosen standard stars, and serves as the basis for differential classification of any other star – and distinguish

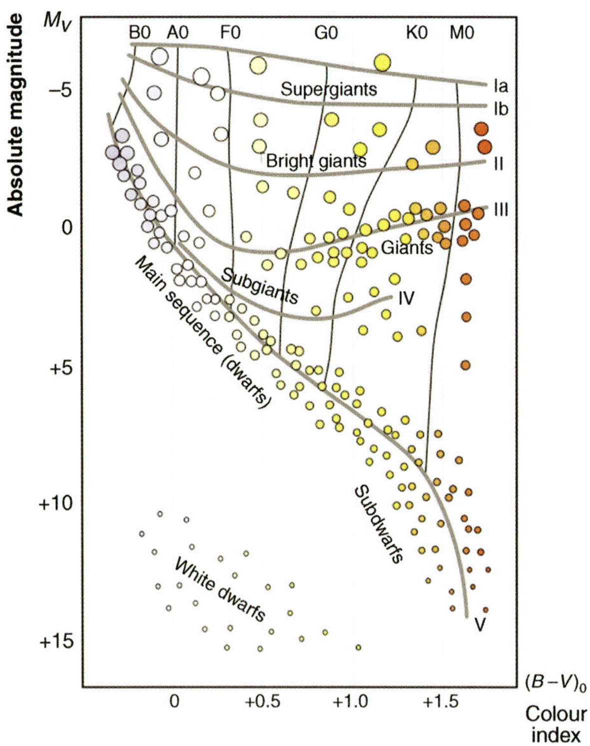

Figure 2.11 *Definition of stellar luminosity classes and their loci in the Hertzsprung–Russell diagram.* (Courtesy of Wikimedia Commons, licensed under the Creative Commons Attribution–Share Alike 3.0 Unported license.)

between supergiants (class I), bright giants (II), normal giants (III), subgiants (IV) and main-sequence ('dwarf') stars (class V), for instance (extended from the optical regime to the near-IR window by Hanson *et al.* 1996). They also include classes VI (subdwarfs) and VII (white dwarfs).

Reylé *et al.* (2006) review the calibration relations for single, M- and T-type cool to ultra-cool subdwarf stars. These are mostly based on near-IR J-band diagnostic spectral features, which can be used to estimate the metallicity, surface gravity and effective temperature, $T_{\rm eff}$. The effective temperature and metallicity can be used to derive the stellar radius from evolutionary models (Reylé *et al.* assumed a generic age of 10 Gyr for their sample subdwarfs), and – through scaling of the model atmospheres to flux-calibrated spectra by a factor R^2/d^2, where R is the stellar radius – a distance can then be derived with an accuracy of better than 10%, although less accurate for subdwarfs with higher effective temperatures, $\geq$3000 K, where radii are poorly constrained.

Stellar spectral types correlate with their surface temperature. Since compact stars are characterized by higher surface gravities than their more diffuse counterparts, they exhibit wider spectral lines because of greater pressure broadening, which implies that we can derive the appropriate stellar radii from the spectral line widths. Using $T_{\rm eff}$ and R, we can obtain the intrinsic luminosity, L, from *Stefan's law*,

$$L = 4\pi R^2 \sigma T_{\rm eff}^4 \tag{2.8}$$

(where σ is the Stefan–Boltzmann constant) and, hence, the absolute magnitude in a given filter, assuming the appropriate bolometric correction. The expected absolute magnitude can then be compared with the star's apparent magnitude, resulting in a 'distance modulus':

$$(m - M)_i = 5\log(d/{\rm pc}) - 5 + A_i, \tag{2.9}$$

where d is the distance sought in the presence of A_i mag of extinction for a given passband i, which remains a major uncertainty in the application of this technique.

This method, generally referred to as **spectroscopic parallax** (although it does not involve measurement of parallaxes as such), works best for main-sequence stars (a typical 10% uncertainty in luminosity results in a 5% uncertainty in distance) because of the relatively steep dependence of luminosity on spectral type. However, for (super)giant stars, the luminosity can only be estimated to within approximately 0.5 mag. Hence, their distance accuracy is roughly 25–30% (e.g. Vacca *et al.* 1996 for single O-type stars), although the uncertainty can be reduced to <10% for statistically significant co-located stellar samples. The uncertainties are smaller for B-type stars because their absolute-magnitude dispersion is lower, particularly in the near-IR range, owing to their longer main-sequence lifetimes.

A variant of this method was originally proposed by Wilson and Bappu (1957). They noticed that the width of *emission core* of the Ca II K absorption line at a wavelength of 3933 Å in cool (G-, K- and M-type) stars tightly correlates with the stars' V-band absolute magnitudes: intrinsically brighter stars exhibit broader absorption lines. Similar correlations exist for the Mg II k line at an ultraviolet wavelength of 2796.34 Å (Cassatella *et al.* 2001) and for the wing emission lines in the extended absorption wings of the Ca^+ K and H lines (see Stencel 1977, 2009; the H line is centred at 3968 Å). The first calibration of the **Wilson–Bappu(–Stencel) effect** was based on independent distance measurements using

Hipparcos parallaxes (Wallerstein *et al.* 1999). Pace *et al.* (2003) obtained an improved calibration relation,

$$M_V = 33.2 - 18.0 \log(W_0), \tag{2.10}$$

where W_0 is the width of the emission core (in km s^{-1}). W_0 can, in principle, be measured out to roughly 7 Mpc, although it is generally used only for bright stars (V $\lesssim$ 15 mag) within a few hundred kpc. The method is affected by the relatively large scatter (~0.6 mag) on the relationship, hence limiting its applicability to stellar groups rather than individual objects (see Pace *et al.* 2003 for an assessment of the possible contributions to this scatter). It may also be subject to a possible metallicity effect for [Fe/H] $\leq$0.4 dex (Pace *et al.* 2003) and contamination due to interstellar extinction, although the residual wing emission lines vary in proportion and are much less affected by the overlying opacity of the stellar atmosphere.

Obtaining high-quality spectra is observationally expensive. Nevertheless, a large-scale, near-IR, high-resolution ($R \equiv \Delta\lambda/\lambda = 20\,000$, where λ is the observational wavelength) survey of Galactic stars, APOGEE (the Apache Point Observatory Galactic Evolution Experiment, part of the Sloan Digital Sky Survey III suite of experiments; SDSS III), will observe 100 000 red giant stars with $H < 13.5$ mag across the full range of the Galactic bulge, bar, disc and halo (e.g. Allende Prieto *et al.* 2008) and, hence, provide an unprecedented Galactic stellar sample for spectroscopic parallax-based distance estimates, which in turn will provide more accurate information on e.g. the Galactic rotation curve and the 3D distribution of dust in the Milky Way. It is currently scheduled for execution between autumn 2011 and spring 2014. A complementary and synergistic, 5-year optical spectroscopic survey of millions of stars in the Milky Way with $13 < g < 20.5$ mag is planned to commence after 2010 using the Large sky Area Multi-Object fiber Spectroscopic Telescope (LAMOST; renamed to the Guoshoujing Telescope) in China (the LAMOST Experiment for Galactic Understanding and Exploration, LEGUE; Newberg *et al.* 2009).

In the meantime, a 'poor man's' version of the spectroscopy parallax method, the **photometric parallax** approach, uses observations of stellar colour to estimate luminosities. Without access to reliable spectral types, the method is predominantly applied to main-sequence stars, which have a reasonably well-defined colour–magnitude relation up to the main-sequence turn-off region, at least for a given metallicity and in the absence of reddening. Applying the same technique to giant stars would result in underestimated distances while the effect for unrecognized multiple systems is similar. The accuracy of the resulting luminosity is much lower than for the equivalent absolute magnitude obtained using the spectroscopic parallax method. However, the colours of many thousands of stars in e.g. star clusters can be observed simultaneously using just a few observations. As a result, a fairly reliable estimate of the distance to the cluster can be obtained.

This technique, which was first applied in this context by Gilmore and Reid (1983; see also Kuijken and Gilmore 1989; Chen *et al.* 2001; Siegel *et al.* 2002), was recently used (Jurić *et al.* 2008) to estimate distances to ~48 million stars with $r < 22$ mag (and $> 5\sigma$ detections in at least either the SDSS g or r filters) in the Milky Way observed as part of the SDSS – covering a large area, 6500 deg^2, at distances spanning the range from 20 to 100 kpc, mostly at high Galactic latitudes – enabling detailed studies of Galactic substructure and discoveries of new streams and stellar overdensities (see also Belokurov *et al.* 2006 for a simple application, resulting in the discovery of the 'Field of Streams'; see Figure 2.12),

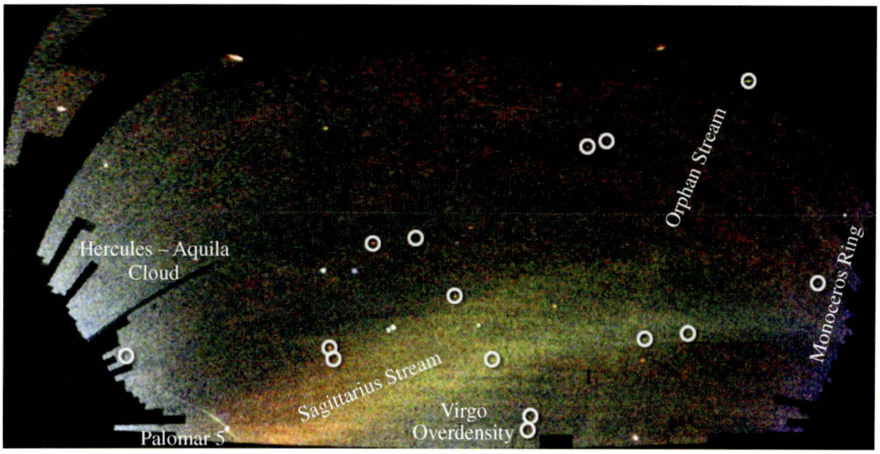

Figure 2.12 *A map of stars in the outer regions of the Milky Way, derived from the SDSS images of the northern sky, shown in a Mercator-like projection. The colour indicates the distance of the stars, while the intensity represents their density on the sky. Structures visible in this map include streams of stars torn from the Sagittarius dwarf galaxy, a smaller 'orphan' stream crossing the Sagittarius streams, the 'Monoceros Ring' which encircles the Milky Way disc, trails of stars being stripped from the GC Palomar 5, and excesses of stars found towards the constellations Virgo and Hercules. Circles enclose new Milky Way companions discovered by the SDSS; two of these are faint GCs, while the others are faint dwarf galaxies.* (Reprinted from V. Belokurov and the Sloan Digital Sky Survey Collaboration (online: www.sdss.org), Field of Streams, Copyright 2006, with permission of the SDSS.)

without the need to rely on *a priori* model assumptions as regards the underlying stellar density distribution, necessarily imposed by having access to only a limited number of lines of sight.

Nevertheless, earlier studies based on smaller numbers of sightlines had already been successful in using a version of the photometric parallax method – although limited to brighter magnitudes and, hence, dominated by giant stars – to detect features like the Monoceros ring (cf. Figure 2.12). For instance, Newberg *et al.* (2002) used some 5 million stars from the SDSS, combined with detailed colour–magnitude diagram analysis, to detect substructures in the Galactic halo, while Rocha-Pinto *et al.* (2003, 2004; see also Majewski *et al.* 2004 for upper main-sequence analysis and a review of previous studies) employed near-IR observations from the Two-Micron All-Sky Survey (2MASS) to select M giants, combined with a colour–magnitude relation appropriate to the core of the Sagittarius dwarf spheroidal galaxy which is currently being cannibalized by the Milky Way, to trace the Monoceros feature in the Galactic Anticentre direction.

A disadvantage of pushing the method down to main-sequence luminosities is its limited applicable distance range compared to that accessible through RR Lyrae (see Section 3.5.5) and M-giant analysis (approximately 20 versus 100 kpc). On the other hand, the number of main-sequence stars is vastly larger than that of other tracers: the number ratio of main-sequence to RR Lyrae stars in the SDSS sample is $\sim$10 000, and even larger for M giants (e.g. Ivezić *et al.* 2005; Jurić *et al.* 2008). This large number of main-sequence stars significantly

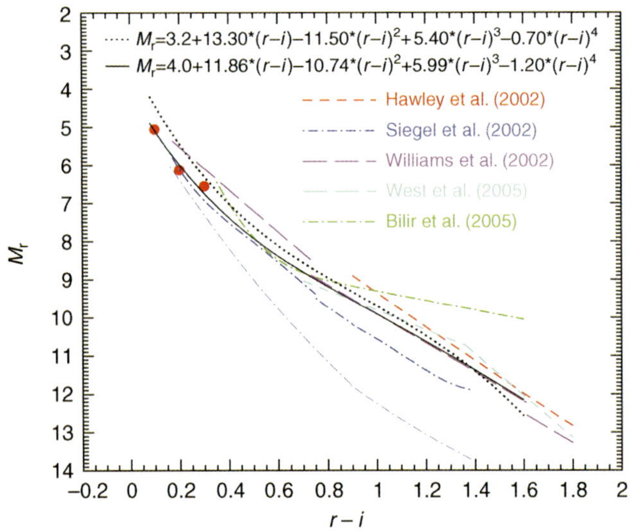

Figure 2.13 *Comparison of photometric parallax relations in the SDSS ugriz system (Jurić et al. 2008). The two relations adopted by Jurić et al. are shown by the dotted ('bright' normalization) and solid ('faint' normalization) lines. The bottom thin line from Siegel et al. (2002) corresponds to low-metallicity stars. The large symbols show SDSS observations of the GC M13.* (Reprinted from M. Juric *et al.*, *Astrophysical Journal*, **673**, The Milky Way tomography with SDSS. I. Stellar number density distribution, p. 864–914, Copyright 2008, with permission of the AAS and M. Juric.)

reduces the Poissonian noise, even when the complete SDSS sample is separated into smaller samples with high spatial resolution.

In the current era of large-scale sky surveys, the potential to map the Milky Way in unprecedented 3D detail is significant. The SDSS, in particular,[9] represents a very attractive source for such studies given its high catalogue completeness and multi-passband CCD photometry to faint flux levels. In addition, because most stars ($\sim$98%) in the database are main-sequence stars (Finlator *et al.* 2000), photometric parallaxes represent a perfectly adequate method to estimate rough distances.

In addition to the small contamination by binary systems and more evolved stars, and the effects of metallicity and extinction variations, a key uncertainty to be dealt with is the choice of photometric parallax relation (see Jurić *et al.* 2008 for a review). The relevant relations available in the literature depend on the methodology used to derive them, such as geometric parallax measurements or fits to GC colour–magnitude sequences, photometric systems, and the absolute magnitude and metallicity ranges for which they are applicable. Most exhibit significant intrinsic scatter, on the order of half a magnitude or more, and they

[9] The Sloan Extension for Galactic Understanding and Exploration, SEGUE (final data release in 2008), will be able to push the distance limit for photometric parallax application to faint main-sequence stars to $\sim$150 kpc, while the **Panoramic Survey Telescope and Rapid Response System** (Pan-STARRS; Kaiser *et al.* 2002), as precursor to the **Large Synoptic Survey Telescope** (LSST; Tyson 2002), will go much beyond SDSS and catalogue 99% of the stars in the northern hemisphere that have ever been observed in visible light, thus providing access to 3D structural studies of the entire Local Group and the nearest galaxies beyond in unprecedented detail.

are not all mutually consistent, not even the relations corresponding to the same cluster (such as the Hyades), which can differ by a few tenths of a magnitude. Based on a careful review of the literature and their own data analysis, Jurić *et al.* (2008) propose a photometric parallax relation applicable to main-sequence stars in the Milky Way observed in the SDSS r and i filters (see Figure 2.13),

$$M_r = 4.0 + 11.86(r - i) - 10.74(r - i)^2 + 5.99(r - i)^3 - 1.20(r - i)^4, \qquad (2.11)$$

which agrees with previously proposed relations at both the blue and red ends (see Karaali *et al.* 2003, 2005 for alternative relations, carefully calibrated with respect to the trigonometric parallax-derived Hyades main sequence). They derive a second relation based on kinematic data and a simple model for the structure of the Galaxy that will match the data only if the derived photometric parallax is correct (Bond *et al.* 2010). Both relations differ at the blue end, $(r - i) \simeq 0.1$ mag, by 0.66 mag in absolute r-band magnitude. This translates into a systematic distance uncertainty of order 18%. These 'faint' and 'bright' normalizations, respectively, straddle the equivalent *Hipparcos* trigonometric-parallax-based dependence.

Precise determination of the relevant stellar metallicities is the other main source of systematic uncertainties in the photometric parallax relation. The main-sequence morphology (i.e. colour and/or temperature) depends on the stellar population's chemical composition: higher metallicity stellar populations have fainter and redder main-sequence turn-offs at a fixed age than lower metallicity populations. Higher metallicity corresponds to increased opacities in the stellar atmospheres because of the presence of electrons from metals. This allows stars to maintain equilibrium at a lower temperature, which in turn translates into redder colours. In addition, **line blanketing** in stellar atmospheres (the change in temperature structure owing to opacities of numerous emission lines, particularly at shorter wavelengths) also contributes to reddening at higher metallicities. Correcting photometric parallaxes for metallicity effects is not straightforward, however (see e.g. Jurić *et al.* 2008, who discuss a range of methods which one can use to overcome metallicity effects), although the SDSS M_r versus $(r - i)$ relation is less sensitive to metallicity than the traditional M_V versus $(B - V)$ relation (Reid *et al.* 2001).

Bibliography

Adams WS and Kohlschütter A 1914 Some spectral criteria for the determination of absolute stellar magnitudes. *Astrophys. J.* **40**, 385–398.

Alexander DR, Brocato E, Cassisi S, Castellani V, Ciacio F and Degl'Innocenti S 1997 A theoretical approach to globular cluster low main sequence stars. *Astron. Astrophys.* **317**, 90–98.

Allende Prieto C, Majewski SR, Schiavon R, Cunha K, Frinchaboy P, Holtzman J, Johnston K, Shetrone M, Skrutskie M, Smith V and Wilson J 2008 APOGEE: the Apache Point Observatory Galactic Evolution Experiment. *Astron. Nachr.* **329**, 1018–1021.

Andersen J 1991 Accurate masses and radii of normal stars. *Astron. Astrophys. Rev.* **3**, 91–126.

Anderson JA 1920 Application of Michelson's interferometer method to the measurement of close double stars. *Contrib. Mount Wilson Obs./Carnegie Inst. Wash.* **185**, 1–13.

Angelov T 1993 On the dynamical masses of visual binaries. *Bull. Astron. Belgr.* **148**, 1–6.

Arenou F and Luri X 1999 Distances and absolute magnitudes from trigonometric parallaxes. In *Harmonizing Cosmic Distance Scales in a Post-Hipparcos Era* (eds Egret D and Heck A), *Astron. Soc. Pac. Conf. Ser.* **167**, 13–32.

Arias EF, Charlot P, Feissel M and Lestrade J-F 1995 The extragalactic reference system of the International Earth Rotation Service, ICRS. *Astron. Astrophys.* **303**, 604–608.

Baize P and Romaní L 1946 Formules nouvelles pour le calcul des parallaxes dynamiques des couples orbitaux. *Ann. d'Astrophys.* **9**, 13–41.

Baraffe I, Chabrier G, Allard F and Hauschildt PH 1995 New evolutionary tracks for very low mass stars. *Astrophys. J.* **446**, L35–L38.

Baraffe I, Chabrier G, Allard F and Hauschildt PH 1997 Evolutionary models for metal-poor low-mass stars. Lower main sequence of globular clusters and halo field stars. *Astron. Astrophys.* **327**, 1054–1069.

Baraffe I, Chabrier G, Allard F and Hauschildt PH 1998 Evolutionary models for solar metallicity low-mass stars: mass–magnitude relationships and color–magnitude diagrams. *Astron. Astrophys.* **337**, 403–412.

Bedin LR, Anderson J, King IR and Piotto G 2001 Color–magnitude diagram and luminosity function of M4 near the hydrogen-burning limit. *Astrophys. J.* **560**, L75–L78.

Belokurov V, Zucker DB, Evans NW, Gilmore G, Vidrih S, Bramich DM, Newberg HJ, Wyse RFG, Irwin MJ, Fellhauer M, Hewett PC, Walton NA, Wilkinson MI, Cole N, Yanny B, Rockosi CM, Beers TC, Bell EF, Brinkmann J, Ivezić Ž and Lupton R 2006 The Field of Streams: Sagittarius and its siblings. *Astrophys. J.* **642**, L137–L140.

Benedict GF, McArthur BE, Fredrick LW, Harrison TE, Slesnick CL, Rhee J, Patterson RJ, Skrutskie MF, Franz OG, Wasserman LH, Jefferys WH, Nelan E, van Altena W, Shelus PJ, Hemenway PD, Duncombe RL, Story D, Whipple AL and Bradley AJ 2002 Astrometry with the Hubble Space Telescope: a parallax of the fundamental distance calibrator δ Cephei. *Astron. J.* **124**, 1695–1705.

Benedict GF, McArthur BE, Feast MW, Barnes TG, Harrison TE, Patterson RJ, Menzies JW, Bean JL and Freedman WL 2007 Hubble Space Telescope Fine Guidance Sensor parallaxes of Galactic Cepheid variable stars: period–luminosity relations. *Astron. J.* **133**, 1810–1827.

Berry A 1899 *A Short History of Astronomy.* New York, Charles Scribner's & Sons.

Bessel FW 1838 Bestimmung der Entfernung des 61sten Sterns des Schwans. *Astron. Nachr.* **16**, 65–96.

Bilir S, Karaali S, Ak S, Yaz E and Hamzaoğlu E 2006 Galactic longitude dependent Galactic model parameters. *New Astron.* **12**, 234–245.

Bochanski JJ, Hawley SL and West AA 2011 The Sloan Digital Sky Survey DR7 spectroscopic M dwarf catalog. II. Statistical parallax analysis. *Astron. J.* **141**, 98.

Boden A and Quirrenbach A 2008 Astrometry with ground-based interferometers. In *A Giant Step: From Milli- to Micro-Arcsecond Astrometry* (eds Jin WJ, Platais I and Perryman MAC), *Proc. Int'l Astron. Union Symp.* **248**, 36–43.

Bond NA, Ivezić Ž, Sesar B, Jurić M, Munn JA, Kowalski A, Loebman S, Roškar R, Beers TC, Dalcanton J, Rockosi CM, Yanny B, Newberg HJ, Allende Prieto C, Wilhelm R, Lee YS, Sivarani T, Majewski SR, Norris JE, Bailer-Jones CAL, Re Fiorentin P, Schlegel D, Uomoto A, Lupton RH, Knapp GR, Gunn JE, Covey KR, Allyn Smith J, Miknaitis G, Doi M, Tanaka M, Fukugita M, Kent S, Finkbeiner D, Quinn TR, Hawley S, Anderson S, Kiuchi F, Chen A, Bushong J, Sohi H, Haggard D, Kimball A, McGurk R, Barentine J, Brewington H, Harvanek M, Kleinman S, Krzesinski J, Long D, Nitta A, Snedden S, Lee B, Pier JR, Harris H, Brinkmann J and Schneider DP 2010 The Milky Way tomography with SDSS. III. Stellar kinematics. *Astrophys. J.* **716**, 1–29.

Brewer JP, Fahlman GG, Richer HB, Searle L and Thompson I 1993 CCD photometry in the globular cluster NGC 3201. *Astron. J.* **105**, 2158–2171.

Brown AGA, Arenou F, van Leeuwen F, Lindegren L and Luri X 1997 Considerations in making full use of the Hipparcos catalogue. In *Proc. ESA Symp. 'Hipparcos – Venice '97'* (ed. Perryman MAC), *ESA Spec. Publ.* **402**, 63–68.

Brunthaler A, Menten KM, Reid MJ, Henkel C, Bower GC and Falcke H 2009 Discovery of a bright radio transient in M82: a new radio supernova? *Astron. Astrophys.* **499**, L17–L20.

Butkevich AG, Berdyugin AV and Teerikorpi P 2005 Statistical biases in stellar astronomy: the Malmquist bias revisited. *Mon. Not. R. Astron. Soc.* **362**, 321–330.

Buzzoni A, Patelli L, Bellazzini M, Fusi Pecci F and Oliva E 2010 Bolometric correction and spectral energy distribution of cool stars in Galactic clusters. *Mon. Not. R. Astron. Soc.* **403**, 1592–1610.

Cassatella A, Altamore A, Badiali M and Cardini D 2001 On the Wilson–Bappu relationship in the Mg II k line. *Astron. Astrophys.* **374**, 1085–1091.

Chabrier G and Baraffe I 1997 Structure and evolution of low-mass stars. *Astron. Astrophys.* **327**, 1039–1053.

Chabrier G and Méra D 1997 Determination of the globular cluster and halo stellar mass functions and stellar and brown dwarf densities. *Astron. Astrophys.* **328**, 83–94.

Chabrier G, Baraffe I and Plez B 1996 Mass–luminosity relationship and lithium depletion for very low mass stars. *Astrophys. J.* **459**, L91–L94.

Chatterjee S, Cordes JM, Vlemmings WHT, Arzoumanian Z, Goss WM and Lazio TJW 2004 Pulsar parallaxes at 5 GHz with the Very Long Baseline Array. *Astrophys. J.* **604**, 339–345.

Chen B, Stoughton C, Smith JA, Uomoto A, Pier JR, Yanny B, Ivezić Ž, York DG, Anderson JE, Annis J, Brinkmann J, Csabai I, Fukugita M, Hindsley R, Lupton R and Munn JA 2001 Stellar population studies with the SDSS. I. The vertical distribution of stars in the Milky Way. *Astrophys. J.* **553**, 184–197.

Cunha MS, Aerts C, Christensen-Dalsgaard J, Baglin A, Bigot L, Brown TM, Catala C, Creevey OL, Domiciano de Souza A, Eggenberger P, Garcia PJV, Grundahl F, Kervella P, Kurtz DW, Mathias P, Miglio A, Monteiro MJPFG, Perrin G, Pijpers FP, Pourbaix D, Quirrenbach A, Rousselet-Perraut K, Teixeira TC, Thévenin F and Thompson MJ 2007 Asteroseismology and interferometry. *Astron. Astrophys. Rev.* **14**, 217–360.

Dambis AK 2009 The kinematics and zero-point of the log $P-\langle M_K \rangle$ relation for Galactic RR Lyrae variables via statistical parallax. *Mon. Not. R. Astron. Soc.* **396**, 553–569.

D'Antona F and Mazzitelli I 1983 Mass–luminosity relation and initial mass function at the faint end of the main sequence. Is there a real deficit of very low-mass stars? *Astron. Astrophys.* **127**, 149–152.

D'Antona F and Mazzitelli I 1996 Stellar models and luminosity functions for the Population II main sequence down to its lower end. *Astrophys. J.* **456**, 329–336.

de Grijs R, Gilmore GF, Johnson RA and Mackey AD 2002 Mass segregation in young compact star clusters in the Large Magellanic Cloud. II. Mass functions. *Mon. Not. R. Astron. Soc.* **331**, 245–258.

Dravins D, Lindegren L, Madsen S and Holmberg J 1997 Astrometric radial velocities from Hipparcos. In *Proc. ESA Symp. 'Hipparcos – Venice '97'* (ed. Perryman MAC), *ESA Spec. Publ.* **402**, 733–738.

Dzib S, Loinard L, Mioduszewski AJ, Boden AF, Rodríguez LF and Torres RM 2010 VLBA determination of the distance to nearby star-forming regions. IV. A preliminary distance to the proto-Herbig AeBe Star EC 95 in the Serpens core. *Astrophys. J.* **718**, 610–619.

Elson RAW, Gilmore GF, Santiago BX and Casertano S 1995 HST observations of the stellar population of the globular cluster ω Cen. *Astron. J.* **110**, 682–692.

ESA 1997 The Hipparcos and Tycho Catalogues. *ESA SP-1200*.

Fantino E 2000 Dinamica di assetto del satellite Hipparcos dell'ESA ed applicazione alla missione astrometrica Tedesca DIVA (Attitude dynamics of the ESA Hipparcos satellite and application to the German astrometric mission DIVA). PhD thesis, Università degli Studi di Padova, Italy.

Feast MW, Laney CD, Kinman TD, van Leeuwen F and Whitelock PA 2008 The luminosities and distance scales of type II Cepheid and RR Lyrae variables. *Mon. Not. R. Astron. Soc.* **386**, 2115–2134.

Ferraro FR, Carretta E, Bragaglia A, Renzini A and Ortolani S 1997 The luminosity function of the globular cluster NGC 6752 with the Hubble Space Telescope: evidence of mass segregation. *Mon. Not. R. Astron. Soc.* **286**, 1012–1022.

Finlator K, Ivezić Ž, Fan X, Strauss MA, Knapp GR, Lupton RH, Gunn JE, Rockosi CM, Anderson JE, Csabai I, Hennessy GS, Hindsley RB, McKay TA, Nichol RC, Schneider DP, Smith JA and York DG 2000 Optical and infrared colors of stars observed by the Two Micron All Sky Survey and the Sloan Digital Sky Survey. *Astron. J.* **120**, 2615–2626.

Fricke W, Schwan H and Lederle T 1988 *Fifth Fundamental Catalogue. I.* Heidelberg Veröff. Astron. Rechen Inst.

Ghez AM, Salim S, Weinberg NN, Lu JR, Do T, Dunn JK, Matthews K, Morris MR, Yelda S, Becklin EE, Kremenek T, Milosavljevic M and Naiman J 2008 Measuring distance and properties of the Milky Way's central supermassive black hole with stellar orbits. *Astrophys. J.* **689**, 1044–1062.

Gilmore G and Reid N 1983 New light on faint stars. III. Galactic structure towards the South Pole and the Galactic thick disc. *Mon. Not. R. Astron. Soc.* **202**, 1025–1047.

Girardi L, Bressan A, Bertelli G and Chiosi C 2000 Evolutionary tracks and isochrones for low- and intermediate-mass stars: from 0.15 to 7 $M_\odot$, and from $Z = 0.0004$ to 0.03. *Astron. Astrophys. Suppl. Ser.* **141**, 371–383.
Gorshanov DL, Shakht NA and Kisselev AA 2006 Observations of the binary star 61 Cyg on the 26 inch refractor at the Pulkovo observatory. *Astrophys.* **49**, 386–396.
Hanson MM, Conti PS and Rieke MJ 1996 A spectral atlas of hot, luminous stars at 2 microns. *Astrophys. J. Suppl. Ser.* **107**, 281–311.
Hawley SL, Jefferys WH, Barnes III TG and Lai W 1986 Absolute magnitudes and kinematic properties of RR Lyrae stars. *Astrophys. J.* **302**, 626–631.
Hawley SL, Covey KR, Knapp GR, Golimowski DA, Fan X, Anderson SF, Gunn JE, Harris HC, Ivezić Ž, Long GM, Lupton RH, McGehee PM, Narayanan V, Peng E, Schlegel D, Schneider DP, Spahn EY, Strauss MA, Szkody P, Tsvetanov Z, Walkowicz LM, Brinkmann J, Harvanek M, Hennessy GS, Kleinman SJ, Krzesinski JLD, Neilsen EH, Newman PR, Nitta A, Snedden SA and York DG 2002 Characterization of M, L, and T dwarfs in the Sloan Digital Sky Survey. *Astron. J.* **123**, 3409–3427.
Hazard C, Sutton J, Argue AN, Kenworthy CN, Morrison LV and Murray CA 1971 Accurate radio and optical positions of 3C273B. *Nat. Phys. Sci.* **233** 89–92.
Heintz WD 1978 Double stars. *Geophys. Astrophys. Monogr.* **15**.
Henderson T 1842 On the parallax of α Centauri, deduced from T. Maclear's observations at the Cape of Good Hope. *Mon. Not. R. Astron. Soc.* **5**, 171–172.
Henry TJ and McCarthy Jr DW 1993 The mass–luminosity relation for stars of mass 1.0 to 0.08 $M_\odot$. *Astron. J.* **106**, 773–789.
Hirota T, Bushimata T, Choi YK, Honma M, Imai H, Iwadate K, Jike T, Kameno S, Kameya O, Kamohara R, Kan-Ya Y, Kawaguchi N, Kijima M, Kim MK, Kobayashi H, Kuji S, Kurayama T, Manabe S, Maruyama K, Matsui M, Matsumoto N, Miyaji T, Nagayama T, Nakagawa A, Nakamura K, Oh CS, Omodaka T, Oyama T, Sakai S, Sasao T, Sato K, Sato M, Shibata KM, Shintani M, Tamura Y, Tsushima M and Yamashita K 2007 Distance to Orion KL measured with VERA. *Publ. Astron. Soc. Jpn* **59**, 897–903.
Høg E 2008 From the Roemer mission to Gaia. In *A Giant Step: From Milli- to Micro-Arcsecond Astrometry* (eds Jin WJ, Platais I and Perryman MAC), *Proc. Int'l Astron. Union Symp.* **248**, 300–302.
Høg E, Fabricius C, Makarov VV, Urban S, Corbin T, Wycoff G, Bastian U, Schwekendiek P and Wicenec A 2000a The Tycho-2 catalogue of the 2.5 million brightest stars. *Astron. Astrophys.* **355**, L27–L30.
Høg E, Fabricius C, Makarov VV, Bastian U, Schwekendiek P, Wicenec A, Urban S, Corbin T and Wycoff G 2000b Construction and verification of the Tycho-2 Catalogue. *Astron. Astrophys.* **357**, 367–386.
Honma M, Bushimata T, Choi YK, Hirota T, Imai H, Iwadate K, Jike T, Kameya O, Kamohara R, Kan-Ya Y, Kawaguchi N, Kijima M, Kobayashi H, Kuji S, Kurayama T, Manabe S, Miyaji T, Nagayama T, Nakagawa A, Oh CS, Omodaka T, Oyama T, Sakai S, Sato K, Sasao T, Shibata KM, Shintani M, Suda H, Tamura Y, Tsushima M and Yamashita K 2007 Astrometry of Galactic star-forming region Sharpless 269 with VERA: parallax measurements and constraint on outer rotation curve. *Publ. Astron. Soc. Jpn* **59**, 889–895.
Hummel CA, Mozurkewich D, Armstrong JT, Hajian AR, Elias II NM and Hutter DJ 1998 Navy Prototype Optical Interferometer observations of the double stars Mizar A and Matar. *Astron. J.* **116**, 2536–2548.
Ivezić Ž, Vivas AK, Lupton RH and Zinn R 2005 The selection of RR Lyrae stars using single-epoch data. *Astron. J.* **129**, 1096–1108.
Johnson HL and Morgan WW 1953 Fundamental stellar photometry for standards of spectral type on the revised system of the Yerkes spectral atlas. *Astrophys. J.* **117**, 313–352.
Jones RV, Carney BW, Storm J and Latham D 1992 The Baade–Wesselink method and the distances to RR Lyrae stars. VII. The field stars SW Andromedae and DX Delphini and a comparison of recent Baade–Wesselink analyses. *Astrophys. J.* **385**, 646–662.
Jurić M, Ivezić Ž, Brooks A, Lupton RH, Schlegel D, Finkbeiner D, Padmanabhan N, Bond N, Sesar B, Rockosi CM, Knapp GR, Gunn JE, Sumi T, Schneider DP, Barentine JC, Brewington HJ,

Brinkmann J, Fukugita M, Harvanek M, Kleinman SJ, Krzesinski J, Long D, Neilsen Jr EH, Nitta A, Snedden SA and York DG 2008 The Milky Way tomography with SDSS. I. Stellar number density distribution. *Astrophys. J.* **673**, 864–914.

Kaiser N, Aussel H, Burke BE, Boesgaard H, Chambers K, Chun MR, Heasley JN, Hodapp K-W, Hunt B, Jedicke R, Jewitt D, Kudritzki R, Luppino GA, Maberry M, Magnier E, Monet DG, Onaka PM, Pickles AJ, Rhoads PHH, Simon T, Szalay A, Szapudi I, Tholen DJ, Tonry JL, Waterson M and Wick J 2002 Pan-STARRS: a Large Synoptic Survey Telescope array. *Proc. SPIE* **4836**, 154–164.

Karaali S, Karataş Y, Bilir S, Ak SG and Hamzaoğlu E 2003 A new procedure for the photometric parallax estimation. *Publ. Astron. Soc. Aust.* **20**, 270–278.

Karaali S, Bilir S and Tunçel S 2005 New colour transformations for the Sloan photometry, and revised metallicity calibration and equations for photometric parallax estimation. *Publ. Astron. Soc. Aust.* **22**, 24–28.

Kepler J 1596 *Mysterium Cosmographicum* (Preface). Tübingen. Full title: *Prodromus dissertationum cosmographicarum, continens mysterium cosmographicum, de admirabili proportione orbium coelestium, de que causis coelorum numeri, magnitudinis, motuumque periodicorum genuinis & proprijs, demonstratum, per quinque regularia corpora geometrica (Forerunner of the Cosmological Essays, Which Contains the Secret of the Universe; on the Marvelous Proportion of the Celestial Spheres, and on the True and Particular Causes of the Number, Magnitude, and Periodic Motions of the Heavens; Established by Means of the Five Regular Geometric Solids).*

Kim MK, Hirota T, Honma M, Kobayashi H, Bushimata T, Choi YK, Imai H, Iwadate K, Jike T, Kameno S, Kameya O, Kamohara R, Kan-Ya Y, Kawaguchi N, Kuji S, Kurayama T, Manabe S, Matsui M, Matsumoto N, Miyaji T, Nagayama T, Nakagawa A, Oh CS, Omodaka T, Oyama T, Sakai S, Sasao T, Sato K, Sato M, Shibata KM, Tamura Y and Yamashita K 2008 SiO maser observations toward Orion-KL with VERA. *Publ. Astron. Soc. Jpn* **60**, 991–999.

Kobayashi H, Kawaguchi N, Manabe S, Shibata KM, Honma M, Tamura Y, Kameya O, Hirota T, Jike T, Imai H and Omodaka T 2008 Phase referencing VLBI astrometry observation system: VERA. In *A Giant Step: From Milli- to Micro-Arcsecond Astrometry* (eds Jin WJ, Platais I and Perryman MAC), *Proc. Int'l Astron. Union Symp.* **248**, 148–155.

Kovalevsky J 1998 On the uncertainty of distances derived from parallax measurements. *Astron. Astrophys.* **340**, L35–L38.

Kraus S, Balega YY, Berger J-P, Hofmann K-H, Millan-Gabet R, Monnier JD, Ohnaka K, Pedretti E, Preibisch T, Schertl D, Schloerb FP, Traub WA and Weigelt G 2007 Visual/infrared interferometry of Orion Trapezium stars: preliminary dynamical orbit and aperture synthesis imaging of the θ^1 Orionis C system. *Astron. Astrophys.* **466**, 649–659.

Kroupa P 2000 Constraints on stellar-dynamical models of the Orion Nebula Cluster. *New Astron.* **4**, 615–624.

Kroupa P and Tout CA 1997 The theoretical mass–magnitude relation of low mass stars and its metallicity dependence. *Mon. Not. R. Astron. Soc.* **287**, 402–414.

Kroupa P, Tout CA and Gilmore G 1990 The low-luminosity stellar mass function. *Mon. Not. R. Astron. Soc.* **244**, 76–85.

Kroupa P, Tout CA and Gilmore G 1993 The distribution of low-mass stars in the Galactic disc. *Mon. Not. R. Astron. Soc.* **262**, 545–587.

Kuijken K and Gilmore G 1989 The mass distribution in the Galactic disc. II. Determination of the surface mass density of the Galactic disc near the sun. *Mon. Not. R. Astron. Soc.* **239**, 605–649.

Layden AC, Hanson RB, Hawley SL, Klemola AR and Hanley CJ 1996 The absolute magnitude and kinematics of RR Lyrae stars via statistical parallax. *Astron. J.* **112**, 2110–2131.

Leggett SK, Allard F, Berriman G, Dahn CC and Hauschildt PH 1996 Infrared spectra of low-mass stars: toward a temperature scale for red dwarfs. *Astrophys. J. Suppl. Ser.* **104**, 117–143.

Lestrade J-F, Preston RA, Jones DL, Phillips RB, Rogers AEE, Titus MA, Rioja MJ and Gabuzda DC 1999 High-precision VLBI astrometry of radio-emitting stars. *Astron. Astrophys.* **344**, 1014–1026.

Lieske JH, Lederle T, Fricke W and Morando B 1977 Expressions for the precession quantities based upon the IAU (1976) system of astronomical constants. *Astron. Astrophys.* **58**, 1–16.

Lindegren L 2005 The astrometric instrument of Gaia: principles. In *The Three-Dimensional Universe with Gaia* (eds Turon C, O'Flaherty KS and Perryman MAC), *ESA Spec. Publ.* **576**, 29–34.

Lindegren L 2010 Gaia: astrometric performance and current status of the project. In *Relativity in Fundamental Astronomy: Dynamics, Reference Frames, and Data Analysis* (eds Klioner SA, Seidelmann PK and Soffel MH), *Proc. Int'l Astron. Union Symp.* **261**, 296–305.
Loinard L, Mioduszewski AJ, Rodríguez LF, González RA, Rodríguez MI, Torres RM 2005 Multi-epoch VLBA observations of T Tauri South. *Astrophys. J.* **619**, L179–L182.
Loinard L, Torres RM, Mioduszewski AJ, Rodríguez LF, González-Lópezlira RA Lachaume R, Vázquez V and González E 2007 VLBA determination of the distance to nearby star-forming regions. I. The distance to T Tauri with 0.4% accuracy. *Astrophys. J.* **671**, 546–554.
Loinard L, Torres RM, Mioduszewski AJ and Rodríguez LF 2008 A preliminary VLBA distance to the core of Ophiuchus, with an accuracy of 4%. *Astrophys. J.* **675**, L29–L32.
Luri X, Mennessier MO, Torra J and Figueras F 1996 *Astron. Astrophys. Suppl. Ser.* **117**, 405–415.
Luri X, Gomez AE, Torra J, Figueras F and Mennessier MO 1998 The LMC distance modulus from Hipparcos RR Lyrae and classical Cepheid data. *Astron. Astrophys.* **335**, L81–L84.
Majewski SR, Kunkel WE, Law DR, Patterson RJ, Polak AA, Rocha-Pinto HJ, Crane JD, Frinchaboy PM, Hummels CB, Johnston KV, Rhee J, Skrutskie MF and Weinberg M 2004 A Two Micron All Sky Survey view of the Sagittarius dwarf galaxy. II. Swope telescope spectroscopy of M giant stars in the dynamically cold Sagittarius tidal stream. *Astron. J.* **128**, 245–259.
Makarov VV 2002 Computing the parallax of the Pleiades from the Hipparcos intermediate astrometry data: an alternative approach. *Astron. J.* **124**, 3299–3304.
Malmquist KG 1920 A study of the stars of spectral type A. *Medd. Lund Astron. Obs. Ser.* **II (22)**, 1–39.
Mantovani F and Kus A (eds) 2004 *The Role of VLBI in Astrophysics, Astrometry and Geodesy. Proc. NATO Adv. Study Inst.* **135**.
Menten KM, Reid MJ, Forbrich J and Brunthaler A 2007 The distance to the Orion Nebula. *Astron. Astrophys.* **474**, 515–520.
Merrill PW 1922 Interferometer observations of double stars. *Contrib. Mount Wilson Obs./Carnegie Inst. Wash.* **240**, 1–13.
Michelson AA and Pease FG 1921 Measurement of the diameter of α Orionis with the interferometer. *Contrib. Mount Wilson Obs./Carnegie Inst. Wash.* **203**, 1–11.
Morgan WW and Keenan PC 1973 Spectral classification. *Annu. Rev. Astron. Astrophys.* **11**, 29–50.
Morgan WW, Keenan PC and Kellman E 1943 An atlas of stellar spectra, with an outline of spectral classification. The University of Chicago Press.
Murata Y, Saito H and Tsuboi M 2008 The VSOP-2 (ASTRO-G) project. In *The Role of VLBI in the Golden Age for Radio Astronomy (Proc. 9th Eur. VLBI Netw. Symp.)*, p. 49.
Murray CA 1983 *Vectorial Astrometry.* Bristol, Adam Hilger.
Newberg HJ, Yanny B, Rockosi C, Grebel EK, Rix H-W, Brinkmann J, Csabai I, Hennessy G, Hindsley RB, Ibata R, Ivezić Ž, Lamb D, Nash ET, Odenkirchen M, Rave HA, Schneider DP, Smith JA, Stolte A and York DG 2002 The ghost of Sagittarius and lumps in the halo of the Milky Way. *Astrophys. J.* **569**, 245–274.
Newberg HJ and US participants in the LAMOST project of China 2009 The LAMOST Spectroscopic Survey of Milky Way Stars (LEGUE). *Bull. Am. Astron. Soc.* **41**, 229–229.
O'Mullane W, Lammers U, Bailer-Jones C, Bastian U, Brown AGA, Drimmel R, Eyer L, Huc C, Katz D, Lindegren L, Pourbaix D, Luri X, Torra J, Mignard F and van Leeuwen F 2007 Gaia data processing architecture. In *Astronomical Data Analysis Software and Systems XVI* (eds Shaw RA, Hill F and Bell DJ), *Astron. Soc. Pac. Conf. Ser.* **376**, 99–108.
Pace G, Pasquini L and Ortolani S 2003 The Wilson–Bappu effect: a tool to determine stellar distances. *Astron. Astrophys.* **401**, 997–1007.
Pavlovskaya ED 1953 *Perem. Zvezdy* **9**, 349.
Perryman MAC 2009 *Astronomical Applications of Astrometry: Ten Years of Exploitation of the Hipparcos Satellite Data.* Cambridge University Press.
Perryman MAC, Lindegren L, Kovalevsky J, Hoeg E, Bastian U, Bernacca PL, Crézé M, Donati F, Grenon M, van Leeuwen F, van der Marel H, Mignard F, Murray CA, Le Poole RS, Schrijver H, Turon C, Arenou F, Froeschlé M and Petersen CS 1997 The HIPPARCOS Catalogue. *Astron. Astrophys.* **323**, L49–L52.

Petrov L, Gordon D, Gipson J, MacMillan D, Ma C, Fomalont E, Walker RC and Carabajal C 2009 Precise geodesy with the Very Long Baseline Array. *J. Geod.* **83**, 859–876.

Piotto G, Cool AM and King IR 1997 A comparison of deep HST luminosity functions of four globular clusters. *Astron. J.* **113**, 1345–1352.

Popowski P and Gould A 1998a Systematics of RR Lyrae statistical parallax. I. Mathematics. *Astrophys. J.* **506**, 259–270.

Popowski P and Gould A 1998b Systematics of RR Lyrae statistical parallax. II. Proper motions and radial velocities. *Astrophys. J.* **506**, 271–280.

Popper DM 1980 Stellar masses. *Annu. Rev. Astron. Astrophys.* **18**, 115–164.

Ratnatunga KU and Upgren AR 1997 Kinematics of common dwarf stars: a maximum-likelihood analysis. *Astrophys. J.* **476**, 811–828.

Reed BC 1984 On the computation of dynamical parallaxes. *J. Astron. Soc. Can.* **78**, 83–85.

Reid IN, van Wyk F, Marang F, Roberts G, Kilkenny D and Mahoney S 2001 A search for previously unrecognized metal-poor subdwarfs in the Hipparcos astrometric catalogue. *Mon. Not. R. Astron. Soc.* **325**, 931–962.

Reid MJ, Menten KM, Zheng XW, Brunthaler A, Moscadelli L, Xu Y, Zhang B, Sato M, Honma M, Hirota T, Hachisuka K, Choi YK, Moellenbrock GA and Bartkiewicz A 2009 Trigonometric parallaxes of massive star-forming regions. VI. Galactic structure, fundamental parameters, and noncircular motions. *Astrophys. J.* **700**, 137–148.

Reylé C, Scholz R-D, Schultheis M, Robin AC and Irwin M 2006 Optical spectroscopy of high proper motion stars: new M dwarfs within 10 pc and the closest pair of subdwarfs. *Mon. Not. R. Astron. Soc.* **373**, 705–714.

Rocha-Pinto HJ, Majewski SR, Skrutskie MF and Crane JD 2003 Tracing the Galactic anticenter stellar stream with 2MASS M giants. *Astrophys. J.* **594**, L115–L118.

Rocha-Pinto HJ, Majewski SR, Skrutskie MF, Crane JD and Patterson RJ 2004 Exploring halo substructure with giant stars: a diffuse star cloud or tidal debris around the Milky Way in Triangulum–Andromeda. *Astrophys. J.* **615**, 732–737.

Sandstrom KM, Peek JEG, Bower GC, Bolatto AD and Plambeck RL 2007 A parallactic distance of 389^{+24}_{-21} parsecs to the Orion Nebula Cluster from Very Long Baseline Array observations. *Astrophys. J.* **667**, 1161–1169.

Saviane I, Piotto G, Fagotto F, Zaggia S, Capaccioli M and Aparicio A 1998 The Galactic globular cluster NGC 1851: its dynamical and evolutionary properties. *Astron. Astrophys.* **333**, 479–496.

Seidelmann PK 1982 1980 IAU nutation: the final report of the IAU working group on nutation. *Celest. Mech.* **27**, 79–106.

Shao M and Colavita MM 1992 Potential of long-baseline infrared interferometry for narrow-angle astrometry. *Astron. Astrophys.* **262**, 353–358.

Siegel MH, Majewski SR, Reid IN and Thompson IB 2002 Star counts redivivus. IV. Density laws through photometric parallaxes. *Astrophys. J.* **578**, 151–175.

Smith Jr H 1987a The calibration problem. I. Estimation of mean absolute magnitude using trigonometric parallaxes. *Astron. Astrophys.* **171**, 336–341.

Smith Jr H 1987b The calibration problem. II. Trigonometric parallaxes selected according to proper motion and the problem of statistical parallaxes *Astron. Astrophys.* **171**, 342–347.

Smith Jr H and Eichhorn H 1996 On the estimation of distances from trigonometric parallaxes. *Mon. Not. R. Astron. Soc.* **281**, 211–218.

Sovers OJ, Fanselow JL and Jacobs CS 1998 Astrometry and geodesy with radio interferometry: experiments, models, results. *Rev. Mod. Phys.* **70**, 1393–1454.

Stencel RE 1977 Emission lines in the wings of Ca II H and K. II. Stellar observations: dependence of line width on luminosity and related topics. *Astrophys. J.* **215**, 176–187.

Stencel RE 2009 The Wilson–Bappu effect: 50 years later. In *The Biggest, Baddest, Coolest Stars* (eds Luttermoser DG, Smith BJ and Stencel RE), *Astron. Soc. Pac. Conf. Ser.* **412**, 251–254.

Strugnell P, Reid N and Murray CA 1986 The luminosity and kinematics of RR Lyrae stars in the solar neighbourhood. I. Statistical parallaxes of RR Lyraes. *Mon. Not. R. Astron. Soc.* **220**, 413–427.

Torres RM, Loinard L, Mioduszewski AJ and Rodríguez LF 2007 VLBA determination of the distance to nearby star-forming regions. II. Hubble 4 and HDE 283572 in Taurus. *Astrophys. J.* **671**, 1813–1819.

Torres RM, Loinard L, Mioduszewski AJ and Rodríguez LF 2009 VLBA determination of the distance to nearby star-forming regions. III. HP Tau/G2 and the three-dimensional structure of Taurus. *Astrophys. J.* **698**, 242–249.

Tout CA, Pols OR, Eggleton PP and Han Z 1996 Zero-age main-sequence radii and luminosities as analytic functions of mass and metallicity. *Mon. Not. R. Astron. Soc.* **281**, 257–262.

Tyson JA 2002 Large Synoptic Survey Telescope: overview. *Proc. SPIE*, **4836**, 10–20.

Vacca WD, Garmany CD and Shull JM 1996 The Lyman-continuum fluxes and stellar parameters of O and early B-type stars. *Astrophys. J.* **460**, 914–931.

van Leeuwen F 2005 Rights and wrongs of the Hipparcos data. A critical quality assessment of the Hipparcos Catalogue. *Astron. Astrophys.* **439**, 805–822.

van Leeuwen F 2007a Hipparcos, the new reduction of the raw data. *Astrophys. Space Sci. Libr.* **350**.

van Leeuwen F 2007b Validation of the new Hipparcos reduction. *Astron. Astrophys.* **474**, 653–664.

van Leeuwen F 2010 Hipparcos, the new reduction. *Space Sci. Rev.* **151**, 209–226.

van Leeuwen F and Fantino E 2003 Dynamic modelling of the Hipparcos attitude. *Space Sci. Rev.* **108**, 537–576.

Wallerstein G, Machado-Pelaez L and Gonzalez G 1999 The CaII–M_v correlation (Wilson–Bappu effect) calibrated by Hipparcos parallaxes. *Publ. Astron. Soc. Pac.* **111**, 335–341.

West AA, Walkowicz LM and Hawley SL 2005 Updated colors for cool stars in the Sloan Digital Sky Survey. *Publ. Astron. Soc. Pac.* **117**, 706–710.

Williams CC, Golimowski DA, Uomoto A, Reid IN, Henry TJ, Dieterich S, Jue SL, Long GM, Neilsen EH, Spahn EY and Walkowicz LM 2002 Colors, magnitudes, and searches for nearby stars using SDSS/USNO photometry. *Bull. Am. Astron. Soc.* **34**, 1292.

Wilson OC and Vainu Bappu MK 1957 H and K emission in late-type stars: dependence of line width on luminosity and related topics. *Astrophys. J.* **125**, 661–683.

Xu Y, Reid MJ, Zheng XW and Menten KM 2006 The distance to the Perseus spiral arm in the Milky Way. *Science* **5757**, 54–57.

Zacharias N, Urban SE, Zacharias MI, Hall DM, Wycoff GL, Rafferty TJ, Germain ME, Holdenried ER, Pohlman JW, Gauss FS, Monet DG and Winter L 2000 The first US Naval Observatory CCD Astrograph Catalog. *Astron. J.* **120**, 2131–2147.

Zacharias N, Finch C, Girard T, Hambly N, Wycoff G, Zacharias MI, Castillo D, Corbin T, DiVittorio M, Dutta S, Gaume R, Gauss S, Germain M, Hall D, Hartkopf W, Hsu D, Holdenried E, Makarov V, Martines M, Mason B, Monet D, Rafferty T, Rhodes A, Siemers T, Smith D, Tilleman T, Urban S, Wieder G, Winter L and Young A 2010 The third US Naval Observatory CCD Astrograph Catalog (UCAC3). *Astron. J.* **139**, 2184–2199.

Zwahlen N, North P, Debernardi Y, Eyer L, Galland F, Groenewegen MAT and Hummel CA 2004 A purely geometric distance to the binary star Atlas, a member of the Pleiades. *Astron. Astrophys.* **425**, L45–L48.

3

From the Milky Way to the Local Group

An experiment is a question which science poses to Nature, and a measurement is the recording of Nature's answer.

– Max Planck (1858–1947), German physicist and 1918 Nobel laureate (physics)

3.1 Basic Stellar Physics as the Key to Understanding Distance Measurements to Local Group Galaxies

Once we reach the volume where the standard geometric methods no longer enable high-accuracy distance estimates, we need to rely on common physical properties of the stars and stellar populations we can probe sufficiently well to map the three-dimensional Universe. With the exception of Section 3.7, where we discuss a number of innovative geometric distance indicators, this chapter deals with the basic stellar physics underlying most methods of distance determination in the volume of space occupied by Local Group galaxies and slightly beyond. Therefore, we first briefly discuss our current understanding of stellar evolution through the Hertzsprung–Russell diagram (HRD), followed by an overview of the make-up of galactic stellar populations. Where appropriate, we also refer to the relevant methods of distance determination associated with a given stellar evolutionary stage or population.

3.1.1 Stellar Evolution Through the Hertzsprung–Russell Diagram

Although we do not have a firm handle on the actual star formation mechanism, despite significant recent as well as past research efforts (but see Bate 2009 and references therein), the details of stellar evolution following the initial molecular cloud collapse are much better understood.

Once protostars have reached approximate **hydrostatic equilibrium**, i.e. when there is a balance between the gravitational force exerted by the protostellar material and the

An Introduction to Distance Measurement in Astronomy, First Edition. Richard de Grijs.

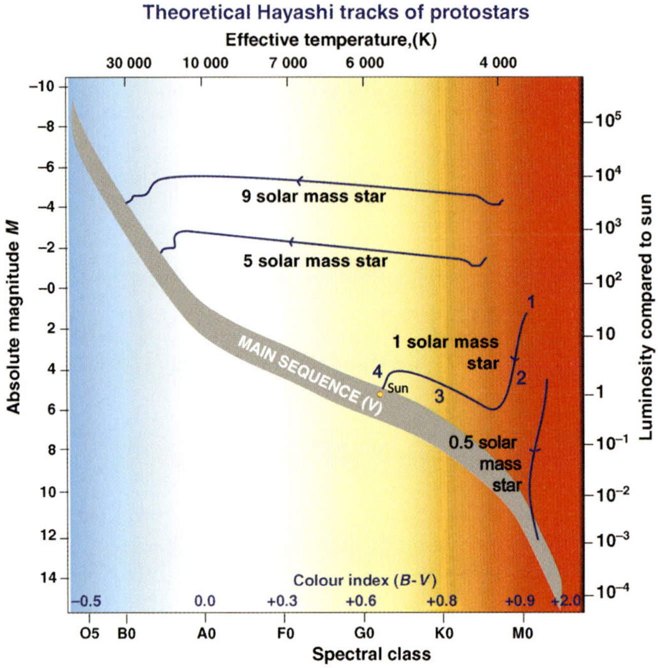

Figure 3.1 *Pre-main-sequence stellar evolutionary tracks in the Hertzsprung–Russell diagram. The four Hayashi tracks show the predicted evolutionary paths for 9, 5, 1 and 0.5 $M_\odot$ stars. Evolutionary stages for solar-mass stars – (1) initial cloud collapse; (2) continuing gravitational collapse; (3) onset of nuclear fusion in the protostellar core; (4) hydrostatic equilibrium.* (Reprinted from R. Hollow, ATOE (online), Star Formation – Theoretical Hayashi tracks of protostars, Copyright 2006, with permission of CSIRO.)

outward thermal pressure, their minimum effective temperature, $T_{\rm eff}$, is given by the **Hayashi boundary** (Hayashi 1961), corresponding to approximately $T_{\rm eff} = 4000$ K. They will continue to contract and cool on the Kelvin–Helmholtz timescale (a theoretical estimate of how long a star would radiate at its current luminosity if the only power source were the conversion of gravitational potential to heat), leading to a decrease in internal thermal pressure and hence gravitational contraction, which in turn results in an increase in the protostellar core temperature. Low-mass stars ($m_* \lesssim 0.5$ $M_\odot$) will remain fully convective and on the close-to-vertical Hayashi track during their **pre-main-sequence** (PMS) evolution, until they reach the main sequence (see Figure 3.1). Higher-mass stars, on the other hand, will reach a stage when radiative energy transport becomes more efficient and convection is overcome, thus leading to radiative equilibrium and stellar evolution along the near-horizontal **Henyey track** (Henyey *et al.* 1955). In particular, stars with masses between 0.4–0.5 and 1.2–1.3 $M_\odot$ – depending on metallicity – develop a radiative core and a convective envelope, while stars with masses in excess of 1.2–1.3 $M_\odot$ are characterized by a convective core and a radiative envelope.

At the onset of hydrogen burning (thermonuclear fusion) in the stellar core (for $m_* > 0.08$ $M_\odot$, depending on metallicity), PMS stars reach the **zero-age main sequence** (ZAMS; see Figure 3.2), a clearly identifiable locus in HRD space where stellar luminosity is a function

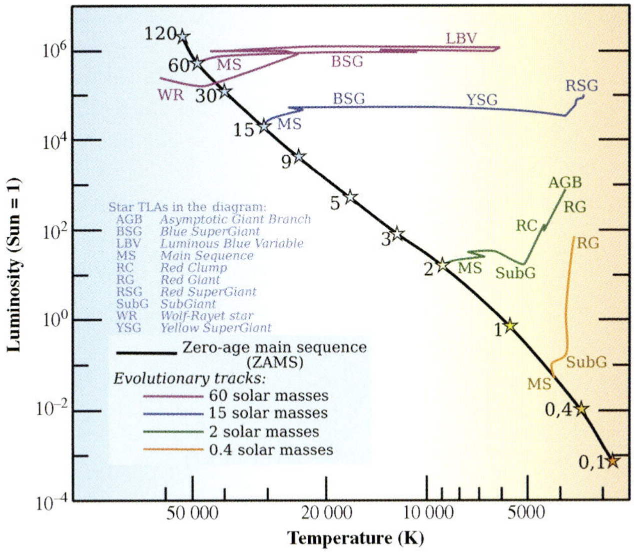

Figure 3.2 *Stellar evolutionary tracks. TLAs: two/three-letter acronyms.* (Courtesy of Wikimedia Commons, licensed under the Creative Commons Attribution–Share Alike 3.0 Unported license.)

of the star's mass, and where stars are in hydrostatic equilibrium and spend most of their lifetimes. They reach this evolutionary stage when the secondary elements of the **proton–proton chain** (see below) attain their equilibrium. This phase is short for a significant fraction of low-mass stars, but becomes of order a billion years for very low-mass stars ($m_* < 0.4\ M_\odot$) because of the long equilibrium time associated with ^{3}He.

The main-sequence morphology (colour, temperature) depends on the stellar population's chemical composition: higher-metallicity, single-age stellar populations (see Section 3.1.2) have fainter and redder main-sequence turn-offs (which refers to the onset of the subgiant branch) at a fixed age than their lower-metallicity counterparts. At higher metallicities, the stellar atmospheric opacity is increased owing to the larger number of electrons from metals. In turn, this allows stars to maintain equilibrium at a lower temperature, thus resulting in redder colours. At the same time, line blanketing in the stellar atmospheres also contributes to producing redder colours for higher-metallicity stars. This increase is driven by the absorption and re-emission of a significant fraction of the photons from the outer layers of the stellar atmosphere back to the inner photosphere. This provides additional energy input and, thus, heats the deeper photospheres, a process referred to as 'backwarming'. This effect increases the local kinetic temperature, leading to stronger photospheric continuum emission and modifies diagnostically important ionization equilibria, which are used for the determination of the effective temperature.

In 1920, Arthur Eddington first suggested that stars obtained their energy from nuclear fusion of hydrogen into helium, based on the precise atomic measurements by Francis William Aston, a British chemist and physicist who won the 1922 Nobel Prize in Chemistry. The exact pathway from hydrogen to helium depends on the stellar mass. For stars with masses $m_* < 1.0$–$1.5\ M_\odot$ (depending on their central temperature and pressure), hydrogen atoms are fused in a number of stages to form helium, which is referred to as the 'proton–proton chain reaction'. Higher-mass, upper main-sequence stars ($m_* > 1.0$–1.3 $M_\odot$) are dominated by fusion reactions in the **CNO cycle** (also known as the Bethe–Weizsäcker cycle; von Weizsäcker 1938; Bethe 1939), where four protons fuse using carbon, nitrogen and oxygen isotopes as catalysts to produce one α particle (identical to a helium nucleus), two positrons and two electron neutrinos. The positrons almost immediately annihilate with electrons, releasing γ rays. The main reactions in the CNO cycle go through

$$^{12}_{6}\mathrm{C} \rightarrow ^{13}_{7}\mathrm{N} \rightarrow ^{13}_{6}\mathrm{C} \rightarrow ^{14}_{7}\mathrm{N} \rightarrow ^{15}_{8}\mathrm{O} \rightarrow ^{15}_{7}\mathrm{N} \rightarrow ^{12}_{6}\mathrm{C}, \tag{3.1}$$

following

$$\begin{array}{lllllllll}
^{12}_{6}\mathrm{C} & + & ^{1}_{1}\mathrm{H} & \rightarrow & ^{13}_{7}\mathrm{N} & + & \gamma & & & + 1.95\ \mathrm{MeV}, \\
^{13}_{7}\mathrm{N} & & & \rightarrow & ^{13}_{6}\mathrm{C} & + & \mathrm{e}^+ & + & \nu_\mathrm{e} & + 2.22\ \mathrm{MeV}, \\
^{13}_{6}\mathrm{C} & + & ^{1}_{1}\mathrm{H} & \rightarrow & ^{14}_{7}\mathrm{N} & + & \gamma & & & + 7.54\ \mathrm{MeV}, \\
^{14}_{7}\mathrm{N} & + & ^{1}_{1}\mathrm{H} & \rightarrow & ^{15}_{8}\mathrm{O} & + & \gamma & & & + 7.35\ \mathrm{MeV}, \\
^{15}_{8}\mathrm{O} & & & \rightarrow & ^{15}_{7}\mathrm{N} & + & \mathrm{e}^+ & + & \nu_\mathrm{e} & + 2.75\ \mathrm{MeV}, \\
^{15}_{7}\mathrm{N} & + & ^{1}_{1}\mathrm{H} & \rightarrow & ^{12}_{6}\mathrm{C} & + & ^{4}_{2}\mathrm{He} & & & + 4.96\ \mathrm{MeV}.
\end{array}$$

Main-sequence stars with $m_* > 1.5$–2.0 $M_\odot$ (depending on metallicity) are dominated by convection in their cores, which is driven by the strong dependence of the CNO-burning efficiency on core temperature. When core convection does not occur, a helium-rich core develops with a mass-dependent size for $m_* > 2$ $M_\odot$, surrounded by an outer convective layer of hydrogen. After the core's hydrogen has been consumed, the star evolves away from the main sequence. Stars with $m_* < 0.5$ $M_\odot$ evolve into white dwarfs, while more massive stars (up to $m_* \sim 6$–10 $M_\odot$) become red giants, fusing the hydrogen in the shell surrounding the helium core because of core contraction. Even more massive stars are predicted to go supernova (SN).

The hydrogen shell-burning process will cause the star's outer atmosphere to expand and consequently decrease in surface temperature, while its luminosity increases rapidly. This process is, as yet, poorly understood. The star is now said to ascend the HRD's red giant branch (RGB). At the same time, the helium core continues to contract until it is supported by **electron degeneracy pressure**. Depending on its mass, the stellar core may ignite helium burning for temperatures of 1–3×10^8 K through the so-called **triple-α process** for $m_* > 0.5$ $M_\odot$, where three α particles are transformed into carbon. In fact, at this stage two helium nuclei fuse together sufficiently rapidly to balance the decay rate of their product (^{8}Be). As such, there are always a few ^{8}Be nuclei present in the stellar core, which can fuse with yet another helium nucleus to form the stable ^{12}Ca isotope:

$$\begin{array}{llllllll}
^{4}_{2}\mathrm{He} & + & ^{4}_{2}\mathrm{He} & \rightarrow & ^{8}_{4}\mathrm{Be} & & - & 0.95\ \mathrm{keV}, \\
^{8}_{4}\mathrm{Be} & + & ^{4}_{2}\mathrm{He} & \rightarrow & ^{12}_{6}\mathrm{C} & + \mathrm{e}^+ + \mathrm{e}^- & + & 7.367\ \mathrm{MeV}.
\end{array}$$

Once the degenerate core reaches a threshold temperature, in low- to intermediate-mass stars ($m_* < 1.8$–2.0 $M_\odot$ for solar metallicity) the entire core will begin helium fusion nearly simultaneously in a 'core **helium flash**'. The cores of more massive stars will reach $\sim 10^8$ K before they are dense enough to be degenerate, implying that helium fusion will begin much more smoothly. Once core helium fusion begins, the star contracts to either the horizontal branch (HB, for metal-poor stars; see Section 3.2.3) or the red clump (more metal-rich and likely younger stars; Chapter 3.2.2). When the helium in the core of sufficiently massive stars is exhausted and has been turned into inert carbon and oxygen, helium-shell burning commences in the form of a shell helium flash, although in many cases hydrogen fusion will also continue in a shell closer to the stellar surface, a process referred to as 'double-shell burning'. (This continues until such stars reach the planetary nebula phase.) This places low- to intermediate-mass stars ($0.6 \lesssim m_*/\mathrm{M}_\odot \lesssim 10$) onto the asymptotic giant branch (AGB). The temperature rises rapidly and leads to a thermal pulse (see, for applications, Section 3.5.3). Thermal pulses occur when AGB stars run out of helium fuel. They derive their energy from fusion of hydrogen in a thin shell surrounding the inert helium shell, but the latter switches back on occasionally, hence leading to **thermal pulsations**. They rapidly release the built-up energy and shed circumstellar gas and dust shells, and allow '**s-process**' reactions (slow neutron capture, a nucleosynthesis process that occurs at relatively low neutron density and intermediate temperatures) to proceed. They also cause mixing of the core material into the outer layers because of convection, which is referred to as '**AGB dredge up**'. Most AGB stars undergo two dredge-up stages. The first dredge up occurs when a low-mass main-sequence star enters the RGB, while stars in the mass range from 4 to 8 $\mathrm{M}_\odot$ undergo a second dredge-up period when core helium fusion ends and convection mixes the products of the CNO cycle, thus leading to oxygen-rich atmospheres. After massive stars evolve onto the AGB, a third dredge-up phase occurs. This happens at roughly the same time as a helium shell flash and leads the convective transport of helium, carbon and s-process elements to the stellar surface, thus forming 'carbon stars'.

More massive stars continue to repeat this cycle, fusing heavier elements in successive, increasingly brief phases. Solar-mass stars will not proceed to fuse carbon; a minimum stellar mass of 6–9 $\mathrm{M}_\odot$ is required for this process. Instead, at the end of the AGB phase, they will eject their outer layers owing to pulsations and strong stellar winds and form planetary nebulae (PNe). Very low-mass stars are not fully convective and may, therefore, not accumulate an inert helium core. As a consequence, they may never become red giants; instead, they are referred to as 'red dwarfs'. Very high-mass stars, on the other hand, become red or blue supergiant stars that wander back and forth horizontally in HRD space. They most likely end their life as Type II SNe (see Section 5.2.2).

Returning to the end phase of low- to intermediate-mass stars, the gases of the PN drift away from the central star (see, for **expansion parallax**, Section 3.7.1), which simultaneously undergoes a two-stage evolution. It first grows hotter as it continues to contract and hydrogen fusion occurs in a shell around the core of carbon and oxygen, and then cools as it radiates away its energy. By this time, fusion ceases, because these stars are typically not sufficiently massive to ignite fusion of heavier elements. Eventually, on timescales of $\sim 10^5$ yr, the central stars will cool down to sufficiently low temperatures that they are no longer able to ionize their increasingly distant expelled outer envelopes, and become carbon oxygen (CO) white dwarfs.

Once stars have reached the white dwarf phase, they consist of electron degenerate matter – mostly composed of carbon and oxygen because of prior stellar evolution – and they slowly release their thermal energy as they cool down. White dwarfs are extremely dense and compact, and supported by electron degeneracy pressure. In HRD space, they will slowly become fainter and redder as they cool down; this can be modelled precisely and is referred to as the **white dwarf cooling sequence** (e.g. Richer *et al.* 2006; Section 3.4).

3.1.2 From Two to Multiple Stellar Populations

Already in the 1920s, Lindblad (1925) argued that the Milky Way seemed to be composed of a set of components with a continuous range of flattening. However, the concept of stellar populations was first introduced by Walter Baade (1944a,b) on the basis of his work on the HRDs of M31, M32 and the dwarf spheroidal companion galaxies in the Local Group, NGC 147, NGC 185 and NGC 205 (see Figure 3.3). He defined as '**Population I**' the stars in the solar neighbourhood, with typical HRDs of similar to those of open star clusters and an overall blue colour index, because the brightest stars are blue. '**Population II**' was better represented by the stars of globular clusters (GCs), the bulge of M31, and the disc and bulge of the Milky Way, with typical HRDs similar to those of GCs, and an overall red colour index, because the brightest stars are red giants.

Modern stellar evolutionary theory tells us that Population I contains young stars: blue main-sequence stars and blue (super)giants are always massive stars, and therefore short-lived. These are mostly found in spiral discs. Population II does not contain young stars; bulges of spiral galaxies and the entire stellar populations of elliptical galaxies are counted among this population. In 1957, the Vatican Symposium on stellar populations defined five stellar populations, with increasing age and decreasing flattening and metal abundance.

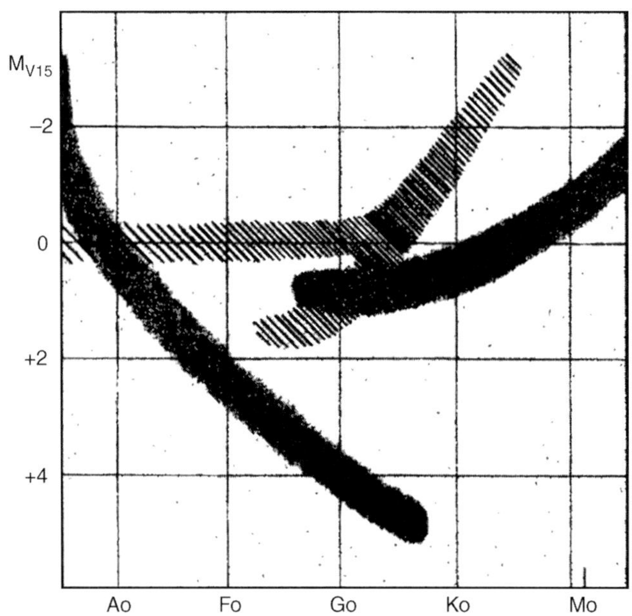

Figure 3.3 *Baade's (1944a) definition of Populations I (shaded) and II (hatched).* (Reprinted from W. Baade, *Astrophysical Journal*, **100**, The resolution of Messier 32, NGC 205, and the central region of the Andromeda nebula, p. 137—146, Copyright 1944, with permission of the AAS.)

These included (i) extreme Population I stars (gas, young stars associated with spiral arms, supergiants, Cepheids, **T Tauri stars**), (ii) older Population I stars (A-type stars, stars with strong emission lines, mostly in the 'thin disc'), (iii) disc population stars (stars in the Galactic nucleus, PNe, novae, RR Lyrae stars with periods less than 0.4 days, weak-line stars), (iv) intermediate Population II stars (high-velocity stars with vertical velocities exceeding 30 km s^{-1}, long-period variables earlier than type M5e, with periods less than 250 days, mostly in the 'thick disc') and (v) halo Population II stars (subdwarfs, GCs with high vertical velocities, RR Lyrae stars with periods longer than 0.4 days).

Given the main distinctions, particularly in age, both Populations I and II have specific secondary distance indicators associated with them. The most important methods for distance estimation to Population I objects include red clump stars (Section 3.2.2), classical Cepheids (Section 3.5.2), Mira variables (Section 3.5.3) and eclipsing binaries (Section 3.7.3), without being exhaustive. The equivalent Population II distance indicators are based on observations of HB and RR Lyrae stars (Sections 3.2.3 and 3.5.5, respectively), white dwarf cooling sequences (Section 3.4) and methods based on ensembles of GCs (e.g. Sections 1.1 and 4.3), among others.

In modern astrophysics, we recognize more diversity in stars than can be described by just two populations. We tend to think in terms of multiple 'stellar populations', which may differ from galaxy to galaxy. The parameters that characterize a population (see Figure 3.4) are as follows:

- *Age*. In particular, how long ago did the last episode of star formation occur?
- *Metallicity*. This is measured from a star's spectrum, using a range of spectral diagnostics or proxies such as the iron or oxygen abundance, so it refers to the surface heavy-element content, i.e. essentially the composition of the gas cloud from which the stars formed.

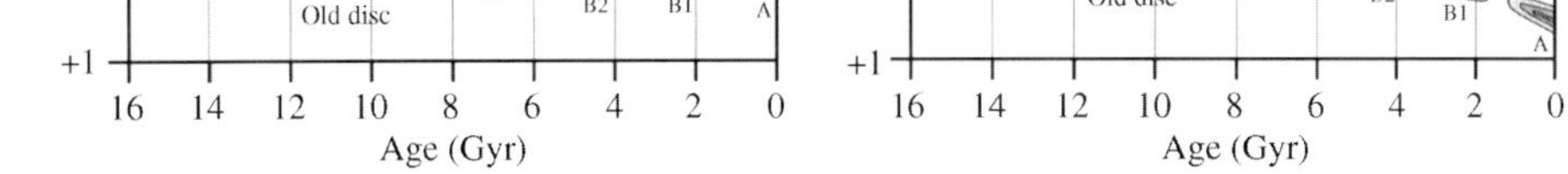

Figure 3.4 *Schematic, composite 'Hodge' population boxes for stellar populations (characterized by age, metallicity and rotation velocity) in the Milky Way (Majewski 1999). The contour levels correspond roughly to levels of 0.5 in terms of instantaneous-to-mean star formation rate (for the disc populations), with the lowest contour at 0.5. Note the use of $|v_{rot}|$ to accommodate the possibility of a retrograde 'young halo'. IP II: intermediate Population II. A, B1, B2, C1 and C2 denote bursts of star formation. See Nissen and Schuster (2009, 2010) for updates on the Galactic disc/halo transition and halo populations.* (Reprinted from S. R. Majewski, Globular Clusters, Stellar Populations and the Formation of the Milky Way, p. 43–107, Copyright 1999, Cambridge University Press.)

- *Kinematics*. How does the population move within a galaxy? Does it have ordered rotation like the disc of a spiral galaxy, or independent, uncorrelated orbits like the GCs? Is it organized as a disc or a spheroid?

Beatrice Tinsley (1980) pioneered the technique of **stellar population synthesis**. It is based on the idea that any stellar population formed in one or more 'single bursts'. This implies that all stars within the population have the same age and the same metallicity, because they formed from the same gas cloud at approximately the same time. The only variable, for any given population, is the stellar **initial mass function**, i.e. the distribution of stellar masses at the time of star formation. Such **'simple' stellar populations** can be used to build up models of star clusters or – using combinations of multiple simple stellar population simultaneously – even galaxies. They are the basis of most of our current understanding and analysis of stellar and galactic star formation histories.

3.2 Open and Globular Cluster Hertzsprung–Russell Diagrams

At distances where individual stellar measurements become cumbersome, if not impossible, with the notable exception of bright variable stars (see Section 3.5), star clusters of at least moderate richness become interesting as hosts of significant numbers of stars with similar properties at essentially the same distance. Their advantages for purposes of distance determination are significant, provided precise calibration can be achieved, thus resulting in the use of various features in their HRDs – or their observational equivalents, the colour–magnitude diagrams – as secondary distance indicators. One of these features, the HB, is intricately linked to the presence of RR Lyrae variable stars, which we will discuss in detail in Section 3.5.5.

3.2.1 Main-Sequence and Subdwarf Fitting

One of the most popular methods of distance determination to resolved star clusters is that of unevolved, zero-age main-sequence fitting. In its simplest form, this reduces to matching the main defining features of an observed cluster HRD at an unknown distance either to their counterparts in a well-calibrated reference HRD or to theoretical **isochrones** (snapshots of theoretical simple stellar populations at a fixed age and chemical composition). Assuming that all observational uncertainties have been dealt with correctly (but see below), the only free parameter in this approach is the luminosity or, alternatively, magnitude, hence a match between both HRDs will result in a distance modulus:

$$(m - M)_0 = 5\log(d/\mathrm{pc}) - 5, \tag{3.2}$$

where m and M are the apparent and the absolute magnitude in a relevant, calibrated filter, respectively, corrected for extinction (indicated by the subscript '0'), and d is the distance sought. Many of the methods used in the nearby Galactic and extragalactic Universe similarly rely on distance moduli to derive the actual, predominantly relative distances.

While this method is robust to first order, the resulting distance accuracy of a generic application is not much better than ~10–15%. As a key rung of the distance ladder, this accuracy is inadequate and additional considerations are required to improve distance determinations based on main-sequence fitting. The underlying assumptions are that (i) extinction effects have been corrected for properly (complications may include deviations from the 'standard' Galactic extinction law and the presence of differential extinction across the face of the cluster of interest; cf. Section 6.1.1), (ii) both the cluster of interest and the reference object have the same physical properties, most notably age and metallicity, (iii) the effects of unresolved binary systems on the main-sequence ridgeline are negligible and (iv) the stellar content and evolutionary processes in both objects are identical.

From an observational point of view, these assumptions imply identical HRD shapes and negligible colour terms (see e.g. An *et al.* 2007b). Shape mismatches introduce systematic uncertainties because of the need for various transformations, particularly colour–$T_{\rm eff}$ calibrations, as do uncertainties in the photometric calibration and the adopted metallicity scale. An *et al.* (2007a,b) attempted to minimize systematic errors by employing a multicolour approach based on the simple main-sequence fitting technique, and achieved significantly improved results. They also quantitatively assessed the effects of unresolved binaries. A 40% binary fraction in their models causes a brightening of the main sequence by ~0.007 mag at optical wavelengths (see also Carretta *et al.* 2000), which is comparable to the photometric accuracy routinely achieved, particularly for *Hubble Space Telescope (HST)*-based observations (see, for an example, Testa *et al.* 2004).

The main-sequence fitting technique is particularly useful if one wants to obtain relative distances between two clusters. After all, a relative distance differential between two identical HRDs results in a simple shift in luminosity or magnitude. *Absolute* cluster distances are more difficult to determine, since their accuracy depends on matching nearby Galactic halo main-sequence subdwarfs – for which reasonably accurate trigonometric parallaxes are available – to cluster main sequences. The main problem associated with this approach is that any given collection of subdwarfs does not automatically give us a 'sequence', because of different distances, parallax uncertainties and metallicities of the sample stars. The problem is to match the nearby subdwarfs to a fiducial main sequence to start with, at the appropriate metallicity, and after proper correction for Lutz–Kelker and Malmquist biases (see Sections 6.1.2 and 6.1.3). In addition, this assumes that field and cluster stars are physically identical.

Subdwarf fitting has been moderately successful in reducing distance uncertainties, particularly on the basis of improved *Hipparcos* parallax measurements (e.g. Reid 1997, 1998; Carretta *et al.* 2000; Grundahl *et al.* 2002; Gratton *et al.* 2003; Kraft and Ivans 2003; An *et al.* 2009; see for significant recent progress VandenBerg *et al.* 2010). Initially, *Hipparcos*-based distances were of order 15% greater – $\Delta(m - M) \sim 0.3$ mag – than those derived from ground-based measurements. However, using a new calibration of the *Hipparcos* data (van Leeuwen 2005, 2007), this discrepancy has now largely disappeared (see also Tabur *et al.* 2009; VandenBerg *et al.* 2010). For low-metallicity GCs, an additional problem is the scarcity of metal-poor local subdwarfs with $[\rm Fe/H] \leq -2$ in the *Hipparcos* Catalogue. This implies that, for a proper calibration, one needs to adjust the colours of the subdwarf calibrators, thus potentially introducing errors in the model atmospheres. As a result of these constraints, the uncertainties are at present minimal only for the nearest subdwarf, Groombridge 1830 (HD 103095), which therefore has a well-determined distance (cf. Gizis 1997) and metallicity. However, do we want to pin the entire GC distance scale and, hence, the age of the Universe on just one star?

3.2.2 Red Clump Stars

Although the technique of main-sequence fitting is popular and has seen marked improvements in accuracy in recent years, its reach is limited because of the intrinsically faint luminosities of most stars in a given cluster. Ideally, one would want to use an intrinsically bright standard candle with many suitable reference objects within reach of accurate, usually *Hipparcos*-based parallax measurements. For this reason, calibration of the absolute magnitude of helium-burning red clump stars (discrepant by ~ 0.5 mag with respect to the zero-age HB; ZAHB) has attracted significant recent attention (see e.g. Paczyński and Stanek 1998; Stanek and Garnavich 1998; see Figure 3.5). Interest has shifted from the conventional optical wavebands to the near-infrared (near-IR) regime (Sarajedini *et al.* 1995: *V*; Seidel *et al.* 1987: *R*; Paczyński and Stanek 1998: *I*; Alves 2000; Alves *et al.* 2002; Grocholski and Sarajedini 2002; Sarajedini *et al.* 2002: *J*, *K*) in an attempt to reduce systematic uncertainties (see also Pietrzyński *et al.* 2010).

Red clump stars are the low- to intermediate-mass analogues of the helium-burning HB stars, and theoretical models predict that their absolute luminosity depends only weakly on age and metallicity. This assumption is the main cause of systematic uncertainties associated with using the red clump as a distance indicator and the dominant reason for a shift to the

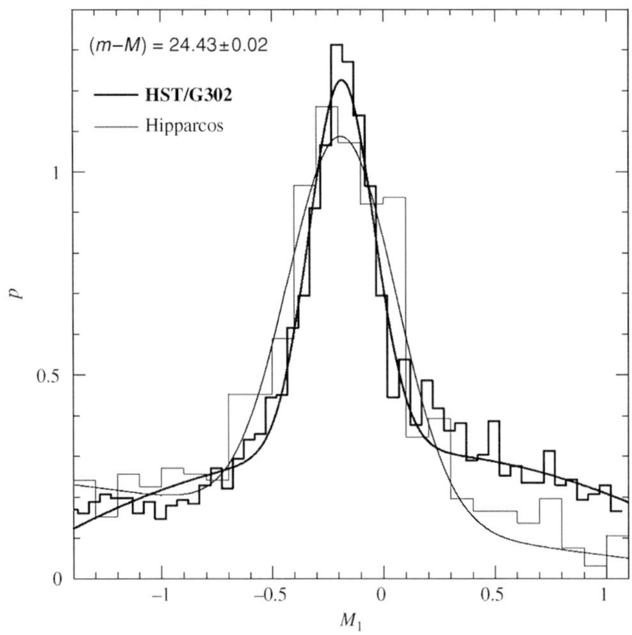

Figure 3.5 Normalized numbers of red clump stars in the solar neighbourhood, based on Hipparcos data, and an M31 field (Stanek and Garnavich 1998) as a function of absolute I-band magnitude (thin solid line and histogram), adopting a distance modulus of $(m - M)_I^0 = 24.43$ mag, which agrees very well with more up-to-date determinations (cf. Section 1.3). (Reprinted from K. Z. Stanek and P. M. Garnavich, *Astrophysical Journal*, **503**, Distance to M31 with the Hubble Space Telescope and Hipparcos red clump stars, L131–L134, Copyright 1998, with permission of the AAS and K. Z. Stanek.)

near-IR wavelength range, where the red clump properties are less sensitive to age and metallicity effects. If one assumes a negligible age and metallicity dependence of the red clump luminosity (but see Pietrzyński *et al.* 2010, particularly in the V and I bands), its apparent magnitude can be used as a simple distance (modulus) indicator, provided adequate extinction corrections are applied (e.g. Paczyński and Stanek 1998; Stanek and Garnavich 1998). Proponents of this approach argue that, given the compactness of the red clump in the *Hipparcos* HRD, this validates the underlying assumption of a roughly constant absolute I-band magnitude, M_I(RC). Stanek and Garnavich (1998) determined that the variance in the I band is only $\sim$0.15 mag.

However, other groups argue that the age and metallicity dependence of the red clump's absolute magnitude is much stronger (e.g. Cole 1998; Sarajedini 1999; Pietrzyński *et al.* 2010). Theoretical models (e.g. Seidel *et al.* 1987; Girardi and Salaris 2001; see also Cole 1998) suggest that M_I(RC) depends on both age and metallicity – although less so than M_V(RC) – becoming fainter as both increase, in good agreement with the observations of Sarajedini (1999). Extending their analysis to the near-IR K band, Grocholski and Sarajedini (2002) show that for ages between approximately 2 and 6 Gyr and $-0.5 \leq$ [Fe/H] ≤ 0 dex, the intrinsic variation in M_K(RC) is minimized (see Figure 3.6; see also Alves 2000), thus suggesting that any lingering uncertainties in age and/or metallicity are unimportant for robust distance determinations. For these parameter ranges,

$$\langle M_K(\mathrm{RC})\rangle = -1.61 \pm 0.04 \text{ mag}. \tag{3.3}$$

Age plays an important role in determining M_K(RC) for younger populations (<2 Gyr), while metallicity variations mainly affect populations older than approximately 2 Gyr (see also Alves 2000; Udalski 2000). Groenewegen (2008b) recently provided an updated calibration in both the I and K bands on the Two-Micron All-Sky Survey (2MASS) photometric system,

$$\langle M_I(\mathrm{RC})\rangle = -0.22 \pm 0.03 \text{ mag}, \tag{3.4}$$

$$\langle M_K(\mathrm{RC})\rangle = -1.54 \pm 0.04 \text{ mag}, \tag{3.5}$$

where he argued that the fainter K-band magnitude is expected on the basis of population corrections, which were required because of selection effects affecting the calibration reference stars compared to his test sample.

It seems, therefore, that the red clump can be used as a reliable distance indicator under certain circumstances. However, with the recent discovery of large numbers of clusters exhibiting multiple main sequences and RGBs, this result has become less secure, particularly for clusters with multiple or more extended red clumps (e.g. Girardi *et al.* 2009). The latter might be a result of helium ignition in stars just massive enough to avoid electron degeneracy settling in their hydrogen-exhausted cores.

Stanek (1995) and Stanek *et al.* (1994, 1997) first used red clump stars to map the Galactic bar (see Cabrera-Lavers *et al.* 2008 for recent updates). Paczyński and Stanek (1998) used the red clump stars observed by the Optical Gravitational Lensing Experiment (OGLE) to estimate the distance to the Galactic Centre. Red clump stars have also been used to determine distances to extragalactic objects. Stanek and Garnavich (1998) used *HST* observations to measure the distance to M31, while Udalski *et al.* (1998) used OGLE data to obtain distances to the Large and Small Magellanic Clouds (LMC and SMC, respectively).

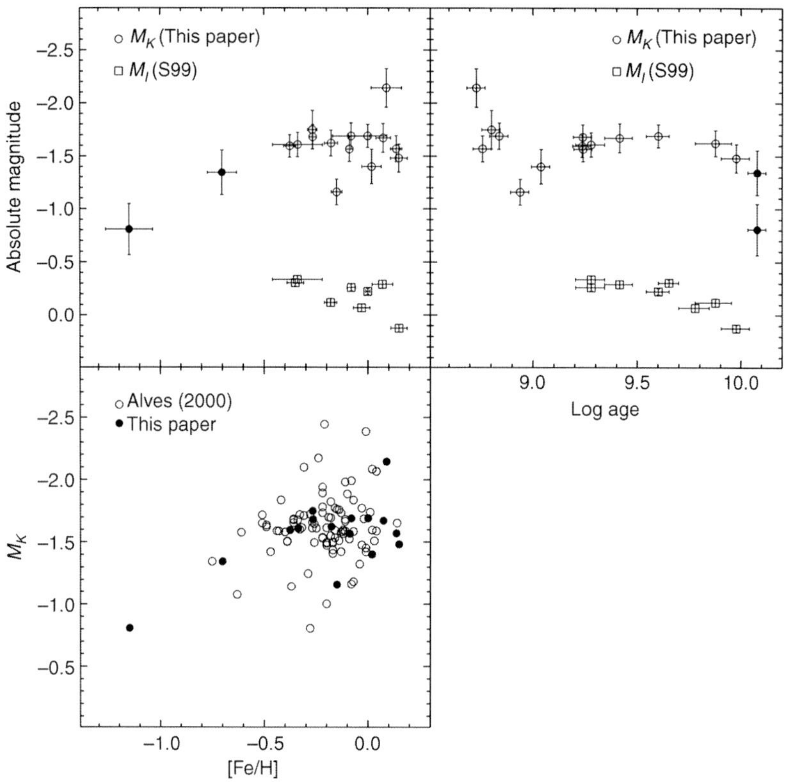

Figure 3.6 *(Top panels) Variation of red clump (RC) absolute magnitude as a function of age (top left) and [Fe/H] (top right). The open and filled circles represent M_K(RC) for 14 open and two GCs (Grocholski and Sarajedini 2002). The open squares are M_I(RC) values. S99: Sarajedini (1999). (Bottom) M_K for Hipparcos solar neighbourhood RC stars (open circles) compared with M_K(RC) for the clusters from the top panels (filled circles). These two data sets show remarkable agreement in their mean K-band magnitudes (Grocholski and Sarajedini 2002).* (Reprinted from A. J. Grocholski and A. Sarajedini, *Astronomical Journal*, **123**, WIYN open cluster study. X. The *K*-band magnitude of the red clump as a distance indicator, p. 1603–1612, Copyright 2002, with permission of the AAS and A. J. Grocholski.)

In addition, they have been helpful in constraining distances to Galactic GCs and to M33 (Kim *et al.* 2002), IC 1613 (Dolphin *et al.* 2001) and the Fornax dwarf spheroidal galaxy (Rizzi *et al.* 2007), to name but a few Local Group galaxies.

3.2.3 The (Zero-Age) Horizontal Branch Level

The luminosities of HB stars are mainly determined by the conductive opacities in the cores of their progenitor RGB stars. The latter are electron degenerate, implying that most of their energy transport occurs through electron conduction. The more efficient the transport of energy away from the helium core, the more difficult it becomes for the core to reach sufficiently high temperatures for the onset of helium burning, which leads to higher helium core masses and, thus, higher luminosities for higher conductivity at the tip of the RGB (TRGB).

Many authors have endeavoured to use 'the' HB level as a standard candle. However, as Ferraro *et al.* (1999) eloquently point out, there is significant controversy about what is meant by 'the' HB level (see also Di Cecco *et al.* 2010). From a theoretical point of view, the ZAHB seems an ideal starting point for our discussion. Because the observed HB, over its full extent from red to extreme blue, cannot be described by a single evolutionary set of models, we consider the ZAHB as the total and core-mass-dependent starting point for stars after helium ignition in their cores, on their way to eventually becoming AGB stars.[1]

From an observational point of view, the ZAHB locus corresponds roughly to the lower envelope of the HB stars in a given coeval and equidistant population in the colour range $0.2 < (B - V) < 0.6$ mag. This is, therefore, not the same as the observed mean HB level ($\langle V_{\rm HB} \rangle$), the mean level of the RR Lyrae stars, $\langle V_{\rm RR} \rangle$, or the 'estimated' ZAHB, as often found in the literature (see, for critical assessments, Lee *et al.* 1990; Ferraro *et al.* 1999). In particular, the difference between the theoretical ZAHB and $\langle V_{\rm RR} \rangle$ is a strong function of HB morphology, i.e. metallicity (Carney *et al.* 1992; Cassisi and Salaris 1997).

Ferraro *et al.* (1999) used theoretical evolutionary tracks to show that the ZAHB level is not the same as the lower envelope of the distribution of HB stars, even when the HB is uniformly populated in the RR Lyrae region. The reason for this is the rapid early evolution away from the ZAHB (cf. Dorman *et al.* 1989): after only 8 Myr ($\sim$8% of their total lifetime on the HB), the stars are already 0.05–0.10 mag brighter than the ZAHB starting line. They subsequently spend 70% of their HB lifetime covering the next 0.1 mag. This implies that the near-ZAHB HB is inherently poorly populated, and the observed lower envelope of the HB will be a poor measure of 'the' HB level because of limited sample size and photometric errors. Ferraro *et al.* (1999) derive an approximate relation to shift $\langle V_{\rm HB} \rangle$ back to the ZAHB,

$$V_{\rm ZAHB} = \langle V_{\rm HB} \rangle + 0.106[{\rm M/H}]^2 + 0.236[{\rm M/H}] + 0.193, \quad (3.6)$$

where [M/H] represents the global metallicity. They also go one step further and use their HB evolutionary models (which are based on a limited mass range for HB stars) to derive the metallicity dependence of the absolute ZAHB level,

$$M_V^{\rm ZAHB} = 1.0005 + 0.3485[{\rm M/H}] + 0.0458[{\rm M/H}]^2, \quad (3.7)$$

or, for $-2.2 < [{\rm M/H}] < -0.4$ dex,

$$M_V^{\rm ZAHB} = 0.23[{\rm M/H}] + 0.94. \quad (3.8)$$

Given that the absolute visual magnitude of the ZAHB is typically fixed at the effective temperature ($\log T_{\rm eff} = 3.83$–3.85) or colour of RR Lyrae stars, $M_V^{\rm ZAHB} \equiv M_V^{\rm ZAHB}({\rm RR})$.

The distance moduli resulting from application of these equations are affected by many uncertainties, including the precise ZAHB level, its zero point and metallicity dependence, and reddening, among others, which implies that the overall associated uncertainty cannot be less than 0.2 mag (Ferraro *et al.* 1999).

However, it has become clear that differences in helium abundance may significantly increase the resulting uncertainties in distance moduli derived on the basis of HB levels (see also Cassisi *et al.* 2007). The HB bolometric luminosity depends on helium abundance, Y, as $dM_{\rm bol}/dY \approx 4.5$ (Catelan 1996). Catelan and de Freitas Pacheco (1996) analysed three

[1] Note that a small fraction of low-mass extreme HB stars evolve immediately to the white dwarf cooling sequence, without going through an AGB phase. These objects are referred to as **AGB manqué stars**.

different chemical enrichment scenarios. Although all three provide essentially the same helium abundance at low metallicities ([Fe/H] < -1 dex), at [Fe/H] ≈ -0.5 dex, ΔY could be $\simeq$0.035 and up to $\sim$0.1 at solar metallicity (see also Catelan 2009). In turn, these differences may lead to changes in HB luminosity by as much as 0.15 and 0.45 mag at [Fe/H] $= -0.5$ and 0.0 dex, respectively (see, for a review, Catelan 2006). This puts the use of the ZAHB level as a standard candle on loose footing when comparing GCs of different metallicities.

3.3 Giants and Supergiants as Standard Candles

Because of their high intrinsic brightness, giant and supergiant stars have attracted significant attention as potential standard candles. With improvements in our understanding of stellar evolution, a number of features associated with the excursions red giant stars make in the HRD can indeed serve as suitable distance indicators. Here, we discuss the TRGB, the blue bump, and distance indicators associated with red and blue supergiants. We discuss Mira variables in Section 3.5.3. The AGB, on the other hand, exhibits too much dispersion in luminosity ($>$0.25 mag), although possibly less at near- and mid-IR wavelengths (see Whitelock *et al.* 2008; Glass *et al.* 2009; Matsunaga *et al.* 2009), to be useful as an accurate standard candle (as exemplified in Davidge 2001, 2002; Davidge and van den Bergh 2001). At one time, the AGB 'bump' (caused by the onset of helium-shell burning, resulting in a temporary slowdown in the star's evolutionary direction; e.g. Gallart 1998) was suggested as a possible distance indicator (Pulone 1992), but the remaining intrinsic uncertainties turned out to be too significant for this feature to be sufficiently reliable. AGB stars collectively stand out in the discs of nearby galaxies and as such dominate – with their red and blue supergiant cousins – the technique of distance determination using surface brightness fluctuations (see Section 4.2).

3.3.1 The Tip of the Red Giant Branch

The TRGB is the maximum absolute luminosity reached by first-ascent red giants, predominantly for ages $\geq$1–2 Gyr. It marks the onset of helium fusion in their degenerate cores. Because the TRGB's absolute bolometric magnitude varies by only $\sim$0.1 mag for a wide range of metallicities and ages (Iben and Renzini 1983; Da Costa and Armandroff 1990; Salaris and Cassisi 1997; Madore *et al.* 2009c), both theoretical predictions and observational evidence strongly support this feature as an accurate distance indicator (but see Cassisi *et al.* 2007 and Sanna *et al.* 2008 for cautionary notes). Its use dates back to 1930, in fact (see Madore and Freedman 1995 for a review).

In recent years, the TRGB's I-band magnitude has become firmly established as a distance indicator for nearby galaxies with well-resolved Population II halo stars (e.g. Lee *et al.* 1993; Madore and Freedman 1995, 1998; Sakai *et al.* 1996; Bellazzini *et al.* 2001; Mager *et al.* 2008; Mould and Sakai 2008; Sanna *et al.* 2008), with similar precision ($\sim$10%: Madore and Freedman 1995) as e.g. Cepheids and RR Lyrae stars (e.g. Lee *et al.* 1993; Section 3.5.5). However, there is a systematic offset of $\sim$0.1 mag between the TRGB and Cepheid distance scales (Tammann *et al.* 2008), while the metallicity dependence of the Cepheid period–luminosity (PL) relation has been calibrated using the TRGB method, thus leading to a circular dependency (Rizzi *et al.* 2007; see also Figure 4 of Sanna *et al.* 2008, for recently updated TRGB calibrations based on scaled solar chemical abundances).

Because the TRGB is more easily observed in the near-IR range and identified as a discontinuity in the stellar luminosity function, recent studies of the Magellanic Clouds, the Galactic bulge and also of local M giants have shifted their focus to these longer wavelengths (e.g. Cioni *et al.* 2000; Kiss and Bedding 2004; Schultheis *et al.* 2004; Tabur *et al.* 2009). In the near-IR regime, interstellar reddening is reduced, but both the colour and luminosity of the TRGB are more sensitive to metallicity than in the I band (Rizzi *et al.* 2007; see also Mager *et al.* 2008). More specifically, the I-band magnitude of the TRGB discontinuity depends only weakly on metallicity for [Fe/H] ≤ -0.7 dex (Iben and Renzini 1983; Lee *et al.* 1993) but at higher metallicities, line-blanketing effects are expected to be more significant (see e.g. Salaris and Cassisi 1998; Bellazzini *et al.* 2001, 2004; Valenti *et al.* 2004; Mager *et al.* 2008; Madore *et al.* 2009c):

$$M_I^{\rm TRGB} = 0.14[{\rm Fe/H}]^2 + 0.48[{\rm Fe/H}] - 3.629 \tag{3.9}$$

(Bellazzini *et al.* 2001, 2004), calibrated using ω Centauri ([Fe/H] $= -1.7 \pm 0.2$ dex) and 47 Tucanae ([Fe/H] $\simeq -0.6$ dex). For ω Cen, Bellazzini *et al.* (2001, 2004) derive

$$M_I^{\rm TRGB} = -4.04 \pm 0.12 \text{ mag.} \tag{3.10}$$

Ferraro *et al.* (2000) published a TRGB calibration in the K band,

$$M_K^{\rm TRGB} = (-0.59 \pm 0.11)[{\rm Fe/H}] - (6.97 \pm 0.15), \tag{3.11}$$

recently updated for [Fe/H] ≤ -0.4 dex and converted to global metallicities (Ferraro *et al.* 2006; see also Tabur *et al.* 2009 for a recent revalidation using revised *Hipparcos* parallaxes) to

$$M_K^{\rm TRGB} = -0.62[{\rm M/H}] - 6.92, \tag{3.12}$$

consistent with an absolute TRGB K-band magnitude in the solar neighbourhood of (Tabur *et al.* 2009)

$$M_K^{\rm TRGB} = -6.85 \pm 0.03 \text{ mag.} \tag{3.13}$$

The most important observational uncertainties affecting the applicability of this method for distance determinations include random photometric errors, sample size, crowding issues and contamination/confusion caused by AGB stars (Renzini 1991; Madore and Freedman 1995). Another potential problem is the relative scarcity of TRGB stars in a given GC-sized stellar population because of the short duration of this evolutionary stage. It is, therefore, crucial to adopt a proper edge-fitting technique to deal with discreteness issues and estimate the precise location of the often sparsely populated tip of the distribution (cf. Madore and Freedman 1995; Sakai *et al.* 1996; Mager *et al.* 2008; Tabur *et al.* 2009). Appropriate techniques include a direct or weighted/smoothed zero-sum (Sobel) kernel edge detector, first used by Lee *et al.* (1993), instead of more ad hoc methods, and since either improved or replaced by alternative approaches (e.g. Madore and Freedman 1995; Sakai *et al.* 1996; Méndez *et al.* 2002; Mager *et al.* 2008; Sanna *et al.* 2008; Madore *et al.* 2009c; and references therein). For a robust application of the TRGB technique, the observed luminosity function must be well populated, with a minimum of ~100 or even 400–500 stars within one magnitude from the tip (Madore and Freedman 1995; Bellazzini *et al.* 2001; Madore *et al.* 2009c; see also Mager *et al.* 2008). Note that edge detection methods often yield multiple peaks, which are not always real, so that *a priori* information needs to be taken

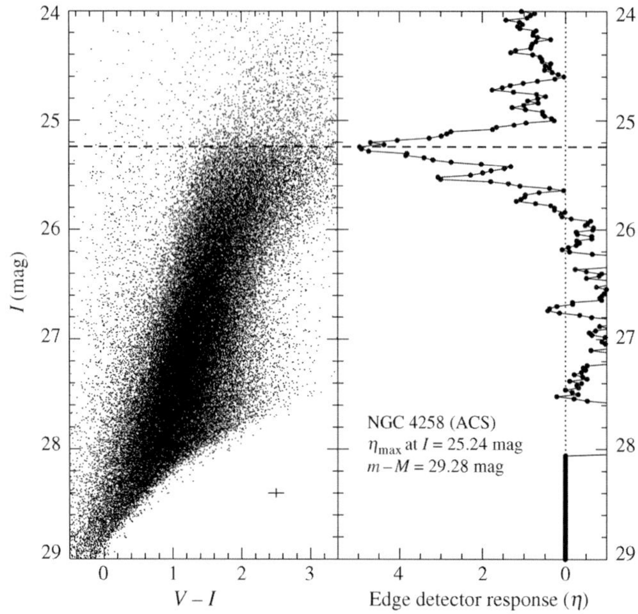

Figure 3.7 *Example of an application of the edge detection technique (Mager et al. 2008). (Left) NGC 4258 colour–magnitude diagram based on HST/Advanced Camera for Surveys (ACS) data. (Right) Value of the edge detector response (η). The maximum η peak gives the measured TRGB apparent magnitude (horizontal dashed line) and, hence, the distance modulus.* (Reprinted from V. Mager *et al.*, *Astrophysical Journal*, **689**, Metallicity-corrected tip of the red giant branch distance to NGC 4258, p. 721–731, Copyright 2008, with permission of the AAS and W. L. Freedman.)

into account to determine the actual TRGB (e.g. Mager *et al.* 2008; Tabur *et al.* 2009; and references therein; see also Figure 3.7).

3.3.2 The Red Giant Branch Bump

During RGB evolution, a narrow hydrogen-burning shell moves towards the outer stellar convection zone. The shell is rather thin in mass. When, after the first dredge up, the convective envelope retreats from the advancing hydrogen-burning shell, a discontinuity in the hydrogen-abundance profile is left and a temporary drop in luminosity results (i.e. evolution through the RGB halts temporarily) when the shell moves from a region of increasing to constant hydrogen abundance (see also Valenti *et al.* 2004). In turn, this leads to the appearance of an overdensity of RGB stars in GC HRDs, referred to as the RGB bump, which was theoretically predicted by Thomas (1967) and Iben (1968) (but see Salaris *et al.* 2007 for complications; see also Di Cecco *et al.* 2010), and convincingly identified for the first time by King *et al.* (1985). The main observational difficulty (in addition to contamination by HB stars) is the often poorly populated status of the RGB bump, requiring sophisticated differential luminosity function fits to derive the apparent magnitude of the RGB bump in GC HRDs, requiring of order 2000 RGB stars in the feature's upper three magnitudes). For up-to-date observational overviews, see Cho and Lee (2002), Valenti *et al.* (2004) and Di Cecco *et al.* (2010).

Its position in luminosity depends on metallicity, helium abundance and stellar mass and, hence, GC age, as well as on other parameters that determine the maximum inward extent of the convection envelope or the position of the hydrogen-burning shell. However, note that Cassisi (2010) and Di Cecco *et al.* (2010) demonstrate – from theoretical and observational perspectives, respectively – that current evolutionary models do not properly predict this evolutionary feature. The overall trend for RGB bump positions is to become brighter with decreasing metallicity (first established by Sweigart 1978), which makes the bump harder to detect in lower-metallicity GCs, because this part of HRD space is poorly populated owing to the rapid rate of stellar evolution at the end of the RGB stage. Yi *et al.* (2001) and Ferraro *et al.* (1999) found that RGB bump luminosities vary as $\Delta M_V/\Delta[\mathrm{Fe/H}] \approx 0.96$ for $-2.3 \leq [\mathrm{Fe/H}] \leq 0.0$ dex and for ages between 7 and 16 Gyr.

Cho and Lee (2002) suggested that in the near-IR regime, GC RGB bumps could be potentially useful standard candles (see Figure 3.8), because they are relatively bright and their position in HRDs is close to the HB (see also Ferraro *et al.* 2000). In addition, Valenti *et al.* (2004) point out that this distance indicator may be of significant interest

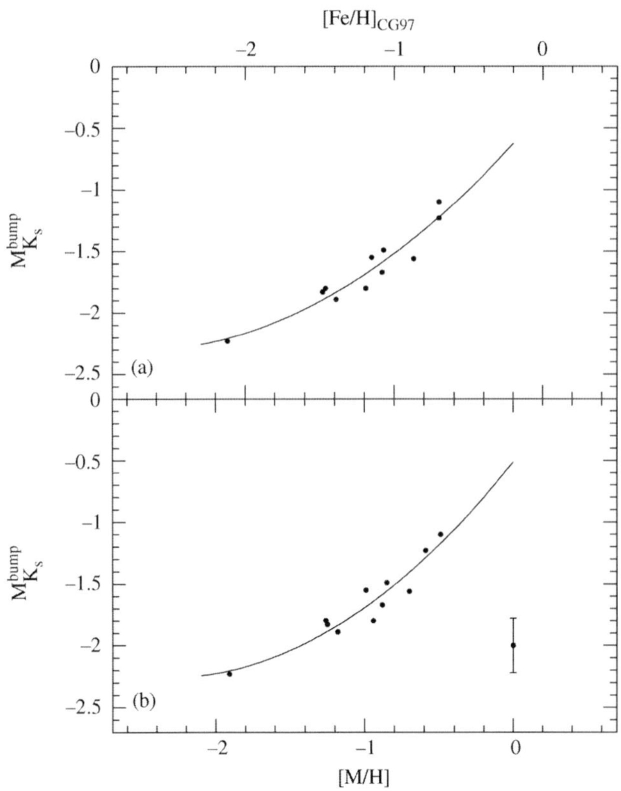

Figure 3.8 *Proposed near-IR calibration of the RGB bump in GCs as a standard candle (Cho and Lee 2002). CG97: Carretta and Gratton (1997).* (Reprinted from D. H. Cho and S. G. Lee, *Astronomical Journal*, **124**, The relation between near-infrared luminosity of red giant branch bumps and metallicity of Galactic globular clusters, p. 977–988, Copyright 2002, with permission of the AAS and D. H. Cho.)

in the era of adaptive-optics-assisted large ground-based telescopes and the *James Webb Space Telescope*, enabling its use as a standard candle in more distant galaxies, out to a few Mpc. The HB level itself, although a good distance indicator in the V band (see Section 3.2.2), is no longer horizontal at longer wavelengths and, therefore, not useful as a distance indicator. The RGB bump as a distance indicator would be affected by observational uncertainties in a quantitatively similar way as the HB in the visible regime, since it has a similar HB metallicity dependence.

On the basis of detailed abundance analysis, Valenti *et al.* (2004) provide blue bump calibration relations in the near-IR J, H and K bands as a function of both [Fe/H] (on the Carretta and Gratton 1997 metallicity scale) and [M/H],

$$M_J^{\rm bump} = 0.66 + 2.03[{\rm Fe/H}]_{\rm CG97} + 0.41[{\rm Fe/H}]^2_{\rm CG97}, \tag{3.14}$$

$$M_H^{\rm bump} = -0.33 + 1.32[{\rm Fe/H}]_{\rm CG97} + 0.21[{\rm Fe/H}]^2_{\rm CG97}, \tag{3.15}$$

$$M_K^{\rm bump} = -0.08 + 1.82[{\rm Fe/H}]_{\rm CG97} + 0.36[{\rm Fe/H}]^2_{\rm CG97} \tag{3.16}$$

and

$$M_J^{\rm bump} = 0.57 + 2.31[{\rm M/H}] + 0.56[{\rm Fe/H}]^2, \tag{3.17}$$

$$M_H^{\rm bump} = -0.38 + 1.53[{\rm M/H}] + 0.38[{\rm Fe/H}]^2, \tag{3.18}$$

$$M_K^{\rm bump} = -0.17 + 2.07[{\rm M/H}] + 0.49[{\rm Fe/H}]^2, \tag{3.19}$$

for $-2.16 \leq [{\rm Fe/H}] \leq -0.38$ and $-1.95 \leq [{\rm M/H}] \leq -0.31$ dex, respectively, in good agreement with earlier determinations.

In addition, the RGB bump position depends moderately on helium abundance (Sweigart 1978; Yun and Lee 1979) and weakly on age (Ferraro *et al.* 1999; Yi *et al.* 2001) for a given metallicity. Based on the theoretical luminosity function analysis of Sweigart and Gross (1978), in the helium-abundance range $Y = 0.1$–0.3 dex, the RGB bump's bolometric luminosity varies almost linearly with [Fe/H] (Yun and Lee 1979). Ferraro *et al.* (1999, 2000) found $\Delta M_V \sim 1.83$ and $\Delta M_{K_s} \sim 1.76$ mag, respectively, so that $\Delta M_{K_s}/\Delta[{\rm Fe/H}] \approx (1.76/1.83)/(\Delta M_V/\Delta[{\rm Fe/H}]) \approx 0.92$ for the appropriate age, metallicity and helium-abundance ranges (Cho and Lee 2002).

Cho and Lee (2002) also derive the variation of the K-band luminosity of the RGB bump as a function of helium abundance ($\Delta M_{K_s}/\Delta Y \approx -4.0$) and age [$\Delta M_{K_s}/\Delta(t/{\rm Gyr}) \approx 0.04$]. Combined with the expected variation in metallicity, they estimate a total observational uncertainty $\Delta M_{K_s} = \pm 0.24$ mag at a given metallicity and for the Galactic GC parameters considered in their work.

3.3.3 Supergiants as Standard Candles

Supergiants are the most luminous 'normal' stars. They are brighter than the Cepheid variables often used for extragalactic distance determinations (cf. Section 3.5.2). They are visible out to galaxies in the Local Group and beyond. As a consequence, their suitability

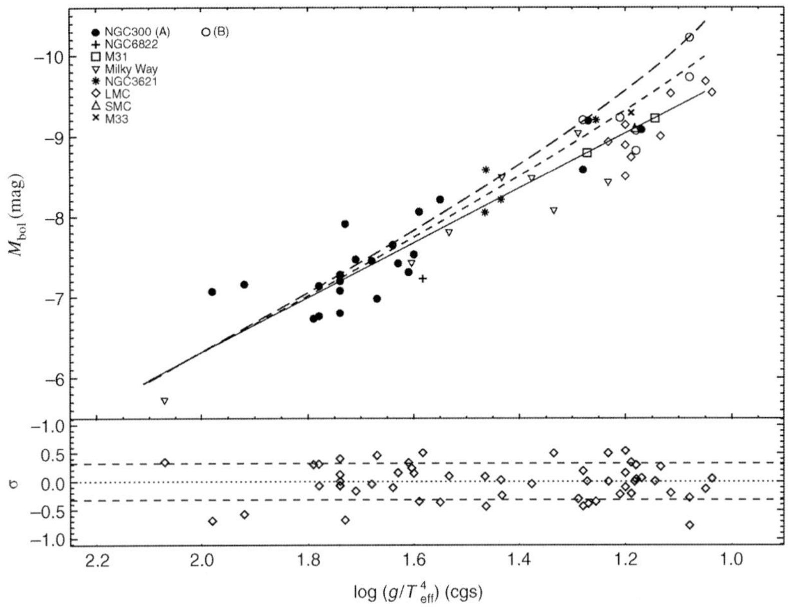

Figure 3.9 *Flux-weighted gravity versus absolute bolometric magnitude (Kudritzki et al. 2008). A (filled circles) and B (open circles): supergiants in NGC 300 and linear regression (solid line). The stellar evolution FGLRs for models with rotation are also overplotted (short-dashed line: Milky Way metallicity; long-dashed line: SMC metallicity).* (Reprinted from R. P. Kudritzki *et al.*, *Astrophysical Journal*, **681**, Quantitative spectroscopy of 24 A supergiants in the Sculptor galaxy NGC 300: flux-weighted gravity–luminosity relationship, metallicity, and metallicity gradient, p. 269–289, Copyright 2008, with permission of the AAS and R. P. Kudritzki.)

for accurate distance determinations has been investigated by many authors, starting with Hubble's (1936) early studies. However, most proposed methods (generally based on detection of the brightest supergiants in a given sample) suffered from uncertainties in distance modulus of ≥ 0.3 mag (e.g. Tully and Wolff 1984; Humphreys 1988; Pierce *et al.* 1992; Shanks *et al.* 1992; Freedman *et al.* 1994; Karachentsev and Tikhonov 1994; Rozanski and Rowan-Robinson 1994; and references therein).

Kudritzki *et al.* (2003) recently developed a promising new *spectroscopic* method for distance determination (see Figure 3.9) using the surprisingly tight relationship between the flux-weighted gravity ($g_F \equiv g/T^4_{\mathrm{eff}}$, where log g and T_{eff} are the stellar surface gravity and effective temperature, respectively) and absolute bolometric magnitude, M_{bol}, of (O-,) B- and A-type blue supergiants (see also Kudritzki *et al.* 2008 and Urbaneja *et al.* 2008 for further validation). Late-B- and early-A-type supergiants offer the greatest potential as extragalactic standard candles (e.g. Kudritzki 1998; Kudritzki *et al.* 1999) because of their high intrinsic luminosities ($-7.0 \geq M_V \geq -9.5$ mag).

The physical basis of the **flux-weighted gravity–luminosity relationship** (FGLR) is the assumption that massive stars evolve through the HRD temperature domain at constant luminosity and mass (Meynet *et al.* 1994; Heger and Langer 2000; Meynet and Maeder 2000), when $g_F =$ constant. This exploits the tight wind momentum (i.e. strength of the radiation-driven winds)–luminosity relationship for blue supergiants (Kudritzki and Puls 2000; Kudritzki and Przybilla 2003; see also Puls *et al.* 2003). If $L \propto M^{\alpha} (\alpha \sim 3)$, as usual,

$$M_{\rm bol} = a \log(g/T_{\rm eff}^4) + b, \tag{3.20}$$

with $a \sim -3.75$. This derivation from first principles leads to a dependency of fundamental stellar parameters (effective temperature and gravity) on absolute magnitude for B- and A-type supergiants that is remarkably close to empirical calibrations (see below).

Kudritzki *et al.* (2003) point out that, compared to the simple assumptions underlying their derivation of the FGLR, the mass-loss history of supergiants and its dependence on stellar angular momentum and metallicity may introduce complications. However, detailed evolutionary calculations (Meynet *et al.* 1994; Meynet and Maeder 2000) indicate, for the relevant luminosity and mass ranges, that the amount of mass lost after leaving the main sequence is small and that differences in mass loss caused by stellar rotation and metallicity have no substantial effects on the theoretical FGLRs derived.

Quantitative spectral analysis of late-B- and early-A-type blue supergiants using state-of-the-art non-LTE (LTE: local thermodynamic equilibrium) radiative-transfer models results in precise determination of the stellar parameters, metallicity, reddening and extinction for all sample stars individually (Urbaneja *et al.* 2005; Przybilla *et al.* 2006; Kudritzki *et al.* 2008). With additional information on stellar photometry, surface temperatures and surface gravities, the de-reddened apparent stellar magnitudes and the tight relationship between flux-weighted gravity, $\log g_F = \log g - 4 \log(T_{\rm eff}/10^4)$, and absolute bolometric magnitude, one can determine a distance using the FGLR (see, for details, U *et al.* 2009 and references therein),

$$M_{\rm bol} = 3.41(\log g_F - 1.5) - 8.02 \tag{3.21}$$

(Kudritzki *et al.* 2008). This calibration is based on a sample of supergiants selected from NGC 300 and seven Local Group galaxies.

FGLR distance determinations require sufficiently large samples, containing at least 10–20 objects (Kudritzki *et al.* 2008; Urbaneja *et al.* 2008), with good multiband photometry and spectra of reasonable quality and resolution ($\leq$5 Å; Kudritzki *et al.* 2003; but see U *et al.* 2009 in relation to uncertainties caused by rotation and atmospheric macroturbulence) to enable quantitative analysis using model atmospheres and radiative-transfer techniques. Although this is, as yet, a novel technique, its prospects as an independent spectroscopic distance indicator are encouraging. At present, the uncertainties in the resulting distance moduli are of order 0.1–0.3 mag (Kudritzki *et al.* 2003; U *et al.* 2009).

Systematic uncertainties affecting the FGLR method are largely related to sample selection criteria. First, in the relevant temperature range we also find brighter unrelated stars, such as luminous blue variables, which are not always easy to distinguish *a priori* from blue supergiants, although they have distinct spectral signatures in the form of strong emission lines. Second, the fraction of observed blue supergiants evolving backwards to the blue after a previous phase as red supergiants is unknown. However, it is important to understand this aspect, since those stars will have lost a significant fraction of their mass as red supergiants and might form an additional sequence below the observed FGLR. Evolutionary calculations indicate that their relative number might depend crucially on metallicity and rotation (Kudritzki and Przybilla 2003).

3.4 White Dwarf Sequences

As the oldest building blocks of the Milky Way containing statistically significant numbers of stars at most evolutionary stages, GCs are the ideal probes to study the properties of the faint ($V \geq 24$ mag) population of white dwarfs. Nevertheless, because of their intrinsically faint magnitudes and crowding in the cluster cores, deep, high-resolution *HST* observations are prerequisite to assess their basic properties (e.g. De Marchi *et al.* 1995; Richer *et al.* 1995, 2006; Cool *et al.* 1996; Renzini *et al.* 1996; Zoccali *et al.* 2001; Sigurdsson *et al.* 2003; Hansen *et al.* 2004, 2007).

The premise of using white dwarfs as standard candles is based on fitting a given GC white dwarf cooling sequence to an appropriate empirical cooling sequence constructed on the basis of local tracers with well-determined trigonometric parallaxes (see Figure 3.10; see also Bono *et al.* 2008b for cross-calibration of white-dwarf-related distance determinations with other well-established methods). This is, in essence, analogous to the classical main-sequence fitting to the local subdwarfs (see Section 3.2.1). However, white dwarf cooling-sequence fitting is more straightforward and relies on fewer assumptions: because of gravitational separation in their atmospheres, it does not involve metallicity determinations and their associated uncertainties, while there are no complications associated with

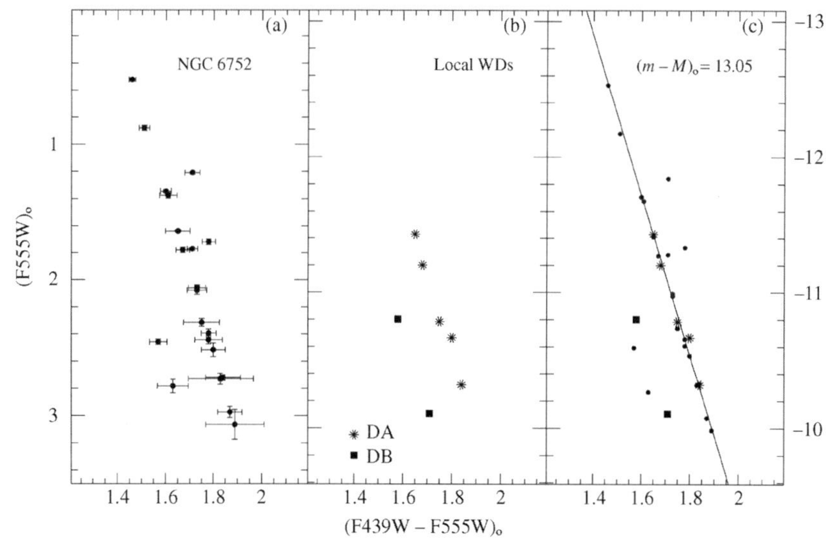

Figure 3.10 *(a) Instrumental HST colour–magnitude diagram for the GC NGC 6752's white dwarfs (Renzini et al. 1996). (b) Instrumental absolute colour–magnitude diagram for the local, calibrating white dwarfs of known trigonometric parallax. DA (hydrogen-rich) and DB (helium-rich) white dwarfs are represented by different symbols. (c) Instrumental colour–magnitude diagram of the cluster and local white dwarfs, with the former having been shifted in magnitude to match the local sequence. This results in a distance modulus for the cluster of* $(m - M)_0 = 13.05$ *mag. The straight line is a linear fit to the cluster white dwarf sequence.* (Reprinted from A. Renzini *et al.*, *Astrophysical Journal*, **465**, The white dwarf distance to the globular cluster NGC 6752 (and its age) with the Hubble Space Telescope, L23–L26, Copyright 1996, with permission of the AAS and A. Renzini.)

convection either. This is so because white dwarfs have virtually metal-free atmospheres: they are either composed of nearly pure hydrogen (called DA) or pure helium (DB). In addition, as pointed out by Renzini *et al.* (1996), white dwarfs are locally much more abundant than subdwarfs and, therefore, accurate trigonometric parallaxes can be obtained for a potentially much larger sample of calibrators. (On the other hand, Hansen *et al.* (2004) point out that using the white dwarf cooling sequence as a distance indicator requires not only a vertical shift in magnitude, but possibly also a shift in the direction of the reddening vector.)

The main uncertainty associated with using white dwarf cooling sequences for distance determination is their dependence in HRD space on the mass of the white dwarf. However, at least in GCs, four independent measurements – the luminosities of the TRGB, the HB, the AGB termination and post-AGB stars – are all very sensitive to the mass of the hydrogen-exhausted core and lead to a very tightly constrained white dwarf mass at the top of the cooling sequence, $M_{\rm WD} = 0.53 \pm 0.02\ {\rm M}_{\odot}$, independent of metallicity (Renzini and Fusi Pecci 1988; see also Cool *et al.* 1996 for NGC 6397; Richer *et al.* 1997 and Hansen *et al.* 2004 for M4). The uncertainty of $0.02\ {\rm M}_{\odot}$ implies an intrinsic uncertainty in the distance modulus of 0.05 mag, so that this method is potentially among the most accurate secondary distance indicators available (see Moehler and Bono 2008 for a review). Of course, this assumes that we can access sufficiently accurate local calibrators, without introducing additional systematic errors. The local calibrators are characterized by a small mass range, $\langle M_{\rm WD} \rangle = 0.59\ {\rm M}_{\odot}$, with a 1σ dispersion of $\sim 0.1\ {\rm M}_{\odot}$ (Bergeron *et al.* 1992; Bragaglia *et al.* 1995), thus leading to an intrinsically narrow local cooling sequence.

Zoccali *et al.* (2001) provide a general equation relating the distance modulus to white dwarf mass and reddening,

$$
\begin{aligned}
(m - M)_V = {} & 13.09 + 3.2E(B - V) - 2[E(B - V) - 0.55] \\
& -2.4(M_{\rm WD}^{\rm cl} - 0.53) + 2.4(\langle M_{\rm WD} \rangle - 0.594), \qquad (3.22)
\end{aligned}
$$

where $\langle M_{\rm WD} \rangle$ is the true mean mass of the local white dwarf sample, $M_{\rm WD}^{\rm cl}$ is the true mass of the cluster white dwarfs and $0.594\ {\rm M}_{\odot}$ is the mean mass of the sampled local DA calibrators, including a correction for a thick hydrogen envelope (cf. Salaris *et al.* 2001).

The overall error budget, including uncertainties in the relative cluster and local white dwarf photometry (and uncertainties in the charge-transfer efficiency of the *HST* detectors, where relevant), extinction, trigonometric parallax and average mass offset of the calibrators as well as their suitability (see Salaris *et al.* 2001), can be as low as ± 0.1 mag in distance modulus (Renzini *et al.* 1996; Salaris *et al.* 2001; Zoccali *et al.* 2001; but see Hansen *et al.* 2004). In turn, this allows more accurate age estimates for those GCs containing a sizeable, observable white dwarf population. Hansen *et al.* (2004) recently argued for a more sophisticated approach to obtain even better constrained age estimates, by fitting the full two-dimensional HRD instead of the one-dimensional white dwarf cooling sequence on its own.

3.5 Period–Density Relations

Starting from Henrietta Leavitt's (1908a,b, 1912) seminal observations of almost 1800 Cepheid variables in the Magellanic Clouds, pulsating variable stars have long been among the best-calibrated and most useful and accurate distance indicators for use at distances beyond those where the standard parallax methods break down.

Assuming virial equilibrium, one can show that the time, P, required for a sound wave to cross a star is

$$P \sim \frac{1}{\sqrt{G\bar{\rho}}}, \tag{3.23}$$

where $\bar{\rho}$ is the mean stellar density. Alternatively,

$$P\left(\frac{\bar{\rho}}{\bar{\rho}_\odot}\right)^{1/2} \equiv Q(P), \tag{3.24}$$

where $Q(P)$ is the star's pulsation constant ($\simeq$1 h for any reasonable stellar make-up), which is a very weak function of the pulsation period if the star's pulsations are gravity driven (Ritter 1879). It can also be shown using simple approximations that, at constant effective temperature, the mean stellar density decreases with increasing luminosity. In the early 1900s, Adams and Joy (1927) and Shapley and Walton (1927) showed that the bright Cepheid variables were characterized by a very narrow range in spectral type or, equivalently, temperature at a given period. This, in turn, led to the establishment of a tight PL relation: more luminous stars are expected to have longer pulsation periods.

The spectral-type/temperature restriction of Cepheids naturally led to the realization that pulsating variables occur only in a narrow '**instability strip**' in the HRD (see Figure 3.11). The instability strip is host to a range of variable stars, including (from top to bottom) classical Cepheids (also known as δ Cephei stars), Mira variables at the top of the AGB, W Virginis stars (intermediate-mass HB stars, also known as 'Population II' Cepheids) and RR Lyrae stars. **SX Phoenicis** and δ Scuti variables (jointly known as **dwarf Cepheids** or ultrashort-period variables) as well as **anomalous Cepheids**, and ZZ Ceti, V777 Herculis and GW Virginis pulsating white dwarfs are also found in the instability strip. On the basis of these simple approximations, all of these stars obey specific period–density relations and may, thus, potentially be used for distance calibration.

3.5.1 The Baade–Wesselink Method

Walter Baade (1926) and Adriaan Wesselink (1946) established an independent method to determine distances to Cepheid variables: they showed that the variation of a star's angular diameter can be inferred from measurements of its pulsation-induced brightness changes on the basis of model atmosphere calculations.

The Baade–Wesselink method, also referred to as the '**pulsation parallax**' in its more recent incarnations (e.g. Di Benedetto 1997, 2008), depends on the fact that stellar pulsation involves physical expansions and contractions, leading to both brightness changes and a periodic variation in the star's radial velocity. A time integration of the radial velocity curve gives the total linear amount by which the star's radius changes; the periodic changes in brightness and the spectrum allow a calculation of the *fraction* by which its radius changes. Thus, one can obtain the star's radius, R. The stellar effective surface temperature, $T_{\rm eff}$, can be obtained from its spectrum, so that its luminosity, $L = 4\pi R^2 \times \sigma T_{\rm eff}^4$ (where σ is the Stefan–Boltzmann constant), leads to an absolute magnitude and, therefore, a distance.

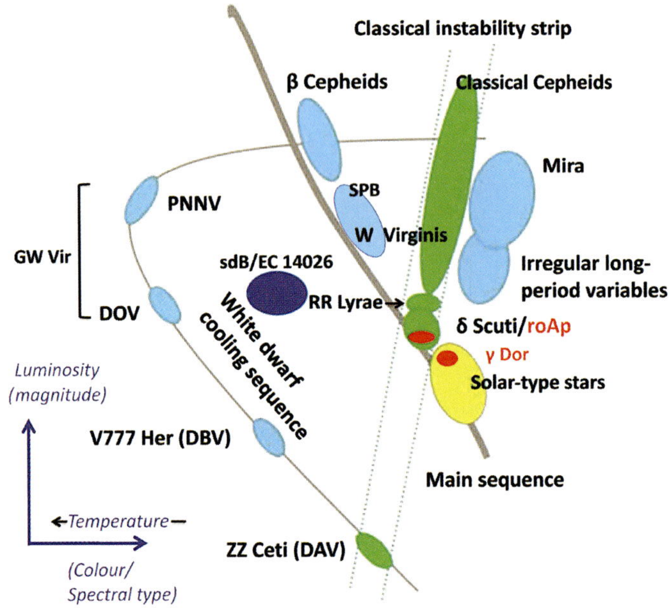

Figure 3.11 Location of the classical instability strip and other variable stars in the HRD. DOV: helium carbon oxygen white dwarfs. DBV: helium white dwarfs. DAV: hydrogen white dwarfs. PNNV: planetary nebula nuclear variables. sdB: pulsating subdwarfs. SPB: slowly pulsating B-type stars. roAP: rapidly oscillating peculiar A-type stars.

The most recent calibration by Beuermann (2006), based on the pioneering Barnes and Evans (1976) relationship and aimed at determining accurate distances to **cataclysmic variables** (binaries in which the radius of the Roche-lobe-filling dwarf secondary star is reasonably well known from Roche geometry), gives

$$S_\lambda = m_\lambda + 5\log(R/\mathrm{R}_0) - 5\log(d/10\ \mathrm{pc}) \quad (3.25)$$

$$= m_\lambda + 5\log\phi_{\mathrm{mas}} + 0.1564, \quad (3.26)$$

where $m_\lambda = -2.5\log(f_\lambda/f_0)$, f_λ and S_λ are, respectively, the star's mean apparent magnitude, flux and surface brightness as a function of wavelength, f_0 is the flux constant at zero wavelength, d is the distance (in pc), R_0 is the distance to the Galactic Centre and ϕ_{mas} is the stellar diameter in milli-arcseconds (mas). S_λ is a function of stellar gravity and metallicity (cf. Beuermann 2006; see also Di Benedetto 2008 and references therein).

Note, however, that the details of this method are highly complex, and involve quite a few assumptions. One of these is that the calibration relations are based on 'spotless' stars, because the presence of star spots will affect the integrated magnitude of one's calibrator stars. Second, the stellar radii used for calibration purposes tend to originate from stellar atmosphere modelling rather than from direct observations. Interferometric observations represent a promising new approach to derive stellar radii directly (see Figure 3.12). The

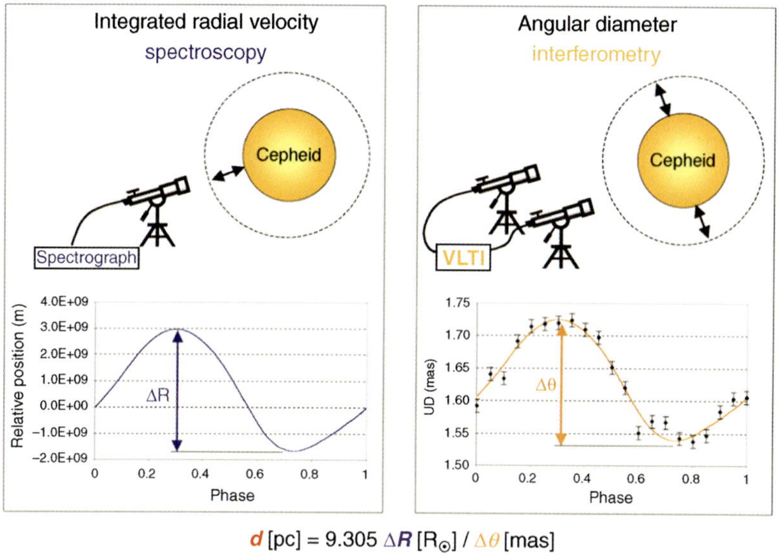

Figure 3.12 *The two observation techniques used for the interferometric version of the Baade–Wesselink method are high-resolution spectroscopy (left) and interferometry (right). The former provides the radial velocity curve over the star's pulsation cycle. When integrated, this, in turn, provides the linear radius variation of the star (in metres). The interferometric observations document variation of the star's angular radius. The ratio of these two quantities gives the distance.* (Reprinted from ESO Images (online), Image eso0432a, Observation Techniques of the Baade–Wesselink Method, Copyright 2004, with permission of ESO.)

first interferometric measurements of stellar radii in the mas range and their changes have now been obtained using the *Very Large Telescope Interferometer* (Ségransan *et al.* 2003; Kervella *et al.* 2004a,b,c,d; see Kervella 2006 for a review) and the *Center for High Angular Resolution Astronomy* array (Berger *et al.* 2006), enabling more precise and independent distance determinations to Cepheid variables. For instance, *l* Carinae's angular diameter of 3.2 mas measured by Kervella's team, combined with a period of 35.5 days, leads to a relative distance accuracy of approximately 5%. Initial distance determinations based on these interferometric measurements are fully consistent with previously obtained geometric distances using *Hubble Space Telescope* (e.g. Benedict *et al.* 2002, 2007) and *Hipparcos* parallaxes and potentially allow the Cepheid distance scale to be established with better than 5% accuracy.

3.5.2 Classical Cepheid Variables

The classical Cepheids are located at the top of the instability strip, which implies that they are the most massive class of variable stars, with masses $\geq$3.5–4.0 $M_\odot$. After stars with masses in excess of $\sim$3.5 $M_\odot$ reach the RGB, their evolutionary tracks will start to move to lower temperatures in the HRD, i.e. they perform a 'blue loop'. For stars more massive than $\sim$5 $M_\odot$, these excursions will cause them to pass into the instability strip and, hence, start to pulsate as classical Cepheids. Given their high masses, classical Cepheids are young stars and, as such, they are usually found in or near regions of active star formation in galaxies, such as in spiral arms.

Many Cepheids pulsate in three different modes, with excited fundamental and first- and second-overtone, as well as mixed modes. This allows two independent mass determinations, on the basis of both our understanding of stellar evolution combined with the star's position in the HRD and pulsation theory. Stellar mass determinations are important for our understanding of the origin and details of the PL relation: the Cepheid PL relation represents the variation with stellar mass of the fundamental pulsation frequency, at least to first order. However, as usual the details are more complex, because stellar structure – and, hence, the pulsation characteristics – also depends on metallicity, so that we should strictly be dealing with a **PL–colour relation** (see Caldwell and Coulson 1986, Sandage and Tammann 2006, Groenewegen 2008a and Bono *et al.* 2010 for a review). PL relations neglect the width in temperature of the instability strip. This assumption is valid in the near-IR regime, but not at optical wavelengths. However, PL–colour relations suffer from both theoretical and observational uncertainties, such as sensitivity to colour–temperature relations and to reddening uncertainties, respectively. Figure 3.13 provides an overview of the observational PL slopes for Milky Way Cepheids as a function of passband, hence illustrating the PL–colour behaviour.

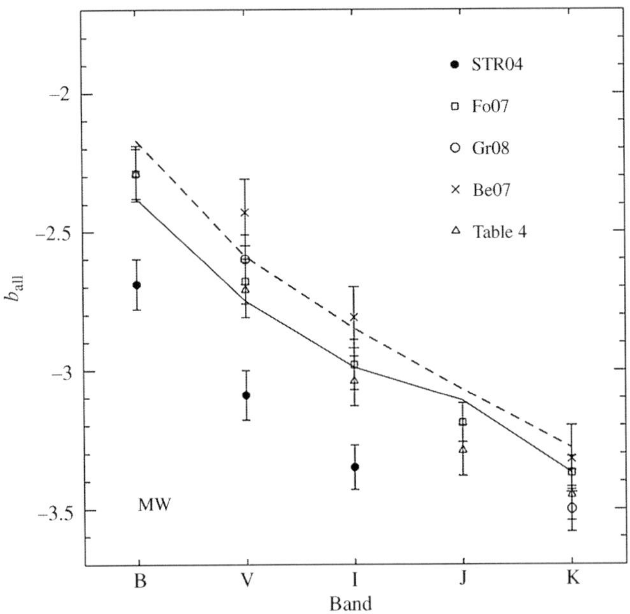

Figure 3.13 *Comparison between observed slopes, b_{all}, of the Galactic PL relations (open symbols) and predicted values for a chemical composition of 12 + log(O/H) = 8.58 and 8.89 dex (solid and dashed lines, respectively; Bono et al. 2010). STR04: Sandage et al. (2004). Fo07: Fouqué et al. (2007). Gr08: Groenewegen (2008a). Be07: Benedict et al. (2007). Table 4: Bono et al. (2010).* (Reprinted from G. Bono *et al.*, *Astrophysical Journal*, **715**, Insights into the Cepheid distance scale, p. 277–291, Copyright 2010, with permission of the AAS and G. Bono.)

The importance of colours/metallicity for the exact slope and zero point of the PL relation naturally implies that the PL relations for the Milky Way and the Magellanic Clouds must be different, given their different stellar populations (e.g. Sandage *et al.* 2004; see also Ngeow and Kanbur 2005; Tanvir *et al.* 2005; Sandage and Tamman 2006, 2008; and references therein; but see Bono *et al.* 2010 for a recent opposing view), which also carries through into different locations of the instability strip in these galaxies. Thus, one cannot use Magellanic Cloud PL calibrations to determine the zero point for Galactic Cepheids obscured by poorly determined amounts of extinction. Yet, in early applications of Cepheid PL relations as distance indicators, this was precisely how one used to proceed. The problem is compounded by uncertain differences among the Local Group galaxies as a function of colour (cf. Sandage and Tammann 2006, 2008; see also Tammann *et al.* 2003). Without corrections for these colour differences, the PL relation is useful only to an accuracy of 0.3 mag, although the metallicity dependence virtually disappears in the near-IR regime, for metallicities between approximately −0.35 and −1.0 dex (cf. Gieren *et al.* 2005; Pietrzyński and Gieren 2006; see also Groenewegen 2008a). In addition, Sandage and Tammann (2006, 2008) argue that it is now well established that the LMC's period–colour relation exhibits a real break at a period of ∼10 days (but see Fouqué *et al.* 2007 and Bono *et al.* 2010 for contrasting results) – not found in the Milky Way – which will, of course, lead to similar differences in the galaxies' PL relations.

Sandage and Tammann (2006) give the current best PL relations for Cepheids in the Milky Way (see also Fouqué *et al.* 2007):

$$M_B = -(2.692 \pm 0.093) \log P - (0.575 \pm 0.107), \quad \text{rms} = 0.25 \text{ mag}, \tag{3.27}$$

$$M_V = -(3.087 \pm 0.085) \log P - (0.914 \pm 0.098), \quad \text{rms} = 0.23 \text{ mag}, \tag{3.28}$$

$$M_I = -(3.348 \pm 0.083) \log P - (1.429 \pm 0.097), \quad \text{rms} = 0.23 \text{ mag}, \tag{3.29}$$

where the period, P, is given in days. The root-mean-square (rms) values per star are consistent with theoretical expectations based on the width of the instability strip. Sandage *et al.* (2004) show that, particularly at the short- and long-period extremes, the LMC and Galactic PL relations differ significantly, e.g. by 0.36 mag at $\log P = 0.5$ ($P = 3$ days) and by 0.16 mag at $\log P = 1.7$ ($P = 50$ days). Recently, Fouqué *et al.* (2007) obtained new Cepheid PL relations for the LMC:

$$M_B = -(2.393 \pm 0.040) \log P + (17.368 \pm 0.009), \quad \text{rms} = 0.27 \text{ mag}, \tag{3.30}$$

$$M_V = -(2.734 \pm 0.029) \log P + (17.052 \pm 0.007), \quad \text{rms} = 0.20 \text{ mag}, \tag{3.31}$$

$$M_I = -(2.957 \pm 0.020) \log P + (16.589 \pm 0.005), \quad \text{rms} = 0.13 \text{ mag}, \tag{3.32}$$

as well as for a range of other optical and near-IR broad-band filters, with the slopes steepening and the scatter decreasing towards longer wavelengths. Recent results for ultralong-period Cepheids (with fundamental-mode periods between 80 and 210 days; first discovered by Leavitt 1908b and referred to as long-period Leavitt variables) suggest that these objects have a very shallow PL relation in optical passbands and, particularly, in the reddening-free **Wesenheit magnitudes** (see Majaess *et al.* 2010 for Wesenheit calibrations for multiple types of variable stars) presently in common use for PL analysis, with a similar scatter as that found for classical Cepheids (Bird *et al.* 2009). As such, they could be even more useful as standard candles than their classical counterparts, out to much greater distances ($\sim$100 Mpc and beyond, compared to $\leq$ 30 Mpc for classical Cepheids) by virtue of their higher luminosities (see also Grieve *et al.* 1985).

On Galactic scales, the main uncertainties in the Cepheid distance calibration depend on the accuracy of distance determination to the Pleiades (which is affected by an uncertainty of up to 0.04 mag in M_V or a 12% uncertainty in its distance; see Section 7.2) and – more importantly – any Galactic metallicity gradient or metallicity differential with respect to this zero-point calibration target (e.g. Feast and Walker 1987; Sandage and Tamman 2006, 2008). The latter could lead to an uncertainty of order 0.1 mag in M_V near solar metallicity, although Sandage and Tamman (2006) point out that the average metallicity of the current sample of Cepheid calibrators is roughly solar, so that this effect may be small or negligible. Any such effects could, in principle, be masked if the short- and long-period variables are distributed differently across the Milky Way, although this does not seem to be the case.

Adopting, possibly incorrectly, either of the local (LMC or Galactic) PL relations as a presumably firmly established rung of the distance ladder could have profound consequences for determination of the Hubble constant on the basis of Cepheid light curves (see Section 4.1). Clearly, this would cause a major impediment for the *HST* Key Project. To mitigate the potentially devastating effects of these calibration uncertainties, Freedman *et al.* (2001) used Galactic and LMC calibrators but applied metallicity corrections that were linear in either [Fe/H] or [O/H], resulting in lingering systematic uncertainties in M_V at the level of 0.2 mag (see also Madore and Freedman 2009 for a discussion of the remaining uncertainties), although distance accuracies of 5% or better have been claimed, either in an absolute sense (e.g. Macri *et al.* 2006; An *et al.* 2007a) or with respect to the LMC (e.g. Gieren *et al.* 2008 for the Local Group galaxy WLM, and other results from the Araucaria Project; Pietrzyński and Gieren 2009).

Recent work has extended the useful wavelength baseline for Cepheid PL relations to the mid-IR range on the basis of *Spitzer Space Telescope* observations (see Figure 3.14; e.g. Freedman *et al.* 2008; Ngeow and Kanbur 2008; Madore *et al.* 2009a,b; Ngeow *et al.* 2009; Marengo *et al.* 2010; see also Freedman and Madore 2010 for a novel approach), although more work remains to be done on characterization of the uncertainties and the effects of metallicity differences.

3.5.3 Mira Variables

Mira variables are in their late evolutionary phases and occupy the instability strip where it crosses the AGB. The prototype long-period variable, Mira (o) Ceti, now lends its name to all large-amplitude ($\Delta V > 2.5$ mag) regular long-period variables with pulsation periods in excess of 80 days.

Stars lose a significant fraction of their total mass during the relatively short AGB phase. In turn, this may lead to instabilities in the helium- and hydrogen-burning shells surrounding the star's degenerate core, thus liberating sufficient energy to sustain a range of stellar pulsation modes. These pulsations cause shocks in the envelope, facilitating the channelling of material to higher atmospheric layers. The temperature is then low enough to allow dust formation, which is ejected by radiation pressure due to the luminous core. Pulsations, probably in the fundamental mode and related to opacity changes (the κ **mechanism**, first

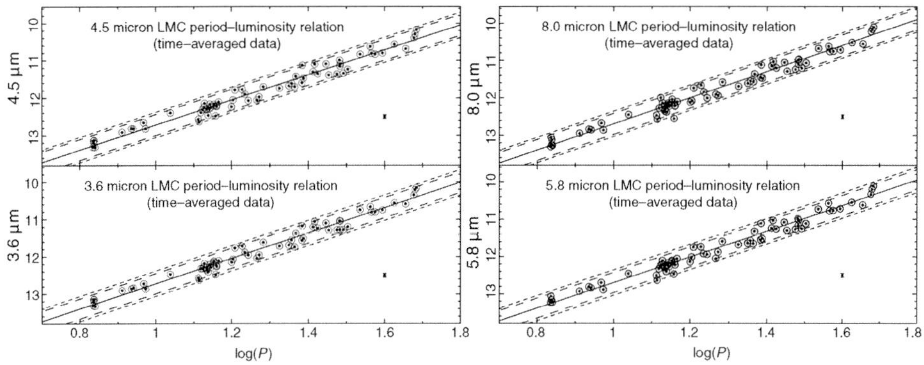

Figure 3.14 *Time-averaged mid-IR Spitzer Space Telescope PL relations for LMC Cepheids (Madore et al. 2009b). The solid line is a weighted least-squares fit to the data. The thick broken and thin dashed lines are two estimated $\pm 2\sigma$ (typically ± 0.28 mag) bounds on the instability strip. The symbol sizes are comparable to the typical photometric errors for mid-range Cepheids, i.e. ± 0.05 mag at $\log(P/\text{days}) = 1.3$. A magnitude error of ± 0.05 mag is shown in the bottom right-hand corner of each figure; at the shortest periods, the errors on a single observation can be as high as ± 0.10 mag.* (Reprinted from B. F. Madore *et al.*, *Astrophysical Journal*, **695**, The Cepheid period–luminosity relation (the Leavitt law) at mid-infrared wavelengths. II. Second-epoch LMC data, p. 988–995, Copyright 2009, with permission of the AAS and B. F. Madore.)

identified by Eddington[2]), affect both the stellar radius and temperature. In combination with the shocks and dust formation, they produce significant photometric variability (e.g. Fraser *et al.* 2005). However, the pulsation mechanism and the relationships between pulsation, mass loss and the ejection of the stellar envelope as the star evolves into a PN are all, as yet, poorly understood *quantitatively*. The main reason for this lack of understanding is that the stellar atmospheres are highly convective, while we do not have a robust theory for pulsations in the presence of convection.

Cook *et al.* (1997) first noted that the variable stars in the MACHO (Massive Compact Halo Objects) database of microlensing events in the LMC occupied five parallel sequences in PL space (see also Fraser *et al.* 2005 for a near-IR view), which were subsequently classified by Wood *et al.* (1999); see Figure 3.15. Mira variables in the Milky Way and the Magellanic Clouds obey the same PL relations (Lebzelter *et al.* 2002), which implies that metallicity differences do not contribute to the observed multiple parallel sequences (but see Groenewegen and Blommaert 2005 for an opposing view). Parallel sequences also exist in the Milky Way (e.g. Bedding and Zijlstra 1998; Glass and Schultheis 2003; Knapp *et al.* 2003; Groenewegen and Blommaert 2005). Fraser *et al.* (2005) suggest that some of the parallel sequences might be caused by amplitude changes caused by colour changes during the variability cycle. In particular, the so-called **SRa (semi-regular type 'a') variables** are very similar to Mira variables yet have amplitudes smaller than 2.5 mag in V.

[2] The κ mechanism implies that the Cepheid pulsation mechanism is a heat engine, based on the opacity of hydrogen and helium (see, for a review, Bono *et al.* 1999). The ionized gas in the star's outer layers is opaque at the faintest part of the pulsation cycle. This leads to a temperature increase because of stellar radiation, hence leading to expansion. The expanding outer atmosphere enables the star to cool – and thus become less ionized and, hence, less opaque – which, in turn, allows radiation to escape. Gravitational attraction by the star's interior mass causes the process to stop and reverse, eventually leading to cyclic behaviour.

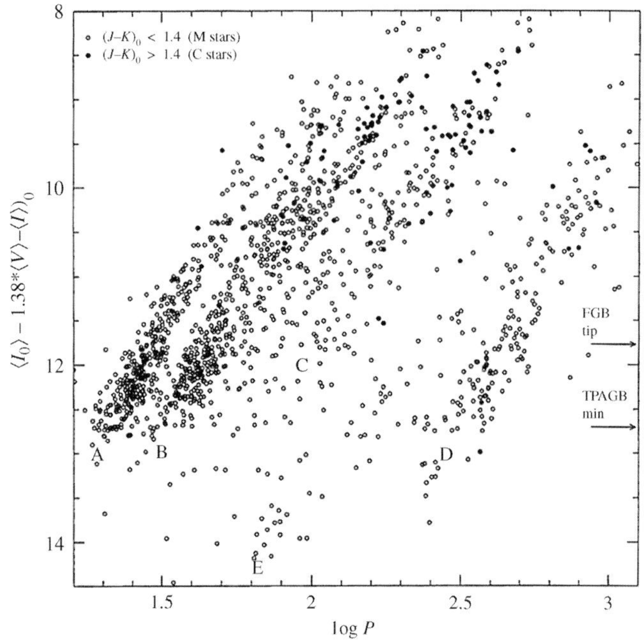

Figure 3.15 *PL relations for Mira variables in the LMC (Wood et al. 1999), based on the* MACHO *database. The five parallel sequences classified by Wood et al. (1999) are indicated. ⟨V⟩, ⟨I⟩: mean magnitudes. FGB: first-ascent giant branch. TPAGB: thermally pulsing AGB.* (Reprinted from P. R. Wood *et al.*, International Astronomical Union Symposium Proceedings, **191**, MACHO observations of LMC red giants: Mira and semi-regular pulsators, and contact and semi-detached binaries, p. 151–158, Copyright 1999, with permission of the IAU.)

The sequence associated with long-period SRa stars may therefore have its origin in the same underlying physics as that for Miras (see also Cioni *et al.* 2001), so that the amplitude cut-off seems arbitrary. The multiple sequences are often associated with radial pulsations caused by various overtone frequencies (e.g. Gautschy and Saio 1996; but see Fraser *et al.* 2005).

The near-IR K-band PL relation of Feast *et al.* (1989) for Mira variables in the LMC, at an adopted distance of 50 kpc, is most often used as a relative distance calibrator (see also Groenewegen *et al.* 2004):

$$M_K = -(3.57 \pm 0.16) \log P + (1.21 \pm 0.39), \text{ rms} = 0.15 \text{ mag}. \qquad (3.33)$$

For the interested reader, Feast *et al.* (1989) provide an historical overview of PL relations and argue that the scatter in the near-IR bands is much smaller than that at optical wavelengths. They provide Mira PL relations in the J, H and K bands. Menzies *et al.* (2008) and Whitelock *et al.* (2009) applied the Mira PL relations to the Local Group galaxies Phoenix and Fornax, respectively, while Rejkuba (2004) showed its applicability out to NGC 5128.

Studies of mostly oxygen-rich Mira variables in the Milky Way have recently attracted renewed interest (e.g. Schultheis *et al.* 2004; Groenewegen and Blommaert 2005; Whitelock *et al.* 2008; Matsunaga *et al.* 2009). Groenewegen and Blommaert (2005) determine a

K-band PL slope of -3.37 ± 0.09 based on OGLE microlensing observations in the direction of the Galactic bulge (see also Matsunaga *et al.* 2005), which agrees to within the uncertainties with the LMC relation of Equation (3.33). The *Spitzer Space Telescope* offers unique opportunities to take this into the mid-IR regime, where preliminary results suggest a very tight Galactic Mira PL relation (Matsunaga *et al.* 2009; see also Yang and Jiang 2010).

3.5.4 W Virginis and Other 'Population II' Cepheids

Turning now to lower masses, once metal-poor HB stars have exhausted helium burning in their cores, a significant fraction migrate to the instability strip. These 'Population II' (or 'Type II') Cepheids are usually referred to after their prototype, W Virginis. They are brighter and less massive than the more common RR Lyrae variables (see Section 3.5.5) at similar metallicities. Most W Virginis stars are located in the Galactic halo, both in the general field and in the old GC population, with a distinct concentration towards the Galactic Centre. These stars have similar periods as the classical Cepheids, yet lower masses. The W Virginis stars form the bulk of the Population II Cepheids at intermediate periods, but we note that at the short-period extreme the class is best represented by **BL Herculis** (BL Her) stars, while at the longest periods they resemble the **RV Tauris** (RV Tau) prototype, which have alternating depths of the minima in their light curves, but a well-defined main pulsation period. The distinctions between these subclasses are somewhat fuzzy and arbitrary, but as a rule of thumb, we can use class separations at periods of 4 and 20 days (e.g. Soszyński *et al.* 2008). In addition, Soszyński *et al.* (2008) define a subclass of peculiar W Virginis (pWV) stars, many of which are binary systems. On the basis of the results of Soszyński *et al.* (2008), Matsunaga *et al.* (2009) suggest that most W Virginis stars occupy a narrower instability strip than the BL Her stars, although there are unresolved issues related to the effects of reddening.

Gingold (1976, 1985) laid the basis for our evolutionary understanding of these stars (see Figure 3.16). Here, BL Her stars evolve from the blue HB to the lower AGB W Vir stars trace out loops blueward of the AGB and RV Tau stars are on their way to a blue post-AGB evolutionary phase. Note, however, that modern stellar evolutionary tracks no longer require these stars to perform blue loops in the HRD. In all cases, these stars are believed to be the direct descendants of hot, low-mass blue HB stars, characterized by little envelope mass (e.g. Catelan 2009). On the basis of updated nonlinear convective models, Di Criscienzo *et al.* (2007) calculated that these BL Her and W Virginis stars pulsate in their fundamental mode and have typical masses between $\sim$0.52 and 0.65 $M_\odot$ (for periods $\geq$0.8 days).

W Virginis stars have at least one claim to notoriety: Edwin Hubble used them to determine the distance to M31, although he believed that he was dealing with classical Cepheids. As a result, he obtained a distance that was significantly too small because the PL relation for these stars is offset with respect to that of the classical Cepheids.

Population II Cepheids define tight PL and Wesenheit relations, particularly at near-IR wavelengths, where the effects of metallicity variations are minimal if not negligible (e.g. Di Criscienzo *et al.* 2007; Matsunaga *et al.* 2009, 2010). This suggests that they may be particularly useful for distance measurements to disc and halo stellar populations (see Matsunaga *et al.* 2009, 2010 for near-IR least-squares PL solutions for both BL Her and W Virginis stars separately and combined).

One caveat of note here is that RV Tau stars in the LMC start to deviate from linear PL relations towards longer periods (log P/days from 1.3 to 1.8), although Galactic GC RV Tau

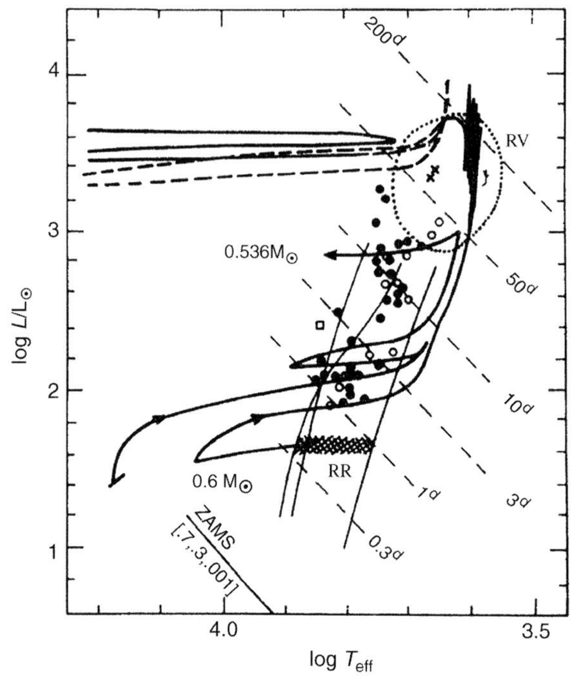

Figure 3.16 *Gingold's (1985) original proposal for the evolution of Population II Cepheids. This HRD shows the positions of RR Lyrae stars ('RR'), Population II Cepheids (solid and open circles, denoting cluster and field variables, respectively), RV Tau stars ('RV': dotted region and crosses) and an anomalous Cepheid (open square). The Population II instability strip fundamental and first-overtone blue edge are shown as thin solid lines; dashed lines represent approximate loci of constant period. ZAMS: zero-age main sequence (for helium abundance $Y = 0.3$ and $Z = 0.01$). Dashed tracks: rapid evolutionary phases.* (Reprinted from R. A. Gingold, *Memorie della Società Astronomica Italiana*, **56**, The evolutionary status of Type II Cepheids, p. 169–191, Copyright 1985, with permission of the Società Astronomica Italiana.)

stars do not (cf. Matsunaga *et al.* 2009; see also Russell 1998; Zsoldos 1998; Matsunaga *et al.* 2006). It has been suggested that these longer-period RV Tau stars ($P > 40$ days) may be on loops from the AGB, similarly to the shorter-period W Virginis stars (e.g. Matsunaga *et al.* 2009), because those for which absolute distance calibrations can be obtained are located well down the giant branch. Matsunaga *et al.* (2009) also speculate that for $P >$ 20 days, Galactic GC RV Tau stars represent the long-period tail of the W Virginis stars, while they represent a distinct stellar population in the LMC. If so, this would introduce significant uncertainties in their absolute and even relative distance estimates.

The key Population II Cepheid often used to place their PL distance moduli on an absolute distance scale is κ Pavonis (κ Pav), the nearest of its class, for which pulsation parallaxes can be used (see Section 3.5.1). However, it now transpires that κ Pav is unusually bright for its pulsation period and is, hence, likely a pWV-type object rather than a normal W Virginis star (see Matsunaga *et al.* 2009). Nevertheless, Di Criscienzo *et al.* (2007) show that the distances implied by Population II Cepheid PL relations are consistent with those derived from RR Lyrae analysis.

As early as 1949, Joy identified a group of variable stars in GCs with periods between 0.8 and 3 days but brighter than the 'normal' RR Lyrae stars by up to one magnitude (see Sandage and Tammann 2006 for an historical overview and early classification confusion), which were subsequently named 'above-HB' (AHB) stars but have since been identified with the more common BL Her variables. Their field prototypes include XX Vir and SW Taurus. The physical importance of these stars is that they trace the transition from the post-HB evolutionary phase to the base of the AGB (Strom *et al.* 1970). Curiously, these stars cross the instability strip some 100$\times$ faster than ordinary RR Lyrae, in $<10^6$ yr (Dorman 1992), which enables their use for studies of secular period changes caused by stellar evolution (Wehlau and Bohlender 1982; Diethelm 1996).

3.5.5 RR Lyrae Stars

RR Lyrae stars, pulsating stars with periods from 0.2 to 0.8–0.9 days, are among the most extensively studied stellar tracers, for reasons ranging from evolutionary and pulsational modelling to understanding fundamental problems in physics. They are also among the most often used distance indicators to old, low-mass stars in the Milky Way and the Local Group of galaxies (see, for a review, Bono 2003). Nevertheless, a number of unsolved problems related to their physical pulsation modes remain, although they are not critical for obtaining reliable distance measurements, including the so-called **Blažhko effect** (amplitude or phase modulations on timescales of typically tens to hundreds of days, occurring in $\sim$50% of fundamental-mode pulsating RR Lyrae stars; e.g. Blažhko 1907; Szeidl 1988; Moskalik and Poretti 2002; Smith *et al.* 2003; Kolenberg *et al.* 2006, 2010), mixed-mode behaviour (combined fundamental, first- and sometimes second-overtone pulsators; e.g. Bono *et al.* 1996; Feuchtinger 1998) and the formation and propagation of the shock front during the pulsation cycle (e.g. Bono *et al.* 1994; Chadid *et al.* 2000).

The empirical basis of using RR Lyrae stars as accurate distance indicators relies on the M_V–[Fe/H] relationship, $M_V = a + b$[Fe/H], which is plagued by problems of its own and, particularly, by calibration issues. Although internal errors in both photometric measurements and metallicity determinations are often of order a few hundredths of a magnitude or iron-abundance dex, uncertainties in the M_V–[Fe/H] calibration are usually in the range from 0.10 to 0.25 mag (e.g. Bono *et al.* 2001; Sandage and Tammann 2006). This suggests that the latter are dominated by systematic uncertainties (see Bono 2003 for a review).

The three main, most popular calibration methods (see, for reviews, Cacciari and Clementini 2003; Sandage and Tamman 2006; Tammann *et al.* 2008) include the use of statistical parallaxes (see Section 2.1.3), the Baade–Wesselink moving-atmosphere method (see Section 3.5.1) and main-sequence fitting (see Section 3.2.1). Sandage and Tammann (2006) review the uncertainties in the slope of the M_V–[Fe/H] relationship and conclude that the most secure determination of the relationship at the time of their review is by Clementini *et al.* (2003) (see Figure 3.17),

$$M_V = (0.840 \pm 0.017) + (0.214 \pm 0.047)[\mathrm{Fe/H}], \qquad (3.34)$$

using an LMC distance modulus of $(m - M)_0 = 18.54$ mag. The latter was derived without the need to use long-period Cepheids for which the PL relation may exhibit a different

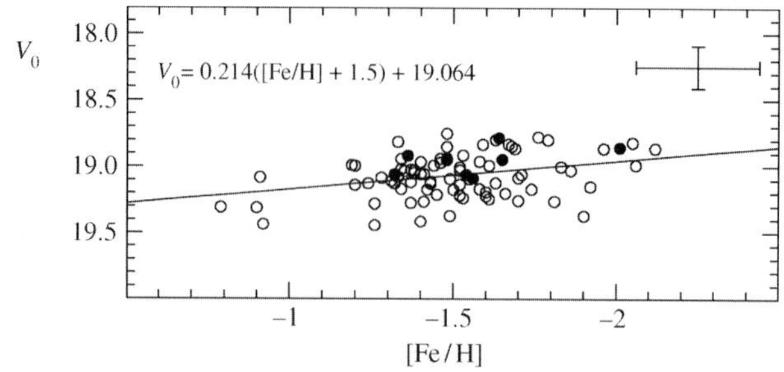

***Figure 3.17** Luminosity–metallicity relation defined by LMC RR Lyrae stars (Clementini et al. 2003). Filled circles mark double-mode pulsators.* (Reprinted from G. Clementini *et al.*, *Astronomical Journal*, **125**, Distance to the Large Magellanic Cloud: the RR Lyrae stars, p. 1309–1329, Copyright 2003, with permission of the AAS and G. Clementini.)

(shallower) slope. Differently phrased, it appears – based on theoretical HB models as a function of evolutionary stage, the pulsation equation (using observed input parameters as a function of metallicity) and semi-empirical observational data – that the slope of the relationship, b, is a function of [Fe/H] and α enhancement: b apparently increases for higher metallicities (cf. Sandage and Tammann 2006 and references therein; particularly Vandenberg *et al.* 2000; Cacciari and Clementini 2003). This trend is particularly noticeable at optical wavelengths and almost disappears in the near-IR K band.

It has, therefore, become customary to represent the M_V–[Fe/H] relation by a parabolic equation, e.g.

$$M_V(\mathrm{RR}) = 1.179 + 0.548([\mathrm{Fe/H}]) + 0.108([\mathrm{Fe/H}])^2 \quad \text{(Catelan et al. 2004)}, \tag{3.35}$$

$$M_V(\mathrm{RR}) = 1.109 + 0.600([\mathrm{Fe/H}]) + 0.140([\mathrm{Fe/H}])^2 \quad \text{(Sandage 2006)}, \tag{3.36}$$

$$M_V(\mathrm{RR}) = 1.576 + 1.068([\mathrm{Fe/H}]) + 0.242([\mathrm{Fe/H}])^2 \quad \text{(Sandage and Tammann 2006)}. \tag{3.37}$$

All of these calibrations are ~0.15 mag brighter than expected for the distances based on statistical parallax measurements, implying lingering systematic effects in either of the methods.

From a physical point of view, the uncertainties in the M_V–[Fe/H] relation (see Bono 2003 for a review) and their dependence on metallicity have part of their origin in the main underlying assumption that RR Lyrae stars are located on the ZAHB. However, field and GC RR Lyrae show a luminosity spread, with the intrinsic width of the ZAHB becoming greater towards more metal-rich Galactic GCs (e.g. Carney *et al.* 1992; Sandage 1993; Bono *et al.* 1995; Cassisi and Salaris 1997), which can largely be traced back to a change in bolometric correction with metallicity and temperature: RR Lyrae stars of the same mass and luminosity but different temperature (pulsation periods) exhibit a spread of ~0.1 V mag (greater for longer wavelengths, implying a broader luminosity range and, hence, a better defined PL relation for wavelengths longward of the I passband;[3] see also Longmore *et al.* 1986, 1990

[3] Because of the effects of these bolometric corrections, in the V band and at shorter wavelengths, the dominant relation is one between periods and metallicities.

for early empirical evidence; Bono 2003). In turn, this implies that the luminosities of RR Lyrae samples drawn from different environments may be affected by different evolutionary histories, particularly at bluer wavelengths (e.g. Demarque *et al.* 2000; Catelan *et al.* 2004), as well as by HB morphology (see below; Caputo *et al.* 1998).

Additional uncertainties are introduced by the intrinsic difference in luminosity between fundamental-mode RR Lyrae stars (RRab, exhibiting high-amplitude nonsinusoidal pulsations) and their nonpulsating counterparts. The discrepancy is greatest ($\approx$0.1 mag) close to the blue edge of the instability strip in the V band, but it disappears in K. Bono (2003) and Catelan *et al.* (2004) provide workable PL–metallicity (Z) relations in the K band (see also Sollima *et al.* 2006):

$$M_K = 0.139 - 2.071(\log P/\text{days} + 0.30) + 0.167 \log Z \quad \text{(Bono 2003)} \tag{3.38}$$

for Galactic GC metallicities $0.0001 \leq Z \leq 0.02$ (solar metallicity), and

$$M_K = -0.597 - 2.353 \log P/\text{days} + 0.175 \log Z. \tag{3.39}$$

3.5.6 Dwarf and Anomalous Cepheids

Anomalous Cepheids occupy parts of parameter space also covered by other types of variables, but they are not easily explained on the basis of differences in stellar evolution. Their periods, 0.8–2 days, overlap with those of RR Lyrae stars, yet the shapes of their light curves are distinctly different from the BL Her or W Virginis-type stars (see below) in the same period range in GCs: they are of lower amplitude and more symmetrical than typical RR Lyrae (and W Virginis)-like light curves, although they are brighter than these comparison objects.

These stars were first discovered in the dwarf spheroidal galaxies in the Local Group (e.g. Baade and Swope 1961; see Zinn and Searle 1976 and Nemec *et al.* 1994 for reviews), but are now also known to exist in some metal-poor Galactic GCs, notably in NGC 5466 and possibly in ω Centauri. Bono (2003) provides the most up-to-date review of the possible evolutionary origin of these stars (see also Fiorentino *et al.* 2006); they are usually associated with classical and type II Cepheids. Their location in PL space at brighter magnitudes than the equivalent RR Lyrae pulsators suggests that they are of higher, intermediate mass (of order 1.5–3.5 $M_\odot$) and $\sim$1 Gyr old (e.g. Caputo and Degli'Innocenti 1995; Castellani and Degli'Innocenti 1995; Bono *et al.* 1997). However, the existence of such 'young', high-mass stars is deemed impossible in the environments where they are found, given the relevant stellar evolutionary timescales involved. As such, it has been suggested that they may be the products of coalesced or mass-transfer binary systems undergoing pulsations because of their loci in the instability strip (see also the review by Nemec *et al.* 1988). Alternatively, they could have been formed in a more recent star formation event. It is, as yet, unclear why this type of variable star seems ubiquitously present in the dwarf spheroidals accompanying the Milky Way but is virtually absent in Galactic GCs, where one would expect such objects to originate from the so-called 'blue straggler' stars. This distinction may, in fact, be of use in further investigations of the origin of the Milky Way's halo (e.g. Wallerstein 2002).

Pulsating variables in the lower part of the instability strip (i.e. on or close to the turn-off of the main sequence; cf. McNamara and Powell 1990; see also Rucinski 2004 for a review of **W Ursae Majoris – W UMa, also known as EW-type – contact binary systems** as distance indicators) used to be referred to as 'dwarf Cepheids'. They have (ultra)short periods of ~80 min to a few hours and small amplitudes (<0.3 mag in the V band). They are now more commonly known by their prototypes, δ Scuti and **AI Velorum**. These stars tend to display simultaneous radial and nonradial pulsations.

Since their discovery in GCs (see Nemec and Mateo 1990a,b and McNamara 1997 for a recent calibration), the δ Scuti and RR Lyrae distance scales could be matched successfully. Therefore, their use for distance determinations is as link between the near-main-sequence variable templates – for which accurate parallax measurements are available – and the brighter RR Lyrae class. Although most δ Scuti stars have high metallicities and move slowly, an important subclass of high-velocity, metal-poor objects is commonly known by its prototypes, SX Phoenicis (SX Phe), **CY Aquarius** and **DY Pegasus**.

McNamara (1997) used his observations of SX Phe stars with *Hipparcos* parallaxes for the current best calibration (but see McNamara (2000) and Sandage and Tamman (2006) for caveats related to Lutz–Kelker-type biases in the trigonometric parallaxes; cf. Section 6.1.2),

$$M_V(\text{SX Phe}) = (-3.725 \pm 0.083)\log(P/\text{days}) - (1.933 \pm 0.087), \tag{3.40}$$

leading, in turn, to a statistical calibration for evolved HB (RR Lyrae) stars in Galactic GCs:

$$M_V(\text{evolved HB}) = (0.29 \pm 0.13)[\text{Fe/H}] + (0.90 \pm 0.06). \tag{3.41}$$

3.6 Novae as Standard Candles

In a number of pioneering studies, Zwicky (1936) and McLaughlin (1939) established that the absolute magnitudes of Galactic novae at maximum brightness correlate with the rate of decline of their light curves, v_d, as (Schmidt 1957)

$$M_\text{max} = a + b\log(t_2), \tag{3.42}$$

where a and b are constants and the decay time t_2 is the period required for the nova's luminosity to decrease by 2 mag below maximum, which is normally used to define the speed class of classical novae. An alternative expression involving t_3, corresponding to a 3 mag drop in brightness, is also in common use, but it is less practically useful for extragalactic novae (cf. Della Valle and Livio 1995; but see Della Valle and Gilmozzi 2002). The rate of decline then follows from $v_\text{d} = 2/t_2$. If only t_3 is known, t_2 follows from $t_2 \simeq (t_3/2.75)^{1.14}$ (Warner 1995).

These observations could, therefore, potentially provide us with accurate distance estimates, if at least the correlation can be calibrated sufficiently robustly *and* the intrinsic variation in M_max were minimal for a given filter. One of the main advantages of using classical novae as standard candles is their intrinsic brightness (they are brighter than Cepheid variables) and relatively frequent occurrence, at least in the nearby extragalactic settings where accurate observations are feasible. Della Valle and Livio (1995) derived a robust calibration of the M_max–v_d relationship (see Figure 3.18), based on the myriad methods used prior to their work, which often relied on completely different zero points and merits. On the basis of a statistically significant sample of novae in M31 and the LMC, they derived

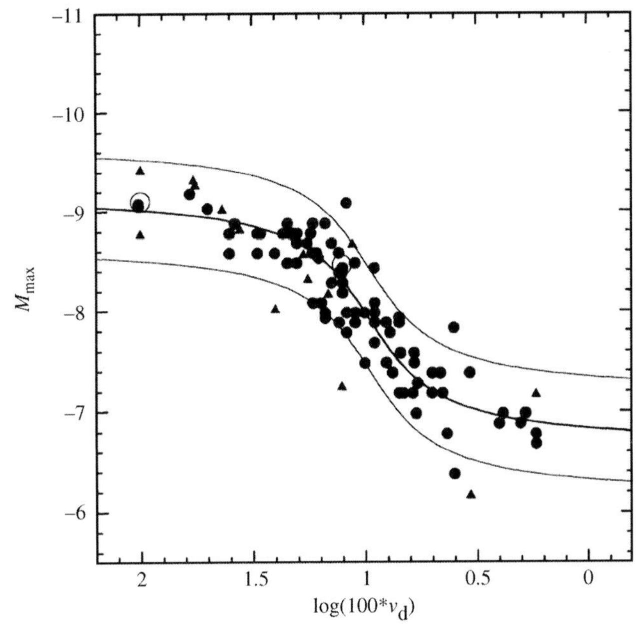

Figure 3.18 *Maximum magnitude versus rate of decline for novae in M31 (solid circles) and the LMC (triangles) (Della Valle and Livio 1995). Open circles: Nova Cygni 1992 and Nova Herculis 1991. The solid curves represent the best fit ±3σ to Equation (3.43).* (Reprinted from M. Della Valle and M. Livio, *Astrophysical Journal*, **452**, The calibration of novae as distance indicators, p. 704–709, Copyright 1995, with permission of the AAS and M. Della Valle.)

an improved calibration relationship,

$$M_{V,\max} = -7.92 - 0.81 \arctan\left(\frac{1.32 - \log(t_2)}{0.23}[\text{rad}]\right), \tag{3.43}$$

for $t_2 \leq 50$ days. For longer decay times, Cohen's (1988) relation applies:

$$M_{V,\max} = 2.41 \log(t_2) - 10.70. \tag{3.44}$$

Adopting Equation (3.43) and applying it to Nova Cygni 1992 (V1974 Cyg) and Nova Herculis 1991 – including careful consideration of the effects of foreground extinction (see also Shafter 1997 for a discussion) – Della Valle and Livio (1995) obtained distances which are consistent with independently determined distances to these objects. They also showed that this relationship yields reliable distances to Virgo cluster galaxies, and as such it may be used as a robust distance indicator.

In the context of the present discussion, it is perhaps more interesting to explore *why* the $M_{\max}$–v_d works reasonably well, from a physical point of view. Novae are a type of cataclysmic variables (see, for a review, Warner 1995), i.e. close binary systems consisting of a low-mass dwarf star that transfers mass to a more massive CO white dwarf companion star. (The physics driving outbursts in more massive oxygen neon magnesium white dwarfs is expected to be similar; Prialnik and Kovetz 1995.) The build-up of hydrogen-rich matter onto the white dwarf's surface may eventually lead to a thermonuclear runaway process and, hence, an eruption in which the accreted material is expelled. This scenario is theoretically consistent with observations of all types of CO white dwarfs evolving into classical novae (Prialnik and Kovetz 1995).

The main physical parameter determining the maximum luminosity of a nova in outburst as well as its rate of decline is the mass of the white dwarf (Livio 1992; Prialnik and Kovetz 1995). Theoretical considerations (e.g. Livio 1992; Prialnik and Kovetz 1995) imply that the strength of the nova outburst also depends on the accretion rate, the white dwarf's temperature, the magnetic field and the chemical composition of the ejected material, among others (Livio 1992):

$$L_{\max}(\mathrm{L}_\odot) = \beta(M_{\mathrm{WD}}, L_{\mathrm{WD}}, B, \dot{M}, X_i) L_{\mathrm{Edd}}(\mathrm{L}_\odot), \tag{3.45}$$

where M_{WD} and L_{WD} are the mass and luminosity of the white dwarf, B is the surface magnetic field strength, $\dot{M}$ is the mass accretion rate and X_i is the chemical composition of the white dwarf's accreted surface layer. L_{Edd} is the white dwarf's Eddington luminosity, i.e. the luminosity required to balance the inward gravitational force and the outward continuum radiation force, assuming hydrostatic equilibrium and spherical symmetry. However (Livio 1992),

$$\beta(M_{\mathrm{WD}}, L_{\mathrm{WD}}, B, \dot{M}, X_i) \approx \beta(M_{\mathrm{WD}}) \simeq (M_{\mathrm{WD}}/0.6\,\mathrm{M}_\odot)^3. \tag{3.46}$$

Nevertheless, the dependence of the white dwarf's maximum luminosity in outburst on such a large number of parameters leads to a relatively large intrinsic scatter of $\sim$1–1.5 mag in the $M_{\max}$–v_{d} relationship. In addition, white dwarf masses in nova systems are often poorly determined, while a significant amount of additional scatter is caused by uncertainties in foreground extinction corrections (e.g. Shafter 1997).

Prialnik and Kovetz (1995) concede that the theoretical state of the art is on shaky ground in the high-luminosity regime (well above the Eddington limit) of nova outbursts and their characteristics (see also Shara 1994 for concerns regarding the empirical $M_{\max}$–v_{d} relationship for the brightest novae).

Della Valle and Livio (1995) point out that, for slow to moderately fast decline rates, the nova's constant bolometric luminosity in the light curve's 'plateau' phase,

$$L(\mathrm{L}_\odot) \simeq 4.6 \times 10^4 (M_{\mathrm{WD}}[\mathrm{M}_\odot] - 0.26) \tag{3.47}$$

(Iben and Tutukov 1989), can also be used as a standard candle, with a scatter of order 1 mag (Livio 1992).

3.7 Geometric Methods

A number of independent *geometric* methods of distance determination allow much needed cross-checks of stellar-population-based distance estimates out to the nearest galaxies, to place the extragalactic distance ladder on a firmer footing. These include, in particular, PN expansion parallaxes, SN light echoes, eclipsing binaries and water maser orbits. We discuss these methods, their merits and uncertainties, in turn.

3.7.1 Planetary Nebula Expansion Parallaxes

Accurate distance estimates to PNe, while abundant in the Milky Way, have proven notoriously difficult to obtain beyond the distance range where ground- and space-based trigonometric parallaxes can currently be used successfully, i.e. to better than 20% accuracy (see e.g. Benedict *et al.* 2003; Hajian 2006; Harris *et al.* 2007). Indirect distance indicators to PNe include spectroscopic parallaxes of the central stars or close companions (see Section 2.2), cluster membership, reddening, angular expansion, and the so-called **Shklovsky (1956) method**. The latter is one of the oldest and simplest statistical distance indicators for use with PNe. It requires knowledge of a PN's absolute Hβ flux and angular size, and is based on the assumptions that every PN has the same ionized mass, is optically thin and can be characterized by a constant electron density and filling factor (cf. Pottasch and Zijlstra 1992). For optically thick PNe, Daub (1982) – with improvements by Cahn *et al.* (1992) – assumed that the constant ionized mass approach of Shklovsky (1956) is still valid, but only for PNe that are optically thin to the Lyman continuum radiation from their central stars (density bounded). For recent reviews of statistical and individual distance estimates to PNe, see Stanghellini *et al.* (2008) and Phillips (2002, 2006); see also Guzmán *et al.* (2009). The uncertainties associated with each of these methods are significant and the results are, at times, mutually inconsistent (see e.g. Pottasch and Acker 1998; Terzian and Teymourian 2005).

Instead, a potentially more accurate and promising method of distance determination to individual PNe is the expansion parallax approach (e.g. Masson 1986, 1989a,b; Gómez *et al.* 1993; Hajian *et al.* 1993, 1995; Hajian and Terzian 1996; Kawamura and Masson 1996; Christianto and Seaquist 1998; Palen *et al.* 2002; Guzmán *et al.* 2006, 2009; Zijlstra *et al.* 2008). This technique measures the angular expansion of a given PN on the plane of the sky, $\dot{\theta}$, and compares the resulting angular velocity to the radial expansion along the line of sight, $v_{\rm exp}$ (see Hajian 2006 for a detailed overview of the method). The angular expansion can be measured using multi-epoch radio interferometry or optical *HST* observations (cf. Hajian 2006; O'Dell *et al.* 2009). The expansion parallax distance is then (Guzmán *et al.* 2006)

$$d = 211 \left(\frac{v_{\rm exp}}{\rm km\ s^{-1}}\right) \left(\frac{\dot{\theta}}{\rm mas\ yr^{-1}}\right)^{-1} \ ({\rm pc}). \tag{3.48}$$

An example for the PN Hen 2–104 in the southern Crab region is given in Figure 3.19 (Santander-García *et al.* 2008).

The systematic uncertainties dominating the resulting distance estimates include the choice of spectroscopic velocity (different ions tend to give different velocities, thus requiring detailed photo-ionization or hydrodynamic modelling to decide on the most suitable tracer; cf. Gesicki *et al.* 1998) and the assumption of linear, spherical expansion (see Mellema 2004; Schönberner *et al.* 2005; Hajian 2006; Zijlstra *et al.* 2008). While corrections are generally applied to take these uncertainties and their effects into account, Mellema (2004) points out that the expansion rate across the sky (which is a pattern velocity) is not necessarily the same as the line-of-sight velocity (which is a material or bulk velocity) in the presence of discontinuities such as shocks and ionization fronts (see also Hajian 2006). The pattern velocity is always higher by 30 $\pm$ 20% than the bulk velocity.

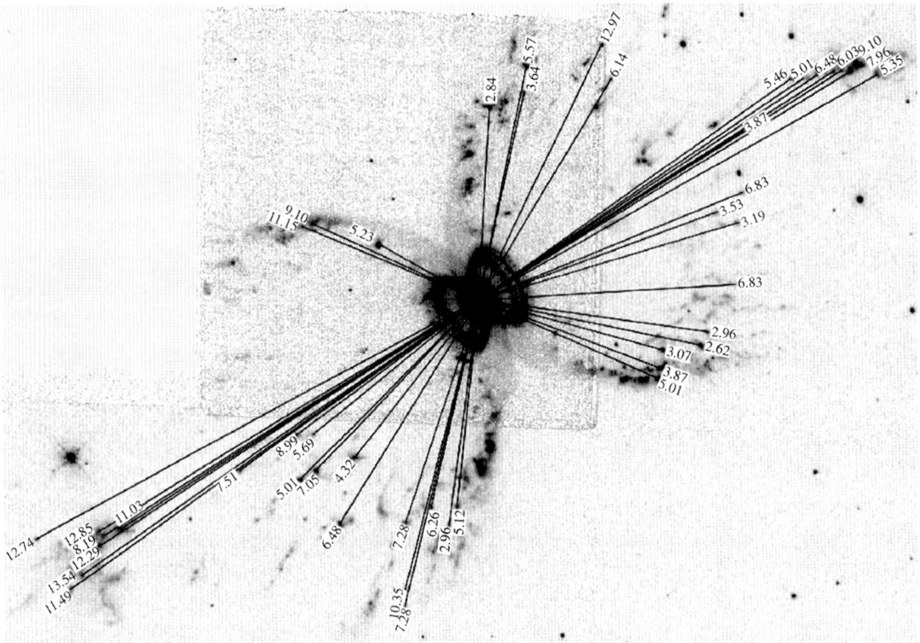

Figure 3.19 *HST [N II] image, along with the radial expansion (in mas yr^{-1}) of the knots of the outer lobes and jets of the PN Henize 2–104, resulting in $d_{\text{outer lobes}} = 3.1 \pm 1.0$ kpc and $d_{\text{jets}} = 3.4 \pm 1.0$ kpc, or $d_{\text{Hen 2-104}} = 3.3 \pm 0.9$ kpc. (Santander-García et al. 2008)* (Reprinted from M. Santander-García *et al.*, *Astronomy and Astrophysics*, **485**, A close-up look at the Southern Crab, p. 117–126, Copyright 2008, with permission of ESO.)

3.7.2 Supernova Light Echoes

When a SN explodes and generates a flash of light, in the generalized, spherically symmetric case, the associated photons propagate in all directions. However, interstellar space is not empty. It contains large numbers of dust clouds that scatter the incident photons out of their direction of propagation. Shklovsky (1964) and van den Bergh (1965) first suggested that, for an observer on Earth, this scattering process will be evident by the faint local brightening of dust clouds near the SN event. The very low surface brightnesses of these 'light echoes' make detection difficult, however. Nevertheless, they provide a rare, direct method for distance determination out to galaxies in and beyond the Local Group. Light echoes have been seen associated with e.g. Cassiopeia A (Wheeler *et al.* 2008), the LMC objects SN 1987A (e.g. Panagia *et al.* 1991; see also Gould and Uza 1998) and SNR 0509–675 (Rest *et al.* 2008; see also Rest *et al.* 2005 for an overview), SN 1993J in M81 (Sugerman and Crotts 2002; Bartel *et al.* 2007), as well as SN 1572 (Tycho Brahe's SN; e.g. Krause *et al.* 2008), SN 1991T in NGC 4527 (Sparks *et al.* 1999) and SN 2003gd in M74 (Sugerman 2005). They have also been associated with outbursts of classical novae and other transient sources in the sky, such as the variable star V838 Monocerotis which underwent a major outburst in 2002 (see Figure 3.20; Henden *et al.* 2002; Bond *et al.* 2003; Crause *et al.* 2005; Sparks *et al.* 2008; and references therein; see, for an X-ray-based approach, Hu *et al.* 2003,

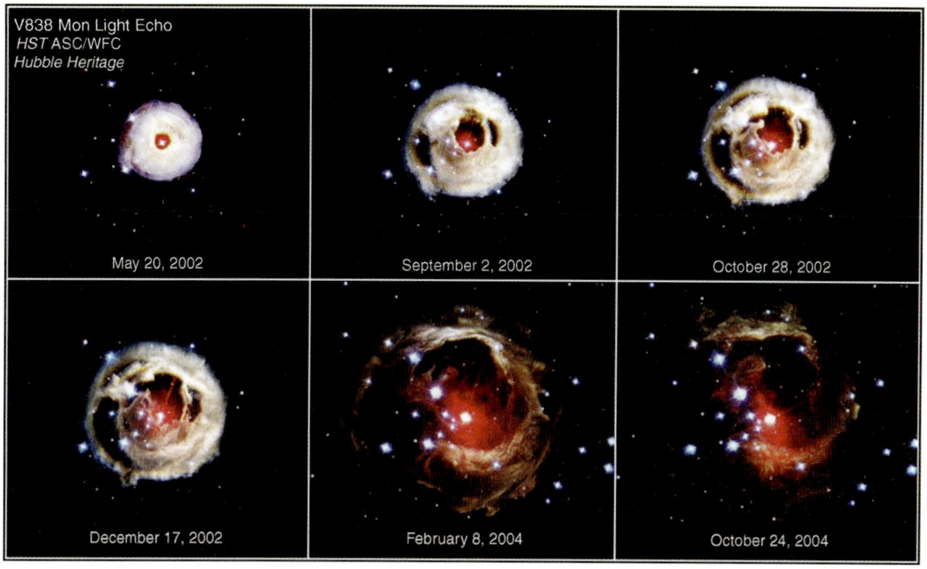

Figure 3.20 *Time sequence of HST images of the light echo around V838 Monocerotis (V838 Mon), taken between May 2002 and October 2004. All six pictures were taken with the Advanced Camera for Surveys using filters sensitive to blue, visible and near-IR wavelengths. The apparent expansion of the light echo, as light from the early 2002 outburst of V838 Mon propagates outwards into the surrounding dust, is clearly shown.* (Courtesy of NASA/ESA, and The Hubble Heritage Team: Space Telescope Science Institute/Association of Universities for Research in Astronomy, Inc.)

based on pioneering work by Trümper and Schönfelder 1973; Thompson *et al.* 2006) as well as the Galactic long-period Cepheid RS Puppis (Kervella *et al.* 2008; but see Bond and Sparks 2009) and type I X-ray bursts[4] (e.g. Thompson *et al.* 2006), such as Cygnus X-3 (Predehl *et al.* 2000; see also Xiang *et al.* 2007 for application to a low-mass X-ray binary system). However, Galactic novae are extremely rare (the most recent was GK Persei 1901: Kapteyn 1902; Perrine 1902; Ritchey 1902), as are light echoes associated with other transient objects. We will therefore focus on SNe in the remainder of this section.

Because of the finite speed of light, light echoes will become evident some time after the event from which they originate. We can, therefore, use the maximum observed angular size of the light echo, combined with the speed of light and accurate measurements of the time delay, to independently obtain a reliable distance to the SN. To obtain the distance to its host galaxy's centre of mass, we can use relative distance measurements connecting the position of the SN with the galaxy's centre.

Let us now consider the geometry of the material involved in some more detail (see Figure 3.21). Given that we generally expect SNe to shed their outer shells isotropically, any large-scale deviation from circular symmetry of the resulting light echo (i.e. a flattening of the ring of scattered light) is most likely caused by inclination effects. If the assumption of spherical

[4] Type I X-ray bursts are characterized by a sharp rise followed by a slow and gradual decline of the luminosity profile.

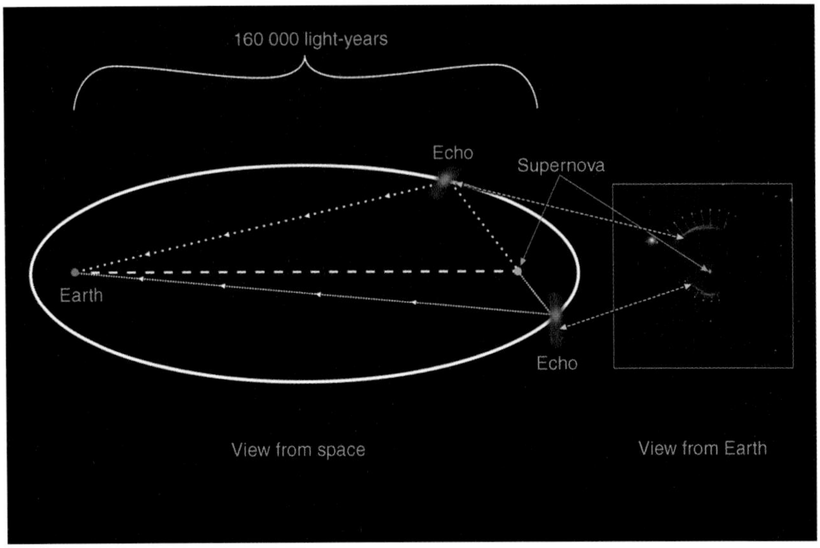

Figure 3.21 *Schematic of the geometry of the light path that creates a SN light echo. An echo occurs when the Earth is at one focus of an imaginary ellipse and the SN remnant is at the other, with dust clouds that happen to be located at the surface of the resulting ellipse. When the light from the SN reaches these dust clouds, it is reflected towards an observer on Earth. The reflection appears as an arc; this arc would be a full, circular slice through the ellipse as viewed from Earth if dust were equally distributed around the full volume of space that the SN light is travelling through.* (Courtesy of P. Marenfeld and the National Optical Astronomy Observatory/Association of Universities for Research in Astronomy/National Science Foundation.)

symmetry holds and the ring is coplanar with the SN, one can obtain the distance as

$$d_{\rm SN} = \frac{c(t_+ + t_-)}{\theta_+}, \tag{3.49}$$

where t_- and t_+ are the delay times of light arriving from the extremities of the ring's minor and major axes, with apparent major- and minor-axis angular diameters θ_+ and θ_-.

However, in practice, one first estimates the ring's inclination, and the associated uncertainties, to derive the optimum error budget. One can derive its inclination on the basis of either the axis ratio or the ratio of time delays, $\eta_t \equiv t_-/t_+$, respectively,

$$i_\theta = \cos^{-1}\frac{\theta_-}{\theta_+} \tag{3.50}$$

and

$$i_t = \frac{\pi}{2} - 2\tan^{-1}\eta_t^{1/2}. \tag{3.51}$$

Although the overall geometry outlined here seems straightforward, reality is usually not so simple. Particularly if the light echoes are resolved, one needs to take a number of additional considerations into account. First, our assumption of circular geometry is most likely an oversimplification. For instance, in the best-studied case of SN 1987A in the LMC, both determinations of t_+ are internally consistent, but the t_- measurements are different by 4σ (Gould 1995; Gould and Uza 1998). Specifically, Gould and Uza (1998) determined $t_+ = 380.7 \pm 6.3$ and 377.8 ± 8.6 days on the basis of the ultraviolet N III] and N IV] emission-line observations of Panagia *et al.* (1991), respectively. The equivalent t_- measurements yielded 87.8 ± 2.7 and 65.6 ± 5.6 days, respectively. Gould (1995) showed that this discrepancy most likely originated from our assumption of circular symmetry, significantly affecting the determination of t_- but not of t_+.

Second, we need to consider whether the emission used to measure delay times at a variety of wavelengths actually originates from the same region(s) in the ring. For instance, for SN 1987A, Plait *et al.* (1995) measured a ring size on the basis of optical [O III] emission lines, while Panagia *et al.*'s (1991) delay times were based on ultraviolet lines. It has been suggested that the ultraviolet lines originated from the ring's inner edge, while the optical lines came from the main body. Gould and Uza (1998) argue that, in this case and using the proper geometry including a finite ring thickness, the ultraviolet light curve could result in an underestimate of the light travel time across the optical ring diameter of up to 7% and, thus, a similar underestimate of the distance. However, they note that this scenario is unlikely given the very similar ionization potentials of O III and N III/N IV, as well as their likely spatial distributions.

It seems, therefore, that the only remaining uncertainty affecting ring size and, hence, distance determinations is the underlying assumption of spherical symmetry. Gould (1994) attempted to include the effects of intrinsically elliptical light echoes. He showed that if the ring is elliptical but $i_t = i_\theta$ within the statistical uncertainties, the resulting distance is an overestimate by a factor of $(1 + 0.4e^4)$, where e is the intrinsic ellipticity. On the basis of independently determined axis ratios assuming an elliptical light echo, Gould and Uza (1998) find for SN 1987A

$$\frac{b}{a} = 0.95 \pm 0.02. \tag{3.52}$$

Sparks (1994, 1996) showed that using polarimetry could circumvent the problem of the spatial origin of a light echo (see also Bond and Sparks 2009). The distribution of linear polarization should depend only on the projected radius from the star and the time since the outburst event. He showed that this applies to light echoes because the degree of linear polarization depends predominantly on the scattering angle and not on the circumstellar density (cf. Bond and Sparks 2009).

Finally, in addition to uncertainties caused by our assumptions on the ring geometry, we also need to consider potential errors caused by a misinterpretation of the underlying physics. Most importantly, we have so far assumed that the fluorescent, scattered emission from the gas commences as soon as the SN's energetic photons hit the gas cloud. It is possible, however, that there is a slight delay in the onset of fluorescent emission as the gas first recombines from highly ionized states, for instance. Neglecting this step will lead to ring size and, hence, distance overestimates. Gould and Uza (1998) consider this carefully for SN 1987A and point out that the response time in the light curves of Sonneborn *et al.* (1997) is clearly a function of ionization state, with the lower states responding later. One should also be careful and consider whether **optical depth** effects may mask some emission

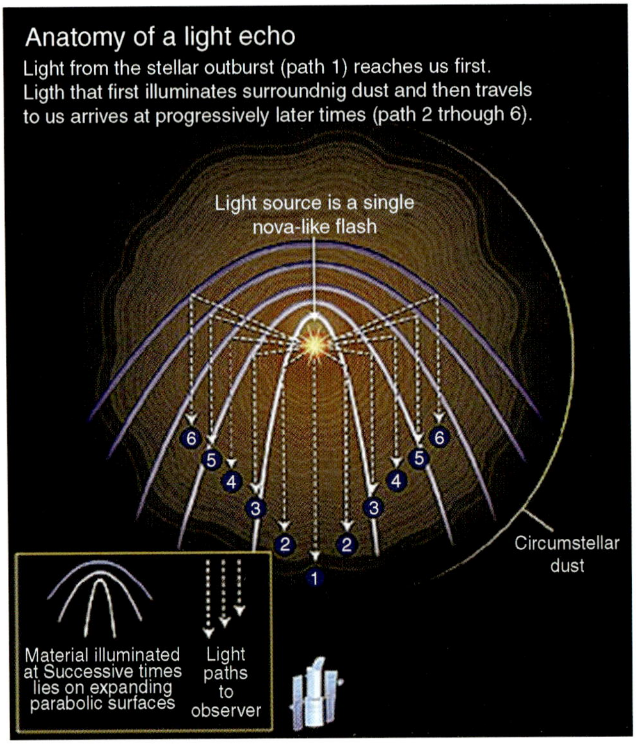

Figure 3.22 *As a SN flash expands at the speed of light, it scatters off and illuminates different regions of a nearby dust cloud, creating delayed echoes across the cloud.* (Courtesy of NASA/ESA and A. Feild/Space Telescope Science Institute.)

and, hence, result in an apparent offset in time delay caused by the emission originating from physically distinct parts of the ring.

Detailed observations of the time variability of nearby SN light echoes seem to suggest that the expanding photon shell travels at superluminal speeds (Bond *et al.* 2003; Krause *et al.* 2008; see Pastorello and Patat 2008 for a review). This effect is caused by simple geometry as the photons propagate to different sections of the circum-SN gas cloud, as shown in Figure 3.22.

3.7.3 Eclipsing Binary Stars

Eclipsing, double-lined detached binary systems offer the potential to derive independent geometric distance estimates, both within the Milky Way (with Algol being the prototypical example) and to galaxies as distant as M81 ($d \sim 3.63 \pm 0.34$ Mpc; Freedman *et al.* 1994). On the basis of photometric and spectroscopic time-sequence monitoring, they provide the most accurate, distance-independent information about stellar masses, radii, luminosities and effective temperatures (cf. Andersen *et al.* 1991; Thompson *et al.* 2010; and references therein). In these systems, the orbital plane of both components is approximately aligned with the observer's line of sight. Therefore, eclipses caused by the stellar orbital motions result in periodic variations in the systems' combined intensity.

The shape of the alternate eclipses – i.e. caused by the primary fully or partially obscuring the secondary component, and vice versa – provides information about the fractional radii of the two stars, $r_i = R_i/a$ $(i = 1, 2)$, where a is the orbital size. The best results are obtained for detached binaries (for small r_i), because this geometry allows the simple assumption of spherical stars, without complications caused by mass transfer, gas streams and/or accretion discs.

We can determine the fractions of the total light attributed to each component, $L_i/L_{\rm tot}$ $(i = 1, 2)$, from light curve analysis. If a third, unresolved companion is present, $L_3/L_{\rm tot} \leq L_2/L_{\rm tot} \leq L_1/L_{\rm tot} > 0$, high-accuracy photometric light curves can also reveal its presence (e.g. Goecking *et al.* 1994; Gatewood *et al.* 1995). In addition, light curve analysis enables derivation of the orbital period, inclination and eccentricity.

On the basis of the radial velocity curves, which provide information on the velocity amplitudes of both binary components, the orbital period and inclination, the size of the orbit and the linear stellar radii can be calculated, in essence using Kepler's third law. The uncertainties involved in all of these geometric estimates combined can be reduced to below the 1% level (e.g. Andersen *et al.* 1991; Paczyński 1997). The only assumptions requiring an understanding of the underlying physics are related to the subsequent derivation of the intrinsic fluxes or surface brightnesses of the two components, F_i $(i = 1, 2)$, using observed photometric or spectroscopic inputs or full spectral energy distributions, or any other suitable approach (e.g. Paczyński 1997; Ribas *et al.* 2002, 2005; Wyithe and Wilson 2002; Wilson 2007; but see Torres *et al.* 2010 for caution regarding uncertainties in the effective-temperature scale). The absolute luminosity of each component in the filter of interest is (Paczyński 1997)

$$L_i = 4\pi R_i^2 F_i. \tag{3.53}$$

The distance to either component can now be derived as a simple distance modulus,

$$d_i = \left(\frac{L_i}{4\pi F_{i,\rm obs}}\right)^{1/2} = \left(\frac{F_i}{F_{i,\rm obs}}\right)^{1/2} R_i, \tag{3.54}$$

where $F_{i,\rm obs}$ is the observed flux for star i. For a more detailed basic description of the method, see Paczyński (1997), Guinan *et al.* (1998) or Wilson (2007, 2008).

Although the basic method has been known and used for well over a century (see, for an extensive historical overview, Kruszewski and Semeniuk 1999), with the recent advent of 8 m-class telescopes, the technique can now be used to estimate distances to and fundamental properties of extragalactic eclipsing binaries, such as in the Magellanic Clouds (e.g. Ostrov *et al.* 2000, 2001; Wyithe and Wilson 2002; Fitzpatrick *et al.* 2003; Harries *et al.* 2003; North *et al.* 2010; and references therein), M31 (e.g. Section 1.3; Ribas and Jordi 2003; Ribas *et al.* 2004; Todd *et al.* 2005; Vilardell *et al.* 2006), M33 (Bonanos *et al.* 2006) or even the M81 group (e.g. Prieto *et al.* 2008), with an accuracy of better than 5% (Andersen *et al.* 1991; Clausen 2004; Torres *et al.* 2010; see Wilson 2008 and North *et al.* 2010 for detailed accounts of the systematic uncertainties involved). A variant of the technique has even been applied successfully to estimate the distance from the Sun to the Galactic Centre, using stellar motions around the Milky Way's central black hole (see Section 1.1; Eisenhauer *et al.* 2003; Ghez *et al.* 2008).

Modern light curve analysis approaches have become very sophisticated in their treatment of potentially detrimental issues (e.g. Prša and Zwitter 2005; Çakırlı and İbanoğlu 2010; Torres *et al.* 2010), including the effects of slight nonsphericity of the component stars (e.g. Wilson 2008), limb darkening (cf. Van Hamme 1993), proximity effects (tides and reflection; cf. Wilson 2008) and unresolved or marginally resolved double lines, in which case use of a two-dimensional cross-correlation technique is preferred. Foreground reddening still needs to be dealt with separately and independently. In fact, these techniques are now so reliable that Wyithe and Wilson (2002; see also Wilson 2004, 2008) confidently propose that semi-detached and overcontact binaries are more useful as direct distance indicators, because for these systems we can use information not available for detached objects. They point out as potential advantages that (i) light curve solutions can be strengthened by taking lobe filling into account, (ii) only single-lined spectra may be needed for radial velocities because the component mass ratio can be determined from photometry for complete eclipses, (iii) nearly all semi-detached binaries have circular orbits, which is not true for detached binaries, (iv) aliasing, here defined as the difficulty in distinguishing solutions with interchanged radii, is much less severe and (v) the condition of complete eclipse, which removes a near degeneracy between inclination and the ratio of the radii, is identified with improved reliability.

New instruments and ever larger telescopes open up exciting prospects in this field. The long baselines offered by optical interferometry, e.g. with the *Very Large Telescope Interferometer*, allow direct measurement of the orbital separation and fundamental stellar parameters, in particular the stellar effective temperatures, of eclipsing binary systems, thus further reducing any remaining uncertainties (e.g. Shabun *et al.* 2008, 2009). In addition, access to IR light curves (eclipse photometry) can improve temperature determinations and, hence, distance and luminosity estimates (e.g. Thompson *et al.* 2010).

3.7.4 Maser-Based Distance Determinations

With the coming of age of very long baseline interferometry, measuring (Galactic) parallaxes and, hence, Galactic structure (e.g. Genzel *et al.* 1981, 1982; Schneps *et al.* 1981; Reid *et al.* 1988a,b, 2009b; Imai *et al.* 2002; Hachisuka *et al.* 2006; Xu *et al.* 2006; Moellenbrock *et al.* 2009), and extragalactic proper motions of bright, compact radio sources, both on the order of milli-arcseconds, has come within reach. In turn, this enables geometric distance determinations to galaxies in the Local Group (e.g. Brunthaler *et al.* 2005, 2008: M33) and even beyond, up to ~45–50 Mpc (Herrnstein *et al.* 1999 and Argon *et al.* 2007: NGC 4258; Braatz *et al.* 2009a,b, 2010: UGC 3789, NGC 6323; Impellizzeri *et al.* 2010: Mrk 1419).

In a number of pioneering studies, maser (microwave amplification by stimulated emission of radiation: the microwave equivalent of lasers) sources have been used as probes of galactic proper motions. In astrophysics, the most common masing molecule is water (H_2O), but hydroxyl radicals (OH), methanol (CH_3OH), formaldehyde (CH_2O) and silicon monoxide (SiO) have also been observed in this context. Masing in star-forming regions occurs if more electrons occupy higher excited states than lower energy levels, a phenomenon known as *population inversion*. In interstellar space, this gives rise predominantly to a strong but narrow (few km s^{-1}) water line at a frequency of 22.235 GHz. Astrophysical maser sources have high brightness temperatures ($\leq 10^{12}$–10^{15} K) and relatively small sizes ($\leq 10^{10}$ km; cf. Reid 1993; Herrnstein *et al.* 1999).

Combined with *a priori* information on a galaxy's inclination with respect to our line of sight and its rotation curve, based on radial velocity measurements, we can construct an accurate, slightly warped 'tilted-ring' model of the galaxy's dynamical structure, usually assuming circular orbits (e.g. Miyoshi *et al.* 1995; Herrnstein *et al.* 1998, 2005; Brunthaler *et al.* 2005; Argon *et al.* 2007). This allows correlation of the angular proper motion measurements with the rotational velocity information obtained in linear units and, thus, provides an independent distance measurement (see also the review in Olling 2007, particularly as regards systematic uncertainties related to noncircular orbital motions).

Herrnstein *et al.* (1999) were the first to apply this approach to the orbital motions of nuclear water masers in the molecular gas disc surrounding the active galactic nucleus of NGC 4258. They detected two types of masers, a stationary high-velocity population on either side of the nuclear disc in the plane of the sky and a population of systemic masers situated on the near edge of the disc (see also Haschick *et al.* 1994; Greenhill *et al.* 1995b; Nakai *et al.* 1995; Argon *et al.* 2007). The latter move over time with respect to a fixed point in the sky (see also Haschick *et al.* 1994; Watson and Wallin 1994; Greenhill *et al.* 1995a,b; Argon *et al.* 2007; Reid *et al.* 2009a), by $\langle \dot{\theta}_{\rm sky} \rangle = 31.5 \pm 1(1\sigma)\mu{\rm as\ yr}^{-1}$ and $\langle \dot{v}_{\rm LOS} \rangle = 9.3 \pm 0.3(1\sigma)$ km s^{-1} yr^{-1} (Herrnstein *et al.* 1999). Combined with other observables from their maser measurements, these proper motion and acceleration (velocity drift) parameters can be linked to the galaxy's geometric distance, d (in Mpc), as

$$\langle \dot{\theta}_{\rm sky} \rangle = 31.5 \left(\frac{d}{7.2}\right)^{-1} \left(\frac{\Omega_{\rm s}}{282}\right)^{1/3} \left(\frac{M_{7.2}}{3.9}\right)^{1/3} \left(\frac{\sin i_{\rm s}}{\sin 82.3^\circ}\right)^{-1} \left(\frac{\cos \alpha_{\rm s}}{\cos 80^\circ}\right)^{-1} \mu{\rm as\ yr}^{-1} \tag{3.55}$$

and

$$\langle \dot{v}_{\rm LOS} \rangle = 9.2 \left(\frac{d}{7.2}\right)^{-1} \left(\frac{\Omega_{\rm s}}{282}\right)^{4/3} \left(\frac{M_{7.2}}{3.9}\right)^{1/3} \left(\frac{\sin i_{\rm s}}{\sin 82.3^\circ}\right)^{-1} {\rm km\ s}^{-1}\ {\rm yr}^{-1}, \tag{3.56}$$

where $\alpha_{\rm s}$ is the disc position angle (east of north) at the mean systemic radius $\langle r_{\rm s} \rangle$, $M_{7.2} = M/d \sin^2 i_{\rm s} \times 10^7$ M$_\odot$ (derived from the rotation curve and evaluated at $d = 7.2$ Mpc and $i_{\rm s} = 82.3^\circ$) and $\Omega_{\rm s} \equiv (GM_{7.2}/\langle r_{\rm s} \rangle^3)^{1/2}$ km s^{-1} mas^{-1} is the projected disc angular velocity at $\langle r_{\rm s} \rangle$ (from the slope of the position–velocity gradient). The values in the denominators are *a priori* estimates for each of the disc parameters (Herrnstein *et al.* 1999).

The resulting *absolute* distance estimate to NGC 4258 is $d = 7.2 \pm 0.3$ Mpc, where the uncertainty includes the results from both equations and the statistical and systematic uncertainties in the input parameters (dominated by a $\sim$5% uncertainty in $\Omega_{\rm s}$), assuming that the masers orbit the nucleus in a thin (Greenhill *et al.* 1995b; Argon *et al.* 2007), circular Keplerian disc (e.g. Miyoshi *et al.* 1995; Herrnstein *et al.* 2005). Additional systematic uncertainties could result from the current weak constraint on the disc's eccentricity – which may increase the overall uncertainty to $\pm$0.5 Mpc, or $d = 7.2 \pm 0.3\,({\rm random}) \pm 0.4\,({\rm systematic})$ Mpc – and possibly contamination by nonkinematic contributions such as travelling density waves (cf. Herrnstein *et al.* 1999). Humphreys *et al.* (2005) and Argon *et al.* (2007) use a significantly expanded data set to reduce the overall uncertainties by approximately a factor of 2 to $\sim$3%. Brunthaler *et al.* (2005, 2008) similarly obtained a slightly less accurate absolute, geometric distance to the Local Group galaxy M33 of $d = 730 \pm 100$ (statistical) $\pm$ 135 (systematic) kpc. When the Square Kilometre Array becomes operational, many more and fainter masers will be detectable out to distances of $\sim$100 Mpc (Fomalont 2005), while promising a distance accuracy of 1% out to 30 Mpc (Fomalont and Reid 2004).

To date, the measured distance to NGC 4258 has been used to establish the zero point of the Cepheid PL relation (Newman *et al.* 2001; Caputo *et al.* 2002; Macri *et al.* 2006; Saha *et al.* 2006; Argon *et al.* 2007; Bono *et al.* 2008a; Humphreys *et al.* 2008; Mager *et al.* 2008; Bresolin 2011), while large-scale efforts are underway to tie maser-based distances with an accuracy of $\sim$3% to an independent determination of the Hubble constant (Humphreys *et al.* 2008; Braatz *et al.* 2009a,b,c; Reid *et al.* 2009a).

3.8 Pulsars: Distance Measurements Outside the 'Classical' Wavelength Range

When a massive star[5] eventually explodes as a SN, its core is compressed while it collapses into a neutron star. Its greatly reduced stellar radius and, hence, its moment of inertia, combined with conservation of angular momentum, results in a stellar remnant characterized by a very high rotation speed. A beam of radiation is emitted from the poles of the magnetic field – mostly in the form of radio or X-ray emission, although optical and/or γ-ray emission has also been reported – along the spinning neutron star's magnetic axis. The latter is not necessarily aligned with the object's rotation axis, which causes the appearance of pulsed electromagnetic radiation, hence the name 'pulsating star' or **pulsar**. A pulsar's spin period slows down over time as electromagnetic power is emitted and because of relativistic pair plasma flows, until it reaches the so-called 'death line', where the radio-pulsar mechanism turns off.

As bright beacons with extremely regular periods, pulsars represent a class of objects to which distances can be determined using observations and physics outside the 'classical' optical/near-IR regime. Most distance determinations to radio pulsars rely on observations of the delay of the pulse time of arrival between two different frequencies (the '**dispersion measure**', DM) and a sophisticated model of the Galactic free-electron density (e.g. Manchester and Taylor 1981; Cordes *et al.* 1991; Taylor and Cordes 1993; Gómez *et al.* 2001; Cordes and Lazio 2002, 2003; and references therein). When they announced the discovery of the first pulsar, LGM-1 ('Little Green Men'-1, later renamed as PSR 1919+21), Hewish *et al.* (1968) suggested that the frequency-dependent variation of pulse arrival time was caused by dispersive signal propagation through the partially ionized interstellar medium

[5] Only a few secure lower limits to the masses of neutron star progenitors have been obtained, of which four for magnetars (strongly magnetized pulsars with magnetic field strengths of $B \sim 10^{15}$ G): (i) a shell of HI around 1E 1048.1−5937 was interpreted as interstellar medium displaced by the wind of a progenitor with initial mass $M_i = 30 - 40\ M_\odot$ (Gaensler *et al.* 2005); (ii) SGR 1900+14 was suggested to be a member of a <10 Myr-old star cluster (Vrba *et al.* 2000), placing a lower limit on the progenitor mass of $M_i \gtrsim 20\ M_\odot$; (iii) SGR 1806−20 was claimed to be a member of a < 4.5 Myr-old star cluster (although see Cameron *et al.* 2005; McClure-Griffiths and Gaensler 2005), providing a limit of $M_i \gtrsim 50\ M_\odot$ (e.g. Fuchs *et al.* 1999; Figer *et al.* 2005); and (iv) Muno *et al.* (2006) derived a minimum mass for CXO J164710.2−455216 in the 4–5 Myr-old star cluster Westerlund 1 of $M_i > 40\ M_\odot$. In addition, Utrobin (2007) derived a mass of the ejecta of SN 1987A of $18.0 \pm 1.5\ M_\odot$, while Graves *et al.* (2005) obtained an upper limit for the luminosity of any survivor of a possible binary system similar to that of F6-type main-sequence stars. They exclude bright young pulsars such as Kes 75 or the Crab pulsar based on optical and X-ray limits.

(ISM): lower-frequency radio waves travel more slowly through such an ISM than their higher-frequency counterparts because of its dispersive nature. For a uniform ISM, the DM would be inversely proportional to the integrated density of free electrons (n_e) along the line of sight (s),

$$\mathrm{DM} = \int_0^{d_p} n_e(s) \mathrm{d}s, \tag{3.57}$$

where d_p is the pulsar's distance. A radio signal propagates through the ISM with group velocity

$$v = c\left(1 - \frac{\nu_p^2}{\nu^2}\right)^{1/2}, \tag{3.58}$$

where c is the speed of light in a vacuum and ν and ν_p are the signal and plasma frequencies, respectively. The latter is expressed as

$$\nu_p = e\left(\frac{n_e}{\pi m}\right)^{1/2}, \tag{3.59}$$

where e and m refer to electron charge and mass, respectively, in conventional SI units. The time delay (Δt, in seconds) caused by free-electron attenuation and scattering in the ISM can be obtained by integrating $(c - v)/c^2$ along the line of sight to a pulsar, which for $\nu \gg \nu_p$ yields (Taylor and Cordes 1993)

$$\Delta t = \frac{e^2}{2\pi m c \nu^2} \int_0^{d_p} n_e \, \mathrm{d}s = \frac{\mathrm{DM}}{241.0 \nu_{\mathrm{GHz}}^2}. \tag{3.60}$$

Therefore, from observations of the frequency drift rate and a model of the number density of free electrons in the ISM, the distance follows naturally. Because the radio frequencies and time-of-arrival differences as a function of frequency can be measured with high precision, most pulsar DMs are known to better than a few percent. Clearly, the dominant uncertainties associated with this approach reside in the accuracy with which we know the electron density distribution and its 3D variation throughout the Milky Way (see e.g. Taylor and Cordes 1993 and Cordes and Lazio 2002 for historical overviews of efforts to improve our understanding of these).

For a first-order distance estimate, we can use the mean electron density over a large fraction of the Galactic disc, $\langle n_e \rangle \approx 0.03$ cm^{-3} (see Cordes and Lazio 2002 and Figure 3.23), combined with our knowledge of the half-thickness of the layer of electrons (~ 1000 pc), a pulsar's DM and its Galactic latitude, b. For an exponentially declining electron-density distribution in the direction perpendicular to the Galactic plane, with a scale height h_e, the pulsar's distance is given by

$$d = \frac{-h_e}{\sin |b|} \ln\left(1 - \frac{\mathrm{DM} \sin |b|}{h_e \langle n_e \rangle}\right), \tag{3.61}$$

while in the limit DM $\sin |b| \rightarrow 0$, Equation (3.61) reduces to

$$d = \mathrm{DM}/\langle n_e \rangle, \tag{3.62}$$

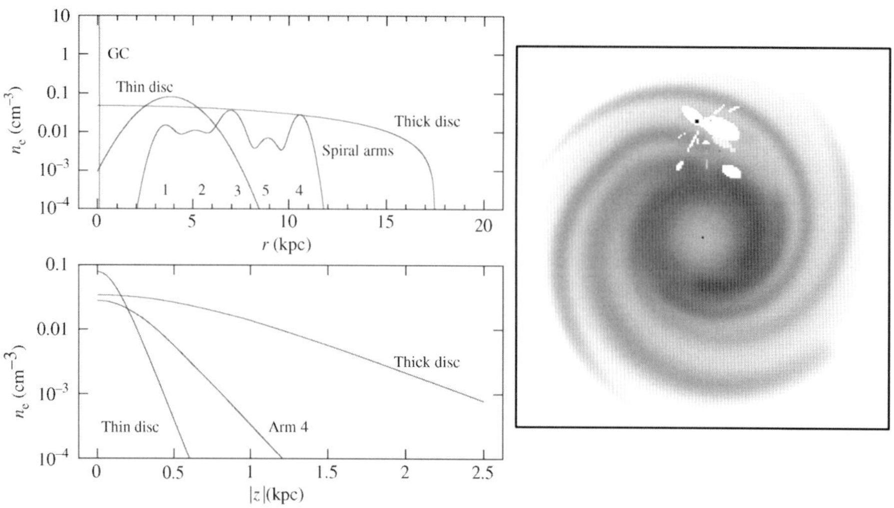

Figure 3.23 *Electron density as a function of (top left) Galactocentric distance along the direction from the Sun to the Galactic Centre for the various large-scale components (the numbers refer to individual spiral arms) and (bottom left) absolute vertical distance from the Galactic plane. The thin-disc profile is for a Galactocentric distance of 3.5 kpc, while the thick-disc component is shown for the solar radius. The spiral arm density is representative of* $(x,y) = (0,10.6)$ *kpc. (right) Top view of the model Galactic electron-density distribution* (30×30 kpc^2)*. The light features near the solar circle represent the local ISM and other 'void' regions; the black dot indicates the position of the Gum Nebula and Vela SN remnant (Cordes and Lazio 2002).* (Reprinted from J. M. Cordes, Unpublished (astro-ph/0207156v3), NE2001. I. A new model for the Galactic distribution of free electrons and its fluctuations, Copyright 2002, with permission of J. M. Cordes.)

which is an adequate approximation for many pulsars. A simple correction to the DM is required if an **HII region** along the line of sight contributes a significant component, where the 'corrected' DM is equivalent to the difference between the measured value and that contributed by the perturbing source along the line of sight.

However, to match direct pulsar observations, combined with independent distance determinations, models of the Galactic free-electron distribution must include more than a simple, one- or two-component (non-)axisymmetric structure consisting of an outer, thick-disc density distribution and an inner, thin-disc feature. The thick-disc component is responsible for the DMs of high-Galactic-latitude GC pulsars (e.g. Reynolds 1989; Taylor and Cordes 1993; Cordes and Lazio 2002, 2003). It has a scale length and height of ~20 and 1 kpc, respectively, and a functional form that truncates at a radius of roughly 20 kpc (cf. Lazio and Cordes 1998a,b). The inner, large-scale Galactic component is best represented by an annular Gaussian ring, similar to the well-known CO molecular ring (Dame *et al.* 1987). It contains the significant contribution from HII regions surrounding hot, massive OB stars, associations and young clusters that are located close (~0.1 kpc) to the Galactic plane.

In addition to these smooth, large-scale components, Ghosh and Rao (1992), Taylor and Cordes (1993) and Gómez *et al.* (2001) already identified the need for spiral arm structure in 3D free-electron models. Cordes and Lazio (2002, 2003) provide its most up-to-date redefinition and refinement, now also including a local (Orion–Cygnus) spiral arm (see also Figure 3.23). Although spiral structure is usually implemented as a smooth, higher-density component, in practice a realistic implementation requires modelling of both regions with higher than ambient density or 'clumps', such as the apparently prominent, nearby Gum Nebula and Vela SN remnant, ionized overdensities in the Cygnus region – including that surrounding the Galactic continuum source Cyg X-3 – as well as weaker enhancements in the spiral arms or specific HII regions that contain molecular masers, and lower-density 'voids' (cf. Cordes and Lazio 2002, 2003).

The Galactic Centre region requires modelling by a separate component, which is predominantly responsible for scattering of Sgr A* and OH maser emission (see Cordes and Lazio 2002), and so does the local, solar neighbourhood ISM. The latter has been studied in great detail using e.g. continuum X-ray observations and interstellar NaI absorption, revealing a structure consisting of (i) a hot, local bubble centred on the Sun, (ii) the North Polar Spur (part of 'Loop I', one of the largest coherent structures on the sky, spanning over 100 degrees, known from nonthermal radio-continuum and X-ray maps and corresponding to a thin bubble of gas that may have originated from a SN explosion), (iii) a local superbubble prominent in the third quadrant – $12^{\rm h} \leq$ Right Ascension (RA) $\leq 18^{\rm h}$ – and (iv) a low-density region in the first quadrant ($0^{\rm h} \leq {\rm RA} \leq 6^{\rm h}$) (cf. Cordes and Lazio 2002, Appendix A, and references therein).

Even the most sophisticated free-electron model cannot include all small-scale perturbing, ionizing features in the Milky Way that may affect the resulting DM measurements. Such features will lead to DM variations (ΔDM) and, hence, distance uncertainties,

$$\Delta d_{\rm p} = \frac{\Delta {\rm DM}}{n_{\rm e}(x, y, z, d_{\rm p})} \tag{3.63}$$

in general or, for objects that are located more than one scale height above the Galactic plane and assuming a plane-parallel model (cf. Cordes and Lazio 2002),

$$\Delta d_{\rm p} = \frac{\Delta {\rm DM}}{n_{\rm e,0}(1 - {\rm DM}/{\rm DM}_{\infty})}, \tag{3.64}$$

where ${\rm DM}_{\infty} = n_{\rm e,0} h_z / \sin |b|$ is the model's maximum DM, for a mid-plane density $n_{\rm e,0}$ and scale height h_z (b: Galactic latitude). Following Prentice and ter Haar (1969) and Bronfman *et al.* (2000), Cordes and Lazio (2002) quote expected ΔDM values ranging from $\sim$3 to 10 and approximately 75 pc cm^{-3} for the ionizing Strömgren spheres around O9/B1 and O5 stars, respectively, to 100 – 200 pc cm^{-3} for OB associations and 300–500 pc cm^{-3} for those very rare pulsars whose lines of sight intersect ultracompact HII regions.

Independent validation of pulsar distances resulting from measurements of their DMs, combined with such ever more sophisticated models of the Galactic free-electron distribution, is crucial. Statistical distance cross-checks can be obtained from 21 cm neutral hydrogen absorption measurements using radio telescopes or from fitting X-ray spectra, combined with a suitable Galactic rotation curve, while individual measurements can be corroborated by association of pulsars with SN remnants (e.g. Hoogerwerf *et al.* 2001; Vlemmings *et al.* 2004) or star clusters (e.g. Fuchs *et al.* 1999; Vrba *et al.* 2000; Figer *et al.* 2005; Muno *et al.* 2006), or through parallax measurements based on **pulse timing** (e.g. van Straten *et al.* 2001; Hotan *et al.* 2006; Verbiest *et al.* 2009) or interferometric techniques (e.g. Chatterjee *et al.* 2001, 2009; Brisken *et al.* 2002); see Lyne and Graham-Smith (2005, their Chapter 4) and the Pulsar Astrometry Project[6] for detailed and up-to-date overviews.

Finally, the pulsar at the centre of the Crab Nebula is often used as a standard candle suitable for normalization purposes in the X- and γ-ray regimes (e.g. Kirsch *et al.* 2005; Jourdain and Roques 2009; Weisskopf *et al.* 2010; Meyer *et al.* 2010), because it is the only hard X-ray source that is both sufficiently bright and steady for this application. For instance, Kirsch *et al.* (2005) showed that its X-ray differential photon spectrum up to energies of 30 keV is well described by a power law of the form

$$\mathrm{d}N/\mathrm{d}E = (9.7 \pm 1.0)E^{-(2.1\pm0.03)} \text{ photons cm}^{-2}\text{ s}^{-1}\text{ keV}^{-1} \tag{3.65}$$

(where N and E represent the number of photons and their energy, respectively), i.e. quantitatively exactly the same as the spectrum published by Toor and Seward (1974) three decades earlier.

However, Wilson-Hodge *et al.* (2010) recently concluded that the Crab pulsar's use as standard candle may be fundamentally unsafe, based on their detailed analysis of X- and γ-ray observations from four independent satellites. They showed that its nebular emission, a combination of synchrotron radiation up to $\sim$100 MeV and a harder, **inverse Compton spectrum** (caused by photons gaining energy through collisions with electrons) up to TeV energies (de Jager *et al.* 1996; Hester 2008), is not constant at a level of $\sim$3.5% yr^{-1}, at least for energies up to 100 keV. This is likely caused by changes in the nebula's shock acceleration or the nebular magnetic field. In addition, the Crab Nebula's integrated flux in the 15–50 keV band has been steadily declining by $\sim$7% since 2008 (becoming larger with increasing energies), while correlated variations on a $\sim$3-year timescale are also reported (cf. Wilson-Hodge *et al.* 2010). As a consequence, use of the Crab Nebula as calibration source at high energies appears compromised at levels better than $\sim$10%.

Bibliography

Adams WS and Joy AH 1927 The relationship of spectral type to period among variable stars. *Proc. Nat'l Acad. Sci. USA* **13**, 391–393.

Alves DR 2000 K-band calibration of the red clump luminosity. *Astrophys. J.* **539**, 732–741.

Alves DR, Rejkuba M, Minniti D and Cook KH 2002 K-band red clump distance to the Large Magellanic Cloud. *Astrophys. J.* **573**, L51–L54.

An D, Terndrup DM, Pinsonneault MH, Paulson DB, Hanson RB and Stauffer JR 2007a The distances to open clusters from main-sequence fitting. III. Improved accuracy with empirically calibrated isochrones. *Astrophys. J.* **655**, 233–260.

An D, Terndrup DM and Pinsonneault MH 2007b The distances to open clusters from main-sequence fitting. IV. Galactic Cepheids, the LMC, and the local distance scale. *Astrophys. J.* **671**, 1640–1668.

An D, Pinsonneault MH, Masseron T, Delahaye F, Johnson JA, Terndrup DM, Beers TC, Ivans II and Ivezić Ž 2009 Galactic globular and open clusters in the Sloan Digital Sky Survey. II. Test of theoretical stellar isochrones. *Astrophys. J.* **700**, 534–544.

Andersen J, Clausen JV, Nordstrom B, Tomkin J and Mayor M 1991 Absolute dimensions of eclipsing binaries. XVII. TZ Fornacis: stellar and tidal evolution in a binary with a fully-fledged red giant. *Astron. Astrophys.* **246**, 99–117.

Argon AL, Greenhill LJ, Reid MJ, Moran JM and Humphreys EML 2007 Toward a new geometric distance to the active galaxy NGC 4258. I. VLBI monitoring of water maser emission. *Astrophys. J.* **659**, 1040–1062.

[6] `http://www.astro.cornell.edu/~shami/psrvlb/`.

Baade W 1926 Über eine Möglichkeit, die Pulsationstheorie der δ Cephei-Veränderlichen zu prüfen. *Astron. Nachr.* **228**, 359–360.
Baade W 1944a The resolution of Messier 32, NGC 205, and the central region of the Andromeda nebula. *Astrophys. J.* **100**, 137–146.
Baade W 1944b NGC 147 and NGC 185, two new members of the Local Group of galaxies. *Astrophys. J.* **100**, 147–150.
Baade W and Swope HH 1961 The Draco system, a dwarf galaxy. *Astron. J.* **66**, 300–347.
Barnes TG and Evans DS 1976 Stellar angular diameters and visual surface brightness. I. Late spectral types. *Mon. Not. R. Astron. Soc.* **174**, 489–502.
Bartel N, Bietenholz MF, Rupen MP and Dwarkadas VV 2007 SN 1993J VLBI. IV. A geometric distance to M81 with the expanding shock front method. *Astrophys. J.* **668**, 924–940.
Bate MR 2009 The importance of radiative feedback for the stellar initial mass function. *Mon. Not. R. Astron. Soc.* **392**, 1363–1380.
Bedding TR and Zijlstra AA 1998 Hipparcos period–luminosity relations for Mira and semiregular variables. *Astrophys. J.* **506**, L47–L50.
Bellazzini M, Ferraro FR and Pancino E 2001 A step toward the calibration of the red giant branch tip as a standard candle. *Astrophys. J.* **556**, 635–640.
Bellazzini M, Ferraro FR, Sollima A, Pancino E and Origlia L 2004 The calibration of the RGB tip as a standard candle. Extension to near infrared colors and higher metallicity. *Astron. Astrophys.* **424**, 199–211.
Benedict GF, McArthur BE, Fredrick LW, Harrison TE, Slesnick CL, Rhee J, Patterson RJ, Skrutskie MF, Franz OG, Wasserman LH, Jefferys WH, Nelan E, van Altena W, Shelus PJ, Hemenway PD, Duncombe RL, Story D, Whipple AL and Bradley AJ 2002 Astrometry with the Hubble Space Telescope: a parallax of the fundamental distance calibrator δ Cephei. *Astron. J.* **124**, 1695–1705.
Benedict GF, McArthur BE, Fredrick LW, Harrison TE, Skrutskie MF, Slesnick CL Rhee J, Patterson RJ, Nelan E, Jefferys WH, van Altena W, Montemayor T, Shelus PJ, Franz OG, Wasserman LH, Hemenway PD, Duncombe RL, Story D, Whipple AL and Bradley AJ 2003 Astrometry with the Hubble Space Telescope: a parallax of the central star of the planetary nebula NGC 6853. *Astron. J.* **126**, 2549–2556.
Benedict GF, McArthur BE, Feast MW, Barnes TG, Harrison TE, Patterson RJ, Menzies JW, Bean JL and Freedman WL 2007 Hubble Space Telescope Fine Guidance Sensor parallaxes of Galactic Cepheid variable stars: period–luminosity relations. *Astron. J.* **133**, 1810–1827.
Berger DH, Gies DR, McAlister HA, ten Brummelaar TA, Henry TJ, Sturmann J, Sturmann L, Turner NH, Ridgway ST, Aufdenberg JP and Mérand A 2006 First results from the CHARA array. IV. The interferometric radii of low-mass stars. *Astrophys. J.* **644**, 475–483.
Bergeron P, Saffer RA and Liebert J 1992 A spectroscopic determination of the mass distribution of DA white dwarfs. *Astrophys. J.* **394**, 228–247.
Bethe HA 1939 Energy production in stars. *Phys. Rev.* **55**, 434–456.
Beuermann K 2006 Barnes–Evans relations for dwarfs with an application to the determination of distances to cataclysmic variables. *Astron. Astrophys.* **460**, 783–792.
Bird JC, Stanek KZ and Prieto JL 2009 Using ultra long period Cepheids to extend the cosmic distance ladder to 100 Mpc and beyond. *Astrophys. J.* **695**, 874–882.
Blažhko S. 1907 Mitteilung über veränderliche Sterne. *Astron. Nachr.* **175**, 325–326.
Bonanos AZ, Stanek KZ, Kudritzki RP, Macri LM, Sasselov DD, Kałużny J, Stetson PB, Bersier D, Bresolin F, Matheson T, Mochejska BJ, Przybilla N, Szentgyorgyi AH, Tonry J and Torres G 2006 The first DIRECT distance determination to a detached eclipsing binary in M33. *Astrophys. J.* **652**, 313–322.
Bond HE and Sparks WB 2009 On geometric distance determination to the Cepheid RS Puppis from its light echoes. *Astron. Astrophys.* **495**, 371–377.
Bond HE, Henden A, Levay ZG, Panagia N, Sparks WB, Starrfield S, Wagner RM, Corradi RLM and Munari U 2003 An energetic stellar outburst accompanied by circumstellar light echoes. *Nature* **422**, 405–408.
Bono G 2003 RR Lyrae distance scale: theory and observations. In *Stellar Candles for the Extragalactic Distance Scale* (eds Alloin D and Gieren W), *Lect. Notes Phys.* **635**, 85–104.

Bono G, Caputo F and Stellingwerf RF 1994 On the application of the Baade–Wesselink method to RR Lyrae stars. *Astrophys. J.* **432**, L51–L54.

Bono G, Castellani V, Degl'Innocenti S and Pulone L 1995 Advanced evolutionary phases of low-mass stars: the role of the original helium. *Astron. Astrophys.* **297**, 115–126.

Bono G, Caputo F, Castellani V and Marconi M 1996 Double-mode RR Lyrae variables: pulsational masses revisited. *Astrophys. J.* **471**, L33–L36.

Bono G, Caputo F, Santolamazza P, Cassisi S and Piersimoni A 1997 Evolutionary scenario for metal-poor pulsating stars. II. Anomalous Cepheids. *Astron. J.* **113**, 2209–2218.

Bono G, Marconi M and Stellingwerf RF 1999 Classical Cepheid pulsation models. I. Physical structure. *Astrophys. J. Suppl. Ser.* **122**, 167–205.

Bono G, Caputo F, Castellani V, Marconi M and Storm J 2001 Theoretical insights into the RR Lyrae K-band period–luminosity relation. *Mon. Not. R. Astron. Soc.* **326**, 1183–1190.

Bono G, Caputo F, Fiorentino G, Marconi M and Musella I 2008a Cepheids in external galaxies. I. The maser-host galaxy NGC 4258 and the metallicity dependence of period–luminosity and period–Wesenheit relations. *Astrophys. J.* **684**, 102–117.

Bono G, Stetson PB, Sanna N, Piersimoni A, Freyhammer LM, Bouzid Y, Buonanno R, Calamida A, Caputo F, Corsi CE, Di Cecco A, Dall'Ora M, Ferraro I, Iannicola G, Monelli M, Nonino M, Pulone L, Sterken C, Storm J, Tuvikene T and Walker AR 2008b On the relative distances of ω Centauri and 47 Tucanae. *Astrophys. J.* **686**, L87–L90.

Bono G, Caputo F, Marconi M and Musella I 2010 Insights into the Cepheid distance scale. *Astrophys. J.* **715**, 277–291.

Braatz III JA, Reid M, Humphreys E, Kuo C, Condon J, Lo KY, Greenhill L, Henkel C, Zaw I and Tilak A 2009a The Megamaser Cosmology Project: the maser distance to UGC 3789. *Bull. Am. Astron. Soc.* **41**, 718.

Braatz JA, Reid MJ, Greenhill LJ, Kuo C-Y, Condon JJ, Lo K-Y and Henkel C 2009b Water masers in AGN accretion disks. In *Approaching Micro-Arcsecond Resolution with VSOP-2: Astrophysics and Technologies* (eds Hagiwara Y, Fomalont E, Tsuboi M and Murata Y), *Astron. Soc. Pac. Conf. Ser.* **402**, 274–279.

Braatz JA, Condon JJ, Henkel C, Lo K-Y and Reid MJ 2009c Cosmology with water-vapor megamasers. *Astro2010: The Astronomy and Astrophysics Decadal Survey*, Science White Papers **23**.

Braatz JA, Reid MJ, Humphreys EML, Henkel C, Condon JJ and Lo KY 2010 The Megamaser Cosmology Project. II. The angular-diameter distance to UGC 3789. *Astrophys. J.* **718**, 657–665.

Bragaglia A, Renzini A and Bergeron P 1995 Temperatures, gravities, and masses for a sample of bright DA white dwarfs and the initial-to-final mass relation. *Astrophys. J.* **443**, 735–752.

Bresolin F 2011 Revisiting the abundance gradient in the maser host galaxy NGC 4258. *Astrophys. J.* **729**, 56.

Brisken WF, Benson JM, Goss WM and Thorsett SE 2002 Very Long Baseline Array measurement of nine pulsar parallaxes. *Astrophys. J.* **571**, 906–917.

Bronfman L, Casassus S, May J and Nyman L-Å 2000 The radial distribution of OB star formation in the Galaxy. *Astron. Astrophys.* **358**, 521–534.

Brunthaler A, Reid MJ, Falcke H, Greenhill LJ and Henkel C 2005 The geometric distance and proper motion of the Triangulum galaxy (M33). *Science* **307**, 1440–1443.

Brunthaler A, Reid MJ, Falcke H, Henkel C and Menten KM 2008 Microarcsecond astrometry in the Local Group. In *A Giant Step: From Milli- to Micro-Arcsecond Astrometry* (eds Jin WJ, Platais I and Perryman MAC), *Proc. Int'l Astron. Union Symp.* **248**, 474–480.

Cabrera-Lavers A, González-Fernández C, Garzón F, Hammersley PL and López-Corredoira M 2008 The long Galactic bar as seen by UKIDSS Galactic Plane Survey. *Astron. Astrophys.* **491**, 781–787.

Cacciari C and Clementini G 2003 Globular cluster distances from RR Lyrae stars. In *Stellar Candles for the Extragalactic Distance Scale* (eds Alloin D and Gieren W), *Lect. Notes Phys.* **635**, 105–122.

Cahn JH, Kaler JB and Stanghellini L 1992 A catalogue of absolute fluxes and distances of planetary nebulae. *Astron. Astrophys. Suppl. Ser.* **94**, 399–452.

Çakırlı Ö and İbanoğlu C 2010 NSVS06507557: a low-mass double-lined eclipsing binary. *Mon. Not. R. Astron. Soc.* **401**, 1141–1150.

Caldwell JAR and Coulson IM 1986 The geometry and distance of the Magellanic Clouds from Cepheid variables. *Mon. Not. R. Astron. Soc.* **218**, 223–246.

Cameron PB, Chandra P, Ray A, Kulkarni SR, Frail DA, Wieringa MH, Nakar E, Phinney ES, Miyazaki A, Tsuboi M, Okumura S, Kawai N, Menten KM and Bertoldi F 2005 Detection of a radio counterpart to the 27 December 2004 giant flare from SGR 1806−20. *Nature* **434**, 1112–1115.

Caputo F and Degl'Innocenti S 1995 HB morphology and age indicators for metal-poor stellar systems with age in the range of 1 to 20 Gyr. *Astron. Astrophys.* **298**, 833–842.

Caputo F, Santolamazza P and Marconi M 1998 The role of chemical composition in RR Lyrae pulsational properties. I. Periods. *Mon. Not. R. Astron. Soc.* **293**, 364–368.

Caputo F, Marconi M and Musella I 2002 The Cepheid period–luminosity relation and the maser distance to NGC 4258. *Astrophys. J.* **566**, 833–837.

Carney BW, Storm J and Jones RV 1992 The Baade–Wesselink method and the distances to RR Lyrae Stars. VIII. Comparisons with other techniques and implications for globular cluster distances and ages. *Astrophys. J.* **386**, 663–684.

Carretta E and Gratton RG 1997 Abundances for globular cluster giants. I. Homogeneous metallicities for 24 clusters. *Astron. Astrophys. Suppl. Ser.* **121**, 95–112.

Carretta E, Gratton RG, Clementini G and Fusi Pecci F 2000 Distances, ages, and epoch of formation of globular clusters. *Astrophys. J.* **533**, 215–235.

Cassisi S 2010 Stellar models: firm evidence, open questions and future developments. In *Stellar Populations: Planning for the Next Decade* (eds Bruzual G and Charlot S), *Proc. Int'l Astron. Union Symp.* **262**, 13–22.

Cassisi S and Salaris M 1997 A critical investigation on the discrepancy between the observational and the theoretical red giant luminosity function 'bump'. *Mon. Not. R. Astron. Soc.* **285**, 593–603.

Cassisi S, Potekhin AY, Pietrinferni A, Catelan M and Salaris M 2007 Updated electron-conduction opacities: the impact on low-mass stellar models. *Astrophys. J.* **661**, 1094–1104.

Castellani V and Degl'Innocenti S 1995 Dwarf spheroidals and the evolution of not-too-old Population II stars. *Astron. Astrophys.* **298**, 827–832.

Catelan M 1996 The distance modulus of the Large Magellanic Cloud. Constraints from RR Lyrae pulsation properties. *Astron. Astrophys.* **307**, L13–L16.

Catelan M 2006 Near-field cosmology with horizontal branch and RR Lyrae stars. *Rev. Mex. Astron. Astrophys.* **26**, 93–96.

Catelan M 2009 Horizontal branch stars: the interplay between observations and theory, and insights into the formation of the Galaxy. *Astrophys. Space Sci.* **320**, 261–309.

Catelan M and de Freitas Pacheco JA 1996 Metal-rich globular clusters of the Galaxy: morphology of the helium-burning 'clump', and the determination of relative ages through the 'ΔV' method. *Publ. Astron. Soc. Pac.* **108**, 166–175.

Catelan M, Pritzl BJ and Smith HA 2004 The RR Lyrae period–luminosity relation. I. Theoretical calibration. *Astrophys. J. Suppl. Ser.* **154**, 633–649.

Chadid M, Gillet D and Fokin AB 2000 Van Hoof effect between metallic lines in RR Lyrae. II. Comparison with purely radiative models. *Astron. Astrophys.* **363**, 568–574.

Chadid M, Benkő JM, Szabó R, Paparó M, Chapellier E, Kolenberg K, Poretti E, Bono G, Le Borgne J-F, Trinquet H, Artemenko S, Auvergne M, Baglin A, Debosscher J, Grankin KN, Guggenberger E and Weiss WW 2010 First CoRoT light curves of RR Lyrae stars. Complex multiplet structure and non-radial pulsation detections in V1127 Aquilae. *Astron. Astrophys.* **510**, A39.

Chatterjee S, Cordes JM, Lazio TJW, Goss WM, Fomalont EB and Benson JM 2001 Parallax and kinematics of PSR B0919+06 from VLBA astrometry and interstellar scintillometry. *Astrophys. J.* **550**, 287–296.

Chatterjee S, Brisken WF, Vlemmings WHT, Goss WM, Lazio TJW, Cordes JM, Thorsett SE, Fomalont EB, Lyne AG and Kramer M 2009 Precision astrometry with the Very Long Baseline Array: parallaxes and proper motions for 14 pulsars. *Astrophys. J.* **698**, 250–265.

Cho D-H and Lee S-G 2002 The relation between near-infrared luminosity of red giant branch bumps and metallicity of Galactic globular clusters. *Astron. J.* **124**, 977–988.

Christianto H and Seaquist ER 1998 Angular expansion measurement of the young and compact planetary nebula Vy 2–2. *Astron. J.* **115**, 2466–2474.

Cioni M-RL, van der Marel RP, Loup C and Habing HJ 2000 The tip of the red giant branch and distance of the Magellanic Clouds: results from the DENIS survey. *Astron. Astrophys.* **359**, 601–614.
Cioni M-RL, Marquette J-B, Loup C, Azzopardi M, Habing HJ, Lasserre T and Lesquoy E 2001 Variability and spectral classification of LMC giants: results from DENIS and EROS. *Astron. Astrophys.* **377**, 945–954.
Clausen JV 2004 Eclipsing binaries as precise standard candles. *New Astron. Rev.* **48**, 679–685.
Clementini G, Gratton R, Bragaglia A, Carretta E, Di Fabrizio L and Maio M 2003 Distance to the Large Magellanic Cloud: the RR Lyrae stars. *Astron. J.* **125**, 1309–1329.
Cohen JG 1988 Nova expansion parallaxes. In *The Extragalactic Distance Scale* (eds van den Bergh S and Pritchet CJ), *Proc. Astron. Soc. Pac. Conf. Ser.* **4**, 114–127.
Cole AA 1998 Age, metallicity, and the distance to the Magellanic Clouds from red clump stars. *Astrophys. J.* **500**, L137–L140.
Cook K, Alcock C, Alves D, Minniti D, Marshall S, Vandehei T, Griest K, Allsman R, Axelrod T, Freeman K, Peterson B, Rodgers A, Pratt M, Becker A, Stubbs C, Tomaney A, Lehner M, Bennett D, Nelson C, Quinn P, Sutherland W and Welch D 1997 The MACHO project: microlensing and variable stars. In *Variable Stars and the Astrophysical Returns of Microlensing Surveys* (eds Ferlet R, Maillard J-P and Raban B), *Proc. Inst. Astrophys. Paris Colloq.* **12**, 17–28.
Cool AM, Piotto G and King IR 1996 The main sequence and a white dwarf sequence in the globular cluster NGC 6397. *Astrophys. J.* **468**, 655–662.
Cordes JM and Lazio TJW 2002 NE2001. I. A new model for the Galactic distribution of free electrons and its fluctuations. Unpublished (astro-ph/0207156v3).
Cordes JM and Lazio TJW 2003 NE2001. II. Using radio propagation data to construct a model for the Galactic distribution of free electrons. Unpublished (astro-ph/0301598v1).
Cordes JM, Weisberg JM, Frail DA, Spangler SR and Ryan M 1991 The Galactic distribution of free electrons. *Nature* **354**, 121–124.
Crause LA, Lawson WA, Menzies JW and Marang F 2005 V838 Mon: light echo evolution and distance estimate. *Mon. Not. R. Astron. Soc.* **358**, 1352–1360.
Da Costa GS and Armandroff TE 1990 Standard globular cluster giant branches in the $(M_I, (V-I)_0)$ plane. *Astron. J.* **100**, 162–181.
Dame TM, Ungerechts H, Cohen RS, de Geus EJ, Grenier IA, May J, Murphy DC, Nyman L-Å and Thaddeus P 1987 A composite CO survey of the entire Milky Way. *Astrophys. J.* **322**, 706–720.
Daub CT 1982 A statistical survey of local planetary nebulae. *Astrophys. J.* **260**, 612–624.
Davidge TJ 2001 The brightest asymptotic giant branch stars in the inner bulge of M31. *Astron. J.* **122**, 1386–1396.
Davidge TJ 2002 The upper asymptotic giant branch of the elliptical galaxy Maffei 1 and comparisons with M32 and NGC 5128. *Astron. J.* **124**, 2012–2022.
Davidge TJ and van den Bergh S 2001 The detection of bright asymptotic giant branch stars in the nearby elliptical galaxy Maffei 1. *Astrophys. J.* **553**, L133–L136.
de Jager OC, Harding AK, Michelson PF, Nel HI, Nolan PL, Sreekumar P and Thompson DJ 1996 Gamma-ray Observations of the Crab Nebula: a study of the synchro-Compton spectrum. *Astrophys. J.* **457**, 253–266.
Della Valle M and Gilmozzi R 2002 Rebirth of novae as distance indicators due to efficient large telescopes. *Science* **296**, 1275–1275.
Della Valle M and Livio M 1995 The calibration of novae as distance indicators. *Astrophys. J.* **452**, 704–709.
De Marchi G, Paresce F and Romaniello M 1995 Low mass stars and white dwarfs in NGC 6397. In *Stellar Populations* (eds van der Kruit PC and Gilmore G), *Proc. Int'l Astron. Union Symp.* **164**, 408–408.
Demarque P, Zinn R, Lee Y-W and Yi S 2000 The metallicity dependence of RR Lyrae absolute magnitudes from synthetic horizontal-branch models. *Astron. J.* **119**, 1398–1404.
Di Benedetto GP 1997 Improved calibration of cosmic distance scale by Cepheid pulsation parallaxes. *Astrophys. J.* **486**, 60–74.

Di Benedetto GP 2008 The Cepheid distance to the Large Magellanic Cloud and NGC 4258 by the surface brightness technique and improved calibration of the cosmic distance scale. *Mon. Not. R. Astron. Soc.* **390**, 1762–1776.

Di Cecco A, Bono G, Stetson PB, Pietrinferni A, Becucci R, Cassisi S, Degl'Innocenti S, Iannicola G, Prada Moroni PG, Buonanno R, Calamida A, Caputo F, Castellani M, Corsi CE, Ferraro I, Dall'Ora M, Monelli M, Nonino M, Piersimoni AM, Pulone L, Romaniello M, Salaris M, Walker AR and Zoccali M 2010 On the $\Delta V_{\rm HB}^{\rm bump}$ parameter in globular clusters. *Astrophys. J.* **712**, 527–535.

Di Criscienzo M, Caputo F, Marconi M and Cassisi S 2007 Synthetic properties of bright metal-poor variables. II. BL Hercules stars. *Astron. Astrophys.* **471**, 893–900.

Diethelm R 1996 Period changes of AHB1 variables. *Astron. Astrophys.* **307**, 803–806.

Dolphin AE, Saha A, Skillman ED, Tolstoy E, Cole AA, Dohm-Palmer RC, Gallagher JS, Mateo M and Hoessel JG 2001 Deep Hubble Space Telescope imaging of IC 1613. I. Variable stars and distance. *Astrophys. J.* **550**, 554–569.

Dorman B 1992 The structure of horizontal-branch models. I. The zero-age horizontal branch. *Astrophys. J. Suppl. Ser.* **80**, 701–724.

Dorman B, Vandenberg DA and Laskarides PG 1989 On the horizontal branch and the helium content of 47 Tucanae. *Astrophys. J.* **343**, 750–759.

Eddington AS 1920 The internal constitution of the stars. *The Observatory* **43**, 341–358.

Eisenhauer F, Schödel R, Genzel R, Ott T, Tecza M, Abuter R, Eckart A and Alexander T 2003 A geometric determination of the distance to the Galactic center. *Astrophys. J.* **597**, L121–L124.

Feast MW and Walker AR 1987 Cepheids as distance indicators. *Annu. Rev. Astron. Astrophys.* **25**, 345–375.

Feast MW, Glass IS, Whitelock PA and Catchpole RM 1989 A period–luminosity–colour relation for Mira variables. *Mon. Not. R. Astron. Soc.* **241**, 375–392.

Ferraro FR, Messineo M, Fusi Pecci F, de Palo MA, Straniero O, Chieffi A and Limongi M 1999 The giant, horizontal, and asymptotic branches of Galactic globular clusters. I. The catalog, photometric observables, and features. *Astron. J.* **118**, 1738–1758.

Ferraro FR, Montegriffo P, Origlia L and Fusi Pecci F 2000 A new infrared array photometric survey of Galactic globular clusters: a detailed study of the red giant branch sequence as a step toward the global testing of stellar models. *Astron. J.* **119**, 1282–1295.

Ferraro FR, Valenti E and Origlia L 2006 An empirical tool to derive metallicity, reddening, and distance for old stellar populations from near-infrared color–magnitude diagrams. *Astrophys. J.* **649** 243–247.

Feuchtinger MU 1998 A nonlinear model for RR Lyrae double mode pulsation. *Astron. Astrophys.* **337**, L29–L33.

Figer DF, Najarro F, Geballe TR, Blum RD and Kudritzki RP 2005 Massive stars in the SGR 1806−20 cluster. *Astrophys. J.* **622**, L49–L52.

Fiorentino G, Limongi M, Caputo F and Marconi M 2006 Synthetic properties of bright metal-poor variables. I. 'Anomalous' Cepheids. *Astron. Astrophys.* **460**, 155–166.

Fitzpatrick EL, Ribas I, Guinan EF, Maloney FP and Claret A 2003 Fundamental properties and distances of Large Magellanic Cloud eclipsing binaries. IV. HV 5936. *Astrophys. J.* **587**, 685–700.

Fomalont E 2005 Phase referencing using more than one calibrator. In *Future Directions in High Resolution Astronomy: the 10th Anniversary of the VLBA* (eds Romney J and Reid M), *Astron. Soc. Pac. Conf. Proc.* **340**, 460–465.

Fomalont E and Reid M 2004 Microarcsecond astrometry using the SKA. *New Astron. Rev.* **48**, 1473–1482.

Fouqué P, Arriagada P, Storm J, Barnes TG Nardetto N, Mérand A, Kervella P, Gieren W, Bersier D, Benedict GF and McArthur BE 2007 A new calibration of Galactic Cepheid period–luminosity relations from *B* to *K* bands, and a comparison to LMC relations. *Astron. Astrophys.* **476**, 73–81.

Fraser OJ, Hawley SL, Cook KH and Keller SC 2005 Long-period variables in the Large Magellanic Cloud: results from MACHO and 2MASS. *Astron. J.* **129**, 768–775.

Freedman WL and Madore BF 2010 A physically-based method for scaling Cepheid light curves for future distance determinations. *Astrophys. J.* **719**, 335–340.

Freedman WL, Hughes SM, Madore BF, Mould JR, Lee MG Stetson P, Kennicutt RC, Turner A, Ferrarese L, Ford H, Graham JA, Hill R, Hoessel JG, Huchra J and Illingworth GD 1994 The Hubble Space Telescope extragalactic distance scale Key Project. I. The discovery of Cepheids and a new distance to M81. *Astrophys. J.* **427**, 628–655.

Freedman WL, Madore BF, Gibson BK, Ferrarese L, Kelson DD, Sakai S, Mould JR, Kennicutt Jr RC, Ford HC, Graham JA, Huchra JP, Hughes SMG, Illingworth GD, Macri LM and Stetson PB 2001 Final results from the Hubble Space Telescope Key Project to measure the Hubble constant. *Astrophys. J.* **553**, 47–72.

Freedman WL, Madore BF, Stetson PB, Hughes SMG, Holtzman JA, Mould JR, Trauger JT, Gallagher III JS, Ballester GE, Burrows CJ, Casertano S, Clarke JT, Crisp D, Ferrarese L, Ford H, Graham JA, Griffiths RE, Hester JJ, Hill R, Hoessel JG, Huchra J, Kennicutt RC, Scowen PA, Sparks B, Stapelfeldt KR, Watson AM and Westphal J 2004 First Hubble Space Telescope observations of the brightest stars in the Virgo galaxy M100 = NGC 4321. *Astrophys. J.* **435**, L31–L34.

Freedman WL, Madore BF, Rigby J, Persson SE and Sturch L 2008 The Cepheid period–luminosity relation at mid-infrared wavelengths. I. First-epoch LMC data. *Astrophys. J.* **679**, 71–75.

Fuchs Y, Mirabel F, Chaty S, Claret A, Cesarsky CJ and Cesarsky DA 1999 ISO observations of the environment of the soft gamma-ray repeater SGR 1806−20. *Astron. Astrophys.* **350**, 891–899.

Gaensler BM, McClure-Griffiths NM, Oey MS, Haverkorn M, Dickey JM and Green AJ 2005 A stellar wind bubble coincident with the anomalous X-ray pulsar 1E 1048.1−5937: are magnetars formed from massive progenitors? *Astrophys. J.* **620**, L95–L98.

Gallart C 1998 Observational discovery of the asymptotic giant branch bump in densely populated color–magnitude diagrams of galaxies and star clusters. *Astrophys. J.* **495**, L43–L46.

Gatewood G, de Jonge JK and Heintz WD 1995 Astrometric studies in the region of Algol. *Astron. J.* **109**, 434–439.

Gautschy A and Saio H 1996 Stellar pulsations across the HR diagram. II. *Annu. Rev. Astron. Astrophys.* **34**, 551–606.

Genzel R, Downes D, Schneps MH, Reid MJ, Moran JM, Kogan LR, Kostenko VI, Matveenko LI and Ronnang B 1981 Proper motions and distances of H_2O maser sources. II. W51 Main. *Astrophys. J.* **247**, 1039–1051.

Genzel R, Becklin EE, Wynn-Williams CG, Moran JM, Reid MJ, Jaffe DT and Downes D 1982 Infrared and radio observations of W51: another Orion-KL at a distance of 7 kiloparsecs? *Astrophys. J.* **255**, 527–535.

Gesicki K, Zijlstra AA, Acker A and Szczerba R 1998 Velocity fields of planetary nebulae. *Astron. Astrophys.* **329**, 265–275.

Ghez AM, Salim S, Weinberg NN, Lu JR, Do T, Dunn JK, Matthews K, Morris MR, Yelda S, Becklin EE, Kremenek T, Milosavljevic M and Naiman J 2008 Measuring distance and properties of the Milky Way's central supermassive black hole with stellar orbits. *Astrophys. J.* **689**, 1044–1062.

Ghosh T and Rao AP 1992 Low-frequency variability and the Galactic distribution of electron-density fluctuation. *Astron. Astrophys.* **264**, 203–216.

Gieren W, Storm J, Barnes III TG, Fouqué P, Pietrzyński G and Kienzle F 2005 Direct distances to Cepheids in the Large Magellanic Cloud: evidence for a universal slope of the period–luminosity relation up to solar abundance. *Astrophys. J.* **627**, 224–237.

Gieren W, Pietrzyński G, Szewczyk O, Soszyński I, Bresolin F, Kudritzki RP, Urbaneja MA, Storm J and Minniti D 2008 The Araucaria project: the distance to the Local Group galaxy WLM from near-infrared photometry of Cepheid variables. *Astrophys. J.* **683**, 611–619.

Gingold RA 1976 The evolutionary status of Population II Cepheids. *Astrophys. J.* **204**, 116–130.

Gingold RA 1985 The evolutionary status of Type II Cepheids. *Mem. Soc. Astron. It.* **56**, 169–191.

Girardi L and Salaris M 2001 Population effects on the red giant clump absolute magnitude, and distance determinations to nearby galaxies. *Mon. Not. R. Astron. Soc.* **323**, 109–129.

Girardi L, Rubele S and Kerber L 2009 Discovery of two distinct red clumps in NGC 419: a rare snapshot of a cluster at the onset of degeneracy. *Mon. Not. R. Astron. Soc.* **394**, L74–L78.

Gizis JE 1997 M-subdwarfs: spectroscopic classification and the metallicity scale. *Astron. J.* **113**, 806–822.

Glass IS and Schultheis M 2003 Period–magnitude relations for M giants in Baade's Window. NGC 6522. *Mon. Not. R. Astron. Soc.* **345**, 39–48.

Glass IS, Schultheis M, Blommaert JADL, Sahai R, Stute M and Uttenthaler S 2009 Mid-infrared period–magnitude relations for AGB stars. *Mon. Not. R. Astron. Soc.* **395**, L11–L15.

Goecking K-D, Duerbeck HW, Plewa T, Kałużny J, Schertl D, Weigelt G and Flin P 1994 The W Ursae Majoris system ER Ori: a multiple star. *Astron. Astrophys.* **289**, 827–836.

Gómez GC, Benjamin RA and Cox DP 2001 A re-examination of the distribution of Galactic free electrons. *Astron. J.* **122**, 908–920.

Gómez Y, Rodríguez LF and Moran JM 1993 Detection of the angular expansion rate and determination of the distance of the planetary nebula NGC 6302. *Astrophys. J.* **416**, 620–622.

Gould A 1994 The ring around supernova 1987A revisited. I. Ellipticity of the ring. *Astrophys. J.* **425**, 51–56.

Gould A 1995 The supernova ring revisited. II. Distance to the Large Magellanic Cloud. *Astrophys. J.* **452**, 189–194.

Gould A and Uza O 1998 Upper limit to the distance to the Large Magellanic Cloud. *Astrophys. J.* **494**, 118–124.

Gratton RG, Fusi Pecci F, Carretta E, Clementini G, Corsi CE and Lattanzi M 1997 Ages of globular clusters from Hipparcos parallaxes of local subdwarfs. *Astrophys. J.* **491**, 749–771.

Gratton RG, Bragaglia A, Carretta E, Clementini G, Desidera S, Grundahl F and Lucatello S 2003 Distances and ages of NGC 6397, NGC 6752 and 47 Tuc. *Astron. Astrophys.* **408**, 529–543.

Graves GJM, Challis PM, Chevalier RA, Crotts A, Filippenko AV, Fransson C, Garnavich P, Kirshner RP, Li W, Lundqvist P, McCray R, Panagia N, Phillips MM, Pun CJS, Schmidt BP, Sonneborn G, Suntzeff NB, Wang L and Wheeler JC 2005 Limits from the Hubble Space Telescope on a point source in SN 1987A. *Astrophys. J.* **629**, 944–959.

Greenhill LJ, Jiang DR, Moran JM, Reid MJ, Lo KY and Claussen MJ 1995a Detection of a subparsec diameter disk in the nucleus of NGC 4258. *Astrophys. J.* **440**, 619–627.

Greenhill LJ, Henkel C, Becker R, Wilson TL and Wouterloot JGA 1995b Centripetal acceleration within the subparsec nuclear maser disk of NGC 4258. *Astron. Astrophys.* **304**, 21–33.

Grieve GR, Madore BF and Welch DL 1985 Leavitt variables: bright variable supergiants and their implications for the distance scale. *Astrophys. J.* **294**, 513–516.

Grocholski AJ and Sarajedini A 2002 WIYN open cluster study. X. The K-band magnitude of the red clump as a distance indicator. *Astron. J.* **123**, 1603–1612.

Groenewegen MAT 2008a Baade–Wesselink distances and the effect of metallicity in classical Cepheids. *Astron. Astrophys.* **488**, 25–35.

Groenewegen MAT 2008b The red clump absolute magnitude based on revised Hipparcos parallaxes. *Astron. Astrophys.* **488**, 935–941.

Groenewegen MAT and Blommaert JADL 2005 Mira variables in the OGLE bulge fields. *Astron. Astrophys.* **443**, 143–156.

Groenewegen MAT, Romaniello M, Primas F and Mottini M 2004 The metallicity dependence of the Cepheid P–L relation. *Astron. Astrophys.* **420**, 655–663.

Grundahl F, Stetson PB and Andersen MI 2002 The ages of the globular clusters M 71 and 47 Tuc from Strömgren *uvby* photometry. Evidence for high ages. *Astron. Astrophys.* **395**, 481–497.

Guinan EF, Fitzpatrick EL, Dewarf LE, Maloney FP, Maurone PA, Ribas I, Pritchard JD, Bradstreet DH and Giménez A 1998 The distance to the Large Magellanic Cloud from the eclipsing binary HV 2274. *Astrophys. J.* **509**, L21–L24.

Guzmán L, Gómez Y and Rodríguez LF 2006 Expansion parallax for the compact planetary nebula M2–43. *Rev. Mex. Astron. Astrophys.* **42**, 127–130.

Guzmán L, Loinard L, Gómez Y and Morisset C 2009 Expansion parallax of the planetary nebula IC 418. *Astron. J.* **138**, 46–49.

Hachisuka K, Brunthaler A, Menten KM, Reid MJ, Imai H, Hagiwara Y, Miyoshi M, Horiuchi S and Sasao T 2006 Water maser motions in W3(OH) and a determination of its distance. *Astrophys. J.* **645**, 337–344.

Hajian AR 2006 Distances to planetary nebulae. In *Planetary Nebulae in Our Galaxy and Beyond* (eds Barlow MJ and Méndez RH), *Proc. Int'l Astron. Union Symp.* **234**, 41–48.
Hajian AR and Terzian Y 1996 Planetary nebulae expansion distances. III. *Publ. Astron. Soc. Pac.* **108**, 258–264.
Hajian AR, Terzian Y and Bignell C 1993 Planetary nebulae expansion distances. *Astron. J.* **106**, 1965–1972.
Hajian AR, Terzian Y and Bignell C 1995 Planetary nebulae expansion distances. II. NGC 6572, NGC 6210, NGC 3242, and NGC 2392. *Astron. J.* **109**, 2600–2607.
Hansen BMS, Richer HB, Fahlman GG, Stetson PB, Brewer J, Currie T, Gibson BK, Ibata R, Rich RM and Shara MM 2004 Hubble Space Telescope observations of the white dwarf cooling sequence of M4. *Astrophys. J. Suppl. Ser.* **155**, 551–576.
Hansen BMS, Anderson J, Brewer J, Dotter A, Fahlman GG, Hurley J, Kalirai J, King I, Reitzel D, Richer HB, Rich RM, Shara MM and Stetson PB 2007 The white dwarf cooling sequence of NGC 6397. *Astrophys. J.* **671**, 380–401.
Harries TJ, Hilditch RW and Howarth ID 2003 Ten eclipsing binaries in the Small Magellanic Cloud: fundamental parameters and Cloud distance. *Mon. Not. R. Astron. Soc.* **339**, 157–172.
Harris HC, Dahn, CC, Canzian B, Guetter HH, Leggett SK, Levine SE, Luginbuhl CB, Monet AKB, Monet DG, Pier JR, Stone RC, Tilleman T, Vrba FJ and Walker RL 2007 Trigonometric parallaxes of central stars of planetary nebulae. *Astron. J.* **133**, 631–638.
Haschick AD, Baan WA and Peng EW 1994 The masering torus in NGC 4258. *Astrophys. J.* **437**, L35–L38.
Hayashi C 1961 Stellar evolution in early phases of gravitational contraction. *Publ. Astron. Soc. Jpn* **13**, 450–452.
Heger A and Langer N 2000 Presupernova evolution of rotating massive stars. II. Evolution of the surface properties. *Astrophys. J.* **544**, 1016–1035.
Henden A, Munari U and Schwartz M 2002 V838 Monocerotis. *Int'l Astron. Union Circ.* **7859**, 1.
Henyey LG, Lelevier R and Levée RD 1955 The early phases of stellar evolution. *Publ. Astron. Soc. Pac.* **67**, 154–160.
Herrnstein JR, Moran JM, Greenhill LJ, Blackman EG and Diamond PJ 1998 Polarimetric observations of the masers in NGC 4258: an upper limit on the large-scale magnetic field 0.2 pc from the central engine. *Astrophys. J.* **508**, 243–247.
Herrnstein JR, Moran JM, Greenhill LJ, Diamond PJ, Inoue M, Nakai N, Miyoshi M, Henkel C and Riess A 1999 A geometric distance to the galaxy NGC 4258 from orbital motions in a nuclear gas disk. *Nature* **400**, 539–541.
Herrnstein JR, Moran JM, Greenhill LJ and Trotter AS 2005 The geometry of and mass accretion rate through the maser accretion disk in NGC 4258. *Astrophys. J.* **629**, 719–738.
Hester JJ 2008 The Crab Nebula: an astrophysical chimera. *Annu. Rev. Astron. Astrophys.* **46**, 127–155.
Hewish A, Bell SJ, Pilkington JDH, Scott PF and Collins RA 1968 Observation of a rapidly pulsating radio source. *Nature* **217**, 709–713.
Hoogerwerf R, de Bruijne JHJ and de Zeeuw PT 2001 On the origin of the O and B-type stars with high velocities. II. Runaway stars and pulsars ejected from the nearby young stellar groups. *Astron. Astrophys.* **365**, 49–77.
Hotan AW, Bailes M and Ord SM 2006 High-precision baseband timing of 15 millisecond pulsars. *Mon. Not. R. Astron. Soc.* **369**, 1502–1520.
Hu J, Zhang SN and Li TP 2003 Distance determination of variable Galactic sources. In *Stellar Astrophysics – A Tribute to Helmut A. Abt* (eds Cheng KS, Leung KC and Li TP), *Astrophys. Space Sci. Libr.* **298**, 55–59.
Hubble E 1936 The luminosity function of nebulae. I. The luminosity function of resolved nebulae as indicated by their brightest stars. *Astrophys. J.* **84**, 158–179.
Humphreys EML, Greenhill LJ, Reid MJ, Beuther H, Moran JM, Gurwell M, Wilner DJ and Kondratko PT 2005 First detection of millimeter/submillimeter extragalactic H_2O maser emission. *Astrophys. J.* **634**, L133–L136.

Humphreys EML, Reid MJ, Greenhill LJ, Moran JM and Argon AL 2008 Toward a new geometric distance to the active galaxy NGC 4258. II. Centripetal accelerations and investigation of spiral structure. *Astrophys. J.* **672**, 800–816.

Humphreys RM 1988 The brightest stars as distance indicators. In *The Extragalactic Distance Scale* (eds van den Bergh S and Pritchet CJ), *Proc. Astron. Soc. Pac. Conf. Ser.* **4**, 103–112.

Iben Jr I 1968 Low-mass red giants. *Astrophys. J.* **154**, 581–595.

Iben Jr I and Renzini A 1983 Asymptotic giant branch evolution and beyond. *Annu. Rev. Astron. Astrophys.* **21**, 271–342.

Iben Jr I and Tutukov AV 1989 Model stars with degenerate dwarf cores and helium-burning shells: a stationary-burning approximation. *Astrophys. J.* **342**, 430–448.

Imai H, Watanabe T, Omodaka T, Nishio M, Kameya O, Miyaji T and Nakajima J 2002 3-D kinematics of water masers in the W 51A region. *Publ. Astron. Soc. Jpn* **54**, 741–755.

Impellizzeri C, Braatz J, Kuo C, Schenker M, Condon J, Henkel C, Reid M and Lo F 2010 Mrk 1419: a new distance determination. *Bull. Am. Astron. Soc.* **42**, 243–243.

Jourdain E and Roques JP 2009 The high-energy emission of the Crab Nebula from 20 keV to 6 MeV with Integral SPI. *Astrophys. J.* **704**, 17–24.

Joy AH 1949 Spectra of the brighter variables in globular clusters. *Astrophys. J.* **110**, 105–116.

Kapteyn JC 1902 Über die Bewegung der Nebel in der Umgebung von Nova Persei. *Astron. Nachr.* **157**, 201–204.

Karachentsev ID and Tikhonov NA 1994 New photometric distances for dwarf galaxies in the Local Volume. *Astron. Astrophys.* **286**, 718–724.

Kawamura J and Masson C 1996 Distances to planetary nebulae BD $+30^{\circ}3639$ and NGC 6572. *Astrophys. J.* **461**, 282–287.

Kervella P 2006 Cepheid distances from interferometry. *Mem. Soc. Astron. It.* **77**, 227–230.

Kervella P, Thévenin F, Morel P, Berthomieu G, Bordé P and Provost J 2004a The diameter and evolutionary state of Procyon A. Multi-technique modeling using asteroseismic and interferometric constraints. *Astron. Astrophys.* **413**, 251–256.

Kervella P, Fouqué P, Storm J, Gieren WP, Bersier D, Mourard D, Nardetto N and du Coudé Foresto V 2004b The angular size of the Cepheid l Carinae: a comparison of the interferometric and surface brightness techniques. *Astrophys. J.* **604**, L113–L116.

Kervella P, Bersier D, Mourard D, Nardetto N, Fouqué P and Coudé du Foresto V 2004c Cepheid distances from infrared long-baseline interferometry. III. Calibration of the surface brightness–color relations. *Astron. Astrophys.* **428**, 587–593.

Kervella P, Thévenin F, Di Folco E and Ségransan D 2004d The angular sizes of dwarf stars and subgiants. Surface brightness relations calibrated by interferometry. *Astron. Astrophys.* **426**, 297–307.

Kervella P, Mérand A, Szabados L, Fouqué P, Bersier D, Pompei E and Perrin G 2008 The long-period Galactic Cepheid RS Puppis. I. A geometric distance from its light echoes. *Astron. Astrophys.* **480**, 167–178.

Kim M, Kim E, Lee MG, Sarajedini A and Geisler D 2002 Determination of the distance to M33 based on the tip of the red giant branch and the red clump. *Astron. J.* **123**, 244–254.

King CR, Da Costa GS and Demarque P 1985 The luminosity function on the subgiant branch of 47 Tucanae. A comparison of observation and theory. *Astrophys. J.* **299**, 674–682.

Kirsch MG, Briel UG, Burrows D, Campana S, Cusumano G, Ebisawa K, Freyberg MJ, Guainazzi M, Haberl F, Jahoda K, Kaastra J, Kretschmar P, Larsson S, Lubinski P, Mori K, Plucinsky P, Pollock AM, Rothschild R, Sembay S, Wilms J and Yamamoto M 2005 Crab: the standard X-ray candle with all (modern) X-ray satellites. *Proc. SPIE* **5898**, 22–33.

Kiss LL and Bedding TR 2004 Red variables in the OGLE–II data base. II. Comparison of the Large and Small Magellanic Clouds. *Mon. Not. R. Astron. Soc.* **347**, L83–L87.

Knapp GR, Pourbaix D, Platais I and Jorissen A 2003 Reprocessing the Hipparcos data of evolved stars. III. Revised Hipparcos period–luminosity relationship for Galactic long-period variable stars. *Astron. Astrophys.* **403**, 993–1002.

Kolenberg K, Smith HA, Gazeas KD, Elmasl A, Breger M, Guggenberger E, van Cauteren P, Lampens P, Reegen P, Niarchos PG, Albayrak B, Selam SO, Özavɑ I and Aksu O 2006 The Blažhko effect of RR Lyrae in 2003–2004. *Astron. Astrophys.* **459**, 577–588.

Kolenberg K, Szabó R, Kurtz DW, Gilliland RL, Christensen-Dalsgaard J, Kjeldsen H, Brown TM, Benkő JM, Chadid M, Derekas A, Di Criscienzo M, Guggenberger E, Kinemuchi K, Kunder A, Kolláth Z, Kopacki G, Moskalik P, Nemec JM, Nuspl J, Silvotti R, Suran MD, Borucki WJ, Koch D and Jenkins JM 2010 First Kepler results on RR Lyrae stars. *Astrophys. J.* **713**, L198–L203.
Kraft RP and Ivans II 2003 A globular cluster metallicity scale based on the abundance of FeII. *Publ. Astron. Soc. Pac.* **115**, 143–169.
Krause O, Tanaka M, Usuda T, Hattori T, Goto M, Birkmann S and Nomoto K 2008 Tycho Brahe's 1572 supernova as a standard Type Ia as revealed by its light-echo spectrum. *Nature* **456**, 617–619.
Kruszewski A and Semeniuk I 1999 Nearby Hipparcos eclipsing binaries for color–surface brightness calibration. *Acta Astron.* **49**, 561–575.
Kudritzki RP 1998 Quantitative spectroscopy of the brightest blue supergiant stars in galaxies. In *Stellar Astrophysics for the Local Group* (eds Aparicio A, Herrero A and Sanchez F), *Proc. VIII Canary Islands Winter School of Astrophysics*. pp. 149–262. Cambridge University Press.
Kudritzki RP and Przybilla N 2003 Blue supergiants as a tool for extragalactic distances. Theoretical concepts. In *Stellar Candles for the Extragalactic Distance Scale* (eds Alloin D and Gieren W), *Lect. Notes Phys.* **635**, 123–148.
Kudritzki RP and Puls J 2000 Winds from hot stars. *Annu. Rev Astron. Astrophys.* **38**, 613–666.
Kudritzki RP, Puls J, Lennon DJ, Venn KA, Reetz J, Najarro F, McCarthy JK and Herrero A 1999 The wind momentum–luminosity relationship of galactic A- and B-supergiants. *Astron. Astrophys.* **350**, 970–984.
Kudritzki RP, Bresolin F and Przybilla N 2003 A new extragalactic distance determination method using the flux-weighted gravity of late B and early A supergiants. *Astrophys. J.* **582**, L83–L86.
Kudritzki RP, Urbaneja MA, Bresolin F, Przybilla N, Gieren W and Pietrzyński G 2008 Quantitative spectroscopy of 24 A supergiants in the Sculptor galaxy NGC 300: flux-weighted gravity–luminosity relationship, metallicity, and metallicity gradient. *Astrophys. J.* **681**, 269–289.
Lazio TJW and Cordes JM 1998a The radial extent and warp of the ionized Galactic disk. I. A VLBA survey of extragalactic sources toward the Anticenter. *Astrophys. J. Suppl. Ser.* **115**, 225–240.
Lazio TJW and Cordes JM 1998b The radial extent and warp of the ionized Galactic disk. II. A likelihood analysis of radio-wave scattering toward the Anticenter. *Astrophys. J.* **497**, 238–252.
Leavitt HS 1908a 1777 variables in the Magellanic Clouds. *Ann. Harvard Coll. Obs.* **60**, 87–108.
Leavitt HS 1908b Ten variable stars of the Algol type. *Ann. Harvard Coll. Obs.* **60**, 109–146.
Leavitt HS and Pickering EC 1912 Periods of 25 variable stars in the Small Magellanic Cloud. *Harvard Coll. Obs. Circ.* **173**, 1–3.
Lebzelter T, Schultheis M and Melchior AL 2002 AGAPEROS: searching for variable stars in the LMC bar. II. Temporal and near-IR analysis of long-period variables. *Astron. Astrophys.* **393**, 573–583.
Lee MG, Freedman WL and Madore BF 1993 The tip of the red giant branch as a distance indicator for resolved galaxies. *Astrophys. J.* **417**, 553–559.
Lee Y-W, Demarque P and Zinn R 1990 The horizontal-branch stars in globular clusters. I. The period-shift effect, the luminosity of the horizontal branch, and the age–metallicity relation. *Astrophys. J.* **350**, 155–172.
Lindblad B 1925 Star-streaming and the structure of the stellar system. *Ark. Mat. Astron. Fys.* **19A** (21), 1–8.
Livio M 1992 Classical novae and the extragalactic distance scale. *Astrophys. J.* **393**, 516–522.
Longmore AJ, Fernley JA and Jameson RF 1986 RR Lyrae stars in globular clusters: better distances from infrared measurements? *Mon. Not. R. Astron. Soc.* **220**, 279–287.
Longmore AJ, Dixon R, Skillen I, Jameson RF and Fernley JA 1990 Globular cluster distances from the RR Lyrae log(period)–infrared magnitude relation. *Mon. Not. R. Astron. Soc.* **247**, 684–695.
Lyne AG and Graham-Smith F 2005 *Pulsar Astronomy*. Cambridge University Press.
Macri LM, Stanek KZ, Bersier D, Greenhill LJ and Reid MJ 2006 A new Cepheid distance to the maser-host galaxy NGC 4258 and its implications for the Hubble constant. *Astrophys. J.* **652**, 1133–1149.
Madore BF and Freedman WL 1995 The tip of the red giant branch as a distance indicator for resolved galaxies. II. Computer simulations. *Astron. J.* **109**, 1645–1652.

Madore BF and Freedman WL 1998 Hipparcos parallaxes and the Cepheid distance scale. *Astrophys. J.* **492**, 110–115.

Madore BF and Freedman WL 2009 Concerning the slope of the Cepheid period–luminosity relation. *Astrophys. J.* **696**, 1498–1501.

Madore BF, Rigby J, Freedman WL, Persson SE, Sturch L and Mager V 2009a The Cepheid period–luminosity relation (the Leavitt law) at mid-infrared wavelengths. III. Cepheids in NGC 6822. *Astrophys. J.* **693**, 936–939.

Madore BF, Freedman WL, Rigby J, Persson SE, Sturch L and Mager V 2009b The Cepheid period–luminosity relation (the Leavitt law) at mid-infrared wavelengths. II. Second-epoch LMC data. *Astrophys. J.* **695**, 988–995.

Madore BF, Mager V and Freedman WL 2009c Sharpening the tip of the red giant branch. *Astrophys. J.* **690**, 389–393.

Mager VA, Madore BF and Freedman WL 2008 Metallicity-corrected tip of the red giant branch distance to NGC 4258. *Astrophys. J.* **689**, 721–731.

Majaess DJ, Turner DG, Lane DJ, Henden A and Krajci T 2010 Anchoring the universal distance scale via a Wesenheit template. *J. Am. Assoc. Variable Star Obs.*, e-print: ejaavso134 (arXiv:1007.2300).

Majewski SR 1999 Stellar populations and the formation of the Milky Way. In: *Globular Clusters*, 10th Canary Islands Winter School of Astrophysics (eds Martínez Roger C, Pérez Fournón I and Sánchez F), pp. 43–107. Cambridge University Press.

Manchester RN and Taylor JH 1981 Observed and derived parameters for 330 pulsars. *Astron. J.* **86**, 1953–1973.

Marengo M, Evans NR, Barmby P, Bono G, Welch DL and Romaniello M 2010 Galactic Cepheids with Spitzer. I. Leavitt law and colors. *Astrophys. J.* **709**, 120–134.

Masson CR 1986 Angular expansion measurement with the VLA. The distance to NGC 7027. *Astrophys. J.* **302**, L27–L30.

Masson CR 1989a The structure of NGC 7027 and a determination of its distance by measurement of proper motions. *Astrophys. J.* **336**, 294–303.

Masson CR 1989b The structures of and distances to BD $+30^{\circ}3639$ and NGC 6572. *Astrophys. J.* **346**, 243–250.

Matsunaga N, Fukushi H and Nakada Y 2005 Mira variables in the Galactic bulge with OGLE–II data. *Mon. Not. R. Astron. Soc.* **364**, 117–125.

Matsunaga N, Fukushi H, Nakada Y, Tanabé T, Feast MW, Menzies JW, Ita Y, Nishiyama S, Baba D, Naoi T, Nakaya H, Kawadu T, Ishihara A and Kato D 2006 The period–luminosity relation for type II Cepheids in globular clusters. *Mon. Not. R. Astron. Soc.* **370**, 1979–1990.

Matsunaga N, Kawadu T, Nishiyama S, Nagayama T, Hatano H, Tamura M, Glass IS and Nagata T 2009 A near-infrared survey of Miras and the distance to the Galactic centre. *Mon. Not. R. Astron. Soc.* **399**, 1709–1729.

Matsunaga N, Feast MW and Soszyński I 2011 Period–luminosity relations of type II Cepheids in the Magellanic Clouds. *Mon. Not. R. Astron. Soc.*, **413**, 223–234.

McClure-Griffiths NM and Gaensler BM 2005 Constraints on the distance to SGR 1806−20 from Hɪ absorption. *Astrophys. J.* **630**, L161–L163.

McLaughlin DB 1939 The light curves of novae. *Pop. Astron.* **47**, 410–418, continued: 481–493, concluded: 538–548.

McNamara DH 1997 Luminosities of SX Phoenicis, large-amplitude Delta Scuti, and RR Lyrae stars. *Publ. Astron. Soc. Pac.* **109**, 1221–1232.

McNamara DH 2000 The high-amplitude δ Scuti stars. In *Delta Scuti and Related Stars* (eds Breger M and Montgomery M), *Astron. Soc. Pac. Conf. Ser.* **210**, 373–380.

McNamara DH and Powell JM 1990 The dwarf Cepheid instability strip. In *Proc. Confrontation Between Stellar Pulsation and Evolution* (eds Cacciari C and Clementini G), pp. 316–319. Astronomical Society of the Pacific.

Mellema G 2004 On expansion parallax distances for planetary nebulae. *Astron. Astrophys.* **416**, 623–629.

Méndez B, Davis M, Moustakas J, Newman J, Madore BF and Freedman WL 2002 Deviations from the local Hubble flow. I. The tip of the red giant branch as a distance indicator. *Astron. J.* **124**, 213–233.

Menzies J, Feast M, Whitelock P, Olivier E, Matsunaga N and da Costa G 2008 Asymptotic giant branch stars in the Phoenix dwarf galaxy. *Mon. Not. R. Astron. Soc.* **385**, 1045–1052.

Meyer M, Horns D and Zechlin H-S 2010 The Crab Nebula as a standard candle in very high-energy astrophysics. *Astron. Astrophys.* **523**, A2.

Meynet G and Maeder A 2000 Stellar evolution with rotation. V. Changes in all the outputs of massive star models. *Astron. Astrophys.* **361**, 101–120.

Meynet G, Maeder A, Schaller G, Schaerer D and Charbonnel C 1994 Grids of massive stars with high mass loss rates. V. From 12 to 120 $M_\odot$ at $Z = 0.001, 0.004, 0.008, 0.020$ and 0.040. *Astron. Astrophys. Suppl. Ser.* **103**, 97–105.

Miyoshi M, Moran J, Herrnstein J, Greenhill L, Nakai N, Diamond P and Inoue M 1995 Evidence for a black hole from high rotation velocities in a subparsec region of NGC 4258. *Nature* **373**, 127–129.

Moehler S and Bono G 2008 White dwarfs in globular clusters. In *White Dwarfs*, *Astrophys. Space Sci. Libr.*, in press (arXiv:0806.4456).

Moellenbrock GA, Claussen MJ and Goss WM 2009 A precise distance to IRAS 00420+5530 via H_2O maser parallax with the VLBA. *Astrophys. J.* **694**, 192–204.

Moskalik P and Poretti E 2002 Frequency analysis of δ Scuti and RR Lyrae stars in the OGLE–I database. In *Radial and Nonradial Pulsations as Probes of Stellar Physics* (eds Aerts C, Bedding TR and Christensen-Dalsgaard J), *Proc. Int'l Astron. Union Colloq.* **185**, 392–395.

Mould J and Sakai S 2008 The extragalactic distance scale without Cepheids. *Astrophys. J.* **686**, L75–L78.

Muno MP, Clark JS, Crowther PA, Dougherty SM, de Grijs R, Law C, McMillan SLW, Morris MR, Negueruela I, Pooley D, Portegies Zwart S and Yusef-Zadeh F 2006 A neutron star with a massive progenitor in Westerlund 1. *Astrophys. J.* **636**, L41–L44.

Nakai N, Inoue M, Miyazawa K, Miyoshi M and Hall P 1995 Search for extremely-high-velocity H_2O maser emission Seyfert galaxies. *Publ. Astron. Soc. Jpn* **47**, 771–799.

Nemec J and Mateo M 1990a Dwarf Cepheids in the Carina dwarf galaxy. In *Proc. Evolution of the Universe of Galaxies* (ed. Kron RG), pp. 134–137. Astronomical Society of the Pacific.

Nemec J and Mateo M 1990b SX Phoenicis stars. In *Proc. Confrontation Between Stellar Pulsation and Evolution* (eds Cacciari C and Clementini G), pp. 64–84. *Astronomical Society of the Pacific.*

Nemec JM, Wehlau A and Mendes de Oliveira C 1988 Variable stars in the Ursa Minor dwarf galaxy. *Astron. J.* **96**, 528–559.

Nemec JM, Linnell Nemec AF and Lutz TE 1994 Period–luminosity–metallicity relations, pulsation modes, absolute magnitudes, and distances for population II variable stars. *Astron. J.* **108**, 222–246.

Newman JA, Ferrarese L, Stetson PB, Maoz E, Zepf SE, Davis M, Freedman WL and Madore BF 2001 A revised Cepheid distance to NGC 4258 and a test of the distance scale. *Astrophys. J.* **553**, 562–574.

Ngeow C-C and Kanbur SM 2005 The linearity of the Wesenheit function for the Large Magellanic Cloud Cepheids. *Mon. Not. R. Astron. Soc.* **360**, 1033–1039.

Ngeow C-C and Kanbur SM 2008 The period–luminosity relation for the Large Magellanic Cloud Cepheids derived from Spitzer archival data. *Astrophys. J.* **679**, 76–85.

Ngeow C, Kanbur S, Ghobrial L, Neilson H and Macri L 2009 IRAC band period–luminosity relations from LMC Cepheids: application to three nearby galaxies. In *Stellar Pulsation: Challenges for Theory and Observation* (eds Guzik JA and Bradley PA), *Proc. Am. Inst. Phys. Conf.* **1170**, 37–39.

Nissen PE and Schuster WJ 2009 The Galactic disk–halo transition. Evidence from stellar abundances. In *The Galaxy Disk in Cosmological Context* (eds Andersen J, Bland-Hawthorn J and Nordström B), *Proc. Int'l Astron. Union Symp.* **254**, 103–108.

Nissen PE and Schuster WJ 2010 Two distinct halo populations in the solar neighborhood. Evidence from stellar abundance ratios and kinematics. *Astron. Astrophys.* **511**, L10.

North P, Gauderon R, Barblan F and Royer F 2010 VLT multi-object spectroscopy of 33 eclipsing binaries in the Small Magellanic Cloud. New distance and depth of the SMC, and a record-breaking apsidal motion. *Astron. Astrophys.* **520**, A74.

O'Dell CR, Henney WJ and Sabbadin F 2009 Tangential motions and spectroscopy within NGC 6720, the Ring Nebula. *Astron. J.* **137**, 3815–3825.

Olling RP 2007 Accurate extragalactic distances and dark energy: anchoring the distance scale with rotational parallaxes. *Mon. Not. R. Astron. Soc.* **378**, 1385–1399.
Ostrov PG, Lapasset E and Morrell NI 2000 Determination of masses and radii of the massive eclipsing binary system HV 2543 in the Large Magellanic Cloud. *Astron. Astrophys.* **356**, 935–942.
Ostrov PG, Morrell NI and Lapasset E 2001 Investigation of the Magellanic eclipsing binary HV 2241. *Astron. Astrophys.* **377**, 972–980.
Paczyński B 1997 Detached eclipsing binaries as primary distance and age indicators. In *The Extragalactic Distance Scale* (ed. Livio M), pp. 273–280. Cambridge University Press.
Paczyński B and Stanek KZ 1998 Galactocentric distance with the Optical Gravitational Lensing Experiment and Hipparcos red clump stars. *Astrophys. J.* **494**, L219–L222.
Palen S, Balick B, Hajian AR, Terzian Y, Bond HE and Panagia N 2002 Hubble Space Telescope expansion parallaxes of the planetary nebulae NGC 6578, NGC 6884, NGC 6891, and IC 2448. *Astron. J.* **123**, 2666–2675.
Panagia N, Gilmozzi R, Macchetto F, Adorf H-M and Kirshner RP 1991 Properties of the SN 1987A circumstellar ring and the distance to the Large Magellanic Cloud. *Astrophys. J.* **380**, L23–L26.
Pastorello A and Patat F 2008 Echo from an ancient supernova. *Nature* **456**, 587–589.
Perrine CD 1902 Observations of the nebulosity about Nova Persei for polarization effects. *Astrophys. J.* **16**, 257–262.
Phillips JP 2002 The distances of planetary nebulae: a scale based upon nearby sources. *Astrophys. J. Suppl. Ser.* **139**, 199–217.
Phillips JP 2006 The problem with reddening distances to planetary nebulae. *Rev. Mex. Astron. Astrophys.* **42**, 229–239.
Pierce MJ, McClure RD and Racine R 1992 High-resolution imaging of Virgo Cluster galaxies. I. The distance based on the brightest stars in NGC 4571. *Astrophys. J.* **393**, 523–529.
Pietrzyński G and Gieren W 2006 The Araucaria project. *Mem. Soc. Astron. It.* **77**, 239–244.
Pietrzyński G and Gieren W 2009 The Araucaria project. Improving the cosmic distance scale. *Rev. Mex. Astron. Astrophys.* **35**, 233–234.
Pietrzyński G, Górski M, Gieren W, Laney D, Udalski A and Ciechanowska A 2010 The Araucaria Project. Population effects on the *V*- and *I*-band magnitudes of red clump stars. *Astron. J.* **140**, 1038–1042.
Plait PC, Lundqvist P, Chevalier RA and Kirshner RP 1995 HST observations of the ring around SN 1987A. *Astrophys. J.* **439**, 730–751.
Planck MK, von Laue M and Gaynor F (transl.) 1949 *Scientific Autobiography and Other Papers*. Philosophical Library.
Pottasch SR and Acker A 1998 A comparison of Hipparcos parallaxes with planetary nebulae spectroscopic distances. *Astron. Astrophys.* **329**, L5–L8.
Pottasch SR and Zijlstra AA 1992 Shklovsky distances to Galactic bulge planetary nebulae. *Astron. Astrophys.* **256**, 251–254.
Predehl P, Burwitz V, Paerels F and Trümper J 2000 Chandra measurement of the geometrical distance to Cyg X-3 using its X-ray scattering halo. *Astron. Astrophys.* **357**, L25–L28.
Prentice AJR and ter Haar D 1969 On Hɪɪ regions and pulsar distances. *Mon. Not. R. Astron. Soc.* **146**, 423–444.
Prialnik D and Kovetz A 1995 An extended grid of multicycle nova evolution models. *Astrophys. J.* **445**, 789–810.
Prieto JL, Stanek KZ, Kochanek CS and Weisz DR 2008 LBT discovery of a yellow supergiant eclipsing binary in the dwarf galaxy Holmberg IX. In *Massive Stars as Cosmic Engines* (eds Bresolin F, Crowther PA and Puls J), *Proc. Int'l Astron. Union Symp.* **250**, 333–338.
Prša A and Zwitter T 2005 A computational guide to physics of eclipsing binaries. I. Demonstrations and perspectives. *Astrophys. J.* **628**, 426–438.
Przybilla N, Butler K, Becker SR and Kudritzki RP 2006 Quantitative spectroscopy of BA-type supergiants. *Astron. Astrophys.* **445**, 1099–1126.
Pulone M 1992 AGB clump as standard candle. *Mem. Soc. Astron. It.* **63**, 485–490.
Puls J, Repolust T, Hoffmann TL, Jokuthy A and Venero ROJ 2003 Advances in radiatively driven wind models. In *A Massive Star Odyssey: From Main Sequence to Supernova* (eds van der Hucht K, Herrero A and Esteban C), *Proc. Int'l Astron. Union Symp.* **212**, 61–69.

Reid IN 1997 Younger and brighter – new distances to globular clusters based on Hipparcos parallax measurements of local subdwarfs. *Astron. J.* **114**, 161–179.
Reid IN 1998 Hipparcos subdwarf parallaxes – metal-rich clusters and the thick disk. *Astron. J.* **115**, 204–228.
Reid MJ 1993 The distance to the center of the Galaxy. *Annu. Rev. Astron. Astrophys.* **31**, 345–372.
Reid MJ, Schneps MH, Moran JM, Gwinn CR, Genzel R, Downes D and Roennaeng B 1988a The distance to the center of the Galaxy: H_2O maser proper motions in Sagittarius B2(N). *Astrophys. J.* **330**, 809–816.
Reid MJ, Gwinn CR, Moran JM and Matthews AH 1988b R_0 from H_2O maser proper motions in Sgr B2–Middle. *Bull. Am. Astron. Soc.* **20**, 1017.
Reid MJ, Braatz JA, Condon JJ, Greenhill LJ, Henkel C and Lo KY 2009a The Megamaser Cosmology Project. I. Very long baseline interferometric observations of UGC 3789. *Astrophys. J.* **695**, 287–291.
Reid MJ, Menten KM, Zheng XW, Brunthaler A and Xu Y 2009b A trigonometric parallax of Sgr B2. *Astrophys. J.* **705**, 1548–1553.
Rejkuba M 2004 The distance to the giant elliptical galaxy NGC 5128. *Astron. Astrophys.* **413**, 903–912.
Renzini A 1991 Magellanic globulars as cosmological tools. In *The Magellanic Clouds* (eds Haynes R and Milne D), *Proc. Int'l Astron. Union Symp.* **148**, 165–169.
Renzini A and Fusi Pecci F 1988 Tests of evolutionary sequences using color–magnitude diagrams of globular clusters. *Annu. Rev. Astron. Astrophys.* **26**, 199–244.
Renzini A, Bragaglia A, Ferraro FR, Gilmozzi R, Ortolani S, Holberg JB, Liebert J, Wesemael F and Bohlin RC 1996 The white dwarf distance to the globular cluster NGC 6752 (and its age) with the Hubble Space Telescope. *Astrophys. J.* **465**, L23–L26.
Rest A, Suntzeff NB, Olsen K, Prieto JL, Smith RC, Welch DL, Becker A, Bergmann M, Clocchiatti A, Cook K, Garg A, Huber M, Miknaitis G, Minniti D, Nikolaev S and Stubbs C 2005 Light echoes from ancient supernovae in the Large Magellanic Cloud. *Nature* **438**, 1132–1134.
Rest A, Matheson T, Blondin S, Bergmann M, Welch DL, Suntzeff NB, Smith RC, Olsen K, Prieto JL, Garg A, Challis P, Stubbs C, Hicken M, Modjaz M, Wood-Vasey WM, Zenteno A, Damke G, Newman A, Huber M, Cook KH, Nikolaev S, Becker AC, Miceli A, Covarrubias R, Morelli L, Pignata G, Clocchiatti A, Minniti D and Foley RJ 2008 Spectral identification of an ancient supernova using light echoes in the Large Magellanic Cloud. *Astrophys. J.* **680**, 1137–1148.
Reynolds RJ 1989 The column density and scale height of free electrons in the Galactic disk. *Astrophys. J.* **339**, L29–L32.
Ribas I and Jordi C 2003 A project to determine an accurate distance to M31 using eclipsing binaries as standard candles. *Rev. Mex. Astron. Astrophys.* **16**, 150–153.
Ribas I, Fitzpatrick EL, Maloney FP, Guinan EF and Udalski A 2002 Fundamental properties and distances of Large Magellanic Cloud eclipsing binaries. III. EROS 1044. *Astrophys. J.* **574**, 771–782.
Ribas I, Jordi C, Vilardell F, Giménez Á and Guinan EF 2004 A program to determine a direct and accurate distance to M31 from eclipsing binaries. *New Astron.* **48**, 755–758.
Ribas I, Jordi C, Vilardell F, Fitzpatrick EL, Hilditch RW and Guinan EF 2005 First determination of the distance and fundamental properties of an eclipsing binary in the Andromeda galaxy. *Astrophys. J.* **635**, L37–L40.
Richer HB, Fahlman GG, Ibata RA, Stetson PB, Bell RA, Bolte M, Bond HE, Harris WE, Hesser JE, Mandushev G, Pryor C and Vandenberg DA 1995 Hubble Space Telescope observations of white dwarfs in the globular cluster M4. *Astrophys. J.* **451**, L17–L20.
Richer HB, Fahlman GG, Ibata RA, Pryor C, Bell RA, Bolte M, Bond HE, Harris WE, Hesser JE, Holland S, Ivanans N, Mandushev G, Stetson PB and Wood MA 1997 White dwarfs in globular clusters: Hubble Space Telescope observations of M4. *Astrophys. J.* **484**, 741–760.
Richer HB, Anderson J, Brewer J, Davis S, Fahlman GG, Hansen BMS, Hurley J, Kalirai JS, King IR, Reitzel D, Rich RM, Shara MM and Stetson PB 2006 Probing the faintest stars in a globular star cluster. *Science* **313**, 936–940.
Ritchey GW 1902 Nebulosity about Nova Persei. Recent photographs. *Astrophys. J.* **15**, 129–131.

Ritter A 1879 *Ann. Phys. Chem. Neue Folge* **8**, 157.

Rizzi L, Held EV, Saviane I, Tully RB and Gullieuszik M 2007 The distance to the Fornax dwarf spheroidal galaxy. *Mon. Not. R. Astron. Soc.* **380**, 1255–1260.

Rozanski R and Rowan-Robinson M 1994 The accuracy of the brightest stars in galaxies as distance indicators. *Mon. Not. R. Astron. Soc.* **271**, 530–552.

Rucinski SM 2004 Contact binary stars of the W UMa type as distance tracers. *New Astron. Rev.* **48**, 703–709.

Russell SC 1998 RV Tau variables in globular clusters: clues on their nature? *Publ. Astron. Soc. Aust.* **15**, 189–193.

Saha A, Thim F, Tammann GA, Reindl B and Sandage A 2006 Cepheid distances to SNe Ia host galaxies based on a revised photometric zero point of the HST WFPC2 and new PL relations and metallicity corrections. *Astrophys. J. Suppl. Ser.* **165**, 108–137.

Sakai S, Madore BF and Freedman WL 1996 Tip of the red giant branch distances to galaxies. III. The dwarf galaxy Sextans A. *Astrophys. J.* **461**, 713–723.

Salaris M and Cassisi S 1997 The ‘tip’ of the red giant branch as a distance indicator: results from evolutionary models. *Mon. Not. R. Astron. Soc.* **289**, 406–414.

Salaris M and Cassisi S 1998 A new analysis of the red giant branch ‘tip’ distance scale and the value of the Hubble constant. *Mon. Not. R. Astron. Soc.* **298**, 166–178.

Salaris M, Cassisi S, García-Berro E, Isern J and Torres S 2001 On the white dwarf distances to Galactic globular clusters. *Astron. Astrophys.* **371**, 921–931.

Salaris M, Held EV Ortolani S, Gullieuszik M and Momany Y 2007 Deep near-infrared photometry of the globular cluster 47 Tucanae. Reconciling theory and observations. *Astron. Astrophys.* **476**, 243–253.

Sandage A 1993 The Oosterhoff period–metallicity relation for RR Lyrae stars at the blue fundamental edge of the instability strip. *Astron. J.* **106**, 687–702.

Sandage A 2006 On the predicted and observed color boundaries of the RR Lyrae instability strip as a function of metallicity. *Astron. J.* **131**, 1750–1765.

Sandage A and Tammann GA 2006 Absolute magnitude calibrations of Population I and II Cepheids and other pulsating variables in the instability strip of the Hertzsprung–Russell diagram. *Annu. Rev. Astron. Astrophys.* **44**, 93–140.

Sandage A and Tammann GA 2008 Temperature differences in the Cepheid instability strip require differences in the period–luminosity relation in slope and zero point. *Astrophys. J.* **686**, 779–784.

Sandage A, Tammann GA and Reindl B 2004 New period–luminosity and period–color relations of classical Cepheids. II. Cepheids in LMC. *Astron. Astrophys.* **424**, 43–71.

Sanna N, Bono G, Stetson PB, Monelli M, Pietrinferni A, Drozdovsky I, Caputo F, Cassisi S, Gennaro M, Prada Moroni PG, Buonanno R, Corsi CE, Degl’Innocenti S, Ferraro I, Iannicola G, Nonino M, Pulone L, Romaniello M and Walker AR 2008 On the distance and reddening of the starburst galaxy IC 10. *Astrophys. J.* **688**, L69–L73.

Santander-García M, Corradi RLM, Mampaso A, Morisset C, Munari U, Schirmer M, Balick B and Livio M 2008 Hen 2–104: a close-up look at the Southern Crab. *Astron. Astrophys.* **485**, 117–126.

Sarajedini A 1999 WIYN open cluster study. III. The observed variation of the red clump luminosity and color with metallicity and age. *Astron. J.* **118**, 2321–2326.

Sarajedini A, Lee Y-W and Lee D-H 1995 Ages for globular clusters with predominantly red horizontal branches. *Astrophys. J.* **450**, 712–721.

Sarajedini A, Grocholski AJ, Levine J and Lada E 2002 *K*-band red clump distances to the Large Magellanic Cloud clusters Hodge 4 and NGC 165. *Astron. J.* **124**, 2625–2632.

Schmidt T 1957 Die Lichtkurven–Leuchtkraft-Beziehung neuer Sterne. *Z. Astrophys.* **41**, 182–201.

Schneps MH, Moran JM, Genzel R, Reid MJ, Lane AP and Downes D 1981 Proper motions and distances of H_2O maser sources. III. W51 North. *Astrophys. J.* **249**, 124–133.

Schönberner D, Jacob R and Steffen M 2005 The evolution of planetary nebulae. III. Internal kinematics and expansion parallaxes. *Astron. Astrophys.* **441**, 573–588.

Schultheis M, Glass IS and Cioni M-R 2004 Late-type giant variables in NGC 6522, LMC and SMC. How do they differ? *Astron. Astrophys.* **427**, 945–958.

Ségransan D, Kervella P, Forveille T and Queloz D 2003 First radius measurements of very low mass stars with the VLTI. *Astron. Astrophys.* **397**, L5–L8.

Seidel E, Da Costa GS and Demarque P 1987 Intermediate-age core helium burning stars and the distance to the Magellanic Clouds. *Astrophys. J.* **313**, 192–199.

Shabun K, Richichi A, Munari U, Siviero A and Paczyński B 2008 The VLTI as a tool to study eclipsing binaries for an improved distance scale. In *A Giant Step: From Milli- to Micro-Arcsecond Astrometry* (eds Jin WJ, Platais I and Perryman MAC), *Proc. Int'l Astron. Union Symp.* **248**, 118–119.

Shabun K, Richichi A, Munari U, Siviero A and Paczyński B 2009 The VLTI as a tool to study eclipsing binaries for an improved distance scale. In *Science with the VLT in the ELT Era* (ed. Moorwood A), *Astrophys. Space Sci. Proc.*, pp. 211–216.

Shafter AW 1997 On the nova rate in the Galaxy. *Astrophys. J.* **487**, 226–236.

Shanks T, Tanvir NR, Major JV, Doel AP, Dunlop CN and Myers RM 1992 High-resolution imaging of Virgo cluster galaxies. II. Detection of the brightest stars in NGC 4523. *Mon. Not. R. Astron. Soc.* **256**, 29P–32P.

Shapley H and Walton ML 1927 Investigations of Cepheid variables. I. The period–spectrum relation. *Harvard Coll. Obs. Circ.* **313**, 1–7.

Shara MM 1994 Masses of white dwarfs in O–Ne–Mg novae: observational constraints, Galactic ^{26}Al, ^{22}Na gamma-rays, and M31 novae. *Astron. J.* **107**, 1546–1550.

Shklovsky IS 1956 Once more on the distances to planetary nebulae and the evolution of their nuclei. *Astron. Zh.* **33**, 222 (English transl. *Astron. Circ.* **306**, 2–4).

Sigurdsson S, Richer HB, Hansen BM, Stairs IH and Thorsett SE 2003 A young white dwarf companion to pulsar B1620–26: evidence for early planet formation. *Science* **301**, 193–196.

Smith HA, Church JA, Fournier J, Lisle J, Gay P, Kolenberg K, Carney BW, Dick I, Peterson RC and Hakes B 2003 The Blažhko effect of RR Lyrae in 1996. *Publ. Astron. Soc. Pac.* **115**, 43–48.

Sollima A, Cacciari C and Valenti E 2006 The RR Lyrae period–K-luminosity relation for globular clusters: an observational approach. *Mon. Not. R. Astron. Soc.* **372**, 1675–1680.

Sonneborn G, Fransson C, Lundqvist P, Cassatella A, Gilmozzi R, Kirshner RP, Panagia N and Wamsteker W 1997 The evolution of ultraviolet emission lines from circumstellar material surrounding SN 1987A. *Astrophys. J.* **477**, 848–864.

Soszyński I, Udalski A, Szymański MK, Kubiak M, Pietrzyński G, Wyrzykowski Ł, Szewczyk O, Ulaczyk K and Poleski R 2008 The Optical Gravitational Lensing Experiment. The OGLE–III catalog of variable stars. II. Type II Cepheids and anomalous Cepheids in the Large Magellanic Cloud. *Acta Astron.* **58**, 293–312.

Sparks WB 1994 A direct way to measure the distances of galaxies. *Astrophys. J.* **433**, 19–28.

Sparks WB 1996 Geometric distance measurement. II. *Astrophys. J.* **470**, 195–200.

Sparks WB, Macchetto F, Panagia N, Boffi FR, Branch D, Hazen ML and della Valle M 1999 Evolution of the light echo of SN 1991T. *Astrophys. J.* **523**, 585–592.

Sparks WB, Bond HE, Cracraft M, Levay Z, Crause LA, Dopita MA, Henden AA, Munari U, Panagia N, Starrfield SG, Sugerman BE, Wagner RM and White RL 2008 V838 Monocerotis: a geometric distance from Hubble Space Telescope polarimetric imaging of its light echo. *Astron. J.* **135**, 605–617.

Stanek KZ 1995 Magnitude offset between lensed stars and observed stars: a new probe of the structure of the Galactic bar. *Astrophys. J.* **441**, L29–L32.

Stanek KZ and Garnavich PM 1998 Distance to M31 with the Hubble Space Telescope and Hipparcos red clump stars. *Astrophys. J.* **503**, L131–L134.

Stanek KZ, Mateo M, Udalski A, Szymański M, Kałużny J and Kubiak M 1994 Color–magnitude diagram distribution of the bulge red clump stars: evidence for the Galactic bar. *Astrophys. J.* **429**, L73–L76.

Stanek KZ, Udalski A, Szymański M, Kałużny J, Kubiak M, Mateo M and Krzemiński W 1997 Modeling the Galactic bar using red clump giants. *Astrophys. J.* **477**, 163–175.

Stanghellini L, Shaw RA and Villaver E 2008 The Magellanic Cloud calibration of the Galactic planetary nebula distance scale. *Astrophys. J.* **689**, 194–202.

Strom SE, Strom KM, Rood RT and Iben Jr I 1970 On the evolutionary status of stars above the horizontal branch in globular clusters. *Astron. Astrophys.* **8**, 243–250.

Sugerman BEK 2005 Discovery of a light echo from SN 2003gd. *Astrophys. J.* **632**, L17–L20.

Sugerman BEK and Crotts APS 2002 Multiple light echoes from SN 1993J. *Astrophys. J.* **581**, L97–L100.

Sweigart AV 1978 Survey of some recent stellar-evolution calculations. In *The HR Diagram: the 100th Anniversary of Henry Norris Russell* (eds Philip AGD and Hayes DS), *Proc. Int'l Astron. Union Symp.* **80**, 333–343.

Sweigart AV and Gross PG 1978 Evolutionary sequences for red giant stars. *Astrophys. J. Suppl. Ser.* **36**, 405–437.

Szeidl B 1988 RR Lyrae stars: beat and Blažhko effect. In *Proc. Multimode Stellar Pulsations* (eds Kovacs G, Szabados L and Szeidl B), pp. 45–66.

Tabur V, Kiss LL and Bedding TR 2009 Hipparcos calibration of the tip of the red giant branch. *Astrophys. J.* **703**, L72–L75.

Tammann GA, Sandage A and Reindl B 2003 New period–luminosity and period–color relations of classical Cepheids. I. Cepheids in the Galaxy. *Astron. Astrophys.* **404**, 423–448.

Tammann GA, Sandage A and Reindl B 2008 Comparison of distances from RR Lyrae stars, the tip of the red giant branch, and classical Cepheids. *Astrophys. J.* **679**, 52–70.

Tanvir NR, Hendry MA, Watkins A, Kanbur SM, Berdnikov LN and Ngeow C-C 2005 Determination of Cepheid parameters by light-curve template fitting. *Mon. Not. R. Astron. Soc.* **363**, 749–762.

Taylor JH and Cordes JM 1993 Pulsar distances and the Galactic distribution of free electrons. *Astrophys. J.* **411**, 674–684.

Terzian Y and Teymourian A 2005 The life and death of planetary nebulae. In *The Initial Mass Function 50 Years Later* (eds Corbelli E, Palle F and Zinnecker H), *Astrophys. Space Sci. Libr.* **327**, 521–526.

Testa V, Chieffi A, Limongi M, Andreuzzi G and Marconi G 2004 The distance to NGC 5904 (M 5) via the subdwarf main sequence fitting method. *Astron. Astrophys.* **421**, 603–612.

Thomas H-C 1967 Sternentwicklung VIII. Der Helium-Flash bei einem Stern von 1. 3 Sonnenmassen. *Z. Astrophys.* **67**, 420–455.

Thompson IB, Kaluzny J, Rucinski SM, Krzeminski W, Pych W, Dotter A and Burley GS 2010 The Cluster AgeS Experiment (CASE). IV. Analysis of the eclipsing binary V69 in the globular cluster 47 Tuc. *Astron. J.* **139**, 329–341.

Thompson TWJ, Rothschild RE and Tomsick JA 2006 Geometric distance determination using type I X-ray bursts. *Astrophys. J.* **650**, 1063–1069.

Tinsley BM 1980 Evolution of the stars and gas in galaxies. *Fundam. Cosmic Phys.* **5**, 287–388.

Todd I, Pollacco D, Skillen I, Bramich DM, Bell S and Augusteijn T 2005 A survey of eclipsing binary stars in the eastern spiral arm of M31. *Mon. Not. R. Astron. Soc.* **362**, 1006–1014.

Toor A and Seward FD 1974 The Crab Nebula as a calibration source for X-ray astronomy. *Astron. J.* **79**, 995–999.

Torres G, Andersen J and Gimenez A 2010 Accurate masses and radii of normal stars: modern results and applications. *Astron. Astrophys. Rev.* **18**, 67–126.

Trümper J and Schönfelder V 1973 Distance determination of variable X-ray sources. *Astron. Astrophys.* **25**, 445–450.

Tully RB and Wolff SC 1984 Mass loss in A and B supergiants and the extragalactic distance scale. *Astrophys. J.* **281**, 67–75.

U V, Urbaneja MA, Kudritzki RP, Jacobs BA, Bresolin F and Przybilla N 2009 A new distance to M33 using blue supergiants and the FGLR method. *Astrophys. J.* **704**, 1120–1134.

Udalski A 2000 The Optical Gravitational Lensing Experiment: red clump stars as a distance indicator. *Astrophys. J.* **531**, L25–L28.

Udalski A, Szymański M, Kubiak M, Pietrzyński G, Woźniak P and Żebruń K 1998 Optical Gravitational Lensing Experiment. Distance to the Magellanic Clouds with the red clump stars: are the Magellanic Clouds 15% closer than generally accepted? *Acta Astron.* **48**, 1–17.

Urbaneja MA, Herrero A, Kudritzki RP, Najarro F, Smartt SJ, Puls J, Lennon DJ and Corral LJ 2005 Blue luminous stars in nearby galaxies: quantitative spectral analysis of M33 B-type supergiant stars. *Astrophys. J.* **635**, 311–335.

Urbaneja MA, Kudritzki RP, Bresolin F, Przybilla N, Gieren W and Pietrzyński G 2008 The Araucaria project: the Local Group galaxy WLM. Distance and metallicity from quantitative spectroscopy of blue supergiants. *Astrophys. J.* **684**, 118–135.

Utrobin V 2007 Supernova 1987A: the ejecta mass and the explosion energy. In *Supernova 1987A, 20 Years After: Supernovae and Gamma-Ray Bursters* (eds Immler S, Weiler K and McCray R), *Am. Inst. Phys. Conf. Proc.* **937**, 25–32.

Valenti E, Ferraro FR and Origlia L 2004 Red giant branch in near-infrared colour–magnitude diagrams. I. Calibration of photometric indices. *Mon. Not. R. Astron. Soc.* **351**, 1204–1214.

VandenBerg DA, Swenson FJ, Rogers FJ, Iglesias CA and Alexander DR 2000 Models for old, metal-poor stars with enhanced α-element abundances. I. Evolutionary tracks and ZAHB loci; observational constraints. *Astrophys. J.* **532**, 430–452.

VandenBerg DA, Casagrande L and Stetson PB 2010 An examination of recent transformations to the $BV(RI)_C$ photometric system from the perspective of stellar models for old stars. *Astron. J.* **140**, 1020–1037.

van den Bergh S 1965 Light echoes from ancient supernovae. *Publ. Astron. Soc. Pac.* **77**, 269–271.

Van Hamme W 1993 New limb-darkening coefficients for modeling binary star light curves. *Astron. J.* **106**, 2096–2117.

van Leeuwen F 2005 Rights and wrongs of the Hipparcos data. A critical quality assessment of the Hipparcos catalogue. *Astron. Astrophys.* **439**, 805–822.

van Leeuwen F 2007 Hipparcos, the new reduction of the raw data. *Astrophys. Space Sci. Libr.* **350**, Springer.

van Straten W, Bailes M, Britton M, Kulkarni SR, Anderson SB, Manchester RN and Sarkissian J 2001 A test of general relativity from the three-dimensional orbital geometry of a binary pulsar. *Nature* **412**, 158–160.

Verbiest JPW, Bailes M, Coles WA, Hobbs GB, van Straten W, Champion DJ, Jenet FA, Manchester RN, Bhat NDR, Sarkissian JM, Yardley D, Burke-Spolaor S, Hotan AW and You XP 2009 Timing stability of millisecond pulsars and prospects for gravitational-wave detection. *Mon. Not. R. Astron. Soc.* **400**, 951–968.

Vilardell F, Ribas I and Jordi C 2006 Eclipsing binaries suitable for distance determination in the Andromeda galaxy. 2006 *Astron. Astrophys.* **459**, 321–331.

Vlemmings WHT, Cordes JM, Chatterjee S 2004 Separated at birth: the origin of the pulsars B2020+28 and B2021+51 in the Cygnus Superbubble. *Astrophys. J.* **610**, 402–410.

von Weizsäcker CF 1938 Über Elementumwandlungen in Innern der Sterne. II. *Phys. Z.* **39**, 633–646.

Vrba FJ, Henden AA, Luginbuhl CB, Guetter HH, Hartmann DH and Klose S 2000 The discovery of an embedded cluster of high-mass stars near SGR 1900+14. *Astrophys. J.* **533**, L17–L20.

Wallerstein G 2002 The Cepheids of Population II and related stars. *Publ. Astron. Soc. Pac.* **114**, 689–699.

Warner B 1995 Systematics of surperoutbursts in dwarf novae. *Astrophys. Space Sci.* **226**, 187–211.

Watson WD and Wallin BK 1994 Evidence from masers for a rapidly rotating disk at the nucleus of NGC 4258. *Astrophys. J.* **432**, L35–L38.

Wehlau A and Bohlender D 1982 An investigation of period changes in cluster BL Herculis stars. *Astron. J.* **87**, 780–791.

Weisskopf MC, Guainazzi M, Jahoda K, Shaposhnikov N, O'Dell SL, Zavlin VE, Wilson-Hodge C and Elsner RF 2010 On calibrations using the Crab Nebula and models of the nebular X-ray emission. *Astrophys. J.* **713**, 912–919.

Wesselink AJ 1946 The observations of brightness, colour and radial velocity of δ Cephei and the pulsation hypothesis. *Bull. Astron. Inst. Neth.* **10**, 91–98.

Wheeler JC, Maund JR and Couch SM 2008 The shape of Cas A. *Astrophys. J.* **677**, 1091–1099.

Whitelock PA, Feast MW and van Leeuwen F 2008 AGB variables and the Mira period–luminosity relation. *Mon. Not. R. Astron. Soc.* **386**, 313–323.

Whitelock PA, Menzies JW, Feast MW, Matsunaga N, Tanabé T and Ita Y 2009 Asymptotic giant branch stars in the Fornax dwarf spheroidal galaxy. *Mon. Not. R. Astron. Soc.* **394**, 795–809.

Wilson RE 2004 Semi-detached binaries as probes of the Local Group. *New Astron. Rev.* **48**, 695–701.

Wilson RE 2007 Eclipsing binary flux units and the distance problem. In *7th Pac. Rim Conf. on Stellar Astrophysics* (eds Kang YW, Lee H-W, Leung K-C, and Cheng K-S), *Astron. Soc. Pac. Conf. Ser.* **362**, 3–14.

Wilson RE 2008 Eclipsing binary solutions in physical units and direct distance estimation. *Astrophys. J.* **672**, 575–589.

Wilson-Hodge CA, Cherry ML, Baumgartner WH, Beklen E, Narayana Bhat P, Briggs MS, Camero-Arranz A, Case GL, Chaplin V, Connaughton V, Finger MH, Gehrels N, Greiner J, Jahoda K, Jenke P, Kippen RM, Kouveliotou C, Krimm HA, Kuulkers E, Meegan CA, Natalucci L, Paciesas WS, Preece R, Rodi JC, Shaposhnikov N, Skinner GK, Swartz D and von Kienlin A 2010 When a standard candle flickers. *Astrophys. J.* **727**, L40–L43.

Wood PR, Alcock C, Allsman RA, Alves D, Axelrod TS, Becker AC, Bennett DP, Cook KH, Drake AJ, Freeman KC, Griest K, King LJ, Lehner MJ, Marshall SL, Minniti D, Peterson BA, Pratt MR, Quinn PJ, Stubbs CW, Sutherland W, Tomaney A, Vandehei T and Welch DL 1999 MACHO observations of LMC red giants: Mira and semi-regular pulsators, and contact and semi-detached binaries. In *Asymptotic Giant Branch Stars* (eds Le Bertre T, Lebre A and Waelkens C), *Proc. Int'l Astron. Union Symp.* **191**, 151–158.

Wyithe JSB and Wilson RE 2002 Photometric solutions for semidetached eclipsing binaries: selection of distance indicators in the Small Magellanic Cloud. *Astrophys. J.* **571**, 293.

Xiang J, Lee JC and Nowak MA 2007 Using the X-ray dust scattering halo of 4U 1624−490 to determine distance and dust distributions. *Astrophys. J.* **660**, 1309–1318.

Xu Y, Reid MJ, Menten KM and Zheng XW 2006 Search for compact extragalactic radio sources near massive star-forming regions. 2006 *Astrophys. J. Suppl. Ser.* **166**, 526–533.

Yang M and Jiang BW 2010 Red supergiant stars in the Large Magellanic Cloud: I. The period–luminosity relation. *Astrophys. J.* **727**, 53.

Yi S, Demarque P, Kim Y-C, Lee Y-W, Ree CH, Lejeune T and Barnes S 2001 Toward better age estimates for stellar populations: the Y^2 isochrones for solar mixture. *Astrophys. J. Suppl. Ser.* **136**, 417–437.

Yun H-Y and Lee S-W 1979 The helium abundances of globular clusters. *J. Kor. Astron. Soc.* **12**, 17–25.

Zijlstra AA, van Hoof PAM and Perley RA 2008 The evolution of NGC 7027 at radio frequencies: a new determination of the distance and core mass. *Astrophys. J.* **681**, 1296–1309.

Zinn R and Searle L 1976 The masses of the anomalous Cepheids in the Draco system. *Astrophys. J.* **209**, 734–747.

Zoccali M, Renzini A, Ortolani S, Bragaglia A, Bohlin R, Carretta E, Ferraro FR, Gilmozzi R, Holberg JB, Marconi G, Rich RM and Wesemael F 2001 The white dwarf distance to the globular cluster 47 Tucanae and its age. *Astrophys. J.* **553**, 733–743.

Zsoldos E 1998 No RV Tauri stars in globular clusters? *Acta Astron.* **48**, 775–788.

Zwicky F 1936 Life–luminosity relation for novae. *Publ. Astron. Soc. Pac.* **48**, 191–194.

4

Reaching Virgo Cluster Distances and Beyond

An ocean traveler has even more vividly the impression that the ocean is made of waves than that it is made of water.

– Sir Arthur Stanley Eddington (1882–1944), English astronomer and physicist

4.1 The Hubble Space Telescope Key Project

The '*Hubble Space Telescope (HST)* Key Project' has perhaps been the most important programme in the entire *HST* portfolio. The project's ultimate aim – as well as one of the main scientific rationales for embarking on the *HST* project as such – was to measure the Hubble constant (Section 5.1) to an accuracy of 10% or better. The Hubble constant, H_0, the expansion rate of the Universe at the present time, relates a galaxy's galactocentric recessional velocity, v, to its distance, d, through $v = H_0 d$ (Hubble 1929), in the absence of so-called 'peculiar motions', which may be caused by e.g. local over- or underdensities such as galaxy clusters and voids.

The Key Project set out to achieve this aim based on a Cepheid period–luminosity calibration (see Section 3.5.2) of independent, secondary distance determination methods at distances between ~60 and 400 Mpc (Freedman *et al.* 2001 and references therein). To minimize the effects of systematic errors, they used Type Ia and II-P supernovae (see Section 5.2), the Tully–Fisher relation (TFR; see Section 4.5), the **Fundamental Plane** (FP) of elliptical galaxies (see Section 4.6) and surface brightness fluctuations (SBFs; see Section 4.2) to cross-check their Cepheid-based distance estimates (cf. Mould *et al.* 2000); see Figure 4.1.

The Hubble constant is of fundamental interest in numerous cosmological and astrophysical contexts (see Section 5.1). The results of this Key Project have profound implications

An Introduction to Distance Measurement in Astronomy, First Edition. Richard de Grijs.

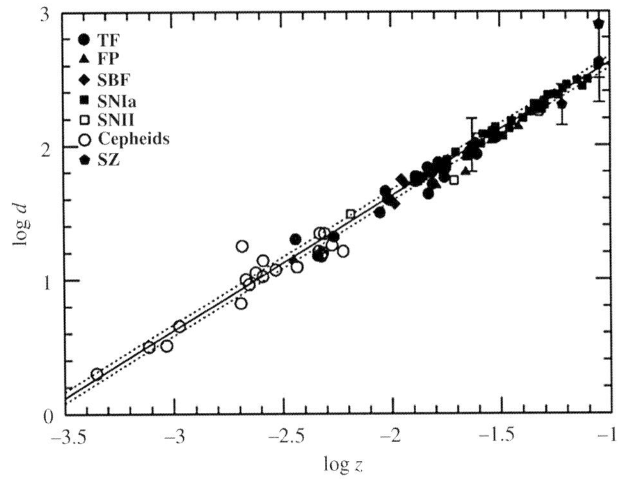

Figure 4.1 **Hubble diagram** *(Freedman et al. 2001): distance (d) versus redshift (z) for Cepheids, the Tully–Fisher relation, Type Ia and II supernovae (SNe), surface brightness fluctuations and the Fundamental Plane calibrated as part of the HST Key Project. Solid black circles are for nearby* ***Sunyaev–Zel'dovich*** *(SZ) clusters with cz < 30 000 km s^{-1} (c is the speed of light in a vacuum and z < 0.1), where the choice of cosmological model does not have a significant effect on the results. The solid line is for $H_0 = 72$ km s^{-1} Mpc^{-1}, with the dashed lines representing ± 10%.* (Reprinted from W. L. Freedman *et al.*, *Astrophysical Journal*, **553**, Final results from the Hubble Space Telescope Key Project to measure the Hubble constant, p. 47–72, Copyright 2001, with permission of the AAS and W. L. Freedman.)

for the extragalactic distance scale. The team collected a significantly increased sample of Cepheid variables in the Large Magellanic Cloud (LMC), properly corrected for the effects of metallicity and incompleteness bias, leading to a new distance modulus to the LMC of 18.50 ± 0.10 mag (50 ± 2 kpc). For the individual secondary distance indicators based on, respectively, Type Ia and Type II-P supernovae, the TFR, SBFs and the FP, they obtained $H_0 = 71 \pm 2 \pm 6, 72 \pm 9 \pm 7, 71 \pm 3 \pm 7, 70 \pm 5 \pm 6$ and $82 \pm 6 \pm 9$ km s^{-1} Mpc^{-1}, where the first and second uncertainties represent the random and systematic errors, respectively. Overall, statistical weighting yields $H_0 = 72 \pm 8$ km s^{-1} Mpc^{-1}. This is remarkably close to the current best determination of the Hubble constant by the 7-year *Wilkinson Microwave Anisotropy Probe* team, combined with baryon acoustic oscillations and measurements of high-z supernovae, of $H_0 = 70.2 \pm 1.4$ km s^{-1} Mpc^{-1}, assuming a Λ **Cold Dark Matter** (ΛCDM) Universe (Komatsu *et al.* 2009, 2011; see also Section 5.1). Both of these results are comfortably within the uncertainties of the recently refined measurements of H_0 based on *HST* observations of Cepheids and Type Ia supernovae, resulting in $H_0 = 74.2 \pm 3.6$ km s^{-1} Mpc^{-1} (Riess *et al.* 2009).

4.2 Surface Brightness Fluctuations

If a galaxy is sufficiently close so that its stars are resolved, we have access to a wealth of observational data, which can help us determine its distance by calibration with respect to primary distance indicators or, more often, Cepheid period–luminosity relations (cf. Freedman *et al.* 2001; Tonry *et al.* 2001; see Section 3.5.2). Even in more distant galaxies ($d \gtrsim 100$ Mpc; cf. Biscardi *et al.* 2008; Blakeslee *et al.* 2009), where individual stars cannot be resolved, the discrete nature of the constituent stars can still be used for accurate ($\ll$10%) distance determinations (Tonry and Schneider 1988; Tonry *et al.* 2001 for ground-based measurements; Blakeslee *et al.* 2009, 2010), and potentially tie near-field stellar distance tracers to the far-field Hubble flow.

Consider a galaxy that is composed of identical stars of luminosity L_*, distributed such that there are n stars per unit area across the galaxy. If we observe this galaxy with an angular resolution $\delta\theta$, each $\delta\theta \times \delta\theta$ resolution element will contain an average of $\bar{N} = n(d\delta\theta)^2$ unresolved stars. The flux from each star is $f_* = L_*/(4\pi d^2)$, so the average total flux from the stars per resolution element (i.e. pixel) is

$$F_* = \bar{N} f_* = nL_*(\delta\theta)^2/(4\pi). \tag{4.1}$$

However, not all pixels will contain exactly the same number of stars: Poissonian fluctuations will introduce a dispersion of $\sqrt{\bar{N}}$, so that the fluctuations in the observed flux per resolution element will be

$$\sigma_{F_*} = \sqrt{\bar{N}} f_* = \frac{n^{1/2}\delta\theta L_*}{4\pi} d^{-1}. \tag{4.2}$$

Thus, the surface brightness fluctuations, normalized by the local galaxy surface brightness (after subtraction of a smooth, large-scale galaxy model and divided by the square root of the mean signal; cf. Tonry and Schneider 1988; Mei *et al.* 2005), scale inversely with the square of the distance. A galaxy at twice the distance would appear twice as smooth,

$$\frac{\sigma_{F_*}}{F_*} = f_* = \frac{L_*}{4\pi d^2}, \tag{4.3}$$

so by comparing the values of f_* among galaxies, one can immediately obtain their relative distances (see Tonry and Schneider 1988 for details). In practice, the SBF method enables one to obtain the spatial power spectrum – i.e. wavenumber, k, versus frequency of occurrence, $P(k)$; the latter includes both the actual signal and noise contributions – of the residual fluctuations after subtraction of the smooth galaxy component (for examples, see Mei *et al.* 2005; Dunn and Jerjen 2006; see Figure 4.2; Blakeslee *et al.* 2010), where the spectrum's amplitude represents the luminosity of the 'fluctuation star'. The SBF pattern corresponds to the Fourier transform of the point spread function, which must be modelled carefully (see e.g. Mei *et al.* 2005; Blakeslee *et al.* 2010).

Although galaxies are not composed of identical stars, in practice the only implication of this assumption is that instead of the fixed luminosity L_*, we need a measurement of some kind of average luminosity $\langle L_* \rangle$, which is equivalent to the luminosity- and relative-number-weighted integral over the luminosity distribution of the constituent stars. If we know the stellar luminosity function of a given galaxy, we can immediately calculate $\langle L_* \rangle$ by integration. In reality, $\langle L_* \rangle$ will be similar to the luminosity of the brightest red giant stars, i.e. this method uses the same stellar population tracer as the tip-of-the-red-giant-branch method discussed in Section 3.3.1, yet reaches significantly beyond that technique. It does not change significantly if the luminosity function of fainter stars varies among galaxies. Since these stars are brightest at red wavelengths, the SBFs will be most readily observable in the optical

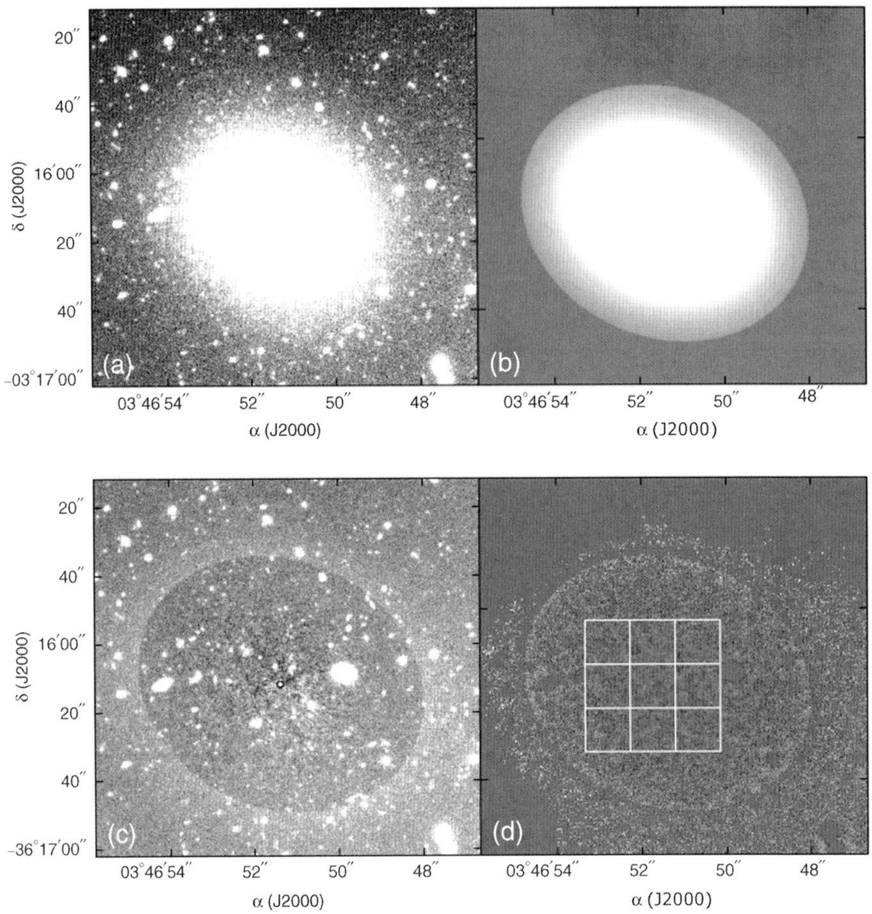

Figure 4.2 *Example application of the SBF technique to Fornax Cluster Catalogue galaxy 318 (Dunn and Jerjen 2006). (a) Original R-band image. (b) Best-fitting model of the galaxy's overall light distribution. (c) Residual after subtraction of the model. (d) Final result after normalization and masking. The fluctuations caused by unresolved stars in the galaxy centre are clearly visible.* (Reprinted from L. P. Dunn and H. Jerjen, *Astronomical Journal*, **132**, First results from SAPAC: toward a three-dimensional picture of the Fornax cluster core, p. 1384–1395, Copyright 2006, with permission of the AAS and H. Jerjen.)

R or I bands (e.g. Ajhar *et al.* 1997; Tonry *et al.* 2001; Dunn and Jerjen 2006; Blakeslee *et al.* 2010; and references therein), where the effects of metallicity and age variations are also minimized for these stars. The value of $\langle L_* \rangle$ in the I band should therefore depend little on the properties of the galaxy's stellar population (but see below). The dependence on stellar population becomes more complicated in the near-infrared (near-IR) regime, where thermally pulsing asymptotic giant branch (AGB) stars are significant contaminants (cf. Jensen *et al.* 2001, 2003; Blakeslee *et al.* 2009). Elliptical galaxies and the bulges of early-type spiral galaxies have fairly consistent stellar populations compared to spiral galaxies, where the stellar population mix usually contains larger dominant age and metallicity ranges, so that the typical average stellar luminosity of the Population II stars (see Section 3.1.2) generally targeted closely approximates a standard candle (cf. Ajhar *et al.* 1997; Tonry *et al.* 1997; Blakeslee *et al.* 2010; and references therein). In practice, however, corrections must be applied to account for variations in age or metallicity among galaxies (see below). The apparent SBF magnitude, $\langle m \rangle$, can thus be linked to the absolute mean magnitude, $\langle M \rangle$ (the standard candle), to yield a distance modulus $\langle m \rangle - \langle M \rangle$ and, hence, a distance.

While we can fairly easily deal with Poissonian noise introduced by the incoming photons (i.e. use longer exposure times), as well as contamination by instrumental readout noise, foreground stars, extinction (for details, see e.g. Ferrarese *et al.* 2006; Blakeslee *et al.* 2010) and superimposed globular clusters, the most troublesome effect is caused by unresolved background objects, of which the luminosity function needs to be modelled and corrected for before SBFs can be used for accurate relative distance measurements (cf. Tonry and Schneider 1988). In addition, the slope of the red giant branch in the Hertzsprung–Russell diagram is a function of metallicity, so that the SBF technique is also expected to be a function of metallicity. Tonry *et al.* (2001) derived a fairly steep metallicity correction for I-band observations,

$$\langle M_I \rangle = -1.74 + 4.5(V - I)_0 - 1.15. \tag{4.4}$$

Clearly, the method works best when applied to the outer regions of galaxies, where the number of stars is lower and Poissonian fluctuations are larger, although the sky brightness places strict limits on the useful range of surface brightnesses. For instance, at a dark, ground-based site, $\mu_{\text{sky},I} \sim 19$ mag arcsec^{-2}, and it is difficult to observe regions of galaxies that are much fainter than this level. At ground-based ~1 arcsecond resolution, this limits accurate I-band photometry to ~20 mag arcsec^{-2}. For a galaxy at $d = 20$ Mpc (e.g. in the Virgo or Fornax clusters; Tonry *et al.* 2001), this surface brightness corresponds to the I-band light from $\langle N \rangle \sim 10^4$ giant stars arcsec^{-2}. Poissonian fluctuations will contribute at a level of ~1%, which implies that this is the level of accuracy required.

With the general availability of *HST* observations for large numbers of early-type galaxies in the Virgo and Fornax clusters (e.g. Côté *et al.* 2004; Jordán *et al.* 2007b), the method's focus has shifted from ground-based I-band data to the use of the *HST*/Advanced Camera for Surveys F850LP filter (z_{850}), with metallicity calibration provided by $(g_{475} - z_{850})$ colours (e.g. Mei *et al.* 2005; Blakeslee *et al.* 2009; for a review of early *HST*-based science, see Blakeslee *et al.* 1999),

$$\langle M \rangle_{z850} = (-2.04 \pm 0.15) + 1.41x + 2.60x^2 + 3.72x^3, \tag{4.5}$$

where $x \equiv (g_{475} - z_{850}) - 1.3$ and the uncertainty in the zero point is dominated by the Cepheid distance calibration and galaxy-type matching. At greater distances, the higher-throughput F814W filter becomes more attractive for SBF applications (e.g. Cantiello *et al.* 2005, 2007; Blakeslee *et al.* 2010), leading to a calibration of

$$\langle M \rangle_{814} = (-1.168 \pm 0.013 \pm 0.092) + (1.83 \pm 0.20)[(g_{475} - I_{814}) - 1.2] \tag{4.6}$$

for $1.06 < (g_{475} - I_{814}) < 1.32$ (AB) mag (Blakeslee *et al.* 2010). The two errors quoted for the relation's zero point represent the statistical and systematic components, respectively, and the root-mean-square (rms) scatter of the data points with respect to Equation (4.6) is 0.029 mag. These two calibrations can be reconciled by a linear relation of the form

(Blakeslee *et al.* 2010, in AB magnitudes)

$$\langle m \rangle_{814} - \langle m \rangle_{850} = (0.72 \pm 0.01) - (0.33 \pm 0.18)[(g_{475} - I_{814}) - 1.2] \tag{4.7}$$

(rms scatter: 0.021 mag), where the zero point of slightly below unity implies that the colour difference is driven by small changes in the underlying stellar population, most likely in the mean age or metallicity (see also Biscardi *et al.* 2008). At present, however, and despite numerous attempts (e.g. Buzzoni 1993; Worthey 1993; Liu *et al.* 2000; Blakeslee *et al.* 2001; Mei *et al.* 2001; Cantiello *et al.* 2003; Mouhcine *et al.* 2005; Raimondo *et al.* 2005; Marín-Franch and Aparicio 2006; Biscardi *et al.* 2008; Cerviño *et al.* 2008; Lee *et al.* 2010; see also González *et al.* 2004; González-Lópezlira *et al.* 2005), the SBF method remains a 'scaling relation' rather than a primary method of distance determination based on well-understood physics associated with some stellar population parameter(s).

To reduce the scatter in the calibration relations, Tonry *et al.* (2001; see also Blakeslee *et al.* 2001, 2002, 2010 and references therein) suggested to use a distance-independent 'fluctuation count', which scales logarithmically with the number of stars in a galaxy and, therefore, with its mass,

$$\langle N \rangle_{\rm f} = \langle m \rangle - m_{\rm tot} = +2.5 \log \left(\frac{L_{\rm tot}}{\langle L \rangle} \right), \tag{4.8}$$

where $L_{\rm tot}/\langle L \rangle$ is a measure of the total luminosity of (part of) a galaxy in units of $\langle L \rangle$, the fluctuation luminosity corresponding to $\langle M \rangle$. The advantage of this approach is that $\langle N \rangle_{\rm f}$ is also independent of Galactic extinction and photometric calibration errors if the same data sets are used to derive the observables. This parameter correlates well with colour for quiescently evolving galaxies, since it is driven by the mass–luminosity relation of early-type galaxies (Blakeslee *et al.* 2001). The resulting scaling relation between $\langle N \rangle_{{\rm f},z}$ (in the z_{850} filter) and $\langle M \rangle_{z_{850}}$ is (Blakeslee *et al.* 2009)

$$\langle M \rangle_{z_{850}} = -1.98 + 0.089(\langle N \rangle_{{\rm f},z} - 18), \tag{4.9}$$

which, using the definition of $\langle N \rangle_{\rm f}$, corresponds to

$$(m - M) \approx 3.58 + 0.91 \langle z \rangle_{850} + 0.09 z_{850,{\rm tot}} \tag{4.10}$$

for a total galactic z_{850} magnitude of $z_{850,{\rm tot}}$. Although the analysis of Blakeslee *et al.* (2009) implies that the resulting scatter in this empirical scaling relation is reduced significantly, it may be affected by systematic environmental and/or type dependences (e.g. Tonry *et al.* 2001; Lisker *et al.* 2008).

4.3 The Globular Cluster Luminosity Function

Objects which have a large spread, σ_M, in the distribution of their absolute magnitudes, M, are not individually useful as standard candles. However, if we observe N such objects, all at the same distance, the average magnitude of the sample will exhibit a smaller scatter, of order $\sigma_M/\sqrt{N}$. If we observe a sufficiently large sample, then the uncertainty in the average magnitude will become small enough for the *average* absolute magnitude to serve as a standard candle. This has been exploited for the 'luminosity functions' (LFs) of both globular clusters and planetary nebulae (PNe; see Section 4.4). The key underlying assumption for practical use of this method of distance determination is that the LFs of the objects do not vary from galaxy to galaxy.

The shape of the **globular cluster luminosity and mass functions** (GCLF, GCMF) in both the Milky Way and many nearby, mostly elliptical and early-type spiral galaxies at distances of up to $\sim$20 Mpc hosting large samples of old globular clusters (GCs) is usually well described on the basis of a Gaussian or lognormal distribution (see Figure 4.3),[1]

$$N(M) \propto \exp[-(M - M_0)^2/2\sigma_M^2], \tag{4.11}$$

where M_0 is the average or 'turnover' magnitude of the distribution.

The GCLF provides strong constraints both on the dynamical evolution of the GC population in a given galaxy (e.g. Gnedin and Ostriker 1997; Vesperini 1998; Fall and Zhang 2001; de Grijs and Parmentier 2007; and references therein) and on the theories of GC formation (see de Grijs and Parmentier 2007 for a review). Much of its analysis was initially motivated by the hope of using the GCLF turnover magnitude or, equivalently, mass as a distance indicator (Harris 2001; Richtler 2003; but see Fritze-von Alvensleben 2004; Villegas *et al.* 2010 for counterarguments), although its usability in dwarf galaxies remains a matter of debate (e.g. Durrell *et al.* 1996; Brodie and Strader 2006; Miller and Lotz 2007; and references therein). Limited, albeit systematic, differences in the detailed shape of the GCLF from one galaxy to another do exist, however (Harris 2001; Brodie and Strader 2006), for instance, variations in the extension of the high-luminosity wing (McLaughlin and Pudritz 1996; Burkert and Smith 2000; Jordán *et al.* 2006; Parmentier and Gilmore 2007).

4.3.1 Elliptical Versus Spiral Galaxy GCLFs

The Milky Way's GCLF is often used as a benchmark for the shape and characteristics, including the turnover magnitude and width, of spiral galaxy GCLFs. It is characterized by a V-band peak magnitude of $M_{0,V} = -7.4 \pm 0.1$ mag, corresponding to a characteristic GC mass of $\langle M_{\rm GC} \rangle \simeq 2 \times 10^5$ $M_\odot$ and a width of $\sigma_{M_V} = 1.15 \pm 0.10$ mag (e.g. Harris 1991, 2001; Harris *et al.* 1998; Richtler 2003). Intriguingly, this shape and these parameters seem to be almost universal among galaxies, since they show only a weak dependence on the size, morphological type, metallicity, and environment of the host galaxy (Ashman *et al.* 1995; Kavelaars and Hanes 1997; Ashman and Zepf 1998; Harris 1999; Whitmore *et al.* 2002; Spitler *et al.* 2008; see also Villegas *et al.* 2010).

For elliptical galaxies in the cores of nearby galaxy clusters, the GCLF turnover magnitude appears to vary by less than $\sim$0.2 mag (Whitmore 1997; see also Whitmore *et al.* 1995; Villegas *et al.* 2010). Their widths are $\sigma_V \simeq 1.4$ mag rather than 1.2 mag as for the Milky Way and M31 (e.g. Barmby *et al.* 2001). This seems consistent with other elliptical galaxies, including NGC 1399 and NGC 5128 (Grillmair *et al.* 1999; Rejkuba 2001), but perhaps not with all (e.g. NGC 720, NGC 4278, NGC 4494; Forbes 1996). However, based on a more detailed comparison, it appears that $M_{0,V}$ is typically $\sim$0.2–0.3 mag fainter for elliptical galaxies in the Virgo cluster than for spiral galaxies in the Local Group. Although the simplest explanation of such a discrepancy would be an error of $\sim$10–15% in the distance

[1] In detail, GCLFs tend to have low-luminosity tails resembling somewhat skewed, evolved Schechter functions instead of symmetrical Gaussian distributions (e.g. Richtler 2003; Jordán *et al.* 2007a; and references therein).

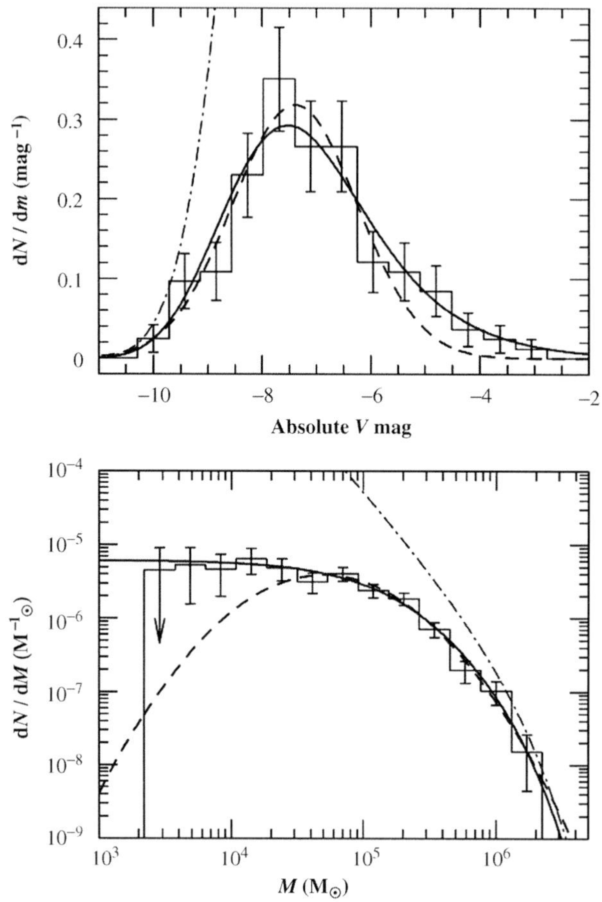

Figure 4.3 *(Top) Fits of a Gaussian (dashed curve) and an evolved Schechter function (solid curve) to the Milky Way GCLF, expressed as the (normalized) number of clusters per unit of absolute V magnitude (Jordán et al. 2007a). The dot-dashed curve is a Schechter function with the same value for the cut-off mass as the solid curve but with the mass-loss parameter set to zero. (Bottom) Corresponding observed GC mass function, dN/dM, and model fits derived from the GCLF assuming a V-band* $M/L_V = 2\ M_\odot/L_{V,\odot}$ *for all clusters.* (Reprinted from A. Jordán et al., *Astrophysical Journal Supplement Series*, **171**, The ACS Virgo cluster survey. XII. The luminosity function of globular clusters in early-type galaxies, p. 101–145, Copyright 2007, with permission of the AAS and A. Jordán.)

modulus towards the Virgo cluster, this is unlikely given that a similar discrepancy has been observed for independent galaxy samples in the Fornax cluster (Blakeslee and Tonry 1996; Kohle *et al.* 1996).

Instead, a potentially more physically realistic explanation of these observations is that GC systems in elliptical and spiral galaxies may have the same *mass* function but different *luminosity* functions. This seems reasonable since the GC systems of elliptical galaxies tend to be slightly redder, i.e. possibly somewhat more metal rich, than those of spiral galaxies (Ashman *et al.* 1995; Whitmore *et al.* 1995). The effect of the population differences required

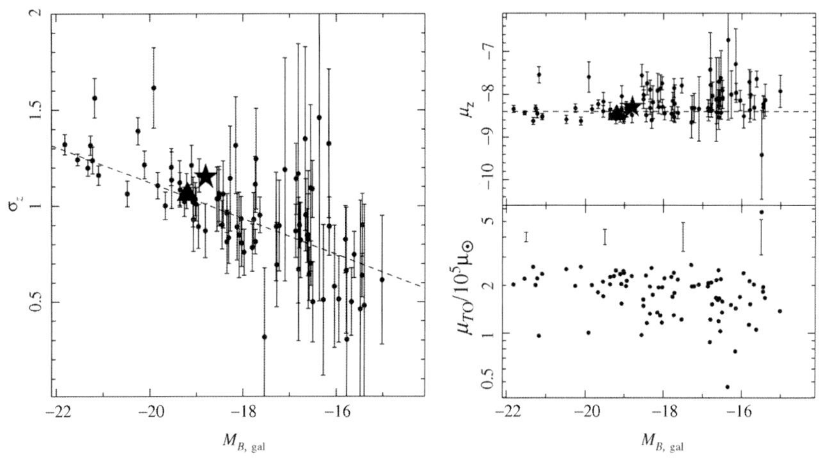

Figure 4.4 *(Left) Gaussian dispersion, σ_z, versus integrated B-band magnitude, $M_{B,gal}$, for the z-band GCLFs of 89 early-type Virgo cluster galaxies from Jordán et al. (2006). The GCLF width varies systematically, being narrower in fainter galaxies. The two anomalously high points at $M_{B,gal} = -21.2$ and -19.9 mag correspond to the galaxies VCC 798 and VCC 2095, both of which have large excesses of faint, diffuse clusters (VCC: Virgo Cluster Catalogue). The large star represents the spheroid luminosity and GCLF dispersion of the Milky Way; the large triangle marks the bulge luminosity and GCLF dispersion of M31. (Right) GCLF turnover magnitude (absolute mean μ_z) and mass versus $M_{B,gal}$.* (Reprinted from A. Jordán *et al.*, *Astrophysical Journal*, **651**, Trends in the globular cluster luminosity function of early-type galaxies, L25–L28, Copyright 2006, with permission of the AAS and A. Jordán.)

for this to work is of order 0.2 mag in the V band, even if part of the colour difference is caused by age differences (see Section 4.3.2).

However, the GCLF accurately reflects the underlying GC mass function only if cluster-to-cluster variations in the integrated cluster mass-to-light ratios (M/L) are small. This requirement is met if the cluster age range is a limited fraction of the mean age of the entire population. For the old GCs in the Milky Way's halo, the range spanned by their visual M/L is limited (i.e. $1 \leq M/L_V \leq 4$, with a mean $\langle M/L_V \rangle \sim 2$ in solar units; Pryor and Meylan 1993; Parmentier and Gilmore 2001; McLaughlin and van der Marel 2005), and the scatter partially reflects variations in the dynamical evolution of individual GCs (see Section 4.3.3).

Alternatively, it has been proposed that $M_{0,V}$ may be brighter for galaxies in low-density environments (e.g. Blakeslee and Tonry 1996) and that more massive galaxies tend to host brighter GCs, on average,[2] and that their GCLF widths are narrower (Jordán *et al.* 2007a; see also Figure 4.4). Spiral galaxies tend to occupy lower-density regions and are characterized by bluer GC systems than their elliptical counterparts, so that this colour difference may indeed imply a correlation with density. Although the dependence of the turnover *magnitude* on integrated galaxy luminosity seems to depend on filter, the mean cluster *mass* of a given GC population depends more clearly on the properties of its host galaxy. Jordán *et al.*

[2] However, the observability of this tendency depends on the filter employed; in the V and g bands, any variation with galaxy luminosity seems minimal if at all observable (cf. Jordán *et al.* 2007a).

(2007a) show that the turnover GC mass in lower-luminosity galaxies ($M_B > -18$ mag) is somewhat lower than for the brightest galaxies (see also Figure 4.4). This implies, therefore, that the stellar population mix might be of sufficient importance to significantly increase the uncertainties associated with the use of the GCLF as a distance indicator.

4.3.2 The Stellar Population Mix

In addition to requiring that the GCLF or GCMF consists of objects of similar M/L, the other crucial stellar population constraint requires that all GCs in a given system are composed of old Population II stars. In turn, this implies a particular stellar population mix characterized by a single metallicity. However, from an observational perspective, many GC systems associated with early-type galaxies exhibit clear bi- or multimodality in their colour distributions (e.g. Larsen *et al.* 2001; Peng *et al.* 2006), despite comprising a single, 'universal' GCLF. It is therefore of crucial importance to (i) understand the origin of the shape of the colour distributions and (ii) then assess whether an *a priori* unknown stellar population mix conspiring to produce the universal GCLF introduces sufficiently significant uncertainties to render a physical understanding of the GCLF universality doubtful at best.

From a theoretical perspective, one would expect metallicity differences among GC systems to impact on the turnover magnitude (e.g. Larsen *et al.* 2001; Girardi *et al.* 2002). In fact, the observed colour distributions are reminiscent of what would be expected for GC systems if they were composed of multiple GC populations, each with slightly different ages and/or metallicities. From the point of view of using the GCLF as a standard candle, it is comforting to learn that the blue population appears universal among different galaxies (Fritze-von Alvensleben 2004; Di Criscienzo *et al.* 2006). The relative importance and precise colour distribution of the redder population appear to be a sensitive function of its host galaxy's properties, however. It has been suggested that these redder, presumably more metal-rich populations formed as a result of violent 'starburst' episodes in their host galaxies, induced by galaxy interactions (e.g. Fritze-von Alvensleben 2004). If so, this would imply that these GCs are younger than their blue-peak counterparts (see also Kundu *et al.* 2005; Puzia *et al.* 2005; Hempel *et al.* 2007a,b; Pipino *et al.* 2007; and references therein).

It is clear, therefore, that for the GCLF to be useful as a secondary distance indicator, one should include only GCs from the blue peak in any bi- or multimodal GC population. Ideally, stellar population differences should be corrected for, so that the most suitable diagnostic for most galactic environments is the GC *mass* function. Nevertheless, despite the uncertainties related to stellar population differences, the GCLF appears universal in the *V* band. Perhaps the most important uncertainty for the application of the GCLF as a standard candle is that – even with the *HST* – for most extragalactic GC systems we do not, or only marginally, reach to turnover magnitude. If the peak magnitude varies, as well as the width of the distribution (e.g. Larsen *et al.* 2001; Richtler 2003; Jordán *et al.* 2006, 2007a; see Figure 4.4), any modelling of the GCLF based on observations of the highest-luminosity GCs only is bound to be affected by significant uncertainties.

4.3.3 GCLF and GCMF Universality Through Dynamical Evolution

The launch of the *HST* and the subsequent discovery – in particular in interacting and merging galaxies – of star clusters with the high luminosities and compact sizes expected for GCs at young ages has prompted renewed interest in the evolution of the cluster LF and MF of massive star clusters, both observationally and theoretically. Since the slope of the observed young cluster LFs is reminiscent of the power-law slope of the high-mass regime of the old GC mass spectrum ($\alpha \sim -2$; McLaughlin 1994), this type of observational evidence has led to the theoretical prediction that *any* initial cluster LF and MF will rapidly be transformed into a Gaussian distribution because of (i) stellar evolutionary fading of the lowest-luminosity and, given a single age, lowest-mass objects to below the detection limit; and (ii) disruption of the low-mass clusters due to both interactions with the gravitational field of the host galaxy and internal two-body relaxation effects, leading to enhanced cluster evaporation (cf. Figure 4.5; e.g. Elmegreen and Efremov 1997; Gnedin and Ostriker 1997; Ostriker and Gnedin 1997; Fall and Zhang 2001; Prieto and Gnedin 2008).

However, from a theoretical point of view, while the preferential removal of the more vulnerable low-mass clusters indeed results in an initial power-law cluster MF being turned into an approximately Gaussian shape (e.g. Okazaki and Tosa 1995; Baumgardt 1998; Vesperini 1998; Fall and Zhang 2001; Prieto and Gnedin 2008), recovering the *present-day* MF after a Hubble time of evolution requires considerable fine-tuning of the models, which is hardly compatible with the near-invariance of the GCMF among large galaxies.

In the currently most popular theoretical model, which starts from a power-law initial cluster mass function, the Gaussian distribution characteristic of old GC populations results

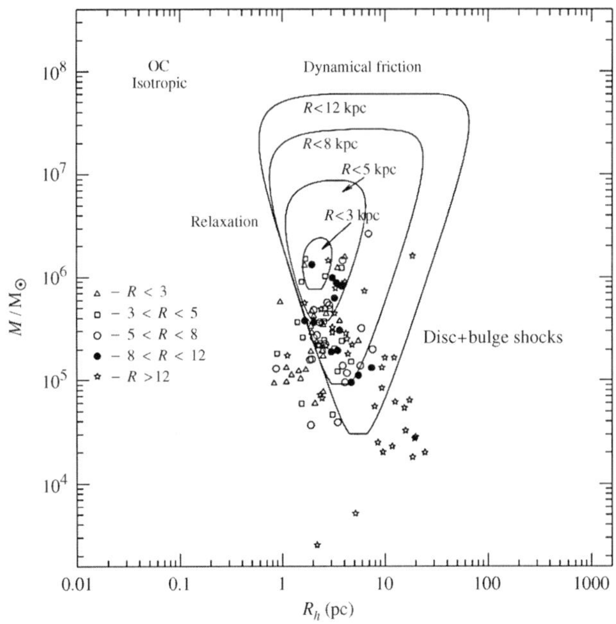

Figure 4.5 *'Vital diagram' for Galactic GCs (Gnedin and Ostriker 1997). The mass–radius plane is restricted by three destruction processes: relaxation, tidal shocks and dynamical friction. Galactic model: Ostriker and Caldwell (1983; OC); kinematic model: isotropic.* (Reprinted from O. Y. Gnedin and J. P. Ostriker, *Astrophysical Journal*, **474**, Destruction of the Galactic globular cluster system, p. 223–255, Copyright 1997, with permission of the AAS and O. Y. Gnedin.)

from evolutionary effects, predominantly the preferential removal of the more vulnerable low-mass clusters (Fall and Zhang 2001). The cluster mass at the turnover of the present-day MF then depends on the age of the cluster system and on the cluster disruption timescale, in the sense that the older the cluster system and/or the shorter the disruption timescale, the higher the expected turnover mass of the cluster system will be. The evolutionary rate of the GCMF turnover is, thus, a function of the initial spatial distribution of the GCs in their host galaxy (Parmentier and Gilmore 2005), of their initial velocity distribution or, equivalently, of the clusters' orbital distribution (Murali and Weinberg 1997; Baumgardt 1998; Baumgardt and Makino 2003; Fall and Zhang 2001), as well as of the circular velocity of the host galaxy (Baumgardt and Makino 2003; see also Gieles *et al.* 2006 and Lamers and Gieles 2006 for the effects of close encounters with giant molecular clouds). Thus, the near-invariance of the GCMF in very different types of galaxies is neither easily understood nor straightforwardly reproduced (Vesperini 2001).

Not only is the observed GC mass at the Gaussian peak universal *among* galaxies, it is also universal *within* galaxies, i.e. the turnover location of the GCMF is constant over a large range of galactocentric distances (e.g. Harris *et al.* 1998 and Kundu *et al.* 1999 – see Figure 4.6 – for M87; Kavelaars and Hanes 1997 and Parmentier and Gilmore 2005 for the Galactic halo; Kavelaars and Hanes 1997 for M31; Spitler *et al.* 2006 for the Sombrero galaxy; but see also Gnedin 1997 for a discussion of the possible role of the statistical methodology used). Yet, because of the higher environmental density in the inner region of any GC system, evolutionary processes proceed at a faster rate there. Therefore, evolutionary models building on the power-law hypothesis, and assuming an isotropic cluster velocity distribution, predict a radial gradient of the mean logarithmic cluster mass, i.e. $\langle \log(M_{\rm cl}[M_\odot]) \rangle$ is higher in the inner than in the outer regions of a given galaxy. The resulting gradient appears too large to be consistent with the observations (Vesperini 2001).

In summary, the initial power-law MF assumption can account for both the present-day turnover GC mass and its near invariance with galactocentric distance, provided that the initial GC velocity distribution is characterized by a strong radial anisotropy that increases as a function of increasing radius, in the sense that the farther from the Galactic Centre a given cluster is located, the higher its orbital eccentricity will be (Fall and Zhang 2001). The net result of this assumption is that the orbits of *all* GCs must have similar perigalactic distances. Because the mass-loss rate of a GC on an eccentric orbit is significantly more sensitive to its perigalactic than to its apogalactic distance (Baumgardt 1998, his Figure 2), the relatively narrow distribution of pericentres causes a nearly invariant GCMF turnover over the entire radial extent of the GC system, regardless of the clusters' loci at any given point in time. In addition, the small anisotropy radius ($\sim$2 kpc) of the velocity distribution or, equivalently, the mean perigalactic distance implies that GC evolutionary processes and, hence, the shift of the GCMF towards higher cluster masses, proceed at a faster rate than in the case of an isotropic velocity distribution. As a consequence, the cluster mass at the peak of the GCMF at an age of 13 Gyr matches the observations, both in the inner and in the outer galactic regions (Fall and Zhang 2001).

Vesperini *et al.* (2003) tested this model against the data of the M87 GC system. Owing to the preferential disruption of GCs on highly eccentric orbits, the initial amount of radial anisotropy in the velocity distribution decreases during the evolution of the cluster population. Yet, the initial radial anisotropy required to reproduce the near constancy of the GCMF turnover mass is so strong that in spite of its steady decrease with time, at an age of 13 Gyr, it will still be significantly stronger than what is inferred from the observed

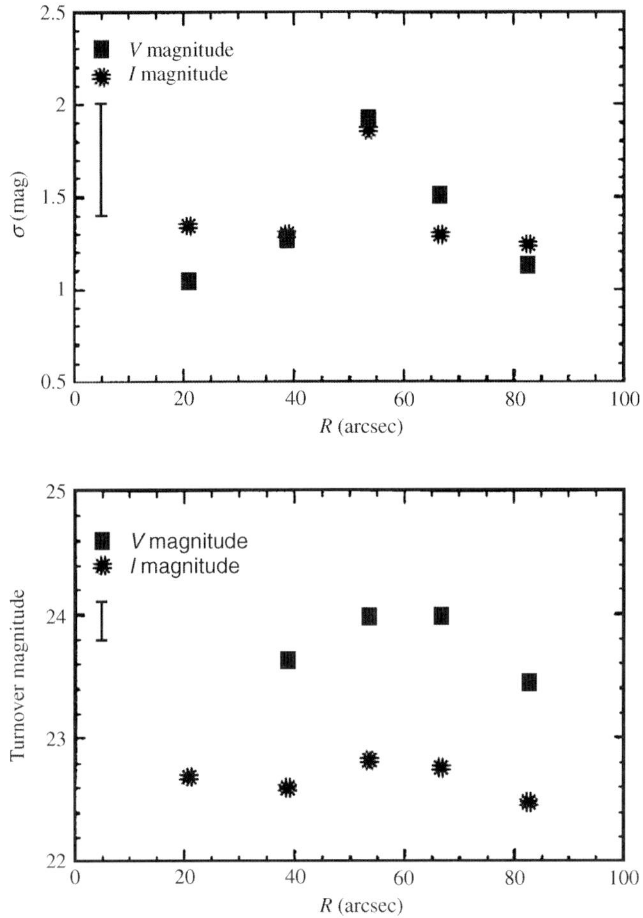

Figure 4.6 *Parameters of the Gaussian curves fitted to the radially constrained GCLFs in M87 (Kundu et al. 1999). (Bottom) Turnover luminosity in V and I as a function of distance. (Top) Corresponding Gaussian dispersions. Both the turnover magnitude and σ are constant within the uncertainties, indicating that the GCLF does not vary with distance.* (Reprinted from A. Kundu *et al.*, *Astrophysical Journal*, **513**, The globular cluster system in the inner region of M87, p. 733–751, Copyright 1999, with permission of the AAS and A. Kundu.)

projected velocity dispersion profile of the M87 GC system (Côté *et al.* 2001). In other words, the lack of any significant radial gradient in the mean logarithmic GC mass, and the observed kinematics of the M87 GC system, cannot be reproduced simultaneously on the basis of this model. The former requires a small initial anisotropy radius (≤ 3 kpc), while the latter requires an anisotropy radius which is greater by at least an order of magnitude.

McLaughlin and Fall (2008) tackled the problem from a new angle by discarding any priors about velocity anisotropy in their model GC system. Instead, they recover the observed insensitivity of the GC turnover mass to galactocentric radius based on a model which incorporates the slow erosion of a GCMF that initially increased towards low masses, such as a power-law distribution, predominantly through cluster evaporation driven by internal two-body relaxation. They conclude that this is not caused by a literal 'universality' of the GCMF turnover mass but originates from a significant variation in this characteristic mass with cluster half-mass density (i.e. the expected outcome of relaxation-driven cluster disruption), combined with significant scatter in the latter as a function of galactocentric distance. The main and as yet unexplained surprise of their result is that the cluster half-mass density does not correlate tightly with perigalactic distance. If it did, one would indeed need highly radially anisotropic orbits to explain the 'universal' GCMF shape.

In another attempt to reconcile the Gaussian MF of old GCs with the power law LF of young massive clusters, Vesperini and Zepf (2003) built on the observed trend between the mass of Galactic GCs and their concentration, i.e. the more massive the cluster, the higher its concentration (van den Bergh 1994). Their results suggest that it may be possible to reproduce, at an age of 13 Gyr, a Gaussian GCMF with a turnover occurring at roughly the observed mass, even if starting from an initial power law. Moreover, dissolution of the low-mass clusters is mostly caused by their low initial concentration. The process depends only weakly on environmental conditions such as the structure of the host galaxy and the cluster orbit. Consequently, their model reproduces the lack of significant radial variation in the mean logarithmic GC mass across the Galactic GC system, even for an isotropic initial velocity distribution. Dissolution of low-concentration clusters may therefore provide the missing link between the power-law LF observed for young massive clusters in violently star-forming environments and the Gaussian MF characteristic of old GC populations, although a detailed study of this effect remains to be done.

4.4 The Planetary Nebulae Luminosity Function

Young PNe provide a second population of objects whose LF is suitable for use as a standard candle (see Ciardullo 2005 for a review). They can be detected easily – even against a background of bright stellar continuum emission from their host galaxy – thanks to their strong [O III]λ5007 Å line emission. Similarly as for GCs, PNe are also found in large numbers in all types of galaxies, which facilitates construction of statistically significant samples.

Observations of PNe through [O III]λ5007 Å filters show that the PNLF varies exponentially with magnitude. However, at high luminosities one encounters a 'cut-off' magnitude, $m_{\rm PN}^{\rm cut}$, beyond which PNe are essentially lacking (see Figure 4.7; Ciardullo *et al.* 1989; see Méndez *et al.* 1993 for an alternative representation):

$$N_{\rm PN}(m) \propto \exp(0.307m) \times (1 - \exp\{3(m_{\rm PN}^{\rm cut} - m)\}). \tag{4.12}$$

Calibration of the absolute value of the cut-off magnitude is generally based on PN observations in the bulge of M31, for which $m_{\rm PN}^{\rm cut} = 19.77 \pm 0.04$ mag, corrected for foreground extinction (Ciardullo *et al.* 1989), in the magnitude system defined by the λ5007 Å bandpass. If we adopt a distance to the M31 bulge of 740 ± 40 kpc (Binney and Merrifield 1998, close to the more up-to-date determinations discussed in Section 1.3), we obtain $M_{\rm PN}^{\rm cut} = -4.45^{+0.02}_{-0.03}$ mag (Ciardullo *et al.* 2002; Ciardullo 2003).[3]

[3] In a recent, preliminary analysis of a new sample of Galactic bulge PNe, Kovacevic *et al.* (2010) find $M_{\rm PN}^{\rm cut} = -4.38 \pm 0.13$ mag.

4.4.1 Applicability

Observations of PN populations in a variety of galaxy types suggest that the PNLF shape and cut-off magnitude are universal and remarkably robust, with a formal uncertainty of 0.05 mag (cf. Figure 4.7), although with a weak metallicity dependence (Dopita *et al.* 1992).

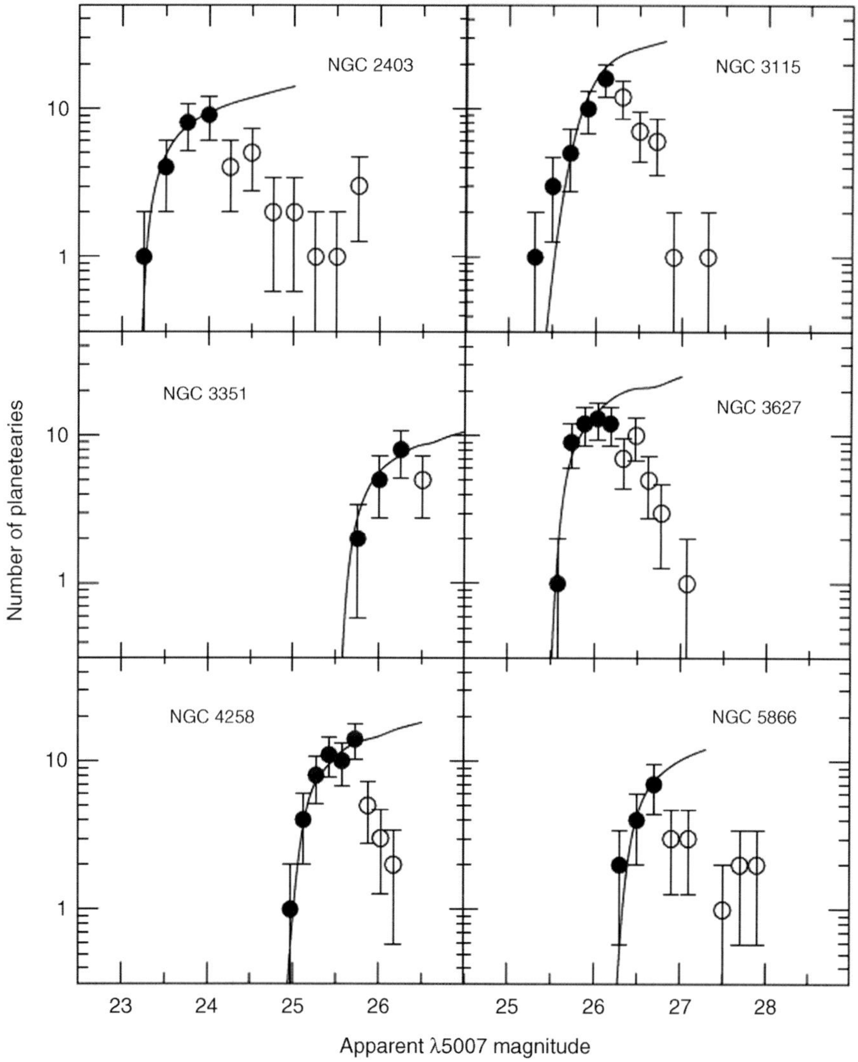

Figure 4.7 *Observed [OIII] PNLFs for samples of PNe in six nearby galaxies (Ciardullo et al. 2002). The curves show the best-fitting empirical PNLFs convolved with the photometric error function and shifted to the most likely distance. The open circles represent points past the completeness limit.* (Reprinted from R. Ciardullo, *Astrophysical Journal*, **577**, Planetary nebulae as standard candles. XII. Connecting the Population I and Population II distance scales, p. 31–50, Copyright 2002, with permission of the AAS and R. Ciardullo.)

These features can, hence, be used as a standard candle (Ciardullo 2005). The sharp cut-off provides an easily recognizable feature in the PNLF, which can be measured accurately, even with limited data: one only needs to observe the bright end of the PNLF. This implies that we can use the PNLF as a secondary distance indicator out to large distances, where only the brightest PNe are detectable.

As opposed to the GCLF, the PNLF works equally well and with similarly high accuracy in both spiral and elliptical galaxies (Jacoby *et al.* 1992; Ciardullo *et al.* 2002; Feldmeier *et al.* 2007; Gerhard *et al.* 2007); it is essentially insensitive to the underlying stellar population: the absolute [OIII]λ5007 Å-band magnitude does not vary among bulge, inner- and outer-disc PNe (Ciardullo 2005). This is of prime importance from a physical point of view, because the PNLF is one of the few accurate (distances to within $\sim$5%) extragalactic standard candles relevant to both Populations I and II out to the large galaxies in the Local Supercluster (cf. Tully 1982), at least until the next-generation telescopes, with mirror diameters $\geq$30 m, become operational. Other relevant techniques include the use of SBFs in galactic bulges – which also often contain Cepheid variables – and of Type Ia supernovae (see Section 5.2.1). Although application of the SBF technique in galactic bulges is difficult, it was originally calibrated based on observations of the dwarf elliptical galaxy M32 (Tonry and Schneider 1988) and also applied to the bulge of M31 (Tonry 1991). This link between the two main stellar population types is crucial for an accurate assessment of the systematic uncertainties still remaining in the extragalactic distance ladder, which are most likely predominantly caused by the uncertain corrections for internal extinction in the calibration galaxies and by colour terms (cf. Section 6.1.1). In fact, Ciardullo (2005) shows that the agreement between PNLF and Cepheid distances (Section 3.5.2) is excellent, with any trends and scatter fully accounted for by the available stellar evolution models and the internal uncertainties. In particular, the stellar and nebular evolution models of Dopita *et al.* (1992) predict exactly the observed trend that for galaxies which are more metal rich than the LMC the absolute [OIII]λ5007 Å-band magnitude is constant, while it fades somewhat for smaller, more metal-poor galaxies.

4.4.2 Physical Basis

The remarkable universality of the PNe cut-off magnitude at high luminosities most likely originates from the contribution of coalesced binary systems to the [OIII]λ5007 Å luminosity of the brightest PNe. Alternative evolutionary paths have also been proposed. These involve younger, higher-mass bipolar or Type I (helium/nitrogen-rich) nebulae (Torres-Peimbert and Peimbert 1997, but see Kovacevic *et al.* 2010 for observational counterarguments) or symbiotic binary stars (Soker 2006; Frankowski and Soker 2009), although this would controversially require that the majority of PNe be ejected by the white dwarf (see Coradi 2003).

The [OIII]λ5007 Å flux emitted by a PN is intimately linked to the ultraviolet flux of its central star, which sensitively depends on the star's mass. At first sight, this would suggest a varying upper luminosity limit as a function of stellar population age because of the initial mass–final mass relation (Weidemann 2000), yet this is contrary to PNe observations. A number of mechanisms are believed to cause a hard upper limit to the forbidden-oxygen flux, all of which place significant emphasis on the rapid evolutionary timescales associated with massive star evolution. These short timescales may not allow certain PNe to attain the low densities required for forbidden-line cooling (e.g. Jacoby 1989; Marigo *et al.* 2004) or may simply not be long enough to generate sufficient amounts of doubly ionized oxygen for collisional excitation (Marigo *et al.* 2004). Alternatively, high-mass stars are associated with massive, dusty circumstellar envelopes, which may not yet have had the time to disperse on the short evolutionary timescales involved (e.g. Meatheringham and Dopita 1991a,b; Ciardullo 2005). We note that these processes by themselves do not give rise to the *universal* high-luminosity PN magnitude cut-off, however.

The principal cause of the high-luminosity cut-off may simply be the enormous luminosities implied for these objects. Taking into account the obscuring effects of the circumstellar envelopes, it is thought that the central massive stars in the brightest PNe emit [O III]λ5007 Å luminosities of $L_{5007} > 6000\ L_\odot$ (cf. Ciardullo 2005). This requires the presence of AGB stars more massive than $0.6\ M_\odot$ (Jacoby 1989; Dopita *et al.* 1992; Vassiliadis and Wood 1994; Blöcker 1995; Stanghellini 1995; but see Schönberner *et al.* 2007 for an opposing view), which in turn must have evolved from progenitors with masses in excess of $2\ M_\odot$ (Weidemann 2000). Such stars are, unfortunately, too sparsely distributed throughout elliptical galaxies – simply because they are too young compared to the main stellar population (cf. Trager *et al.* 2000; Ciardullo *et al.* 2005) – to provide a viable emission mechanism. Alternatively, binary stellar evolution formed through the coalescence of $\sim 1\ M_\odot$ stars (i.e. blue stragglers), which are numerous in elliptical galaxies, could produce sufficient ultraviolet and [O III]λ5007 Å flux to produce the highest-luminosity PNe.

However, Schönberner *et al.* (2007) point out that most models for post-AGB evolution used to date considered only optically thick nebular emission properties (but see Marigo *et al.* 2004). These authors suggest that the constant maximum forbidden-oxygen PN luminosity may be caused by a balanced interplay of the evolutionary timescale of the central star on the one hand and the change of the nebula's optical depth with time on the other. They argue, supported by observations of a Magellanic Cloud PNe sample, that while most PNe are certainly optically thin, those populating the bright end of the PNLF must at least be partially optically thick, with central stellar masses of $\geq 0.6\ M_\odot$. Their hydrodynamical models imply that the shells around more massive central stars remain close to optically thick, thus leading to [O III]λ5007 Å at or just above the observed cut-off magnitude. Alternatively, PNe hosting less massive central stars will become optically thin for Lyman continuum photons. For these objects, their maximum luminosity will be reached around the optically thick/thin transition period, and they will subsequently be fainter and hence not contribute to the high-luminosity cut-off.

4.5 The Tully–Fisher Relation

Perhaps the most widely used secondary distance indicator for spiral galaxies is the Tully–Fisher relation (Tully and Fisher 1977). Its basic premise starts from very simple and straightforward assumptions, yet despite the inherently large uncertainties associated with this approach, the method has stood the test of scientific scrutiny remarkably well.

The underlying idea is that galaxies behave like self-gravitating systems. For such systems, the **virial theorem** implies that the average speed with which test masses (e.g. stars in a galaxy) orbit the system's centre increases with the system's mass, $M_{\rm virial}$:

$$M_{\rm virial} = \eta \frac{R_{\rm eff}\sigma_0^2}{G}, \tag{4.13}$$

where $R_{\rm eff}$ is the system's **effective (∼half-mass) radius**, σ_0 its central velocity dispersion (for spiral galaxies, the rotational velocity, $v_{\rm c}$, is normally used), G the usual Newtonian gravitational constant and η a proportionality constant, usually assumed to be in the range $7.5 \le \eta \le 10$ (but see Fleck *et al.* 2006; Kouwenhoven and de Grijs 2008).

We might also make the reasonable assumption that a galaxy's luminous mass increases proportionally to its virial mass, which essentially implies that we assume a fairly constant M/L among spiral galaxies (see Section 4.5.1). Combined, these two basic assumptions lead us to suggest that the intrinsic luminosity, L, of a galaxy should increase with the amplitude of the mean orbital velocities as

$$L \propto \frac{v_{\rm c}^4}{I_{\rm eff}(M/L)^2}, \tag{4.14}$$

where

$$I_{\rm eff} = \frac{L}{\pi R_{\rm eff}^2} \tag{4.15}$$

is an expression for the surface brightness. Thus, if the product of a galaxy's surface brightness and its $(M/L)^2$ is constant, we retrieve the TFR,

$$L \propto v_{\rm c}^4. \tag{4.16}$$

In other words, the stellar and dark matter distributions are somehow linked.

To understand the physical basis of the TFR, we need to consider the full picture of galaxy formation and evolution. The relationship between luminosity and rotation velocity is, in fact, made up of the combination of three contributing scaling relations: one between the total amount of matter and internal dynamics, the second between total and luminous mass, and the third between luminosity and mass, i.e. the M/L. The relationship between the amount of luminous matter and a galaxy's rotational velocity is the outcome of the galaxy formation process, while that between the luminosity and mass originates from its star formation history. Although both galaxy formation and the history of star formation in individual galaxies are physically understood to at least some degree of confidence, the bottle neck in our full physical understanding of the TFR resides in the relationship between luminous and total mass in a given galaxy and, hence, in the nature of the mysterious dark matter making up a significant fraction of the mass of most spiral galaxies (cf. Gurovich *et al.* 2010).

4.5.1 Wavelength Dependence

Our simple derivation in the previous section, based on straightforward assumptions, naturally results in a TFR slope of approximately 4. If our assumptions of a roughly constant M/L and a constant luminous-to-dark matter fraction hold, the 'standard' CDM paradigm imposes an initial hierarchy of density fluctuations with a power-law slope of $\simeq$2, which in turn predicts a TFR slope of 4, irrespective of the dark matter distribution (cf. Rhee 2004; Gurovich *et al.* 2010; and references therein).

Observational studies using a variety of galaxy samples tend to conclude that a TFR slope of ∼4 is generally found on the basis of near-IR data (e.g. Aaronson *et al.* 1979; Peletier and

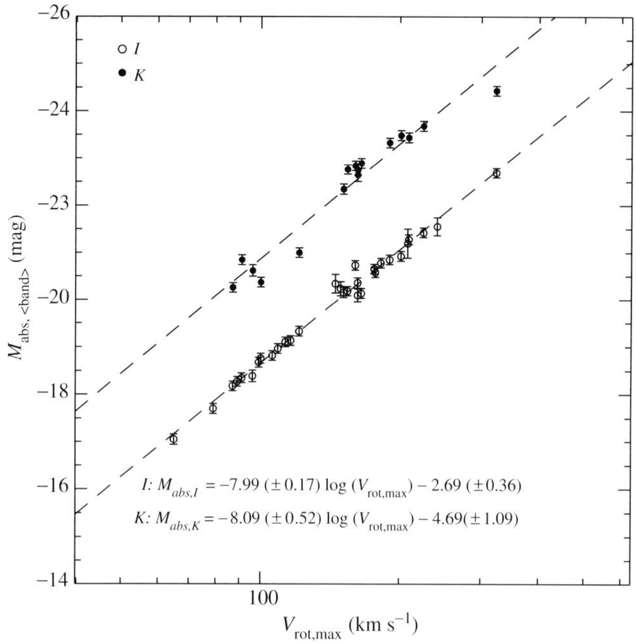

Figure 4.8 *I- and K-band TFRs for a sample of edge-on spiral galaxies (de Grijs and Peletier 1999).*

Willner 1993; Rhee 1996, 2004; de Grijs and Peletier 1999; see Figure 4.8), with shallower slopes resulting from optical observations (cf. Mathewson *et al.* 1992; Pierce and Tully 1992; de Grijs and Peletier 1999; Rhee 2004). This wavelength dependence of the TFR (see Pierce and Tully 1992 for a review), combined with the existence of colour–magnitude relations (CMRs) within and among galaxies (e.g. de Jong 1995; de Grijs and Peletier 1999; see also Section 4.7), implies that the TFR is not a *fundamental* relation (cf. Rhee 2004). In fact, the more fundamental relation of interest for our physical understanding of the TFR is that between a galaxy's luminous mass and the internal dynamics, $M_{\rm lum} - v_{\rm c}$ (see also Rhee 2004; Figure 4.9). It is plausible that this relation reflects the initial conditions of the galaxy formation era.

Spatial variations in the initial conditions for galaxy formation would be manifested as scatter in the $M_{\rm lum} - v_{\rm c}$ relation through a range in dark-to-luminous matter ratios. After all, if the collapse factor varies among galaxies, this would be reflected in variations in the mass ratio of the dark and luminous matter. Different gravitational potentials associated with a range of galaxy haloes – caused by variations in their formation times – would increase the scatter in the halo mass versus dynamics relationship (see Eisenstein and Loeb 1996) and hence the scatter in the $M_{\rm lum} - v_{\rm c}$ relation. In other words, the intrinsic scatter in the $M_{\rm lum} - v_{\rm c}$ relation may give us an insight into the initial conditions of galaxy formation. Unfortunately, the observational scatter in all systematic studies undertaken to date is dominated by random errors.

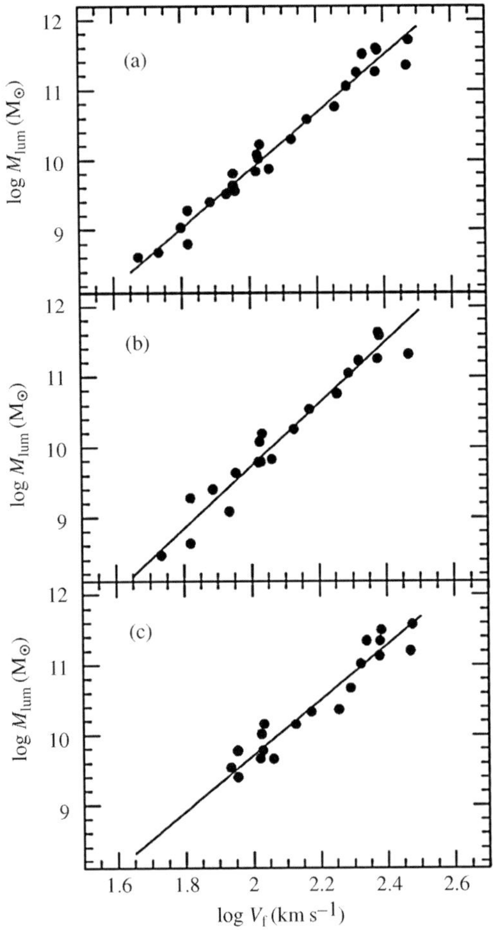

Figure 4.9 *Luminous mass–rotation velocity relation for three different disc models (Rhee 2004). v_f: rotational velocity determined at the flat part of the rotation curve.* (Reprinted from M. H. Rhee, *Journal of the Korean Astronomical Society*, **37**, On the physical basis of the Tully–Fisher relation, p. 15–39, Copyright 2004, with permission of the KAS.)

4.5.2 The Scatter in the Tully–Fisher Relation

Although the TFR slope, particularly at near-IR wavelengths, seems to be understood – and predicted! – fairly well on the basis of the standard CDM cosmology, most models overpredict the scatter on the relationship. This significant discrepancy gives rise to the fundamental question as to whether the small observed scatter may be left over from the cosmological initial conditions or, alternatively, caused by feedback due to gas dynamics and star formation in galaxies (e.g. Eisenstein and Loeb 1996).

Alternatively, there may be a much more straightforward explanation of the small observed scatter. The uncertainties in the individual observables are not fully independent, but often correlated. For instance, as pointed out by Rhee (2004), the errors in v_c and L are linked through the inclination correction for v_c and the inclination-dependent extinction correction for L. If one were to ignore the correlated nature of these uncertainties, this would result in an underestimate of the total error budget.

On the other hand, the small scatter may be physical: a local TFR, relating the local rotational velocity and the integrated luminosity inside the corresponding radius, suggests that the halo-regulated star formation model (Wyse and Silk 1989), which builds upon the Schmidt–Kennicutt star formation law, may be feasible. Here, the local star formation rate is not only driven by the gas surface density but also depends on the local angular frequency, $v(r)/r$, so that the local star formation rate and the halo gravitational potential are linked through the galaxy's rotation curve. This may result in a correlation between v_c and the local luminosity (cf. Madore and Woods 1987; Kennicutt 1989; see also Petrou 1981 and Zaritsky 1992 for related correlations). Rhee (2004) uses these arguments to propose the existence of a well-defined $M_{\rm lum}-L$ relation if the local and global properties of galaxies are physically dependent upon each other. He also suggests that this would naturally explain the wavelength dependence of the TFR. If this scenario is correct, the TFR is a result of the feedback associated with gas dynamics and star formation instead of the imprint of the cosmological initial conditions of galaxy formation (see also Silk 1997 versus Mo *et al.* 1998; Steinmetz and Navarro 1999).

One of the most important assumptions in our derivation of the TFR from first principles is that of a roughly constant M/L among a wide range of spiral galaxies. However, studies using optical and near-IR observations of spiral galaxies (e.g. de Jong 1995) indicate large differences in the $M_{\rm lum}/L$ ratios both within and among galaxies. In addition, late-type galaxies (Scd and later) tend to be more metal poor and younger, on average, than their earlier-type counterparts. This implies significantly varying $M_{\rm lum}/L$ ratios as a function of wavelength, hence likely giving rise to a substantial fraction of the intrinsic scatter in the TFR.

The situation becomes somewhat less clear-cut at distances beyond the nearest large galaxy clusters. At intermediate redshifts ($z \sim 0.5$–1) and particularly in bluer passbands, the scatter on the observed TFR is much greater than expected from our knowledge of the local relation, particularly at the low-luminosity/velocity extreme (e.g. Böhm *et al.* 2004; Conselice *et al.* 2005). This could be caused either by observational incompleteness or by luminosity evolution (see Böhm and Ziegler 2007). In the latter case, distant low-luminosity galaxies would be characterized by significantly lower M/L ratios than in the local Universe, while higher-luminosity galaxies would not exhibit strong M/L evolution.

On the other hand, it is now well established that at least some of the increased scatter in the TFR at intermediate redshifts can be attributed to kinematic peculiarities, from non-circular, nonordered motions to galaxy mergers (e.g. Kannappan and Barton 2004; Flores *et al.* 2006; Kassin *et al.* 2007; Puech *et al.* 2008; and references therein). In particular, Weiner *et al.* (2006) and Kassin *et al.* (2007) showed convincingly that once corrected for systematic deviations from circular rotation, the distant TFR exhibits a significantly decreased dispersion, and no evidence for evolution from intermediate redshifts to the present time in zero point, slope or intrinsic dispersion (see also Flores *et al.* 2006 for the latter). However, Puech *et al.* (2008) report evolution in the zero point of the near-IR (K-band) TFR, suggesting an average brightening by 0.66 ± 0.14 mag of $z \sim 0.6$ galaxies compared to their local counterparts. They attribute this to a possible change in the dominant galaxy type, from quiescent spirals in the local Universe to starburst and luminous IR galaxies at intermediate redshifts. They also infer from their data that a significant fraction of the progenitors of present-day spiral galaxies have roughly doubled their mass over the past ∼6 Gyr, hence leading to a shift in the TFR zero-point calibration.

4.6 Distance Indicators Specific to Elliptical Galaxies

A relation analogous to the TFR exists for elliptical galaxies, first established in the B filter by Faber and Jackson (1976) for a sample of local elliptical and S0-type galaxies (cf. Figure 4.10),

$$L \propto \sigma_0^4, \tag{4.17}$$

which follows naturally from Equations (4.13) and (4.16). However, the scatter on the relationship of ∼0.6 mag is much greater than expected from simple measurement errors (e.g. Dressler *et al.* 1987). This implies, therefore, that distance determinations based on the **Faber–Jackson relation** are inaccurate by ∼25%. It also implies that, while mass is an important driver of a galaxy's random motions, it is not the only parameter of importance (cf. Nigoche-Netro *et al.* 2010). In fact, the properties of elliptical galaxies and the

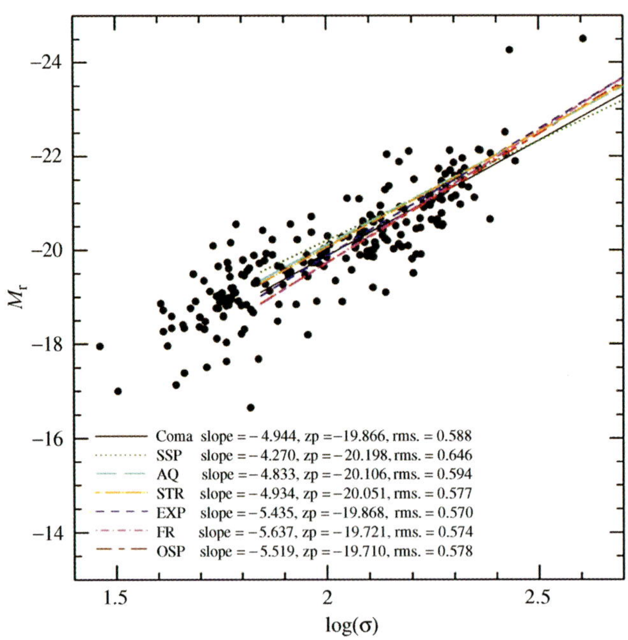

Figure 4.10 *Coma cluster Faber–Jackson relation compared to various model predictions (Allanson et al. 2009).* (Reprinted from S. P. Allanson *et al.*, *Astrophysical Journal*, **702**, The star formation histories of red-sequence galaxies, mass-to-light ratios and the Fundamental Plane, p. 1275–1296, Copyright 2009, with permission of the AAS and M. Hudson.)

bulges of S0 galaxies also depend on their effective radii through their **effective surface brightnesses** ($\Sigma_{\rm eff}$).

We can derive the latter from first principles, assuming that elliptical galaxies are good approximations to virialized systems, so that (see Equation 4.13)

$$\sigma_{\rm eff}^2 = \frac{GM_{\rm ell}}{R_{\rm eff}}, \tag{4.18}$$

where $\sigma_{\rm eff}$ is the mean velocity dispersion within $R_{\rm eff}$ and $M_{\rm ell}$ is the galaxy's mass. Combined with the definition of the effective surface brightness,

$$\Sigma_{\rm eff} = \frac{L/2}{\pi R_{\rm eff}^2}, \tag{4.19}$$

we obtain

$$L\,\Sigma_{\rm eff}(M_{\rm ell}/L)^2 \propto \sigma_{\rm eff}^4. \tag{4.20}$$

Thus, if the assumption holds that both the surface brightness and the M/L ratios of elliptical galaxies are near constant, we obtain the Faber–Jackson relation. Clearly, the large observed scatter implies that one or both of these parameters are not sufficiently invariant to warrant the use of the Faber–Jackson relation as an accurate secondary distance estimator (see also Nigoche-Netro *et al.* 2010 for a discussion of selection biases).

In fact, Binney and Merrifield (1998, their Figure 4.43) show that there is significant real scatter on any of the two-parameter correlations defining the properties of elliptical galaxies. The nonnegligible importance of a second driving parameter, most likely $R_{\rm eff}$, suggests that elliptical galaxies occupy a 3D plane in parameter space. By combining the virial theorem and the definition of the effective surface brightness – Equations (4.13) and (4.19), respectively – we obtain

$$R_{\rm eff} = \left(\frac{\rm constant}{2\pi}\right)\left(\frac{M_{\rm ell}}{L}\right)^{-1}\sigma_0^2\,\Sigma_{\rm eff}^{-1}. \tag{4.21}$$

If $(M_{\rm ell}/L)$ is roughly constant, Equation (4.21) describes a 3D plane, referred to as the *Fundamental Plane* of elliptical galaxies (Djorgovski and Davis 1987; Dressler *et al.* 1987; cf. Figure 4.11).

Surprisingly, the FP is much better defined than perhaps expected from first principles: the scatter perpendicular to the plane is very small, which in turn implies that $M_{\rm ell}/L$ varies by less than ~10% among elliptical galaxies. The observed FP is best represented by

$$R_{\rm eff} \propto \sigma_0^{1.4\pm0.15}\,\Sigma_{\rm eff}^{-0.9\pm0.1}. \tag{4.22}$$

This small but significant deviation or 'tilt' observed from the 'virial' FP is consistent with the assumption of virial equilibrium if

$$\left(\frac{2\pi}{\rm constant}\right)\left(\frac{M_{\rm ell}}{L}\right) \propto M_{\rm ell}^{0.2} \propto L^{0.25}, \tag{4.23}$$

which implies an underlying regularity among elliptical galaxies in dynamical structure (σ_0), stellar $M_{\rm ell}/L$, dark matter content inside $R_{\rm eff}$ and the slope of the stellar initial

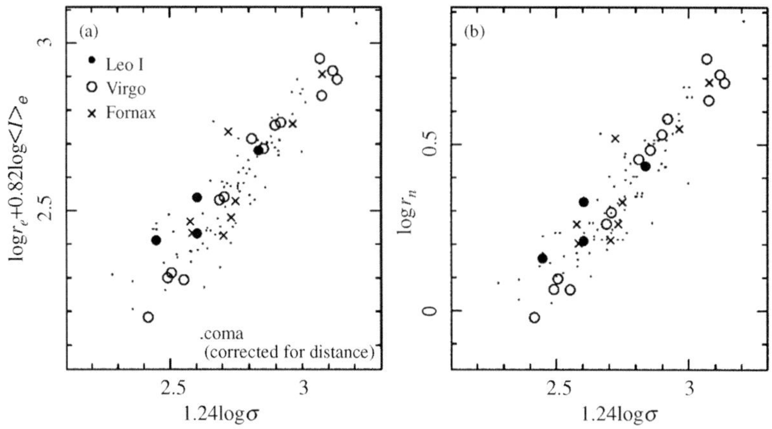

Figure 4.11 *(a) Fundamental plane and (b) D_n–σ relations in the Coma, Leo I, Virgo and Fornax clusters, where distance effects have been removed (Kelson et al. 2000).* (Reprinted from D. D. Kelson *et al.*, *Astrophysical Journal*, **529**, The Hubble Space Telescope Key Project on the extragalactic distance scale. XXVII. A derivation of the Hubble constant using the Fundamental Plane and D_n–σ relations in Leo I, Virgo, and Fornax, p. 768–785, Copyright 2000, with permission of the AAS and D. D. Kelson.)

mass function, among others. Nevertheless, the tilt in the FP suggests either a systematic variation in the stellar M/L or in central dark matter content, or a 'structural nonhomology', i.e. a variation in stellar dynamics as a function of host galaxy mass (e.g. Prugniel and Simien 1996; Trujillo *et al.* 2004; Cappellari *et al.* 2006; La Barbera *et al.* 2008; Allanson *et al.* 2009). As such, the FP is an important tool for studies of galaxy evolution on cosmic timescales. On the basis of *independent* mass and velocity measurements from gravitational lensing observations, Bolton *et al.* (2007) claim that the mass dynamical structure of early-type galaxies (i.e. their stellar dynamics) is independent of mass. They then suggest that this implies that the tilt in the FP is caused by systematic M_{total}/L variations, where they do not distinguish between luminous and dark mass.

It has been shown, both observationally and through detailed stellar population analysis (e.g. Thomas *et al.* 2005; Allanson *et al.* 2009), that at least some of the tilt in the FP is likely caused by local variations in and evolution of the M/L. The FP's zero point is a function of redshift or approximate age of the stellar population and environment (e.g. Treu *et al.* 2001, 2002; van de Ven *et al.* 2003; van der Wel *et al.* 2004), in the sense that at intermediate redshifts ($z \sim 0.4$–1) early-type galaxies are brighter than expected from the local FP relation for a given effective radius and velocity dispersion, although the scatter on the FP remains almost constant. The modest evolution of the FP's zero point and its tightness up to intermediate redshifts is consistent with passive evolution and an old age of the stellar populations of early-type galaxies (e.g. Treu *et al.* 2002; see also van der Wel *et al.* 2004), with (quenched) star formation redshifts for the bulk of the stellar populations of $z \approx 2$–3 (e.g. Kelson *et al.* 1997; van de Ven *et al.* 2003; van Dokkum and Stanford 2003; van der Wel *et al.* 2004; Harker *et al.* 2006), possibly enhanced by more recent bursts of star formation involving $\lesssim$30% of the stellar mass in lower-mass galaxies ('**downsizing**': Trager *et al.* 2000; van der Wel *et al.* 2004).

The tilt decreases systematically with increasing wavelength (Pahre *et al.* 1998; see also Allanson *et al.* 2009), while the average rate of luminosity evolution, which depends on galaxy mass, increases with redshift (Treu *et al.* 2001, 2002, 2005a,b; van de Ven *et al.* 2003; van der Wel *et al.* 2004). However, Bolton *et al.* (2007) suggest that most of the derived variation in $M_{\rm total}/L$ may be caused by varying dark matter fractions (see also Allanson *et al.* 2009). They propose an analogous, tighter plane in mass space, defined by $R_{\rm eff}$, σ_0 and the surface mass density within $R_{\rm eff}/2$, which eliminates the FP's tilt almost entirely and no longer includes a dependence on the galaxy's luminosity. The latter is interesting from the point of view of using this alternative 'mass plane' as a potential distance indicator since it allows the plane to be tracked across cosmic time without the need to account for stellar population changes, provided it can be calibrated locally.

For purposes of distance determination, one commonly uses a specific 2D, 'edge-on' projection of the FP provided by the $D_n-\sigma$ **relation** (Dressler *et al.* 1987; see e.g. Figure 4.11). D_n is the diameter within which $I_{\rm eff} = 20.75\mu_B$ (B-band surface brightness), although the exact surface brightness value is unimportant. It follows that $D_n/R_{\rm eff}$ is larger for higher surface brightness galaxies than for those characterized by lower surface brightnesses, in the sense that, from an integration of the $R^{1/4}$ de Vaucouleurs surface brightness profile,

$$D_n \propto R_{\rm eff} I_{\rm eff}^{0.8}, \tag{4.24}$$

or, because $\Sigma_{\rm eff} = -2.5 \log I_{\rm eff}$,

$$D_n \propto \sigma_0^{1.4} \Sigma_{\rm eff}^{0.07}. \tag{4.25}$$

Since the latter equation exhibits only a weak dependence on $\Sigma_{\rm eff}$, this is essentially a relationship between D_n and σ_0. For the Virgo cluster, Dressler *et al.* (1987) derived

$$D_n = 2.05 \left(\frac{\sigma}{100 \text{ km s}^{-1}} \right)^{1.33} \text{kpc}, \tag{4.26}$$

assuming a distance to the centre of the Virgo cluster of 16 Mpc. Note that the resulting distances are uncertain by ~10–15%, although the relationship is very powerful for larger galaxy samples located at roughly the same distance, such as galaxy clusters (but see Gudehus 1991 for a discussion of systematic biases affecting this relationship).

The three-parameter FP is not a unique solution for constraining the full parameter space occupied by elliptical galaxies. Bender *et al.* (1992) suggested an alternative plane with 'edge-on' projections driven by physical parameters, with axes (cf. Figure 4.12)

$$\kappa_1 = \frac{\log(\sigma_0^2 R_{\rm eff})}{\sqrt{2}} \propto \log M_{\rm ell}, \tag{4.27}$$

$$\kappa_2 = \frac{\log(\sigma_0^2 I_{\rm eff}^2 / R_{\rm eff})}{\sqrt{6}} \propto \log(I_{\rm eff}(M_{\rm ell}/L)^{1/3}) \quad \text{and} \tag{4.28}$$

$$\kappa_3 = \frac{\log(\sigma_0^2/(I_{\rm eff}^2 R_{\rm eff}))}{\sqrt{3}} \propto \log(M_{\rm ell}/L). \tag{4.29}$$

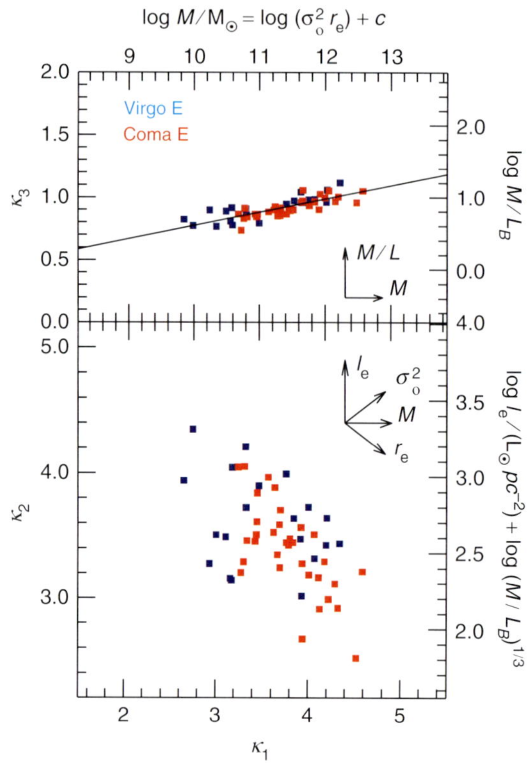

Figure 4.12 Distribution of Virgo and Coma cluster elliptical galaxies in κ **space** *(Bender et al. 1992), defined by the central velocity dispersion (σ_0^2), surface brightness – log $I_e = -0.4\ (\Sigma_e - 27)$ – and effective radius (r_e).* (Reprinted from R. Bender et al., *Astrophysical Journal*, **399**, Dynamically hot galaxies. I. Structural properties, p. 462–477, Copyright 1992, with permission of the AAS and R. Bender.)

This alternative coordinate system is the most physically meaningful orthogonal system in which the FP is projected exactly edge-on. κ_1 is a simple measure of galaxy size, while κ_3 is designed to project the FP edge-on in combination with κ_1 (Bender *et al.* 1992); consequently, the κ_1/κ_2 projection results in approximately a face-on view of the FP. This forces κ_3 to be some combination of log $M_{\rm ell}$ and log($M_{\rm ell}/L$), because $M_{\rm ell}/L \propto L^{0.2}$ is the FP's defining property. κ_2 represents an orthogonal axis to both the κ_1 and κ_3 axes. Bender *et al.* (1992) showed that the tightest projection results from the κ_1/κ_3 relationship, suggesting that elliptical galaxies are characterized by a narrow range in M/L, which correlates weakly with galaxy type – except for dark-matter-dominated, low-luminosity dwarf spheroidal galaxies – and/or mass (a factor of $\sim$3 increase over five orders of magnitude in M/L). Bender *et al.* (1992) conclude that the smooth variation with galaxy mass of a number of key physical properties along the κ-space FP is consistent with a merger history, in which the more massive galaxies were systematically more stellar and less gaseous than their lower-mass counterparts.

The similarity of the M/L ratios and dynamical structures of elliptical galaxies suggests that their stellar populations, ages and dark matter properties are very uniform, with M/L ratios of $\sim$10–20 (in solar units) and a scatter of less than $\sim$10%. The former are consistent with the absence of dark matter within $R_{\rm eff}$, while the latter places strong constraints on the ages ($\sim$10–13 Gyr; cf. van Dokkum and Franx 1996) and metallicities of elliptical galaxies.

4.7 The Colour–Magnitude Relation

The tightness of the CMR for *early-type* galaxies – first established by Baum (1959) and de Vaucouleurs (1961) – makes it potentially useful as a distance indicator, as first suggested by Sandage (1972). Visvanathan and Griersmith (1977) extended the range of galaxy types from elliptical/S0 (E/S0) galaxies to early-type spirals (S0/a to Sab). They found, within the errors, exactly the same optical CMR for the early-type Virgo cluster spirals as had been found for E/S0 galaxies, but with a larger excess *scatter* (see also Visvanathan and Sandage 1977; Griersmith 1980; Bower *et al.* 1992b; Peletier and de Grijs 1998).

Despite the good agreement between the elliptical and spiral galaxy CMRs, later-type spiral galaxies occupy a different region in the colour–magnitude diagram (e.g. Tully *et al.* 1982; Mobasher *et al.* 1986; Peletier and de Grijs 1998; see Figure 4.13). Griersmith (1980) also noticed that differences in zero point of the CMRs seem to follow a systematic trend along the Hubble sequence: colours become systematically bluer for later Hubble types

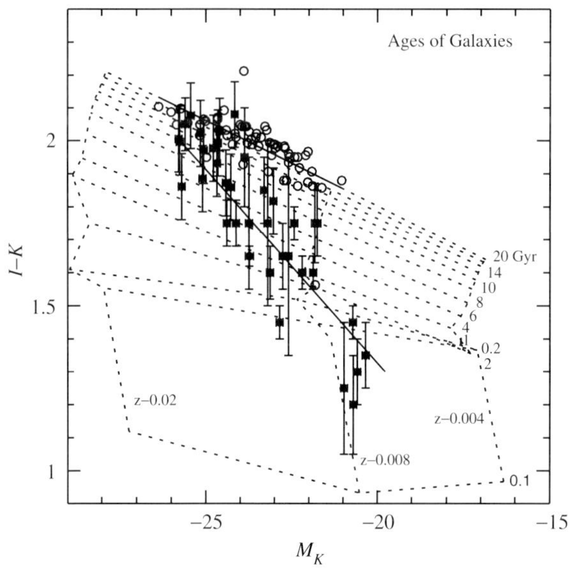

Figure 4.13 *$I - K$ versus M_K (colour–magnitude) relation for elliptical/S0 galaxies from Bower et al. (1992a,b: open circles) and spiral discs (filled symbols), compared to standard simple stellar population models for which metallicities and ages are given (Peletier and de Grijs 1998).*

(see also Peletier and de Grijs 1998), although the slopes for the individual galaxy types are the same within the errors compared to each other and to E/S0 galaxies.

Many attempts have been made to unambiguously determine the universality of the relationship (for early attempts, see e.g. Visvanathan and Griersmith 1977; Visvanathan and Sandage 1977; Griersmith 1980; Aaronson *et al.* 1981; Bower *et al.* 1992a,b, 1998). In general, the ultraviolet–optical CMR *for early-type galaxies* was found to be universal and independent of environment to a high degree (for a review see Ellis *et al.* 1997). However, a number of studies (e.g. Burstein 1977; Faber 1977; Larson *et al.* 1980; Aaronson *et al.* 1981; Abraham *et al.* 1996; van Dokkum *et al.* 1998) subsequently claimed that non-negligible environmental effects were playing a role in the observational evidence. Bower *et al.*'s (1992a,b) observations of Virgo and Coma cluster galaxies support a universal CMR for cluster E/S0 galaxies, over the entire wavelength range, although the dispersion is expected to be considerable. Bower *et al.*'s (1992b) photometry allows the CMR for early-type cluster galaxies to be used to estimate distances to an accuracy of ~20% per galaxy.

For the E/S0 galaxies in the Virgo cluster, the change of colour with absolute magnitude is greatest in the ultraviolet and decreases significantly towards redder wavelengths (Visvanathan and Sandage 1977 and references therein). However, the effect shows up again in the optical/near-IR regime, in the sense that $V - K$ colours are bluer for intrinsically fainter galaxies (e.g. Aaronson *et al.* 1981; Tully *et al.* 1982). The usefulness of near-IR observations of spiral galaxies in the field for measuring extragalactic distances (de Grijs and Peletier 1999) depends on whether *spiral* galaxies in the field follow a universal CMR in the near-IR and, if so, whether the scatter is sufficiently small for a useful application of the relation (see Figure 4.14).

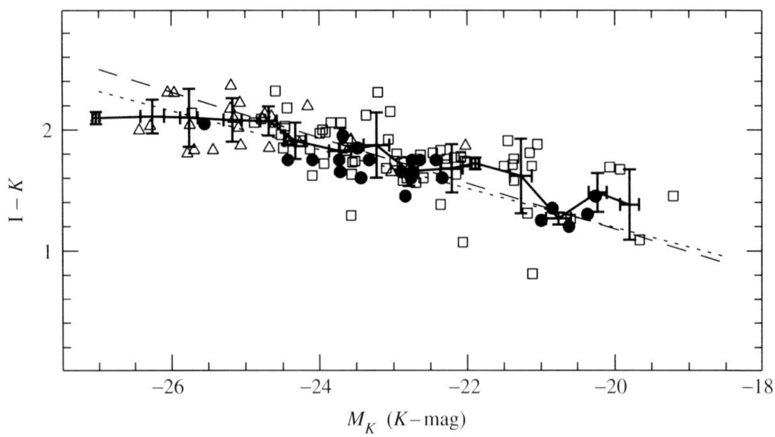

Figure 4.14 *Composite CMR (de Grijs and Peletier 1999) for various observational galaxy samples. The best-fitting composite relation is indicated by the dashed line; the near-IR CMR of de Grijs and Peletier (1999) is indicated by the dotted line. The solid line was obtained by binning the data points in M_K bins of 0.5 mag; the error bars indicate the dispersion in the data points in each bin.*

The main advantage of near-IR observations compared to observations in blue passbands is their relative insensitivity to contamination by the presence of young stellar populations and dust. Absorption corrections for dust in external galaxies, which are largest in the blue, are difficult and controversial. Peletier and de Grijs (1998) found that the scatter in the near-IR CMR for field spiral galaxies can entirely be explained by observational uncertainties. Moreover, they showed that the slope of the relation is steeper for spirals than for ellipticals. We can explain this if we assume that the CMRs for spiral and elliptical galaxies are intrinsically different, in the sense that the stars in spiral galaxies are younger than those in ellipticals and the fraction of young stars in a spiral galaxy (i.e. its 'age') is determined solely by the galaxy's luminosity and not by its environment. The CMR formed by elliptical galaxies, on the other hand, is generally attributed to changes in metallicity.

For distance determinations to individual spiral galaxies (de Grijs and Peletier 1999), we have to take into account the scatter in and the shallow slope of the CMR, as well as the observational errors. By obtaining high-quality observations of and improved distances to moderately or highly inclined galaxies, the accuracy of distance determinations based on CMR analysis can be enhanced by reducing the observational scatter in the apparent K-band magnitudes to a lower limit of $\sim$0.02 mag. The accuracy will thus be limited by the intrinsic dispersion ($\sim$0.5 mag), resulting in a maximum accuracy of distance determinations to spiral galaxies of $\sim$25%. Note that this accuracy approaches the 20% accuracy of Bower *et al.* (1992b) for early-type galaxies, for which the intrinsic scatter in the CMR is significantly smaller.

From a physical point of view, the origin of the intrinsic scatter in the CMR is most interesting. The intrinsic dispersion in the CMR for field spiral galaxies is most likely caused by the nonnegligible effects of different degrees of active star formation, star formation histories, ages, metallicities and extinction. The intrinsic dispersion places strong constraints on the age spread allowed for the dominant fraction of the stellar populations in the galaxies determining the CMR. For early-type galaxies, which have a star formation history dominated by an early, enhanced star formation episode followed by more quiescent evolution, the intrinsic age spread must be $\leq$4 Gyr (Bower *et al.* 1998), although we cannot set tight limits on the subsequent lower level of star formation. In this context, perhaps the most interesting conclusion one can draw from the amount of scatter on the early-type CMR is that the bulk of these galaxies must have been in place $\sim$5 Gyr ago.

Second, the small scatter on the CMR can also help constrain the predominance of galaxy merger activity: unequal-mass mergers lead to a change in CMR slope and an associated increase in the scatter on the relationship. Bower *et al.* (1998) conclude that, in a realistic scenario in which large galaxies form hierarchically, the CMR persists through a large number of merging stages, provided that massive galaxies preferentially merge with systems of similar mass. In general, these constraints on galaxy evolutionary histories suggest that the bulk (>50%) of the present-day stellar populations in galaxies must have formed by $z \sim 1$, and that these galaxies must have been already old enough when undergoing their last merger(s) so that the effects of stellar evolution are small enough to not affect the CMR slope and scatter significantly, except at the reddest extreme (e.g. Hogg *et al.* 2004; Skelton *et al.* 2009). Hogg *et al.* (2004) use the Sloan Digital Sky Survey database to show that the CMR is also essentially independent of environment or galaxy density (see Figure 4.15).

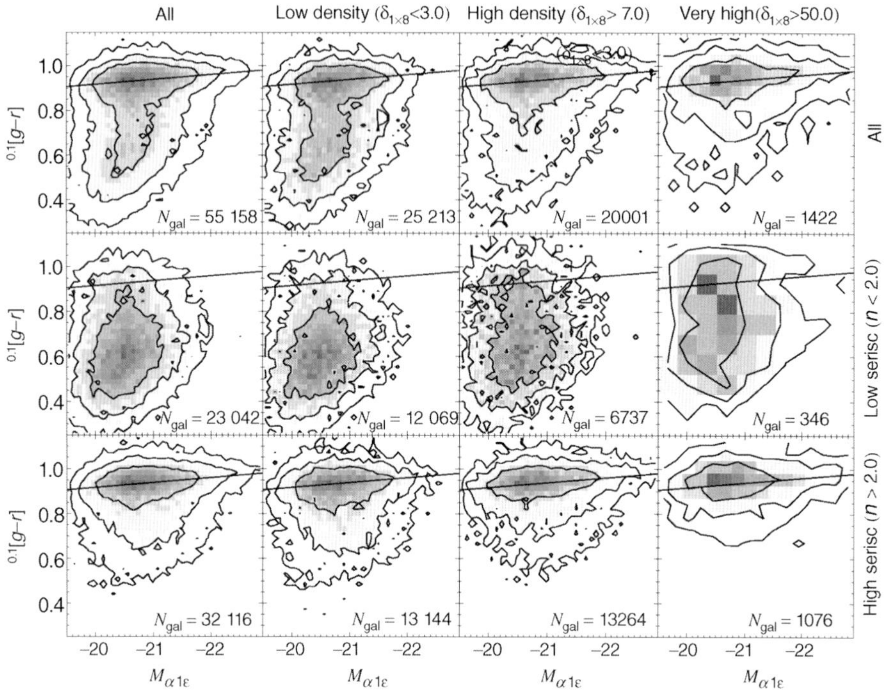

Figure 4.15 *Colour–magnitude relations for Sloan Digital Sky Survey galaxies in the redshift range $0.08 < z < 0.12$ (Hogg et al. 2004). Columns: subsamples cut in overdensity $\delta_{1\times8}$ (a mean density environment has $\delta_{1\times8} = 0$). Rows: subsamples cut in Sérsic index n (n = 1: exponential disc; n = 4: de Vaucouleurs $r^{1/4}$ profile). In each panel, the grey scale monotonically represents the abundance of sample galaxies in the two-dimensional space of colour and magnitude and the contours enclose 52.0, 84.3 and 96.6% of the sample. Overplotted on all panels is the same straight solid line showing the best-fitting CMR from the bottom left-hand panel.* (Reprinted from D. W. Hogg *et al.*, *Astrophysical Journal*, **601**, The dependence on environment of the color–magnitude relation of galaxies, L29–L32, Copyright 2004, with permission of the AAS and D. W. Hogg.)

4.8 HII Regions as Distance Indicators?

Because of their large sizes and high luminosities, giant HII regions are attractive targets as potential extragalactic distance indicators. To this aim, several attempts have been made to use their sizes, velocity dispersions or luminosities, possibly as a function of the properties of or their position in their host galaxies (e.g. Gum and de Vaucouleurs 1953; Sérsic 1960; Sandage and Tammann 1974; Kennicutt 1979, 1984, 1988; de Vaucouleurs 1979; van den Bergh 1980; Sisteró 1988; see also the review of Shields 1990).

Although a number of studies have tried to confirm the proposed empirical relations among luminosity, radius and velocity dispersion, no agreement has been found (see, for reviews, Shields 1990; Fuentes-Masip *et al.* 2000). The dispersion in the results may be caused by observational limitations (Gallagher and Hunter 1983; Hippelein 1986), nonuniform HII region sample selection criteria (Roy *et al.* 1986; Arsenault and Roy 1988), different approaches to estimating HII region radii (e.g. Sandage and Tammann 1974; Melnick 1977; Gallagher and Hunter 1983; Relaño *et al.* 2005) or different treatments of the regions' line profiles and asymmetries (see Relaño *et al.* 2005 for a review).

Rozas *et al.* (1998) published one of the most complete studies, covering the entire HII region population for a single galaxy (M100), but they could not confirm the proposed $L \propto \sigma^4$ relation, which would be expected if the regions were in virial equilibrium and light traces mass: see the arguments leading to Equation (4.16). Instead, they obtained a lower envelope (in σ) in the $L-\sigma$ diagram which they suggested is formed by those HII regions that are close to virial equilibrium. In addition, Terlevich and Melnick (1981) found a scatter in their $L-\sigma$ relation which they suggested might originate from the different metallicities of the HII regions in their sample. In view of these developments, it appears that the observational uncertainties and intrinsic physical conditions in giant HII regions prevent their use as reliable standard candles.

Bibliography

Aaronson M, Huchra J and Mould J 1979 The infrared luminosity/HI velocity–width relation and its application to the distance scale. *Astrophys. J.* **229**, 1–13.

Aaronson M, Persson SE and Frogel JA 1981 The infrared color–magnitude relation for early-type galaxies in Virgo and Coma. *Astrophys. J.* **245**, 18–24.

Abraham RG, Smecker-Hane TA, Hutchings JB, Carlberg RG, Yee HKC, Ellingson E, Morris S, Oke JB and Rigler M 1996 Galaxy evolution in Abell 2390. *Astrophys. J.* **471**, 694–719.

Ajhar EA, Lauer TR, Tonry JL, Blakeslee JP, Dressler A, Holtzman JA and Postman M 1997 Calibration of the surface brightness fluctuation method for use with the Hubble Space Telescope. *Astron. J.* **14**, 626–634.

Allanson SP, Hudson MJ, Smith RJ and Lucey JR 2009 The star formation histories of red-sequence galaxies, mass-to-light ratios and the Fundamental Plane. *Astrophys. J.* **702**, 1275–1296.

Arsenault R and Roy JR 1988 Correlations between integrated parameters and Hα velocity widths in giant extragalactic HII regions. A new appraisal. *Astron. Astrophys.* **201**, 199–207.

Ashman KM and Zepf SE 1998 *Globular Cluster Systems*. Cambridge University Press.

Ashman KM, Conti A and Zepf SE 1995 Globular cluster systems as distance indicators: metallicity effects on the luminosity function. *Astron. J.* **110**, 1164–1170.

Barmby P, Huchra JP and Brodie JP 2001 The M31 globular cluster luminosity function. *Astron. J.* **121**, 1482–1496.

Baum WA 1959 Population inferences from star counts, surface brightness and colors. *Publ. Astron. Soc. Pac.* **71**, 106–117.

Baumgardt H 1998 The initial distribution and evolution of globular cluster systems. *Astron. Astrophys.* **330**, 480–491.

Baumgardt H and Makino J 2003 Dynamical evolution of star clusters in tidal fields. *Mon. Not. R. Astron. Soc.* **340**, 227–246.

Bender R, Burstein D and Faber SM 1992 Dynamically hot galaxies. I. Structural properties. *Astrophys. J.* **399**, 462–477.

Binney J and Merrifield M 1998 *Galactic Astronomy*. Princeton University Press.

Biscardi I, Raimondo G, Cantiello M and Brocato E 2008 Optical surface brightness fluctuations of shell galaxies toward 100 Mpc. *Astrophys. J.* **678**, 168–178.

Blakeslee JP and Tonry JL 1996 Globular clusters in Fornax: does M_0 depend on environment? *Astrophys. J.* **465**, L19–L22.

Blakeslee JP, Davis M, Tonry JL, Dressler A and Ajhar EA 1999 A first comparison of the surface brightness fluctuation survey distances with the galaxy density field: implications for H_0 and Ω. *Astrophys. J.* **527**, L73–L76.

Blakeslee JP, Vazdekis A and Ajhar EA 2001 Stellar populations and surface brightness fluctuations: new observations and models. *Mon. Not. R. Astron. Soc.* **320**, 193–216.

Blakeslee JP, Lucey JR, Tonry JL, Hudson MJ, Narayanan VK and Barris BJ 2002 Early-type galaxy distances from the Fundamental Plane and surface brightness fluctuations. *Mon. Not. R. Astron. Soc.* **330**, 443–457.

Blakeslee JP, Jordán A, Mei S, Côté P, Ferrarese L, Infante L, Peng EW, Tonry JL and West MJ 2009 The ACS Fornax cluster survey. V. Measurement and recalibration of surface brightness fluctuations and a precise value of the Fornax–Virgo relative distance. *Astrophys. J.* **694**, 556–572.

Blakeslee JP, Cantiello M, Mei S, Côté P, Barber DeGraaff R, Ferrarese L, Jordán A, Peng EW, Tonry JL and Worthey G 2010 Surface brightness fluctuations in the Hubble Space Telescope ACS/WFC F814W bandpass and an update on galaxy distances. *Astrophys. J.* **724**, 657–668.

Blöcker T 1995 Stellar evolution of low- and intermediate-mass stars. II. Post-AGB evolution. *Astron. Astrophys.* **299**, 755–769.

Böhm A and Ziegler BL 2007 Evolution of field spiral galaxies up to redshifts $z = 1$. *Astron. J.* **668**, 846–852.

Böhm A, Ziegler BL, Saglia RP, Bender R, Fricke KJ, Gabasch A, Heidt J, Mehlert D, Noll S and Seitz S 2004 The Tully–Fisher relation at intermediate redshift. *Astron. Astrophys.* **420**, 97–114.

Bolton AS, Burles S, Treu T, Koopmans LVE and Moustakas LA 2007 A more fundamental plane. *Astrophys. J.* **665**, L105–L108.

Bower RG, Lucey JR and Ellis RS 1992a Precision photometry of early-type galaxies in the Coma and Virgo clusters: a test of the universality of the colour–magnitude relation. I. The data. *Mon. Not. R. Astron. Soc.* **254**, 589–600.

Bower RG, Lucey JR and Ellis RS 1992b Precision photometry of early-type galaxies in the Coma and Virgo clusters: a test of the universality of the colour–magnitude relation. II. Analysis. *Mon. Not. R. Astron. Soc.* **254**, 601–613.

Bower RG, Kodama T and Terlevich A 1998 The colour–magnitude relation as a constraint on the formation of rich cluster galaxies. *Mon. Not. R. Astron. Soc.* **299**, 1193–1208.

Brodie JP and Strader J 2006 Extragalactic globular clusters and galaxy formation. *Annu. Rev. Astron. Astrophys.* **44**, 193–267.

Burkert A and Smith GH 2000 A proposed functional form for the upper mass spectrum of old globular cluster systems. *Astrophys. J.* **542**, L95–L98.

Burstein D 1977 Discussion following 'The chemical composition of old stellar populations (Faber SM).' In *The Evolution of Galaxies and Stellar Populations* (eds Tinsley BM and Larson RB), pp. 191–198. Yale University Observatory, New Haven, CT.

Buzzoni A 1993 Statistical properties of stellar populations and surface-brightness fluctuations in galaxies. *Astron. Astrophys.* **275**, 433–450.

Cantiello M, Raimondo G, Brocato E and Capaccioli M 2003 New optical and near-infrared surface brightness fluctuation models: a primary distance indicator ranging from globular clusters to distant galaxies? *Astron. J.* **125**, 2783–2808.

Cantiello M, Blakeslee JP, Raimondo G, Mei S, Brocato E and Capaccioli M 2005 Detection of radial surface brightness fluctuations and color gradients in elliptical galaxies with the Advanced Camera for Surveys. *Astrophys. J.* **634**, 239–257.

Cantiello M, Blakeslee J, Raimondo G, Brocato E and Capaccioli M 2007 Surface brightness fluctuations from archival ACS images: a stellar population and distance study. *Astrophys. J.* **668**, 130–149.

Cappellari M, Bacon R, Bureau M, Damen MC, Davies RL, de Zeeuw PT, Emsellem E, Falcón-Barroso J, Krajnović D, Kuntschner H, McDermid RM, Peletier RF, Sarzi M, van den Bosch RCE and van de Ven G 2006 The SAURON project. IV. The mass-to-light ratio, the virial mass estimator and the Fundamental Plane of elliptical and lenticular galaxies. *Mon. Not. R. Astron. Soc.* **366**, 1126–1150.

Cerviño M, Luridiana V and Jamet L 2008 On surface brightness fluctuations: probabilistic and statistical bases. I. Stellar population and theoretical surface brightness fluctuations. *Astron. Astrophys.* **491**, 693–701.

Ciardullo R 2003 Planetary nebulae as extragalactic distance indicators. In *Planetary Nebulae: Their Evolution and Role in the Universe* (eds Kwok S, Dopita M and Sutherland R), *Proc. Int'l Astron. Union Symp.* **209**, 617–624.

Ciardullo R 2005 Planetary nebulae and the extragalactic distance scale. In *Planetary Nebulae as Astronomical Tools* (eds Szczerba R, Stasińska G and Górny SK), *AIP Conf. Proc.* **804**, 277–283.

Ciardullo R, Jacoby GH, Ford HC and Neill JD 1989 Planetary nebulae as standard candles. II. The calibration in M31 and its companions. *Astrophys. J.* **339**, 53–69.

Ciardullo R, Feldmeier JJ, Jacoby GH, Kuzio de Naray R, Laychak MB and Durrell PR 2002 Planetary nebulae as standard candles. XII. Connecting the Population I and Population II distance scales. *Astrophys. J.* **577**, 31–50.

Ciardullo R, Sigurdsson S, Feldmeier JJ and Jacoby GH 2005 Close binaries as the progenitors of the brightest planetary nebulae. *Astrophys. J.* **629**, 499–506.

Conselice CJ, Bundy K, Ellis RS, Brichmann J, Vogt NP and Phillips AC 2005 Evolution of the near-infrared Tully–Fisher relation: constraints on the relationship between the stellar and total masses of disk galaxies since $z \sim 1$. *Astrophys. J.* **628**, 160–168.

Coradi RLM 2003 Large-scale ionized outflows from symbiotic stars: a real link with planetary nebulae? In *Symbiotic Stars Probing Stellar Evolution* (eds Corradi RLM, Mikolajewska R and Mahoney TJ), *Astron. Soc. Pac. Conf. Ser.* **303**, 393–406.

Côté P, McLaughlin DE, Hanes DA, Bridges TJ, Geisler D, Merritt D, Hesser JE, Harris GLH and Lee MG 2001 Dynamics of the globular cluster system associated with M87 (NGC 4486). II. Analysis. *Astrophys. J.* **559**, 828–850.

Côté P, Blakeslee JP, Ferrarese L, Jordán A, Mei S, Merritt D, Milosavljević M, Peng EW, Tonry JL and West MJ 2004 The ACS Virgo Cluster Survey. I. Introduction to the survey. *Astrophys. J. Suppl. Ser.* **153**, 223–242.

de Grijs R and Parmentier G 2007 The long-term survival chances of young massive star clusters. *Chin. J. Astron. Astrophys.* **7**, 155–186.

de Grijs R and Peletier RF 1999 Spiral galaxy distance indicators based on near-infrared photometry. *Mon. Not. R. Astron. Soc.* **310**, 157–167.

de Jong RS 1995 Spiral galaxies. The light and color distribution in the optical and near-infrared. PhD thesis, University of Groningen, the Netherlands.

de Vaucouleurs G 1961 Integrated colors of bright galaxies in the *U, B, V* system. *Astrophys. J. Suppl. Ser.* **5**, 233–289.

de Vaucouleurs G 1979 Velocity dispersions in HII regions as distance indicators. *Astron. Astrophys.* **79**, 274–276.

Di Criscienzo M, Caputo F, Marconi M and Musella I 2006 RR Lyrae-based calibration of the globular cluster luminosity function. *Mon. Not. R. Astron. Soc.* **365**, 1357–1366.

Djorgovski S and Davis M 1987 Fundamental properties of elliptical galaxies. *Astrophys. J.* **313**, 59–68.

Dopita MA, Jacoby GH and Vassiliadis E 1992 A theoretical calibration of the planetary nebulae cosmic distance scale. *Astrophys. J.* **389**, 27–38.

Dressler A, Lynden-Bell D, Burstein D, Davies RL, Faber SM, Terlevich R and Wegner G 1987 Spectroscopy and photometry of elliptical galaxies. I. A new distance estimator. *Astrophys. J.* **313**, 42–58.

Dunn LP and Jerjen H 2006 First results from SAPAC: toward a three-dimensional picture of the Fornax cluster core. *Astron. J.* **132**, 1384–1395.

Durrell PR, Harris WE, Geisler D and Pudritz RE 1996 Globular cluster systems in dwarf elliptical galaxies. II. The Virgo cluster. *Astron. J.* **112**, 972–990.

Eddington AS 1929 *The Nature of the Physical World*, p. 242. Transcript of his 1927 Gifford lecture at the University of Edinburgh. University of Michigan Press, 1981.

Eisenstein DJ and Loeb A 1996 Can the Tully–Fisher relation be the result of initial conditions? *Astrophys. J.* **459**, 432–437.

Ellis RS, Smail I, Dressler A, Couch WJ, Oemler Jr A, Butcher H and Sharples RM 1997 The homogeneity of spheroidal populations in distant clusters. *Astrophys. J.* , **483**, 582–596.

Elmegreen BC and Efremov YuN 1997 A universal formation mechanism for open and globular clusters in turbulent gas. *Astrophys. J.* **480**, 235–245.

Faber SM 1977 The chemical composition of old stellar populations. In *The Evolution of Galaxies and Stellar Populations* (eds Tinsley BM and Larson RB), pp. 157–198. Yale University Observatory, New Haven, CT.

Faber SM and Jackson RE 1976 Velocity dispersions and mass-to-light ratios for elliptical galaxies. *Astrophys. J.* **204**, 668–683.
Fall SM and Zhang Q 2001 Dynamical evolution of the mass function of globular star clusters. *Astrophys. J.* **561**, 751–765.
Feldmeier JJ, Jacoby GH and Phillips MM 2007 Calibrating Type Ia supernovae using the planetary nebula luminosity function. I. Initial results. *Astrophys. J.* **657**, 76–94.
Ferrarese L, Côté P, Jordán A, Peng EW, Blakeslee JP, Piatek S, Mei S, Merritt D, Milosavljević M, Tonry JL and West MJ 2006 The ACS Virgo Cluster Survey. VI. Isophotal analysis and the structure of early-type galaxies. *Astrophys. J. Suppl. Ser.* **164**, 334–434.
Fleck J-J, Boily CM, Lançon A and Deiters S 2006 On the mass of dense star clusters in starburst galaxies from spectrophotometry. *Mon. Not. R. Astron. Soc.* **369**, 1392–1406.
Flores H, Hammer F, Puech M, Amram P and Balkowski C 2006 3D spectroscopy with VLT/GIRAFFE. I. The true Tully–Fisher relationship at $z \sim 0.6$. *Astron. Astrophys.* **455**, 107–118.
Forbes DA 1996 Globular cluster luminosity functions and the Hubble constant from WFPC imaging: galaxies in the Coma I cloud. *Astron. J.* **112**, 1409–1414.
Frankowski A and Soker N 2009 Comparing symbiotic nebulae and planetary nebulae luminosity functions. *Astrophys. J.* **703**, L95–L98.
Freedman WL, Madore BF, Gibson BK, Ferrarese L, Kelson DD, Sakai S, Mould JR, Kennicutt Jr RC, Ford HC, Graham JA, Huchra JP, Hughes SMG, Illingworth GD, Macri LM and Stetson PB 2001 Final results from the Hubble Space Telescope Key Project to measure the Hubble constant. *Astrophys. J.* **553**, 47–72.
Fritze-von Alvensleben U 2004 Secondary globular cluster populations. *Astron. Astrophys.* **414**, 515–530.
Fuentes-Masip O, Muñoz-Tuñón C, Castañeda HO and Tenorio-Tagle G 2000 On the size and luminosity versus velocity dispersion correlations from the giant HII regions in the irregular galaxy NGC 4449. *Astron. J.* **120**, 752–762.
Gallagher JS and Hunter DA 1983 Gas kinematics and the structure of extragalactic giant and supergiant HII regions. *Astrophys. J.* **274**, 141–151.
Gerhard O, Arnaboldi M, Freeman KC, Okamura S, Kashikawa N and Yasuda N 2007 The kinematics of intracluster planetary nebulae and the ongoing subcluster merger in the Coma cluster core. *Astron. Astrophys.* **468**, 815–822.
Gieles M, Portegies Zwart SF, Baumgardt H, Athanassoula E, Lamers HJGLM, Sipior M and Leenaarts J 2006 Star cluster disruption by giant molecular clouds. *Mon. Not. R. Astron. Soc.* **371**, 793–804.
Girardi L, Bertelli G, Bressan A, Chiosi C, Groenewegen MAT, Marigo P, Salasnich B and Weiss A 2002 Theoretical isochrones in several photometric systems. I. Johnson–Cousins–Glass, HST/WFPC2, HST/NICMOS, Washington, and ESO Imaging Survey filter sets. *Astron. Astrophys.* **391**, 195–212.
Gnedin OY 1997 On the evolution of the globular cluster luminosity function: the differences. *Astrophys. J.* **487**, 663–666.
Gnedin OY and Ostriker JP 1997 Destruction of the Galactic globular cluster system. *Astrophys. J.* **474**, 223–255.
González RA, Liu MC and Bruzual A G 2004 Infrared surface brightness fluctuations of Magellanic star clusters. *Astrophys. J.* **611**, 270–293.
González-Lópezlira RA, Albarrán MY, Mouhcine M, Liu MC, Bruzual A G and de Batz B 2005 Near-infrared surface brightness fluctuations and optical colours of Magellanic star clusters. *Mon. Not. R. Astron. Soc.* **363**, 1279–1289.
Griersmith D 1980 Photometric properties of bright early-type spiral galaxies. III. The color–absolute magnitude relation and the analysis of fully corrected colors. *Astron. J.* **85**, 1295–1311.
Grillmair CJ, Forbes DA, Brodie JP and Elson RAW 1999 Hubble Space Telescope imaging of the globular clusters in the Fornax cluster: color and luminosity distributions. *Astron. J.* **117**, 167–180.
Gudehus DH 1991 Systematic bias in cluster galaxy data, affecting galaxy distances and evolutionary history. *Astrophys. J.* **382**, 1–18.
Gum CS and de Vaucouleurs G 1953 Large ring-like HII regions as extragalactic distance indicators. *The Observatory* **73**, 152–155.

Gurovich S, Freeman K, Jerjen H, Puerari I and Staveley-Smith L 2010 The slope of the baryonic Tully–Fisher relation. *Astron. J.* **140**, 663–676.

Harker JJ, Schiavon RP, Weiner BJ and Faber SM 2006 Population synthesis models for late buildup of the red sequence. *Astrophys. J.* **647**, L103–L106.

Harris WE 1991 Globular cluster systems in galaxies beyond the Local Group. *Annu. Rev. Astron. Astrophys.* **29**, 543–579.

Harris WE 1999 Globular cluster systems: formation models and case studies. In *Globular Clusters* (eds Martinez-Roger C, Pérez-Fournón I and Sánchez F), pp. 325–355. Cambridge University Press.

Harris WE 2001 Globular cluster systems. In *Star Clusters* (eds Labhardt L and Binggeli B), *Saas-Fee Advanced Course* **28**, pp. 223–408. Springer.

Harris WE, Harris GLH and McLaughlin DE 1998 M87, globular clusters, and galactic winds: issues in giant galaxy formation. *Astron. J.* **115**, 1801–1822.

Hempel M, Kissler-Patig M, Puzia TH and Hilker M 2007a The age structure of globular cluster systems in early-type galaxies and its dependence on host galaxy properties. *Astron. Astrophys.* **463**, 493–502.

Hempel M, Zepf S, Kundu A, Geisler D and Maccarone TJ 2007b Near-infrared observations of globular clusters in NGC 4472, NGC 4594, NGC 3585, and NGC 5813 and implications for their ages and metallicities. *Astrophys. J.* **661**, 768–778.

Hippelein HH 1986 Turbulent gas motions in giant HII regions. II. The luminosity–velocity dispersion relation. *Astron. Astrophys.* **160**, 374–384.

Hogg DW, Blanton MR, Brinchmann J, Eisenstein DJ, Schlegel DJ, Gunn JE, McKay TA, Rix H-W, Bahcall NA, Brinkmann J and Meiksin A 2004 The dependence on environment of the color–magnitude relation of galaxies. *Astrophys. J.* **601**, L29–L32.

Hubble E 1929 A relation between distance and radial velocity among extra-galactic nebulae. *Proc. Nat'l Acad. Sci. USA* **15**, 168–173.

Jacoby GH 1989 Planetary nebulae as standard candles. I. Evolutionary models. *Astrophys. J.* **339**, 39–52.

Jacoby GH, Branch D, Ciardullo R, Davies RL, Harris WE, Pierce MJ, Pritchet CJ, Tonry JL and Welch DL 1992 A critical review of selected techniques for measuring extragalactic distances. *Publ. Astron. Soc. Pac.* **104**, 599–662.

Jensen JB, Tonry JL, Thompson RI, Ajhar EA, Lauer TR, Rieke MJ, Postman M and Liu MC 2001 The infrared surface brightness fluctuation Hubble constant. *Astrophys. J.* **550**, 503–521.

Jensen JB, Tonry JL, Barris BJ, Thompson RI, Liu MC, Rieke MJ, Ajhar EA and Blakeslee JP 2003 Measuring distances and probing the unresolved stellar populations of galaxies using infrared surface brightness fluctuations. *Astrophys. J.* **583**, 712–726.

Jordán A, McLaughlin DE, Côté P, Ferrarese L, Peng EW, Blakeslee JP, Mei S, Villegas D, Merritt D, Tonry JL and West MJ 2006 Trends in the globular cluster luminosity function of early-type galaxies. *Astrophys. J.* **651**, L25–L28.

Jordán A, McLaughlin DE, Côté P, Ferrarese L, Peng EW, Mei S, Villegas D, Merritt D, Tonry JL and West MJ 2007a The ACS Virgo cluster survey. XII. The luminosity function of globular clusters in early-type galaxies. *Astrophys. J. Suppl. Ser.* **171**, 101–145.

Jordán A, Blakeslee JP, Côté P, Ferrarese L, Infante L, Mei S, Merritt D, Peng EW, Tonry JL and West MJ 2007b The ACS Fornax Cluster Survey. I. Introduction to the survey and data reduction procedures. *Astrophys. J. Suppl. Ser.* **169**, 213–224.

Kannappan SJ and Barton EJ 2004 Tools for identifying spurious luminosity offsets in Tully–Fisher studies: application at low redshift and implications for high redshift. *Astron. J.* **127** 2694–2710.

Kassin SA, Weiner BJ, Faber SM, Koo DC, Lotz JM, Diemand J, Harker JJ, Bundy K, Metevier AJ, Phillips AC, Cooper MC, Croton DJ, Konidaris N, Noeske KG and Willmer CNA 2007 The stellar mass Tully–Fisher relation to $z = 1.2$ from AEGIS. *Astrophys. J.* **660**, L35–L38.

Kavelaars JJ and Hanes DA 1997 A comparison of the globular cluster luminosity functions of the inner and outer halo of the Milky Way and M31. *Mon. Not. R. Astron. Soc.* **285**, L31–L34.

Kelson DD, van Dokkum PG, Franx M, Illingworth GD and Fabricant D 1997 Evolution of early-type galaxies in distant clusters: the Fundamental Plane from Hubble Space Telescope imaging and Keck spectroscopy. *Astrophys. J.* **478**, L13–L16.

Kelson DD, Illingworth GD, Tonry JL, Freedman WL, Kennicutt Jr RC, Mould JR, Graham JA, Huchra JP, Macri LM, Madore BF, Ferrarese L, Gibson BK, Sakai S, Stetson PB, Ajhar EA, Blakeslee JP, Dressler A, Ford HC, Hughes SMG, Sebo KM and Silbermann NA 2000 The Hubble Space Telescope Key Project on the extragalactic distance scale. XXVII. A derivation of the Hubble constant using the Fundamental Plane and $D_n - \sigma$ relations in Leo I, Virgo, and Fornax. *Astrophys. J.* **529**, 768–785.

Kennicutt Jr RC 1979 HII regions as extragalactic distance indicators. III. Application of HII region fluxes and galaxy diameters. *Astrophys. J.* **228**, 704–711.

Kennicutt Jr RC 1984 Structural properties of giant HII regions in nearby galaxies. *Astrophys. J.* **287**, 116–130.

Kennicutt Jr RC 1988 Properties of HII region populations in galaxies. I. The first-ranked HII regions. *Astrophys. J.* **334**, 144–158.

Kennicutt Jr RC 1989 The star formation law in galactic disks. *Astrophys. J.* **344**, 685–703.

Kohle S, Kissler-Patig M, Hilker M, Richtler T, Infante L and Quintana H 1996 The distance of the Fornax cluster based on globular cluster luminosity functions. *Astron. Astrophys.* **309**, L39–L42.

Komatsu E, Dunkley J, Nolta MR, Bennett CL, Gold B, Hinshaw G, Jarosik N, Larson D, Limon M, Page L, Spergel DN, Halpern M, Hill RS, Kogut A, Meyer SS, Tucker GS, Weiland JL, Wollack E and Wright EL 2009 Five-year Wilkinson Microwave Anisotropy Probe observations: cosmological interpretation. *Astrophys. J. Suppl. Ser.* **180**, 330–376.

Komatsu E, Smith KM, Dunkley J, Bennett CL, Gold B, Hinshaw G, Jarosik N, Larson D, Nolta MR, Page L, Spergel DN, Halpern M, Hill RS, Kogut A, Limon M, Meyer SS, Odegard N, Tucker GS, Weiland JL, Wollack E and Wright EL 2011 Seven-year Wilkinson Microwave Anisotropy Probe (WMAP) observations: cosmological interpretation. *Astrophys. J. Suppl. Ser.* **192**, 18.

Kouwenhoven MBN and de Grijs R 2008 The effect of binaries on the dynamical mass determination of star clusters. *Astron. Astrophys.* **480**, 103–114.

Kovacevic AV, Parker QA, Jacoby GH and Miszalski B 2010 A new O[III]λ500Å Galactic bulge planetary nebula luminosity function. In *Proc. Asymmetric Planetary Nebulae V* (ed Zijlstra A), p. E44 (arXiv:1010.1655). Palo Alto, CA, Ebrary.

Kundu A, Whitmore BC, Sparks WB, Macchetto FD, Zepf SE and Ashman KM 1999 The globular cluster system in the inner region of M87. *Astrophys. J.* **513**, 733–751.

Kundu A, Zepf SE, Hempel M, Morton D, Ashman KM, Maccarone TJ, Kissler-Patig M, Puzia TH and Vesperini E 2005 The ages of globular clusters in NGC 4365 revisited with deep HST observations. *Astrophys. J.* **634**, L41–L44.

La Barbera F, Busarello G, Merluzzi P, de la Rosa IG, Coppola G and Haines CP 2008 The SDSS-UKIDSS Fundamental Plane of early-type galaxies. *Astrophys. J.* **689**, 913–918.

Lamers HJGLM and Gieles M 2006 Clusters in the solar neighbourhood: how are they destroyed? *Astron. Astrophys.* **455**, L17–L20.

Larsen SS, Brodie JP, Huchra JP, Forbes DA and Grillmair CJ 2001 Properties of globular cluster systems in nearby early-type galaxies. *Astron. J.* **121**, 2975–2998.

Larson RB, Tinsley BM and Caldwell CN 1980 The evolution of disk galaxies and the origin of S0 galaxies. *Astrophys. J.* **237**, 692–707.

Lee H-C, Worthey G and Blakeslee JP 2010 Effects of α-element enhancement and the thermally pulsing asymptotic giant branch on surface brightness fluctuation magnitudes and broadband colors. *Astrophys. J.* **710**, 421–431.

Lisker T, Grebel EK and Binggeli B 2008 Virgo cluster early-type dwarf galaxies with the Sloan Digital Sky Survey. IV. The color–magnitude relation. *Astron. J.* **135**, 380–399.

Liu MC, Charlot S, Graham JR 2000 Theoretical predictions for surface brightness fluctuations and implications for stellar populations of elliptical galaxies. *Astrophys. J.* **543**, 644–668.

Madore BF and Woods D 1987 The luminosity–velocity diagram: a new distance estimator for spiral galaxies. *Astrophys. J.* **323**, L25–L29.

Marigo P, Girardi L, Weiss A, Groenewegen MAT and Chiosi C 2004 Evolution of planetary nebulae. II. Population effects on the bright cut-off of the PNLF. *Astron. Astrophys.* **423**, 995–1015.

Marín-Franch A and Aparicio A 2006 Surface-brightness fluctuations in stellar populations. IAC-star models for the optical and near-IR wavelengths. *Astron. Astrophys.* **450**, 979–992.

Mathewson DS, Ford VL and Buchhorn M 1992 A southern sky survey of the peculiar velocities of 1355 spiral galaxies. *Astrophys. J. Suppl. Ser.* **81**, 413–659.

McLaughlin DE 1994 An analytical study of the globular-cluster luminosity function. *Publ. Astron. Soc. Pac.* **106**, 47–55.

McLaughlin DE and Fall SM 2008 Shaping the globular cluster mass function by stellar-dynamical evaporation. *Astrophys. J.* **679**, 1272–1287.

McLaughlin DE and Pudritz RE 1996 The formation of globular cluster systems. I. The luminosity function. *Astrophys. J.* **457**, 578–597.

McLaughlin DE and van der Marel RP 2005 Resolved massive star clusters in the Milky Way and its satellites: brightness profiles and a catalog of fundamental parameters. *Astrophys. J. Suppl. Ser.* **161**, 304–360.

Meatheringham SJ and Dopita MA 1991a Optical spectroscopy of Magellanic Cloud planetary nebulae. *Astrophys. J. Suppl. Ser.* **75**, 407–424.

Meatheringham SJ and Dopita MA 1991b Optical spectroscopy of Magellanic Cloud planetary nebulae. II. *Astrophys. J. Suppl. Ser.* **76**, 1085–1098.

Mei S, Kissler-Patig M, Silva DR and Quinn PJ 2001 High K-band surface brightness fluctuations in NGC 1427 and NGC 720. *Astron. Astrophys.* **376**, 793–802.

Mei S, Blakeslee JP, Tonry JL, Jordán A, Peng EW, Côté P, Ferrarese L, Merritt D and Milosavljević M and West MJ 2005 The ACS Virgo Cluster Survey. IV. Data reduction procedures for surface brightness fluctuation measurements with the Advanced Camera for Surveys. *Astrophys. J. Suppl. Ser.* **156**. 113–125.

Melnick J 1977 Velocity dispersions in giant HII regions. Relation with their linear diameters. *Astrophys. J.* **213**, 15–17.

Méndez RH, Kudritzki RP, Ciardullo R and Jacoby GH 1993 The bright end of the planetary nebula luminosity function. *Astron. Astrophys.* **275**, 534–548.

Miller BW and Lotz JM 2007 The globular cluster luminosity function and specific frequency in dwarf elliptical galaxies. *Astrophys. J.* **670**, 1074–1089.

Mo HJ, Mao S and White SDM 1998 The formation of galactic discs. *Mon. Not. R. Astron. Soc.* **295**, 319–336.

Mobasher B, Ellis RS and Sharples RM 1986 A complete galaxy redshift sample. IV. Optical and infrared colour–luminosity relations. *Mon. Not. R. Astron. Soc.* **223**, 11–37.

Mouhcine M, González RA and Liu MC 2005 New near-infrared surface brightness fluctuation models. *Mon. Not. R. Astron. Soc.* **362**, 1208–1218.

Mould JR, Huchra JP, Freedman WL, Kennicutt Jr RC, Ferrarese L, Ford HC, Gibson BK, Graham JA, Hughes SMG, Illingworth GD, Kelson DD, Macri LM, Madore BF, Sakai S, Sebo KM, Silbermann NA and Stetson PB 2000 The Hubble Space Telescope Key Project on the extragalactic distance scale. XXVIII. Combining the constraints on the Hubble constant. *Astrophys. J.* **529**, 786–794.

Murali C and Weinberg MD 1997 Evolution of the Galactic globular cluster system. *Mon. Not. R. Astron. Soc.* **291**, 717–731.

Nigoche-Netro A, Aguerri JAL, Lagos P, Ruelas-Mayorga A, Sánchez LJ and Machado A 2010 The Faber–Jackson relation for early-type galaxies: dependence on the magnitude range. *Astron. Astrophys.* **516**, A96.

Okazaki T and Tosa M 1995 The evolution of the luminosity function of globular cluster systems. *Mon. Not. R. Astron. Soc.* **274**, 48–60.

Ostriker JP and Caldwell JAR 1983 A model for the Galaxy with rising rotational velocity. In *Kinematics, Dynamics and Structure of the Milky Way* (ed. Shuter WLH), pp. 249–257. Dordrecht, Reidel.

Ostriker JP and Gnedin OY 1997 The globular cluster luminosity function as a distance indicator: dynamical effects. *Astrophys. J.* **487**, 667–671.

Pahre MA, de Carvalho RR and Djorgovski SG 1998 Near-infrared imaging of early-type galaxies. IV. The physical origins of the Fundamental Plane scaling relations. *Astron. J.* **116**, 1606–1625.

Parmentier G and Gilmore G 2001 The self-enrichment of galactic halo globular clusters. The mass–metallicity relation. *Astron. Astrophys.* **378**, 97–101.

Parmentier G and Gilmore G 2005 New light on the initial mass function of the Galactic halo globular clusters. *Mon. Not. R. Astron. Soc.* **363**, 326–336.

Parmentier G and Gilmore G 2007 The origin of the Gaussian initial mass function of old globular cluster systems. *Mon. Not. R. Astron. Soc.* **377**, 352–372.

Peletier RF and de Grijs R 1998 The scatter in the near-infrared colour–magnitude relation in spiral galaxies. *Mon. Not. R. Astron. Soc.* **300**, L3–L6.

Peletier RF and Willner SP 1993 The infrared Tully–Fisher relation in the Ursa Major cluster. *Astrophys. J.* **418**, 626–637.

Peng EW, Jordán A, Côté P, Blakeslee JP, Ferrarese L, Mei S, West MJ, Merritt D, Milosavljević M and Tonry JL 2006 The ACS Virgo Cluster Survey. IX. The color distributions of globular cluster systems in early-type galaxies. *Astrophys. J.* **639**, 95–119.

Petrou M 1981 On the local mass to light ratio in galaxies. *Mon. Not. R. Astron. Soc.* **196**, 933–942.

Pierce MJ and Tully RB 1992 Luminosity-line width relations and the extragalactic distance scale. I. Absolute calibration. *Astrophys. J.* **387**, 47–55.

Pipino A, Puzia TH and Matteucci F 2007 The formation of globular cluster systems in massive elliptical galaxies: globular cluster multimodality from radial variation of stellar populations. *Astrophys. J.* **665**, 295–305.

Prieto JL and Gnedin OY 2008 Dynamical evolution of globular clusters in hierarchical cosmology. *Astrophys. J.* **689**, 919–935.

Prugniel P and Simien F 1996 The Fundamental Plane of early-type galaxies: stellar populations and mass-to-light ratio. *Astron. Astrophys.* **309**, 749–759.

Pryor C and Meylan G 1993 Velocity dispersions for Galactic globular clusters. In *Structure and Dynamics of Globular Clusters* (eds Djorgovski SG and Meylan G), *Proc. Astron. Soc. Pac. Conf. Ser.* **50**, 357–372.

Puech M, Flores H, Hammer F, Yang Y, Neichel B, Lehnert M, Chemin L, Nesvadba N, Epinat B, Amram P, Balkowski C, Cesarsky C, Dannerbauer H, di Serego Alighieri S, Fuentes-Carrera I, Guiderdoni B, Kembhavi A, Liang YC, Östlin G, Pozzetti L, Ravikumar CD, Rawat A, Vergani D, Vernet J and Wozniak H 2008 IMAGES. III. The evolution of the near-infrared Tully–Fisher relation over the last 6 Gyr. *Astron. Astrophys.* **484**, 173–187.

Puzia TH, Kissler-Patig M, Thomas D Maraston C, Saglia RP, Bender R, Goudfrooij P and Hempel M 2005 VLT spectroscopy of globular cluster systems. II. Spectroscopic ages, metallicities, and [α/Fe] ratios of globular clusters in early-type galaxies. *Astron. Astrophys.* **439**, 997–1011.

Raimondo G, Brocato E, Cantiello M and Capaccioli M 2005 New optical and near-infrared surface brightness fluctuation models. II. Young and intermediate-age stellar populations. *Astron. J.* **130**, 2625–2646.

Rejkuba M 2001 Deep VLT search for globular clusters in NGC 5128: color–magnitude diagrams and globular cluster luminosity function. *Astron. Astrophys.* **369**, 812–825.

Relaño M, Beckman JE, Zurita A, Rozas M and Giammanco C 2005 The internal dynamical equilibrium of HII regions: a statistical study. *Astron. Astrophys.* **431**, 235–251.

Rhee MH 1996 A physical basis of the Tully–Fisher relation. PhD thesis, University of Groningen, the Netherlands.

Rhee MH 2004 On the physical basis of the Tully–Fisher relation. *J. Kor. Astron. Soc.* **37**, 15–39.

Richtler T 2003 The globular cluster luminosity function: new progress in understanding an old distance indicator. In *Stellar Candles for the Extragalactic Distance Scale* (eds Alloin D and Gieren W), *Lect. Notes Phys.* **635**, 281–305.

Riess AG, Macri L, Casertano S, Sosey M, Lampeitl H, Ferguson HC, Filippenko AV, Jha SW, Li W, Chornock R and Sarkar D 2009 A redetermination of the Hubble constant with the Hubble Space Telescope from a differential distance ladder. *Astrophys. J.* **699**, 539–563.

Roy J-R, Arsenault R and Joncas G 1986 Hα velocity widths of giant extragalactic HII regions. *Astrophys. J.* **300**, 624–638.

Rozas M, Sabalisck N, Beckman JE and Knapen JH 1998 Internal turbulence, viriality, and density bounding of the most luminous HII regions in the spiral galaxy M100. *Astron. Astrophys.* **338**, 15–26.

Sandage A 1972 Absolute magnitudes of E and S0 galaxies in the Virgo and Coma clusters as a function of $U - B$ color. *Astrophys. J.* **176**, 21–30.

Sandage A and Tammann GA 1974 Steps toward the Hubble constant. I. Calibration of the linear sizes of extragalactic HII regions. *Astrophys. J.* **190**, 525–538.

Schönberner D, Jacob R, Steffen M and Sandin C 1997 The evolution of planetary nebulae. IV. On the physics of the luminosity function. *Astron. Astrophys.* **473**, 467–484.

Schönberner D, Jacob R, Steffen M and Sandin C 2007 The evolution of planetary nebulae. IV. On the physics of the luminosity function. *Astron. Astrophys.* **473**, 467–484.

Sérsic JL 1960 The HII regions in galaxies. *Z. Astrophys.* **50**, 168–177.

Shields GS 1990 Extragalactic HII regions. *Annu. Rev. Astron. Astrophys.* **28**, 525–560.

Silk J 1997 Feedback, disk self-regulation, and galaxy formation. *Astrophys. J.* **481**, 703–709.

Sisteró RF 1988 Statistics of HII regions as distance indicators. *Astrophys. Lett. Commun.* **27**, 41–48.

Skelton RE, Bell EF and Somerville RS 2009 The effect of dry mergers on the color–magnitude relation of early-type galaxies. *Astrophys. J.* **699**, L9–L12.

Soker N 2006 Observed planetary nebulae as descendants of interacting binary systems. *Astrophys. J.* **645**, L57–L60.

Spitler LR, Larsen SS, Strader J, Brodie JP, Forbes DA and Beasley MA 2006 Hubble Space Telescope ACS wide-field photometry of the Sombrero galaxy globular cluster system. *Astron. J.* **132**, 1593–1609.

Spitler LR, Forbes DA and Beasley MA 2008 Extending the baseline: spitzer mid-infrared photometry of globular cluster systems in the Centaurus A and Sombrero galaxies. *Mon. Not. R. Astron. Soc.* **389**, 1150–1162.

Stanghellini L 1995 Evolutionary tests of the planetary nebula luminosity function. *Astrophys. J.* **452**, 515–521.

Steinmetz M and Navarro JF 1999 The cosmological origin of the Tully–Fisher relation. *Astrophys. J.* **513**, 555–560.

Terlevich R and Melnick J 1981 The dynamics and chemical composition of giant extragalactic HII regions. *Mon. Not. R. Astron. Soc.* **195**, 839–851.

Thomas D, Maraston C, Bender R and Mendes de Oliveira C 2005 The epochs of early-type galaxy formation as a function of environment. *Astrophys. J.* **621**, 673–694.

Tonry JL 1991 Surface brightness fluctuations: a bridge from M31 to the Hubble constant. *Astrophys. J.* **373**, L1–L4.

Tonry J and Schneider DP 1988 A new technique for measuring extragalactic distances. *Astron. J.* **96**, 807–815.

Tonry JL, Blakeslee JP, Ajhar EA and Dressler A 1997 The SBF survey of galaxy distances. I. Sample selection, photometric calibration, and the Hubble constant. *Astrophys. J.* **475**, 399–413.

Tonry JL, Dressler A, Blakeslee JP, Ajhar EA, Fletcher AB, Luppino GA, Metzger MR and Moore CB 2001 The SBF survey of galaxy distances. IV. SBF magnitudes, colors, and distances. *Astrophys. J.* **546**, 681–693.

Torres-Peimbert S and Peimbert M 1997 Planetary nebulae of type I revisited. In *Planetary Nebulae* (eds Habing HJ and Lamers HJGLM), *Proc. Int'l Astron. Union Symp.* **180**, 175–183.

Trager SC, Faber SM, Worthey G and González JJ 2000 The stellar population histories of early-type galaxies. II. Controlling parameters of the stellar populations. *Astron. J.* **120**, 165–188.

Treu T, Stiavelli M, Bertin G, Casertano S and Møller P 2001 The properties of field elliptical galaxies at intermediate redshift. III. The Fundamental Plane and the evolution of stellar populations from $z \sim 0.4$ to $z = 0$. *Mon. Not. R. Astron. Soc.* **326**, 237–254.

Treu T, Stiavelli M, Casertano S, Møller P and Bertin G 2002 The evolution of field early-type galaxies to $z \sim 0.7$. *Astrophys. J.* **564**, L13–L16.

Treu T, Ellis RS, Liao TX and van Dokkum PG 2005a Keck Spectroscopy of distant GOODS spheroidal galaxies: downsizing in a hierarchical Universe. *Astrophys. J.* **622**, L5–L8.

Treu T, Ellis RS, Liao TX, van Dokkum PG, Tozzi P, Coil A, Newman J, Cooper MC and Davis M 2005b The assembly history of field spheroidals: evolution of mass-to-light ratios and signatures of recent star formation. *Astrophys. J.* **633**, 174–197.

Trujillo I, Burkert A and Bell EF 2004 The tilt of the Fundamental Plane: three-quarters structural nonhomology, one-quarter stellar population. *Astrophys. J.* **600**, L39–L42.

Tully RB 1982 The Local Supercluster. *Astrophys. J.* **257**, 389–422.

Tully RB and Fisher JR 1977 A new method of determining distances to galaxies. *Astron. Astrophys.* **54**, 661–673.

Tully RB, Mould JR and Aaronson M 1982 A color–magnitude relation for spiral galaxies. *Astrophys. J.* **257**, 527–537.
van den Bergh S 1980 Can the Hubble constant be determined from HII region diameters? *Astrophys. J.* **235**, 1–3.
van den Bergh S 1994 Central concentration and luminosity of Galactic globular clusters. *Astrophys. J.* **435**, 203–206.
van der Wel A, Franx M, van Dokkum PG and Rix H-W 2004 The Fundamental Plane of field early-type galaxies at $z = 1$. *Astrophys. J.* **601**, L5–L8.
van de Ven G, van Dokkum PG and Franx M 2003 The Fundamental Plane and the evolution of the M/L ratio of early-type field galaxies up to $z \sim 1$. *Mon. Not. R. Astron. Soc.* **344**, 924–934.
van Dokkum PG and Franx M 1996 The Fundamental Plane in CL 0024 at $z = 0.4$: implications for the evolution of the mass-to-light ratio. *Mon. Not. R. Astron. Soc.* **281**, 985–1000.
van Dokkum PG and Stanford SA 2003 The Fundamental Plane at $z = 1.27$: first calibration of the mass scale of red galaxies at redshifts $z > 1$. *Astrophys. J.* **585**, 78–89.
van Dokkum PG, Franx M, Kelson DD and Illingworth GD 1998 Luminosity evolution of early-type galaxies to $z = 0.83$: constraints on formation epoch and Ω. *Astrophys. J.* **504**, L17–L21.
Vassiliadis E and Wood PR 1994 Post-asymptotic giant branch evolution of low- to intermediate-mass stars. *Astrophys. J. Suppl. Ser.* **92**, 125–144.
Vesperini E 1998 Evolution of the mass function of the Galactic globular cluster system. *Mon. Not. R. Astron. Soc.* **299**, 1019–1039.
Vesperini E 2001 Evolution of globular cluster systems in elliptical galaxies. II. Power-law initial mass function. *Mon. Not. R. Astron. Soc.* **322**, 247–256.
Vesperini E and Zepf SE 2003 Effects of the dissolution of low-concentration globular clusters on the evolution of globular cluster systems. *Astrophys. J.* **587**, L97–L100.
Vesperini E, Zepf SE, Kundu A and Ashman KM 2003 Modeling the dynamical evolution of the M87 globular cluster system. *Astrophys. J.* **593**, 760–771.
Villegas D, Jordán A, Peng EW, Blakeslee JP, Côté P, Ferrarese L, Kissler-Patig M, Mei S, Infante L, Tonry JL and West MJ 2010 The ACS Fornax Cluster Survey. VIII. The luminosity function of globular clusters in Virgo and Fornax early-type galaxies and its use as a distance indicator. *Astrophys. J.* **717**, 603–616.
Visvanathan N and Griersmith D 1977 Absolute magnitude–color relation for early type spirals. *Astron. Astrophys.* **59**, 317–328.
Visvanathan N and Sandage A 1977 The color–absolute magnitude relation for E and S0 galaxies. I. Calibration and tests for universality using Virgo and eight other nearby clusters. *Astrophys. J.* **216**, 214–226.
Weidemann V 2000 Revision of the initial-to-final mass relation. *Astron. Astrophys.* **363**, 647–656.
Weiner BJ, Willmer CNA, Faber SM, Harker J, Kassin SA, Phillips AC, Melbourne J, Metevier AJ, Vogt NP and Koo DC 2006 A survey of galaxy kinematics to $z \sim 1$ in the TKRS/GOODS-N field. II. Evolution in the Tully–Fisher relation. *Astrophys. J.* **653**, 1049–1069.
Whitmore BC 1997 Globular clusters as distance indicators. In *The Extragalactic Distance Scale* (ed. Livio M), *STScI Conf. Ser.*, pp. 254–272. Cambridge University Press.
Whitmore BC, Sparks WB, Lucas RA, Macchetto FD and Biretta JA 1995 Hubble Space Telescope observations of globular clusters in M87 and an estimate of H_0. *Astrophys. J.* **454**, L73–L76.
Whitmore BC, Schweizer F, Kundu A and Miller BW 2002 The luminosity functions of old and intermediate-age globular clusters in NGC 3610. *Astron. J.* **124**, 147–157.
Worthey G 1993 The dependence of the brightness fluctuation distance indicator on stellar population age and metallicity. *Astrophys. J.* **409**, 530–536.
Wyse RFG and Silk J 1989 Star formation rates and abundance gradients in disk galaxies. *Astrophys. J.* **339**, 700–711.
Zaritsky D 1992 The radial distribution of oxygen in disk galaxies. *Astrophys. J.* **390**, L73–L76.

5

From Nearby Galaxy Clusters to Cosmological Distances

In the beginning the Universe was created. This has made a lot of people very angry and been widely regarded as a bad move.

– Douglas Adams (1952–2001), British author

One geometry cannot be more true than another; it can only be more convenient. Geometry is not true, it is advantageous.

– Robert Pirsig (born 1928), American author

5.1 Cosmological Redshifts

5.1.1 Determination of the Current Expansion Rate of the Universe

In 1929, Edwin Hubble first published what was to become a major yardstick for measurements of cosmological distances (Hubble 1929a), now known as the **Hubble law**. He combined his own distance (d) determinations to six 'extragalactic nebulae' in the Local Group, derived based on the period–luminosity relationship for Cepheid variables (see Section 3.5.2) obtained from photographic-plate observations at the 100-inch Hooker telescope (Mt Wilson Observatory, California, USA), with published radial velocity measurements (v_r), corrected for the solar motion through the Milky Way. Assuming a constant upper limit to the 'brightest blue stars' – now known to be HII regions – in those galaxies, he expanded his sample with an additional 18 objects located as far away as the Virgo cluster. This data set allowed him to construct the first version of what is now commonly known as

An Introduction to Distance Measurement in Astronomy, First Edition. Richard de Grijs.

a *Hubble diagram* (see Figure 4.1), relating (galactic) distances to their radial velocities,[1]

$$v_{\rm r}(= cz) = {\rm H}_0 d, \qquad (5.1)$$

where c and z are the speed of light in a vacuum and an object's redshift, i.e. the scale of the Universe at a given distance with respect to current scales: because of the large-scale 'Hubble expansion', the observed wavelength, λ, of light from a distant object appears to be stretched, to an observed wavelength $\lambda(1+z)$. The Hubble law in Equation (5.1) is applicable in any **Friedmann–Lemaître–Robertson–Walker (FLRW)** universe[2] for redshifts $z \ll 1$; at greater redshifts, one needs to take into account the cosmology-specific densities of both baryonic and dark matter, and dark energy (a large, negative pressure; see Section 5.1.3). In fact, accurate measurements of the Hubble diagram in the form of the so-called distance–redshift relation may, therefore, help to discriminate between models and constrain cosmological parameters (e.g. Linder 2008).

The slope of the linear relationship in the diagram, $\Delta v_{\rm r}/\Delta d$, corresponds to the Hubble constant, $\rm H_0$, which is usually expressed in units of km s^{-1} Mpc^{-1}. Physically, the Hubble constant corresponds to the current expansion rate of the Universe. Its inverse, $\rm H_0^{-1}$, which has units of time, is referred to as the **Hubble time**. To first order, it corresponds to the current age of the Universe, t_0. The most up-to-date value for the Hubble constant determined to date – based on the 7-year data of the *Wilkinson Microwave Anisotropy Probe (WMAP)* and baryon acoustic oscillations (BAOs; see Section 5.3.3) – is $\rm H_0 = 70.2 \pm 1.4$ km s^{-1} Mpc^{-1} (Komatsu *et al.* 2011), so that $t_0 = 13.76 \pm 0.11$ Gyr (but see below). We emphasize that $\rm H_0$ and t_0 refer to the *current* expansion rate and age of the Universe: in a uniformly expanding Universe, the Hubble constant is a function of time, $H(t)$. Its accurate determination has occupied generations of astrophysicists because of large and lingering systematic uncertainties that affected and continue to hamper the observations (see also Section 5.3.4). Significant recent research efforts have led to an unprecedented accuracy of $\rm H_0$ measurements, largely thanks to concerted programmes using the *Hubble Space Telescope (HST)*, one of whose primary mission goals was determination of $\rm H_0$ to an accuracy of better than 10% (cf. Freedman *et al.* 2001; Sandage *et al.* 2006). The quest to determine $\rm H_0$ to ever higher accuracy continues. We will discuss the main physical motivations behind this endeavour in Section 5.3.4.

5.1.2 Redshift Surveys and Peculiar Velocities

From Hubble's law, it follows that redshift is the primary cosmological distance indicator. Galaxy redshift distributions based on large-scale redshift surveys – including the Center for Astrophysics (CfA) Redshift Surveys (e.g. Geller and Huchra 1989), the Two-Degree-Field Galaxy Redshift Survey (e.g. Colless *et al.* 2001) and the redshift survey associated with the Sloan Digital Sky Survey (SDSS; e.g. Berlind *et al.* 2006: see Figure 5.1) – have,

[1] This result had been anticipated by Lemaître (1927) based on a mathematical model of an expanding Universe, but because his article was published in French in the *Annals of the Scientific Society of Brussels*, it had not attracted much attention from his colleagues at the time.

[2] The so-called Standard Model of modern cosmology was developed independently by the Soviet cosmologist Alexander Alexandrovich Friedmann (1922, 1924), Belgian Roman-catholic priest, physicist and astronomer Monsignor Georges Henri Joseph Édouard Lemaître (1931, 1933), American mathematician and physicist Howard Percy Robertson (1935, 1936a,b) and British mathematician Arthur Geoffrey Walker (1937), who derived the exact solution of **Einstein's field equations** of general relativity.

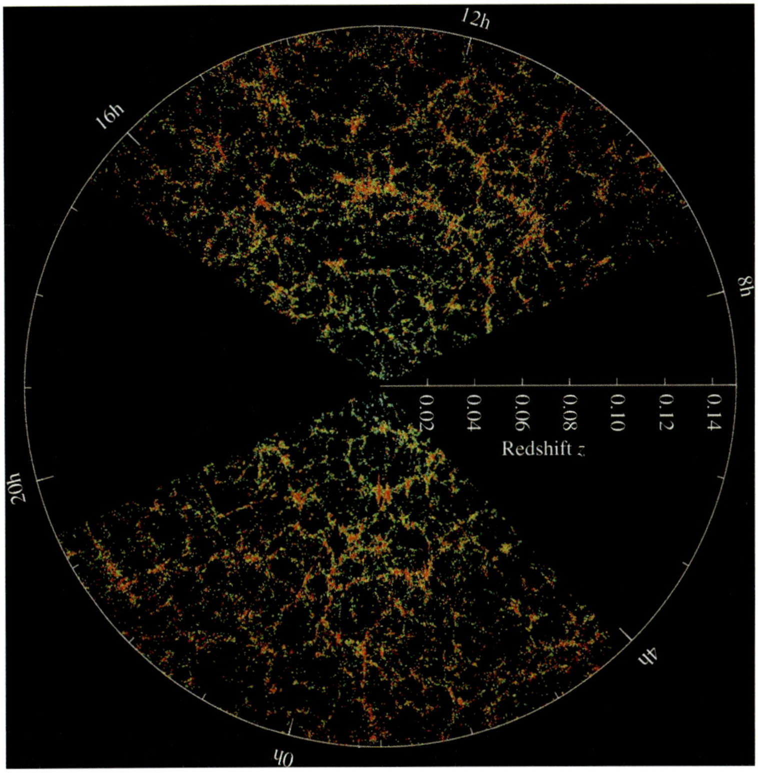

Figure 5.1 *Slices covering the declination range* $-1.25 \leq \delta \leq 1.25$ *degrees through the* SDSS *redshift survey's 3D map of the distribution of galaxies. Earth is at the centre, and each point represents a galaxy. Galaxies are coloured according to the ages of their stars (redder colours correspond to older average ages). Galaxy clusters show a characteristic expansion in the redshift direction known as the 'Finger-of-God' effect.* (Reprinted from M. Blanton and the Sloan Digital Sky Survey Collaboration (online: www.sdss.org), Galaxy Map, Copyright 2008, with permission of the SDSS.)

therefore, long been used as proxies of the true, three-dimensional spatial distributions of large numbers of galaxies. On large scales, this has led to our improved understanding of the large-scale structure of the Universe, which consists of an hierarchical '**cosmic web**' of galaxy clusters and superclusters interspersed with regions referred to as 'voids'. Some of the largest-scale galaxy-populated structures in the Universe known to date are the so-called Great Wall (also known as the Coma Wall or CfA2 Great Wall; Geller and Huchra 1989; Ramella *et al.* 1992), a very large-scale filamentary structure dominated by the Coma galaxy cluster, with an aspect ratio (length with respect to depth) of at least 20, the Sloan Great Wall (Gott *et al.* 2005), which is nearly three times longer than the Great Wall, and the Pisces–Cetus Supercluster Complex (e.g. Haynes and Giovanelli 1986; Tully 1986; Bonometto *et al.* 1993), which includes the Local or Virgo Supercluster.

On the other hand, a 'supervoid' appears to exist in the Galactic southern hemisphere, coinciding with the intrinsic cosmic microwave background (CMB) or *WMAP* 'Cold Spot' (e.g. Vielva *et al.* 2004), a ~10° region on the sky that is some 70 μK colder than the average CMB temperature of approximately 2.7 K (1σ root-mean-square (rms) CMB temperature variations are ≈18 μK; cf. Wright 2004). This is highly improbable (probability $P < 1.85\%$) under the currently favoured cosmological model (e.g. Cruz *et al.* 2007; Inoue *et al.* 2010; and references therein). This supervoid could cause the cold spot (e.g. Inoue and Silk 2006, 2007), although recent optical observations (Bremer *et al.* 2010; Granett *et al.* 2010) appear to rule out the existence of any sufficiently large supervoids at $0.35 < z < 1$ (see also Inoue *et al.* 2010). The cold spot was originally discovered as a significant deviation from the 'Gaussian' random-field probability density function generally used to describe the nature of the primordial CMB density fluctuations. Study of the Gaussianity of the CMB is one of the most powerful approaches to understanding the nature of these density fluctuations: by estimating the probability distribution from the measured CMB temperature fluctuations, models predicting primordial density fluctuations can be rejected or accepted at a certain level of significance.

However, on smaller scales, this one-to-one correspondence between spectroscopic or **photometric redshift** and distance breaks down because of the effects of so-called *peculiar motions*. On scales of galaxy groups and clusters, the motions of the member galaxies are not solely driven by the universal expansion of the Universe. Instead, they are affected by mutual gravitational attractive forces, which in turn cause random deviations of several hundred to up to 1000 km s^{-1} from their general recessional velocities. This superimposed Doppler shift could be either a blue- or a redshift along the line of sight. This leads to the so-called 'Finger-of-God' effect, an elongation of galaxy clusters in redshift space. A related, second-order effect in redshift space is caused by coherent, infalling motions of galaxies during cluster assembly. Instead of an extension along the line of sight, one would observe an apparent flattening of the cluster structure (Kaiser 1987).

From our vantage point in the Milky Way, we need to take these effects into account, given that the Milky Way is subject to peculiar motions related to the dominant, local gravitational potential provided by the Local Group as well as the Virgo cluster of galaxies (causing Local Group motion of $\gtrsim$ 400 km s^{-1}; e.g. Davis and Peebles 1983; Dressler 1984; Lucey *et al.* 2005; Kocevski and Ebeling 2006; Lavaux *et al.* 2010), in whose outskirts we live. In addition, a gravitational overdensity exists at the centre of the Hydra–Centaurus supercluster – known as the **Great Attractor** (e.g. Dressler *et al.* 1987; Lynden-Bell *et al.* 1988; Burstein 1990; Mould *et al.* 2000; Tonry *et al.* 2000) – which may be associated with the Norma galaxy cluster and supercluster (e.g. Woudt *et al.* 2008). It affects the motions of galaxies by up to ±700 km s^{-1} with respect to the local *Hubble flow* – i.e. the distance regime in which the individual motions of galaxies do not create significant perturbations in their **cosmological redshifts** – over a region hundreds of Mpc across. Beyond this significant overdensity of matter, an even more massive mass concentration, the Shapley Supercluster (Hudson *et al.* 2004) has been discovered at a roughly threefold greater distance, which appears to dominate the 'local' velocity field, including that of the Local Group as a whole. The models proposed to correct for local peculiar motions are broadly similar, but differ in detail. For instance, Mould *et al.* (2000) use a simple linear infall model with three attractors – the Local and Shapley Superclusters and the Great Attractor – while Tonry *et al.* (2000) do not consider the Shapley concentration but, instead, include a **mass quadrupole component**, which provides a measure of how far from spherically symmetric the mass distribution is.

The key question in this context is where the 'smooth Hubble flow' starts, i.e. where redshifts of field galaxies become reasonably reliable proxies of their distances. Riess *et al.* (2009a) suggest a minimum redshift of $z = 0.023$ ($d \sim 100$ Mpc), given recent controversy about the possible presence of a local '**Hubble bubble**' characterized by an increased outflow of $\sim 5\%$ within the local void in the range $0.01 < z < 0.023$ (cf. Zehavi *et al.* 1998; Jha *et al.* 2007; see also Section 5.2.1). At $d \sim 100$ Mpc, the Hubble flow velocity is around 7000 km s^{-1} and peculiar velocities will typically amount to a $\sim 5\%$ contribution.

5.1.3 The Prevailing Cosmological Model

A number of model parameters relate cosmological distances, redshifts and times. To understand the relevance of cosmology to the measurement of distances at $z \sim 1$ and beyond, it is essential to review some aspects of distance measurement in general relativistic cosmological models.

5.1.3.1 The Friedmann–Lemaître–Robertson–Walker Metric

In general relativity, the *metric* is the formula which gives the distance between neighbouring points in four-dimensional spacetime. Spacetime is a mathematical model that combines space and time into a single continuum. This metric is, therefore, the fundamental property of any cosmological model, since it defines the geometry of spacetime and its evolution with cosmic epoch. The FLRW metric can be written as

$$ds^2 = -c^2 dt^2 + a^2(t)\left[\frac{dx^2}{1 - kx^2} + x^2\left(d\theta^2 + \sin^2\theta\, d\phi^2\right)\right], \tag{5.2}$$

where (x, θ, ϕ) are spherical, **comoving coordinates**, $a(t)$ is the cosmic scale factor defining the size of the Universe at time t relative to that at the present time, t_0, and k determines the geometry of the Universe ($k < 0$ implies an open geometry, $k = 0$ flat and $k > 0$ closed). If we define $k \in [-1, 0, 1]$, $a(t_0)$, the scale factor at the present time, is a free parameter. Alternatively, if $a(t_0) \equiv 1$, as often assumed, k is a free, nondiscrete parameter. Note that the metric's space and time parts must have a relative minus sign.

Perhaps the most intuitive way to define an astronomical distance in terms of Equation (5.2) is to integrate ds along a radial line, assuming d$t = 0$. If we omit the expansion factor $a(t)$, we get the *comoving distance*, r,

$$r = \int_0^x \frac{dx}{\sqrt{1 - kx^2}}. \tag{5.3}$$

In the convention in which k is continuous, this integral gives

$$r = \begin{cases} \dfrac{1}{\sqrt{|k|}} \sinh^{-1}\left(x\sqrt{|k|}\right) & k < 0, \\ x & k = 0, \\ \dfrac{1}{\sqrt{k}} \sin^{-1}\left(x\sqrt{k}\right) & k > 0. \end{cases} \tag{5.4}$$

It is convenient to rewrite the FLRW metric in terms of r instead of x:

$$\mathrm{d}s^2 = -c^2\mathrm{d}t^2 + a^2(t)\left[\mathrm{d}r^2 + x_k^2(r)\left(\mathrm{d}\theta^2 + \sin^2\theta\,\mathrm{d}\phi^2\right)\right], \tag{5.5}$$

where

$$x_k(r) = \begin{cases} \dfrac{1}{\sqrt{|k|}}\sinh\left(r\sqrt{|k|}\right) & k < 0, \\ r & k = 0, \\ \dfrac{1}{\sqrt{k}}\sin\left(r\sqrt{k}\right) & k > 0. \end{cases} \tag{5.6}$$

The comoving distance is the radial distance from the origin to an object at coordinates (x, θ, ϕ), measured when $a(t) = 1$. We can incorporate the expansion of the Universe into this definition by multiplying by $a(t)$: this gives the **proper distance**,

$$d_{\mathrm{P}} = a(t)r. \tag{5.7}$$

A cosmological model is specified by the behaviour of $a(t)$. To calculate the proper distance in a given cosmological model, note that for a photon, which travels at the speed of light, the spacetime interval $\mathrm{d}s$ is always zero. Therefore, we have

$$c\,\mathrm{d}t = a(t)\,\mathrm{d}r, \tag{5.8}$$

which can be integrated to give

$$r = c\int_{t_{\mathrm{e}}}^{t_{\mathrm{o}}} \frac{1}{a(t)}\mathrm{d}t, \tag{5.9}$$

where t_{e} and t_{o} are, respectively, the times of emission and observation. The latter is usually the present, so the upper limit of the integral is normally t_0, which can be set to 1 if we operate under the assumption that $a(t) = 1$.

Equation (5.9) allows us to derive an expression for the cosmological redshift. Consider two successive wave crests emitted from a source at times t_{e} and $t_{\mathrm{e}} + \lambda_{\mathrm{e}}/c$, where λ_{e} is the emitted wavelength. They arrive at the observer at times t_{o} and $t_{\mathrm{o}} + \lambda_{\mathrm{o}}/c$, where λ_{o} is the observed wavelength. Since the comoving distance does not change with time and assuming the source has no peculiar velocity with respect to the observer, Equation (5.9) gives

$$r = c\int_{t_{\mathrm{e}}}^{t_{\mathrm{o}}} \frac{1}{a(t)}\mathrm{d}t = c\int_{t_{\mathrm{e}}+\lambda_{\mathrm{e}}/c}^{t_{\mathrm{o}}+\lambda_{\mathrm{o}}/c} \frac{1}{a(t)}\mathrm{d}t. \tag{5.10}$$

The interval between $t_{\mathrm{e}} + \lambda_{\mathrm{e}}/c$ and t_{o} is common to both integrals and can be eliminated, leaving

$$c\int_{t_{\mathrm{e}}}^{t_{\mathrm{e}}+\lambda_{\mathrm{e}}/c} \frac{1}{a(t)}\mathrm{d}t = c\int_{t_{\mathrm{o}}}^{t_{\mathrm{o}}+\lambda_{\mathrm{o}}/c} \frac{1}{a(t)}\mathrm{d}t. \tag{5.11}$$

But these integrals are now over the very short time intervals λ_{e}/c and λ_{o}/c, respectively, so we can reasonably assume that $a(t)$ does not change significantly in either case. Hence, we can take it out of the integral, leaving a trivial integral over $\mathrm{d}t$. The result is

$$\frac{\lambda_{\mathrm{e}}}{a\left(t_{\mathrm{e}}\right)} = \frac{\lambda_{\mathrm{o}}}{a\left(t_{\mathrm{o}}\right)}. \tag{5.12}$$

For the case where the time of observation is the present, $a(t_o) = 1$ and we have

$$\frac{\lambda_o}{\lambda_e} = 1 + z = \frac{1}{a(t_e)}. \tag{5.13}$$

The *cosmological redshift* is a direct measure of how much the Universe has expanded since the light was emitted: a source with redshift 1 is observed as it was when the scale factor $a(z = 1) = \frac{1}{2}a(z = 0)$ or equivalently, for $a(t) = 1$, when the Universe was half its present age. Note that although it is common to interpret the cosmological redshift as a Doppler shift, this is misleading: neither the source nor the observer is moving relative to the local coordinate system. The cosmological redshift is instead caused by the expansion of spacetime, not by motion through spacetime, as is the case with the Doppler shift.

Considerations about the geometry and expansion rate of the Universe are of importance for our understanding of how the effects of redshift may affect distances measured to objects beyond the nearest galaxy clusters (see Figure 5.2). Cosmological distance measurements are usually based on either a **luminosity distance** (a *standard candle* method), defined as

$$d_L = \sqrt{\frac{L}{4\pi f}}, \tag{5.14}$$

where f and L are the object's observed flux and intrinsic luminosity in its rest frame, or an **angular diameter distance** (a **standard ruler** method), defined as

$$d_A = \frac{d}{\theta}, \tag{5.15}$$

where d and θ are, respectively, the object's linear and angular sizes (in radians).

From the form of the FLRW metric, Equations (5.2) and (5.5), it is clear that the surface area of a sphere of comoving proper radius r is $4\pi x_k^2$ (not, in general, $4\pi r^2$). In addition, each photon received from the source has had its energy reduced by a factor $(1 + z)$ because

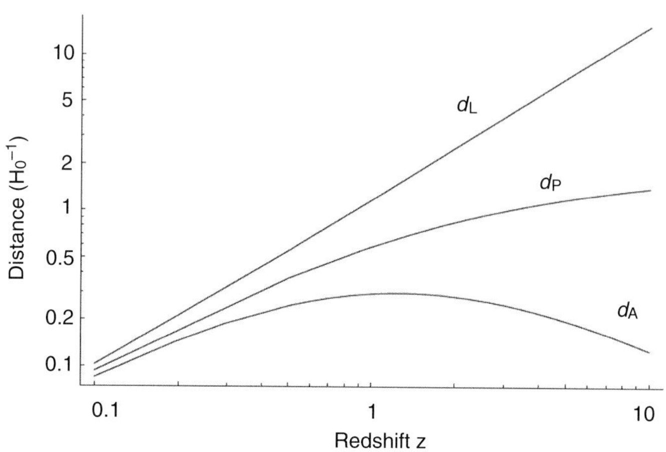

Figure 5.2 *Comparison of the redshift behaviour of proper, luminosity and angular-diameter distances (d_P, d_L and d_A, respectively).*

of the cosmological redshift, and an argument exactly analogous to that used to derive the redshift shows that the rate of reception of photons is also reduced by the same factor. Therefore, the received flux is given by

$$f = \frac{L}{4\pi x_k^2 (1+z)^2}. \tag{5.16}$$

Comparing this with the definition of $d_{\rm L}$ in Equation (5.14), it follows that

$$d_{\rm L} = x_k (1+z) \tag{5.17}$$

in the observer's frame.

Now consider an object of linear size ℓ at comoving distance r, oriented perpendicular to the observer's line of sight. Assuming that $\ell \ll r$, we can set $\mathrm{d}t = \mathrm{d}r = \mathrm{d}\phi = 0$ in Equation (5.5) and write

$$\ell = a(t_{\rm e})\, x_k \,\mathrm{d}\theta = \frac{x_k \,\mathrm{d}\theta}{1+z}, \tag{5.18}$$

using Equation (5.13) to express $a(t_{\rm e})$ in terms of z. Comparing this with the definition of $d_{\rm A}$ in Equation (5.15) gives

$$d_{\rm A} = \frac{x_k}{1+z}, \tag{5.19}$$

again in the observer's frame. At a given redshift z, these two distance measurements are related through Etherington's (1933) reciprocity relation, also known as the '**distance duality relation**' in astronomy,

$$d_{\rm L} = (1+z)^2 d_{\rm A}. \tag{5.20}$$

This is always true. It is valid for all cosmological models based on the **Riemannian geometry** – which includes a broad range of standard and special geometries – and does not depend on either Einstein's field equations or the nature of the matter–energy content. The only requirement is that source and observer are connected by null geodesics in Riemannian spacetime and that the number of photons is conserved. Therefore, it is valid for spatially homogeneous and isotropic as well as anisotropic cosmologies, and also for inhomogeneous cosmological models (Ellis 2007), so that comparisons between angular diameter and luminosity distances are straightforward (see, for validation of the distance duality relation, Basset and Kunz 2004; Uzan *et al.* 2004; De Bernardis *et al.* 2006; Holanda *et al.* 2010). The relationship between $d_{\rm L}$, $d_{\rm A}$ and $d_{\rm P}$ is less simple, because both $d_{\rm L}$ and $d_{\rm A}$ are defined in terms of x_k rather than r. In general, therefore, this relationship depends on the geometry of the Universe. Only for a flat Universe can we set $x_k = r$ and, hence, have simple relations between $d_{\rm L}$, $d_{\rm A}$ and $d_{\rm P}$.

The reason that these distinctions have only become important at this point in our discussion can be clearly seen if we take $z \ll 1$. In this case,

$$\frac{1}{\sqrt{k}} \sin\left(\sqrt{k} r\right) \simeq r \simeq \frac{1}{\sqrt{|k|}} \sinh\left(\sqrt{|k|} r\right), \tag{5.21}$$

so $x_k(r) \simeq r$ in all cases. Also, the factor $(1 + z) \simeq 1$. Therefore, for sufficiently small values of z, all definitions of distance converge to the comoving distance r, and we can simply refer to 'distance' without qualification. This has been the case for all distance indicators considered in earlier chapters.

5.1.3.2 *Cosmological Model Parameters*

As can be seen from Equation (5.13), redshift is essentially a measure of $a(t_e)$, the expansion parameter at the time of emission of the light. For a specified cosmological model, in which the form of $a(t)$ is known, z therefore gives t_e, and hence the light travel or **look-back time** $t_0 - t_e$, which is directly related – via Equation (5.9) – to the comoving distance r. Therefore, it is entirely valid to regard spectroscopic or photometric redshift as a distance indicator (see, for recent improvements in precision of the latter, Mandelbaum *et al.* 2008 and references therein). However, it cannot be directly compared to luminosity distance or angular diameter distance without assuming a specific cosmological model.

The Hubble Constant. The Hubble constant is merely a measure of the current expansion rate, so that constraining its current rate of change provides an indication as to whether the expansion of the Universe is accelerating or decelerating. In turn, the predominant behaviour of the Universe on very large scales depends on the mean density of baryonic and dark matter and energy, since mutual gravitation counteracts the expansion's momentum. A rapidly expanding Universe, i.e. a larger value of H_0, with a lower mass–energy density, ρ, implies that the universal expansion will continue forever. The expansion of a homogeneous, isotropic Universe is described mathematically by the **Friedmann equation**, which in turn is derived from Einstein's field equations for general relativity,

$$H^2(t) = \left(\frac{\dot{a}}{a}\right)^2 = \frac{8\pi G}{3} \sum_i \rho_i(t) - \frac{kc^2}{a^2} + \frac{\Lambda}{3}, \tag{5.22}$$

where G is the Newtonian gravitational constant, $\dot{a}(t)$ is the rate of change of the cosmic scale factor defining the size of the Universe at time t relative to that at the present time, $\rho_i(t)$ are the individual components of the mass–energy density and Λ is the cosmological constant, which can be interpreted as the energy density of the vacuum and is, therefore, sometimes included as a component of ρ. The Friedmann equation can be rewritten as

$$\Omega_{\text{total}} - 1 = \Omega_k = \frac{k}{a^2 H^2}. \tag{5.23}$$

In this context, the **cosmological density parameter** Ω_M is the sum of the mass–energy densities in matter. It is defined as the fraction of ρ in the Universe to the critical value, $\rho_0 \equiv 3H_0^2/(8\pi G)$, which will asymptotically bring the expansion of the Universe to a halt at an infinite time in the future, so

$$\Omega_M = \frac{8\pi G}{3H^2}\rho_0. \tag{5.24}$$

In addition, this Ω can be related to the topology of the Universe, which is determined by Ω_{tot}, Ω_M, and the mass–energy density contained in any possible cosmological constant, Ω_Λ, a type of energy which is characterized by a large, negative pressure and is referred to as *dark energy*. The prevailing cosmological model – denoted ΛCDM, referring to a cold dark matter-dominated Universe which includes a cosmological constant – favours a globally flat Universe, $\Omega_{\rm tot} = 1$, with a subcritical matter density, $\Omega_{\rm M} < 1$, and the remainder comprised of vacuum energy, $\Omega_\Lambda = 1 - \Omega_{\rm M} > 0$.

In such a cosmology, the nonzero terms of the Friedmann equation can be written as

$$\dot{a} = {\rm H}_0 \sqrt{\left[\frac{\Omega_{\rm M}}{a} + (1 - \Omega_{\rm M})\, a^2\right]}. \tag{5.25}$$

This can be integrated using a trigonometric substitution to produce the *parametric equations*

$$a(\theta) = \left(\frac{\Omega_{\rm M}}{1 - \Omega_{\rm M}}\right)^{1/3} \tan^{2/3}\theta \tag{5.26}$$

and

$$t(\theta) = \frac{2}{3H_0}\frac{1}{\sqrt{1 - \Omega_{\rm M}}} \ln|\sec\theta + \tan\theta|, \tag{5.27}$$

where θ is a dummy variable arising from the substitution. These expressions do not yield a neat algebraic expression for $d_{\rm P}$. However, they are fairly straightforward to integrate numerically.

Accelerated Expansion. Riess *et al.* (1998) and Perlmutter *et al.* (1999) independently found that **Type Ia supernovae** (SNe; see Section 5.2.1) at $z \sim 0.5$ appear to be approximately 10% fainter than their local counterparts, which has since been interpreted as evidence for accelerated expansion of the Universe (see Figure 5.3; see, for reviews, Filippenko 2005; Frieman *et al.* 2008). Parameter constraints representing the present, concordance model of the Universe, based on the 7-year *WMAP* observations (e.g. Komatsu *et al.* 2011; see also Section 5.2.1), BAO and independent constraints on ${\rm H}_0$, include $(\Omega h^2, \Omega_{\rm b}, \Omega_\Lambda) = (0.1352 \pm 0.0036, 0.0458 \pm 0.0016, 0.725 \pm 0.016)$, where $h = {\rm H}_0/100$. This corresponds to a Universe that contains approximately 27.5% of matter, $\Omega_{\rm M}$ – of which ~4.6% ordinary **baryons**,[3] $\Omega_{\rm b}$, and the remainder made up of cold and hot dark matter – and 72.5% of dark energy.

The expansion rate of the Universe is given by

$$H^2(z)/{\rm H}_0^2 = \Omega_{\rm M}(1+z)^3 + \Omega_\Lambda(1+z)^{3(1+w)}, \tag{5.28}$$

where $w = P/(\rho c^2)$ is the ratio of the pressure and density of the dark energy, also referred to as the *equation of state*. Current observations are consistent with $w = -1$ within the observational uncertainties and inconsistent with cosmological models that do not include dark energy, i.e. without accelerated expansion. For ordinary matter, $w = 0$, while for radiation, $w = \frac{1}{3}$.

[3] Composite particles consisting of three quarks, including protons and neutrons.

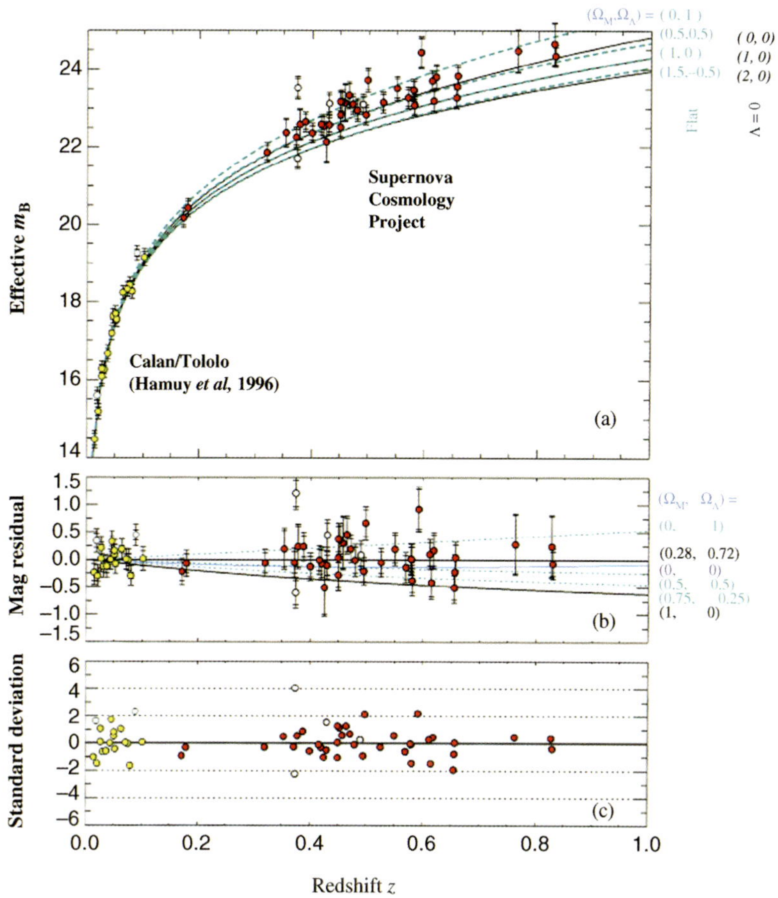

Figure 5.3 *(a) Hubble diagram and (b, c) residuals for 60 Type Ia supernovae (SNe; Perlmutter et al. 1999) corrected for the SN Ia light curve width–luminosity relation. The inner and outer error bars represent the uncertainties due to measurement errors and after accounting for an additional, intrinsic luminosity dispersion of 0.17 mag, respectively. The open circles indicate SNe not included in their final fit, while the parameters determining the various curves are indicated in the figure. (b) The solid curves are the theoretical predictions for* $(\Omega_M, \Omega_\Lambda) = (0,1)$, $(0.5,0.5)$ *and* $(0.75,0.25)$ *(top to bottom) for a set of flat cosmological models and the solid curve at the bottom is for* $(\Omega_M, \Omega_\Lambda) = (1, 0)$. *The middle solid curve is for* $(\Omega_M, \Omega_\Lambda) = (0,0)$. *(c) Uncertainty-normalized residuals from the best-fitting flat cosmology,* $(\Omega_M, \Omega_\Lambda) = (0.28, 0.72)$. (Reprinted from S. Perlmutter et al., *Astrophysical Journal*, **517**, Measurements of Ω and Λ from 42 high-redshift supernovae, p. 565–586, Copyright 1999, with permission of the AAS and S. Perlmutter.)

Returning now to the age of the Universe implied by the current value of the Hubble constant, in an expanding Universe the exact relation between these parameters also depends on the nature of the mass–energy content of the Universe:

$$t_0 = \int_0^\infty \frac{dz}{(1+z)H(z)} = H_0^{-1} \int_0^\infty \frac{dz}{(1+z)[\Omega_M(1+z)^3 + \Omega_\Lambda(1+z)^{3(1+w)}]^{1/2}}. \quad (5.29)$$

This implies that if one were to discard the presence of dark energy, the resulting t_0 would be an underestimate. For instance, for a matter-dominated flat Universe with $\Omega_M = 1.0$ and $\Omega_\Lambda = 0.0$, $t_0 = 9.3$ Gyr for $h = 0.70$ (Freedman and Madore 2010).

5.2 Supernovae as Beacons

SNe have provided key contributions to our improved understanding of extragalactic, cosmological distance measurements since early modern astronomy. Shapley (1919) used them to advance his 'Island Universes' hypothesis. He argued that objects such as SN 1885A (S Andromedae) in the Andromeda galaxy, the first known extragalactic SN, would have had an absolute visual magnitude $M = -16$ mag, which was 'out of the question'. Hubble (1929b) reported on 'a mysterious class of exceptional novae which attain luminosities that are respectable fractions of the total luminosities of the systems in which they appear', which Baade and Zwicky (1934) subsequently called 'supernovae' because of their high luminosities. In turn, their extraordinary brightnesses, combined with the observed homogeneity of the then-available sample of SNe, as evidenced by the small scatter in the Hubble diagram (see e.g. Kowal 1968 for early indications), prompted Wilson (1939) to suggest that they could potentially be used for cosmological distance determinations and derivation of other fundamental parameters at high redshifts.

Based on a sample of 14 objects, Minkowski (1941) 'provisionally' divided them into Type I SNe, which were characterized by an absence of hydrogen absorption lines in their optical spectra, and **Type II** objects, whose spectra showed evidence for Hα and other Balmer absorption lines. On the basis of near-infrared (near-IR) observations, Elias *et al.* (1985) divided SNe I into two subclasses, which they dubbed Ia and Ib, and observed that it was the Type Ia subclass which appeared to be a standard candle and, thus, a potential distance indicator. The present-day, most favoured classification of SNe (see, for a review, Filippenko 1997) still includes Type I subclasses Ia and Ib/Ic, which exhibit strong and weak Si II λ6150 Å absorption in their spectra, respectively. This absorption line is caused by the blueshifted Si II $\lambda\lambda$6347,6371 Å feature. Alternative discriminants between Type Ia and Type Ib/c SNe include differences in the strength or presence of Si II λ4130 Å (Coil *et al.* 2000), Fe II λ4555 Å and Mg II λ4481 Å (Barris *et al.* 2004) absorption features.

Type Ib and Ic SNe are caused by the core collapse of massive stars, such as Wolf–Rayet stars. They have lost their outer hydrogen layer. Hence, they are also called 'stripped core-collapse SNe'. Type Ic SNe are thought to additionally have lost most of their helium content. This is particularly evidenced by a lack of the λ5876 Å helium absorption line in their spectra, which is present in nonionized form in the spectra of Type Ib SNe. Very massive stars, with initial masses $\geq$140 $M_\odot$, are thought to end their lives violently as **'pair-instability' SNe**. Fraley (1968) predicted that the core temperatures of such objects will become so high that photons spontaneously form **electron–positron pairs**. Because of the subsequent reduction of the photon pressure supporting the star's outer layers, a collapse followed by ignition of their oxygen core is triggered, which vaporizes the star. A possible example of this type of SN is SN2007bi, one of the most luminous SNe known (Gal-Yam *et al.* 2009; Young *et al.* 2010). Whereas its light curve and spectral analysis suggest that it may be a pair-instability SN, its host galaxy's metallicity is not as low as expected for the stellar models to produce highly carbon–oxygen-rich cores in 140–260 $M_\odot$ stars (Heger and Woosley 2002).

Type II SNe, resulting from the iron core collapse of massive progenitor stars (zero-age main-sequence mass $\gtrsim 8\ M_\odot$; e.g. Burrows 2000) with significant hydrogen abundances (i.e. red supergiants; cf. Smartt 2009), have been further subdivided into classes II-P (comprising some 50% of all SNe II; Cappellaro *et al.* 1999; Botticella *et al.* 2008; Leaman *et al.* 2010) and II-L, indicating the presence or absence of a 'plateau' in their light curves. When such a star explodes with a significant fraction of its initial hydrogen-rich envelope intact, it exhibits a light curve which is characterized by an optically thick phase of ~100 days of nearly constant luminosity. When all hydrogen in the envelope has recombined, a sudden drop in luminosity of 2–3 mag is observed, followed by an exponential decline caused by the radioactive decay of ^{56}Co into ^{56}Fe (e.g. Kirshner and Kwan 1974; Nadyozhin 2003; Utrobin 2007; Bersten and Hamuy 2009). Type II-L SNe, on the other hand, display a linear decrease in the magnitudes of their light curves following an initial rapid decline (see Figure 5.4).

Type II SNe can also be classified based on their spectral appearance. Although most Type II SNe exhibit broad **P Cygni**-type absorption features on the order of thousands of km s^{-1}, Type IIn SNe have relatively narrow features, which are thought to be caused by interactions with a dense circumstellar medium. In addition, the term Type IIb is reserved for those SNe that show a combination of features normally associated with both Types II and Ib, i.e. they have hydrogen lines near maximum light but later evolve to resemble SNe Ib. Any Type II SN that cannot be classified easily into one of these subcategories is usually referred to as 'peculiar', i.e. IIpec. Table 5.1 provides a quick overview of the current SN classification criteria, while Figure 5.5 shows representative optical SN spectra for the various modern classes. Of all SN subclasses, only Type Ia and II-P SNe show promising

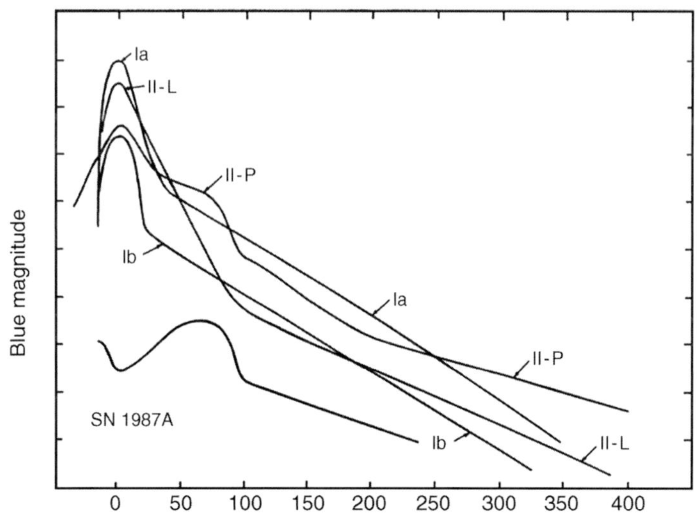

Figure 5.4 *Schematic light curves for SNe of Types Ia, Ib, II-L, II-P, and SN 1987A. The curve for SNe Ib includes SNe Ic as well and represents an average (Wheeler and Harkness 1990).* (Reprinted from J. C. Wheeler and R. P. Harkness, *Reports on Progress in Physics*, **53**, Type I supernovae p. 1467–1557, Copyright 1990, with permission of the Institute of Physics and J. C. Wheeler.)

Table 5.1 *Supernova classification*

	Diagnostic features		Type	Mechanism
No H	Strong Si II		Ia	Thermonuclear
No H	Weak Si II	He I	Ib	Core collapse
		No He I	Ic	
H	At early times only		IIb	
H	Plateau in light curve		II-P	
	Linear decline		II-L	
	Narrow emission lines		IIn	
	Unclassifiable		IIpec	

features for potential use as reliable distance tracers. None of the other core-collapse-type SNe (e.g. Types Ib, Ic, II-L or IIn) have been proposed as standard candles, because their absolute magnitudes vary significantly, nor are they characterized by any other parameter that can be calibrated reliably.

5.2.1 Type Ia Supernovae

Type Ia SNe most likely result from thermonuclear runaway explosions of white dwarfs in accreting binary systems (Whelan and Iben 1973). Although white dwarfs are stars at the end stages of their life cycles which normally die a quiet death by slowly radiating away their remaining heat, carbon oxygen (^{12}C/^{16}O) white dwarfs can continue violent nuclear fusion if their temperatures increase sufficiently. (Oxygen neon magnesium white dwarfs continue to collapse and cannot reignite nuclear fusion.)

A slowly rotating CO white dwarf which accretes matter from a binary companion in a cataclysmic variable system (see Section 3.6) cannot exceed the **Chandrasekhar mass limit**. The maximum nonrotating mass ($\sim$1.38 $M_\odot$, revised upwards since Chandrasekhar 1931) which can be supported against gravitational collapse is determined by the electron degeneracy pressure caused by the compression of the electron-degenerate matter, which in turn increases the electrons' kinetic energy. This increase is determined by either the **Pauli exclusion principle** – prohibiting two electrons to occupy the same quantum state – or the **Heisenberg uncertainty principle**: since their positions are reasonably well determined because of the high electron density in a degenerate white dwarf core, some electrons must have large kinetic energies.

The increasing pressure and density caused by the increasing mass due to either accretion or collisional – respectively, single- or **double-degenerate** – evolution (the product of the latter is referred to as a **super-Chandrasekhar white dwarf**; e.g. SN 2003fg: Howell *et al.* 2006; SN 2006gz: Hicken *et al.* 2007; SN 2009dc: Tanaka *et al.* 2010 and references therein; see also Gilfanov and Bogdan 2010; van Kerkwijk *et al.* 2010) raise the core temperature, and as the white dwarf approaches the Chandrasekhar limit to within $\sim$1%, carbon fusion is somehow ignited. This is followed by oxygen fusion, and a subsonic 'deflagration flame front' originates. Fusion further increases the white dwarf's core

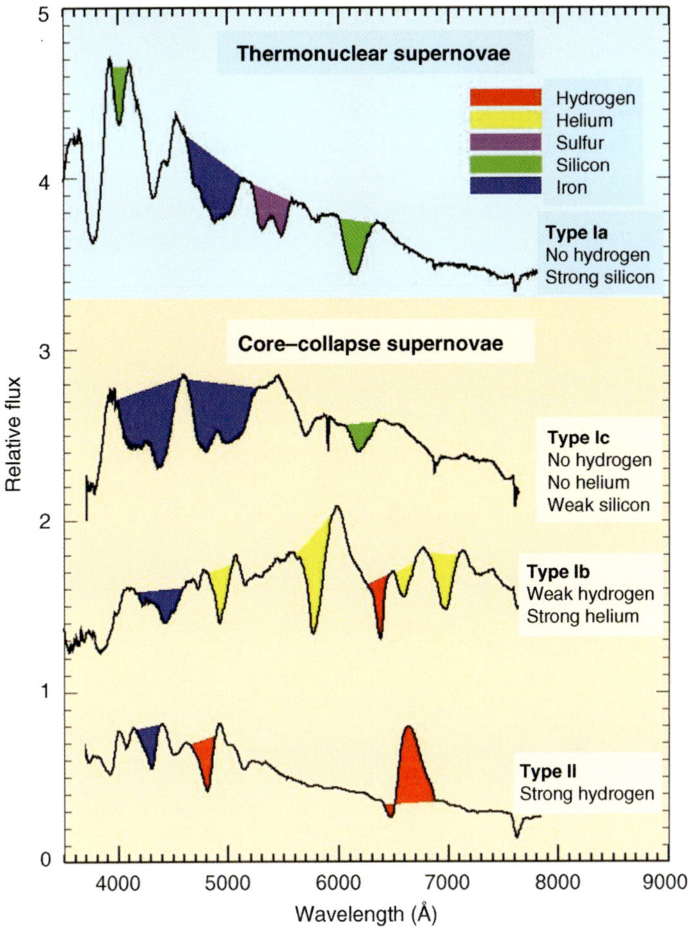

Figure 5.5 *Schematic optical SN spectra highlighting the similarities and differences among the various subclasses.* (Reprinted from D. Kasen, Supernova types (Online: http://supernova.lbl.gov), Copyright 2003, with permission of D. Kasen.)

temperature, eventually – via **Rayleigh–Taylor instabilities** and turbulent interactions – leading to a supersonic detonation wave and, thus, a thermonuclear runaway explosion (e.g. Khokhlov *et al.* 1993; Hillebrandt and Niemeyer 2000; Gamezo *et al.* 2003). This generates a flash with an absolute V-band magnitude of $M_V \simeq -19.3$.

The details of this ignition process are unclear. The general understanding is that the observed, exponentially declining light curves of SNe Ia (see Wood-Vasey *et al.* 2007 for an historical overview) are caused by radioactive decay of ^{56}Ni via ^{56}Co to ^{56}Fe, with the small range of peak brightnesses being due to variations in ^{56}Ni abundance. Nevertheless, despite the lack of a firm understanding of the underlying physics, SNe Ia are presently among the lowest-dispersion, highest-precision (7–10% accuracy; e.g. Hicken *et al.* 2009) distance tracers to objects at sizeable redshifts.[4] This is possible by virtue of the tight relationship between their absolute magnitude at peak brightness, their colour and their rate of decline (cf. Pskovskii 1984; Phillips 1993; Perlmutter *et al.* 1997, 1999; Goldhaber *et al.* 2001; Leibundgut 2001; Knop *et al.* 2003; see also Jha *et al.* 2007 for a sophisticated approach to minimize this multiparameter domain), which is further improved at near-IR wavelengths (e.g. Wood-Vasey *et al.* 2008). Jha *et al.* (2007) provide a useful relationship between the intrinsic absolute V-band magnitude at the time of B-band maximum light, and the shape of the light curve, also known as the '**stretching–luminosity correlation**' (see also Figure 5.6 for an application), which can be used for statistical distance determinations,

$$M_V(t = 0) = -19.504 + 0.736\Delta + 0.182\Delta^2 + 5\log\left(\frac{\mathrm{H}_0}{65\ \mathrm{km\ s^{-1} Mpc^{-1}}}\right)\ \mathrm{mag}. \tag{5.30}$$

Here, Δ (sometimes referred to as Δm_{15}^B) is the light curve shape parameter. It is defined as the B-band magnitude difference between the intensity at peak brightness and that 15 days post-maximum (Phillips 1993; see also Jha *et al.* 2007). In terms of a Hubble diagram, in the form of a magnitude–redshift relation, they provide

$$M_V^0 - 5\log \mathrm{H}_0 = m_V^0 - 25 - 5\log\left\{c(1+z)\int_0^z [\Omega_\mathrm{M}(1+z')^3 + \Omega_\Lambda]^{1/2}\mathrm{d}z'\right\}, \tag{5.31}$$

which is valid for a flat Universe, with $\Omega_\mathrm{M} + \Omega_\Lambda = 1$. Based on their sample of 60 SNe Ia, Jha et al. (2007) derive $M_V^0 - 5\log h_{65} = -19.504 \pm 0.018$ mag, where the Hubble constant is expressed in units of 65 km s^{-1} Mpc^{-1}. Although different techniques produce relative luminosity distances that scatter by $\sim$0.10 mag for individual SNe Ia (e.g. Tonry *et al.* 2003; Sandage *et al.* 2006; see also the discussion in and recent results of Foley and Kasen 2011), the magnitude of the scatter is uncorrelated with redshift.

In fact, the high-precision distance determinations attainable with observations of SNe Ia formed the basis of the notion that the Universe has been undergoing accelerated expansion since $z \sim 0.5$ (Riess *et al.* 1998; Perlmutter *et al.* 1999). Clearly, to make such a claim, it is important to verify that the spectra of high-z SNe Ia are the same as those of their nearby counterparts, a fundamental question that has triggered a significant level of research activity (e.g. Coil *et al.* 2000; Hook *et al.* 2005; Howell *et al.* 2005; Lidman *et al.* 2005; Matheson *et al.* 2005; Blondin *et al.* 2006; Kessler *et al.* 2009; Foley and Kasen 2011). Foley *et al.* (2005) confirmed that spectra of both distant and nearby SNe Ia evolve similarly over their lifetimes, at least to the extent to which we can determine this with our current methods.

A number of technical issues must be addressed in using SNe Ia as distance indicators, particularly K**-corrections** and the effects of extinction, as well as important calibration issues. In the following, we will highlight the state of the art regarding these issues and also point out where improvements are required.

***K*-corrections.** SNe Ia are observed at considerable distances and, therefore, at significant redshifts. As the data consist primarily of apparent magnitudes measured in standard ground-based photometric passbands (e.g. *ugriz* in the SDSS, *ugriBVYJH* in the Carnegie Supernova

[4] Because of the similarity of their spectra to those of Type Ib and Ic SNe, the latter must be removed carefully from any sample aimed at determining extragalactic distances (cf. Homeier 2005).

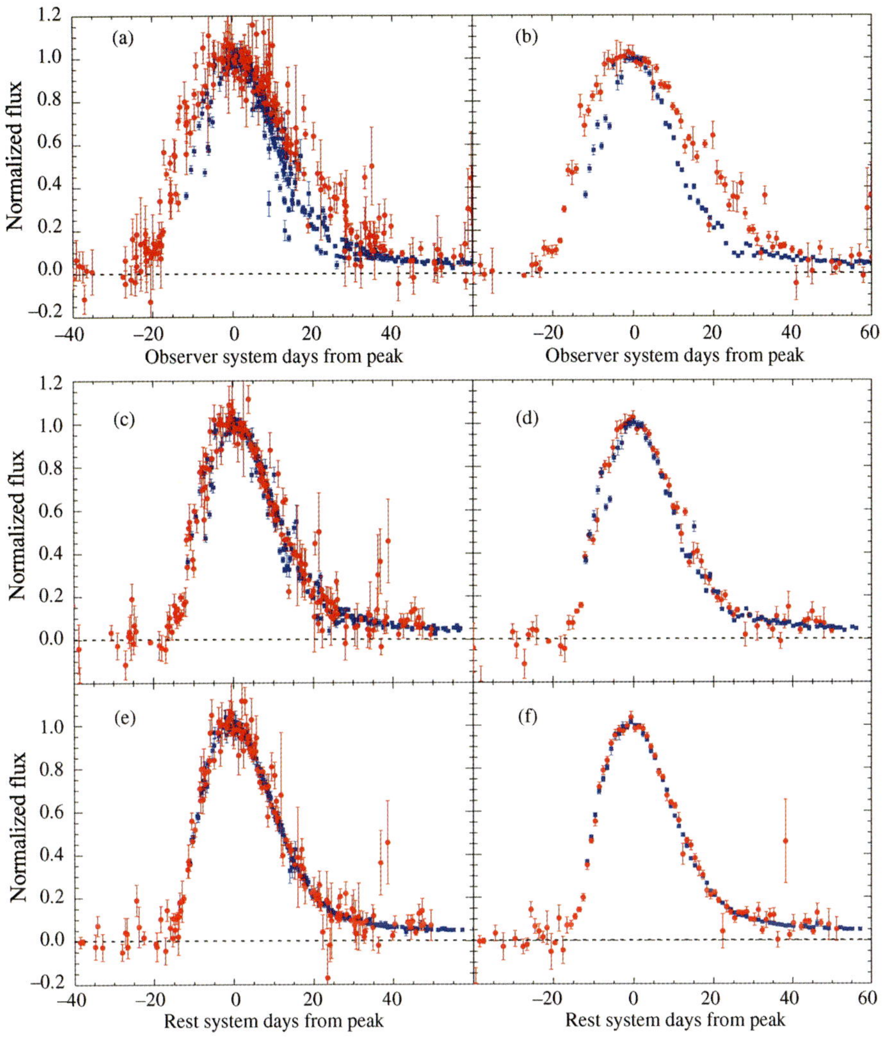

Figure 5.6 *Illustrative example of the concept of a 'composite light curve', constructed by linearly compressing or expanding the time axis for each SN such that all low- and high-z data points can be plotted on a single curve (Goldhaber et al. 2001). The left- and right-hand panels show a data point for each night's observation of each SN and 1-day averages over all SNe, respectively, for two different samples (indicated in blue and red). (a, b) B-band photometry points in the observer system displaced to $t = 0$ at light maximum, and normalized to unit intensity at $t = 0$. There are no corrections for stretch or width; times of observation relative to maximum light are used. (c, d) After transformation of the time axis from the observer frame to the rest frame by dividing by the appropriate $1 + z$ factor for each data point of each SN. (e, f) The timescale for each point is also divided by the fitted stretch factor s. By this stage, essentially all of the dispersion has been removed, and the corrected points fall on a common curve at the level of the measurement uncertainty, of typically 2–4%.* (Reprinted from G. Goldhaber et al., *Astrophysical Journal*, **558**, Timescale stretch parameterization of Type Ia supernova *B*-band light curves, p. 359–368, Copyright 2001, with permission of the AAS and D. E. Groom, Lawrence Berkeley National Laboratory.)

Project), the part of the SN spectrum which falls within each passband will vary with redshift. Because their spectral energy distributions (SEDs) are not flat, this will create a systematic shift of observed flux with redshift, even for quite small redshifts (Jha *et al.* 2007 report U-band corrections of order 0.1 mag for $z = 0.03$). For higher redshifts, cross-filter K-corrections are needed to account for the fact that at $z \gtrsim 0.1$, a particular region of the SED will fall into different passbands: for example, at redshifts of order 0.2, ground-based V-band observations would approximately correspond to the rest-frame B band.

Calculating K-corrections for SNe Ia is not simple: the correction is a function of redshift, time since maximum light, intrinsic luminosity (since for SNe Ia luminosity is correlated with colour), extinction in the host galaxy, and extinction in the Milky Way. The basic approach is to define a set of template rest-frame spectra which can then be artificially redshifted to derive the K-correction for observed SNe. Ideally, one would like full spectroscopic coverage of a number of different SNe Ia to average out any individual idiosyncrasies over the whole duration of the event from before maximum light until at least 100 days post-maximum. This is not very practical. Instead, recipes have been developed to allow a more limited set of templates to cover a much wider range of observed SNe, and/or to apply a colour-based procedure rather than relying on full spectroscopic information.

Nugent *et al.* (2002) note that the K-correction for SN spectra is driven by the continuum rather than by specific spectral features, and that the intrinsic correlation between M_V and $B - V$ observed in SNe Ia of different peak luminosities is quite similar to standard interstellar reddening laws. Therefore, they apply corrections based on the standard Galactic reddening law (Cardelli *et al.* 1989; cf. Section 6.1.1) to adjust the colours of observed SNe to match a template spectrum. The physical motivation for this procedure is that if the SED is continuum-dominated, as SNe Ia are at early epochs, then essentially any slowly varying function of wavelength should be satisfactory for relatively small adjustments, while at late times the colour variation is largely caused by extinction, so the use of a reddening law is clearly appropriate (although see e.g. Hsiao *et al.* 2007 for a more sophisticated approach based on using intrinsic colours).

Thus, in general, the prescription for K-corrections for SNe Ia is to take a library of spectral templates (e.g. Hsiao *et al.* 2007), adjust or 'warp' the closest template to match the photometric colours of the target object and then use this to determine the K-correction. Clearly, the quality of the template spectra is a critical determinant in avoiding systematic errors (cf. Foley *et al.* 2008).

Host Galaxy Extinction. The problem of disentangling intrinsic colour differences from differences in reddening is nontrivial. One approach to achieve this is to construct samples for which the host galaxy extinction is expected to be small, as for e.g. early-type host galaxies or SNe which are on the outskirts of their host galaxy. This does not, however, preclude the presence of local circumstellar or intergalactic dust. One can select the bluest SNe, on the grounds that reddening never makes objects appear bluer and, therefore, the bluest objects are likely to be comparatively unreddened. The problem with this strategy is that there is known variation in the intrinsic colours of SNe Ia. Therefore, a sample of exceptionally blue objects probably is a sample of intrinsically exceptionally blue objects. Thus, using these objects as a baseline for reddening corrections will likely overestimate the correction. One way to avoid this problem is to select a subset of the available data for which intrinsic variations are small.

Phillips *et al.* (1999) find a dispersion of 0.06 mag in the $B - V$ evolution of the light curve at late times ($t = 30 - 90$ days) for six SNe Ia with a wide range of intrinsic luminosities. Their 'unreddened' test sample was selected on the basis of location (early-type hosts or not in the arms or disc of a spiral host) and absence of interstellar spectral lines from NaI or CaII (see also Jha *et al.* 2007). Having established that the $B - V$ colour evolution at late times is essentially independent of the intrinsic brightness or colour of the SN at maximum light, one can then use it as a baseline to determine host galaxy extinction. The problem with this method is that it is difficult to apply to high-z SNe, where the late-time light curve may be too faint to be well observed.

Another approach, used by the Carnegie Supernova Project (Freedman *et al.* 2009), is to construct a 'reddening-free' magnitude: a linear combination of magnitude and colour which is designed to cancel out the effect of reddening, specifically

$$w^i_{BV} = i - R^i_{BV}(B - V), \tag{5.32}$$

where

$$R^i_{BV} = \frac{A_i}{E(B-V)} = \frac{R_i}{R_B - R_V}, \tag{5.33}$$

i is a photometric band and R_i is the ratio of total to selective absorption, defined by a given reddening law. This method is commonly used in studying Cepheid period–luminosity relations (see Section 3.5.2).

Host galaxy extinction, either directly as A_i or in the form of R parameters, is normally included as a 'nuisance' parameter in the multidimensional fits used in modern analyses to extract physically useful results such as distance moduli. In recent years, the values of R_V extracted from such fits have tended to be lower than the canonical Milky Way value of 3.1. For example, Freedman *et al.* (2009) obtain $R_V = 1.74 \pm 0.27$ (statistical) $\pm\, 0.10$ (systematic), while Kessler *et al.* (2009) quote 2.18 ± 0.14 (statistical) $\pm\, 0.48$ (systematic). Jha *et al.* (2007) adopt an average extragalactic value of $R_V \sim 2.5$. However, extinction effects and their distribution in the SNe Ia host galaxies contribute at a fairly low level to the systematic uncertainties in the resulting Hubble constant. Varying the extinction law from $R_V \equiv A_V/E(B-V) = 1.5 - 3.1$ results in an associated variation in H_0 of 0.2 km s^{-1} Mpc^{-1} (Riess *et al.* 2009a). In addition, assuming no extinction at all, or leaving its value as a free parameter introduces a full uncertainty range in the value of H_0 of 0.8 km s^{-1} Mpc^{-1}. Although a better understanding is clearly desirable to minimize systematic errors in future high-precision cosmological studies, there is no reason to believe that current results are seriously biased.

Host Galaxy Light. SN light curves will be contaminated by background light from their host galaxies. This becomes progressively more important as redshift increases, because the effective linear size of a pixel at the host galaxy distance will increase, at least until the maximum angular diameter distance is reached. However, this occurs at a redshift greater than those of most SNe Ia. Hence, the ratio of SN luminosity to host galaxy light decreases with increasing redshift, as well as with increasing time since maximum light.

Host galaxy light is normally removed by subtracting a template image of the host galaxy before (if available) or long after the SN explosion. Since any noise in the template image will introduce a systematic error in the host-subtracted light curve, template images need to be of good quality: for example, the Carnegie Supernova Project (Contreras *et al.* 2010) requires the template image to be 3 mag deeper than the SN images, and obtained under seeing conditions at least as good as the best images from the SN light curve.

Tonry *et al.* (2003) avoided the systematic error introduced by subtraction of a common template by employing the difference between different images of the SN, instead of the difference between the SN images and a standard template. There are $\frac{1}{2}N(N-1)$ such differences, producing an $N \times N$ matrix of flux differences and a corresponding matrix of errors, which can be used to determine the flux scale. A post- or pre-SN image is still required to establish the zero point, but the errors in subtracting the galaxy background are no longer strongly correlated among images. However, this procedure is fairly complex and the benefits are limited, unless the available background galaxy images have low signal-to-noise ratios or are compromised by poor seeing.

Time Dilation. The rate of decline of the SN light curve is a crucial parameter in calibrating its peak luminosity. However, because SNe Ia are seen at significant redshifts, **time dilation** produces a similar effect: the spectrum of a SN at redshift z is 'stretched' by a factor $(1+z)$ compared to that of a nearby comparison object. This effect must be corrected for by rescaling the time axis before applying the various relative calibration algorithms discussed below.

It might be argued that there is an element of circularity in using a correction derived from general relativity to correct data that will be used to probe general relativistic cosmological models. This concern has been addressed by Blondin *et al.* (2008), who used the known *spectroscopic* evolution of SNe Ia to derive rest-frame ages for SN spectra independent of the observed time past maximum light. Their study of 35 spectra of 13 SNe Ia at redshifts between 0.28 and 0.62 clearly showed the expected $(1+z)$ dilation effect.

Relative Calibration of SN Ia Peak Brightness. Phillips (1993) quantified the rate of light curve decline using the parameter Δm_{15}^{B}, the difference in the B filter between the magnitude at the peak and 15 days post-maximum. Phillips *et al.* (1999) updated this analysis with better data and host galaxy extinction corrections using the late-time $B-V$ colour as their baseline, hence deriving an estimate of the 'true', de-reddened, value of Δm_{15}^{B}, which they compared with the absolute B, V and I magnitudes for a sample of 41 low-z SNe Ia.

Direct measurement of Δm_{15}^{B} can be difficult based on noisy data. A solution to this is to develop a family of templates using well-measured SNe, and then estimate Δm_{15}^{B} by fitting the templates to the data in question. This procedure uses all available data instead of just two points. A natural extension to this idea is to use the templates directly to determine the peak absolute magnitude: this is the 'Multi-colour Light Curve Shape' (MLCS) calibration method (Riess *et al.* 1996). Its principle is straightforward. Well-measured SNe are used to produce a set of template light curves in several passbands and/or colours. A 'training set' of SNe with well-measured relative distances is used to establish an empirical relationship between the shapes of the template curves and the peak absolute magnitude. This relationship is then used to determine the absolute magnitudes of SNe at unknown distances by fitting their light curves to the templates (cf. Jha *et al.* 2007).

Inspection of SN light curves, whether real data as in Hamuy *et al.* (1996) or idealized templates as in Jha *et al.* (2007), shows that the overall *shape* of the light curve in U, B and

V does not change much: the different curves for different peak luminosities seem related by a simple scaling factor (see Figure 5.6). The Supernova Cosmology Project (Perlmutter *et al.* 1997, 1999) implemented such a scaling factor, correcting their peak magnitudes by the simple relation

$$m_B^{\rm corr} = m_B + \alpha\,(s - 1)\,, \tag{5.34}$$

where s is the 'stretch factor' required to map a standard template onto the SN light curve and α is a fitted constant: Perlmutter *et al.* (1999) use $\alpha = 0.6$. This relation has the considerable merit of simplicity, but it does not work well for rest-frame passbands redder than V, since in these filters the overall shape of the light curve changes as the peak luminosity changes. It also requires corrections for host galaxy extinction to be made *a priori*, since there is no scope to fit for them as is done in MLCS methods.

Tonry *et al.* (2003) compare results from MLCS, Δm_{15}^B, and a 'Bayesian Adapted Template Method', which performs a likelihood fit of the target SN light curve to a library of well-observed local SNe. Their comparison table for eight SNe shows that the different methods are satisfactorily concordant, with MLCS appearing to yield the smallest formal error estimates.

A rather different approach, used by the Supernova Legacy Search, is presented in Guy *et al.* (2007) [SALT-2]. They fit the flux $F(\mathrm{SN}, t, \lambda)$ using a principal-components decomposition:

$$F(\mathrm{SN}, t, \lambda) = x_0 \times [M_0\,(t, \lambda) + x_1 M_1\,(t, \lambda) + \cdots] \times \exp\,(c\;\mathrm{CL}(\lambda))\,, \tag{5.35}$$

where x_0 is the flux normalization, M_0 represents the average spectrum, M_1 describes the most important source of variability with respect to the average and any higher M_i would describe subsidiary sources of variability. The exponential factor is a colour correction, in exponential form because the fit is done in flux space and not in magnitude: c is the colour offset at time t compared to the colour at maximum light, $c = (B - V)_{\rm max} - (B - V)$ and $\mathrm{CL}(\lambda)$ is a colour correction law, modelled as a third-order polynomial. The components M_i and the colour law $\mathrm{CL}(\lambda)$ are determined from a training sample, while the parameters x_i and c are properties of the individual SN.

Kessler *et al.* (2009) use both SALT-2 and an MLCS variant in an analysis studying the cosmological parameters $\Omega_{\rm M}$, Ω_Λ and w. They find that, while broadly similar, the results from the two models diverge when applied to higher-z SNe. The cause of the discrepancy seems to be partly the different treatment of colour corrections and partly a difference in the results of training, particularly in the rest-frame U band, where there is a lack of good low-z data. The large surveys currently underway should provide both models with better, more consistent training samples, which may well allow resolution of the discrepancy.

Absolute Calibration. Type Ia SNe are secondary distance indicators. They are not geometric, and, given the continuing uncertainty over the exact nature(s) of the progenitor systems, there is no reliable theoretical prediction of their peak luminosity. Therefore, although they can be used for *relative* distance measurements, e.g. studies of deviations from the linear Hubble law, they require calibration for application to *absolute* distance determinations, e.g. the slope of the Hubble law.

Unfortunately, one of the key uncertainties associated with using SNe Ia for calibration of distances beyond the nearest galaxies and clusters of galaxies is the relative paucity of galaxies for which both SNe Ia light curves and Cepheid-based distances are available (e.g. Sandage *et al.* 2006; see also Wood-Vasey *et al.* 2007 for a comprehensive review of the systematic uncertainties affecting SNe Ia-based distance determinations). This means that the number of well-measured, recent examples in nearby galaxies with secure distances is very small: for example, the *HST* Key Project calibration of SNe Ia (Freedman *et al.* 2001) relied on only seven SNe, of which only three are ideal calibrators, i.e. observed before maximum, with CCD or photo-electric – not photographic – photometry, in a location where significant host galaxy extinction is unlikely, and spectroscopically typical. The shortcomings of this calibration are shown by the fact that Saha *et al.* (2001) and Freedman *et al.* (2001), using the same calibration set, report values for H_0 of 58.7 ± 6.3 and $71 \pm 2 \pm 6$ km s^{-1} Mpc^{-1}, respectively, which, while not formally inconsistent if the errors are unrealistically treated as independent, surely indicate the existence of a problem.

Riess *et al.* (2009a,b) recently provided a careful recalibration of Cepheid and SNe Ia distance determinations to six nearby galaxies based on optical and near-IR *HST* observations, using the independently verified geometric distance determination to the maser galaxy NGC 4258 as their anchor (see Section 3.7.4). Near-IR observations of Cepheids have the advantage of reducing both extinction effects and any dependence on Cepheid chemical composition. Riess *et al.* (2009a) derive a value for H_0 with an uncertainty of $<5\%$, $H_0 = 74.2 \pm 3.6$ km s^{-1} Mpc^{-1}, where the quoted uncertainty includes the statistical and systematic errors (cf. an 11% uncertainty resulting from the *HST* Key Project; Section 4.1).

On the other hand, Sandage *et al.* (2006) and Tammann *et al.* (2008) independently pursued a recalibration programme of SNe Ia, Cepheids, RR Lyrae variables and 'tip-of-the-red-giant-branch' (TRGB) distances, and reported $H_0 = 62.3 \pm 4.0$ km s^{-1} Mpc^{-1} for 62 SNe Ia with $3000 < v_{\rm CMB} < 20\,000$ km s^{-1} (recessional velocities with respect to the CMB rest frame). They included the uncertainties in the Cepheid and TRGB zero points in their result. Riess *et al.* (2009a,b) compared the methodology used in the latter article with their own and argue that the Tammann *et al.* (2008) result is plagued by significant cross-calibration issues affecting the use of different detectors, their inclusion of less accurate photographic plate observations and of more heavily reddened Cepheid variables. If correct, this appears to point at either the need for a reassessment of the uncertainties associated with the Tammann *et al.* (2008) work or a global reassessment of the underlying assumptions and resulting uncertainties.

Selection Bias. Apart from biases introduced by rejecting observed but reddened SNe from the analysis sample, there is an intrinsic bias against such events, because they are fainter than unreddened examples and their detection is, therefore, less likely close to the magnitude limits of the particular study (e.g. Wood-Vasey *et al.* 2007). In addition to this reddening-dependent effect, a pure Malmquist bias (see Chapter 6.1.3), i.e. a tendency to preferentially detect SNe that are unusually bright *for their light curve shape* would also introduce a systematic error. Preferentially detecting brighter SNe is not a problem if they are correctly identified as bright on the basis of a slower decline. The effect of such a bias depends on the intrinsic dispersion of the objects in question: clearly, no bias would be introduced if all SNe with the same decline rate had exactly the same peak brightness. While this is unrealistic, studies of low-z samples indicate that the dispersion after correcting for light curve shape is small, so the effect is not dramatic and can be corrected for if the low-z sample is representative of the unbiased sample at high redshift.

The Local Velocity Field. Another potentially important systematic uncertainty affecting the reliability of SNe Ia as suitable distance indicators beyond the nearest galaxies which also host alternative distance tracers is the minimum distance or redshift appropriate for measuring the smooth Hubble flow. Riess *et al.* (2009a) suggest a minimum redshift of $z = 0.023$, given recent controversy about the possible presence of a local 'Hubble bubble' characterized by an increased outflow of $\sim 5\%$ within the local void in the range $0.01 < z < 0.023$ (cf. Zehavi *et al.* 1998; Jha *et al.* 2007). However, Conley *et al.* (2007) provided convincing evidence that the claim of a local bubble may have been caused by the nature of and, particularly, the amount of extinction affecting the local SNe Ia. Nevertheless, coherent large-scale flows could easily induce a distance bias at low redshifts (Cooray and Caldwell 2006; Hui and Greene 2006), potentially affecting the resulting Hubble constant by $\Delta H_0 = +1.0$–1.2 km s^{-1} Mpc^{-1}, depending on the precise treatment of the extinction (Riess *et al.* 2009a). The latter authors suggest that one could, as alternative, only use local SNe Ia affected by $A_V < 0.5$ mag to avoid uncertain extinction corrections, in which case H_0 would be 0.7 km s^{-1} Mpc^{-1} larger than resulting from leaving out SNe Ia measurements at $z < 0.023$ altogether.

Interesting prospects await this field. In particular, the planned satellite mission *WFIRST* (Wide Field Infrared Survey Telescope; cf. Gehrels 2010), a 1.5 m wide-field IR imaging instrument, combined with low-resolution spectroscopic capabilities, will employ a combination of high-z SNe, BAO (see Section 5.3.3) and weak lensing to constrain the parameters of dark energy.

5.2.2 Type II-P Supernovae

5.2.2.1 The Expanding Photosphere Method

A range of methods have been proposed and investigated to obtain distance estimates to Type II – and particularly Type II-P – SNe, including a Baade–Wesselink-type (see Section 3.5.1) **expanding photosphere method** (EPM; Kirshner and Kwan 1974; Schmidt *et al.* 1992, 1994; Hamuy 2001; Jones *et al.* 2009) and, more recently, the so-called synthetic spectral-fitting expanding atmosphere method (SEAM; e.g. Baron *et al.* 2004; Dessart *et al.* 2008). Both methods relate the SN's angular and physical sizes, assume a spherically symmetric expanding photosphere (see Leonard *et al.* 2001) which radiates as a diluted blackbody and rely on high-signal-to-noise photometry and spectroscopy.

Early versions of the EPM simply assumed a blackbody spectrum: there was little alternative at the time because the calculations necessary to produce a realistic model of the expanding SN were not feasible with 1970s computer technology. However, the limitations of this approximation became apparent during the 1980s (Hershkowitz *et al.* 1986a,b): while the results presented were still of limited precision, they were able to conclude that "the flux is diluted below that of a blackbody at the same colour temperature" (Hershkowitz *et al.* 1986b). This led to the introduction of a 'dilution factor' (e.g. Hamuy 2001), ζ_λ, which represents a correction factor relating the true luminosity of the SN to that of a blackbody of the same colour temperature. The definition of ζ_λ is such that the emitted flux per unit area of the photosphere is given by

$$F_{\lambda,\text{emitted}} = \zeta_\lambda^2 \pi B_\lambda(T). \tag{5.36}$$

The SN's angular size, θ, can be derived as (Jones *et al.* 2009)

$$\theta = \frac{R}{d} = \sqrt{\frac{(1+z)f_\lambda}{\pi \zeta_{\lambda'}^2 B_{\lambda'}(T) 10^{-0.4[A(\lambda)+A'(\lambda')]}}}, \tag{5.37}$$

where R is the photospheric radius, d the distance and f_λ the observed flux density at an observed wavelength λ. $B_{\lambda'}(T)$ is the Planck blackbody function in the SN's rest frame characterized by a colour temperature T. In addition, $\lambda' = \lambda/(1+z)$ in the SN's rest frame, and $A(\lambda)$ and $A'(\lambda')$ are the relevant extinction values. For low redshifts ($z \ll 1$), Equation (5.37) reduces to

$$\theta = \frac{R}{d} = \sqrt{\frac{f_\lambda 10^{0.4A(\lambda)}}{\pi \zeta_\lambda^2 B_\lambda(T)}} \tag{5.38}$$

and $R = vt$ after an expansion time $t > 1$ day for an expansion velocity v.

The continuum opacity of the expanding photosphere is dominated by electron scattering (Eastman *et al.* 1996; Dessart and Hillier 2005) and is, thus, essentially 'grey' (i.e. no wavelength dependence). Hence, the radius of the photosphere, defined as the radius at which the optical depth is $\frac{2}{3}$, is well defined, at least in the optical and near-IR regimes, and there is no need to recalculate θ for each passband. Line-blanketing effects in the SN atmosphere also cause deviations from the blackbody curve and are included in ζ_λ. Detailed calculations show that for colour temperatures above 8000–9000 K, ζ_λ is approximately independent of temperature and is equal to 0.4–0.5, depending on the colour used (cf. Jones *et al.* 2009). Below this temperature, ζ_λ climbs rapidly with decreasing temperature. Unfortunately, although the models are detailed, they are not currently in agreement: the two standard calculations, Eastman *et al.* (1996) and Dessart and Hillier (2005), while in agreement as to general shape, differ by $\sim$15% throughout, with Dessart and Hillier (2005) yielding higher values, which would imply brighter SNe. Although the computational model atmosphere code used by Dessart and Hillier (2005) is somewhat more sophisticated than that of Eastman *et al.* (1996), it is not clear that the differences can account for such a significant discrepancy. Therefore, until this issue is resolved, it introduces a systematic error of approximately 15% into distances calculated using EPM.

For a given set of passbands S, an effective angular radius $\theta\zeta_S$ and a colour temperature T_S can be obtained by fitting a Planck function to the observed broad-band magnitudes. The expected magnitude for a given temperature T and passband X must be calculated by convolving the Planck function – corrected for absorption and redshifted down by a factor $(1+z)$ – with the transmission function of the filter $S_X(\lambda)$ (see Appendix B of Hamuy 2001 for details). There are two unknowns, $\theta\zeta_S$ and T_S, so measurements in at least two passbands are required. Calculations using model SN atmospheres can be used to determine ζ_S once T_S is known, thereby retrieving the true angular radius θ.

The energy released in a SN explosion, $E \sim 10^{51}$ erg $= 10^{44}$ J, is much greater than the gravitational binding energy of a typical red supergiant progenitor, $U \sim 10^{49}$ erg $= 10^{42}$ J. Therefore, one can neglect gravitational effects on the ejecta and assume that the envelope is expanding freely at constant speed v. In addition, the initial radius of the progenitor, R_0, is extremely small compared to the radius rapidly reached by the expanding photosphere. Expansion velocities are of order 10^4 km s^{-1}, so in one day the photosphere radius reaches $\sim 10^9$ km, compared to a typical red supergiant radius of $\sim 10^8$ km. Therefore, it is reasonable to assume that the radius at time t is given by constant outward expansion from an initial point,

$$R(t) \simeq \frac{v\,(t - t_0)}{1 + z}, \tag{5.39}$$

where t_0 is the time of the explosion and the factor $(1 + z)$ allows for time dilation. Since $\theta = R/d_\mathrm{A}$, the angular diameter distance is then given by

$$d_\mathrm{A} = \frac{R}{\theta} = \frac{v\,(t - t_0)}{\theta\,(1 + z)}. \tag{5.40}$$

This appears straightforward, but the problem is that we need the *photospheric* expansion velocity, while the velocity we measure will necessarily be derived from Doppler shifts of spectral features generated outside the photosphere. Therefore, it is not immediately obvious that the measured velocity will be a good match to the desired photospheric velocity, and some prescription is needed as to which lines should be used.

Kirshner and Kwan (1974) suggested, from very general considerations regarding the available area of the photosphere, that the red edge of the absorption trough in a P Cygni line profile should be a good approximation to the photospheric velocity. They used the hydrogen Balmer lines, which are prominent and unmistakable. However, most subsequent studies (e.g. Schmidt *et al.* 1992, 1994) preferred weak metal absorption lines such as Fe II $\lambda\lambda$5169, 5018, 4924 Å and Sc II $\lambda\lambda$5526, 5658 Å, because weak lines are likely produced close to the photosphere. More sophisticated model atmosphere calculations suggest that the velocities measured even from weak lines do not necessarily reproduce the true photospheric velocity and, conversely, the velocities measured from optically thick lines such as the Balmer lines do not necessarily overestimate it. In addition, the weak lines are difficult to observe at very early epochs, when the spectrum is dominated by the Balmer lines, and this is the time at which the photospheric radius is most well defined and the EPM should be most applicable. Therefore, Dessart and Hillier (2005, 2006) suggest using the Balmer lines with appropriate corrections from synthetic spectra constructed from model atmospheres. This is the approach taken by Jones *et al.* (2009), who conclude that Hβ is the best line to use, because it is easily identified, not blended with other species and present throughout the period of interest. Comparing the ratio of Hα and Hβ velocities from their 12 SNe II-P with synthetic spectra, they show that both models agree well with the data at high expansion velocities ($v > 7000$ km s^{-1}), but application of the Dessart and Hillier (2005) models yields too low a ratio at lower velocities. (Eastman *et al.* 1996 seems to be better, but does not cover the low-velocity parameter space very well.) The measurement of $R(t)$ from Equation (5.39) is also sensitive to errors in t_0. However, if multiple measurements are available, the dependence on t_0 can be removed by a simple linear fit (e.g. Jones *et al.* 2009).

The requirements for an EPM measurement are in principle fairly simple. It requires at least two and preferably more simultaneous photometric and spectroscopic measurements of the SN, concentrating on early epochs, as well as reliable theoretical models from which correction factors can be obtained for the measured flux (compared to blackbody) and velocity. The first item is required to provide the basic data. Although not particularly demanding, it is rather specific, and largely precludes the use of archive data and general-purpose SN surveys, which are unlikely to have multiple contemporaneous spectral and photometric measurements.

The main problem, however, lies with the second item. Theoretical models are available, but the two in common use disagree significantly, for reasons that are not understood. Consequently, the most recent application of the EPM (Jones *et al.* 2009) depressingly finds a factor of 2 variation in the value of H_0 derived from 12 SNe II-P, from 52.4 ± 4.3 km s^{-1} Mpc^{-1} for the Dessart and Hillier (2005) model atmospheres and the *VI* filter set to 100.5 ± 8.4 km s^{-1} Mpc^{-1} for the Eastman *et al.* (1996) model atmospheres and the *BVI* filter set. However, for a given choice of filter set and model atmosphere, the dispersion of the values about the best-fitting line was quite small, $\sim$15%, indicating that the method may be useful if the issues with the theoretical calculations could be resolved.

Distances derived from the EPM are, in principle, very desirable, since the method is geometric and entirely independent of Cepheid calibration. It also has the merit of being comparatively insensitive to extinction corrections: as pointed out by Eastman and Kirshner (1989), this is because extinction affects both the measured flux (reducing it, and hence leading to an overestimate of the distance) and the measured colour temperature (reducing that too, hence leading to an underestimate of the intrinsic luminosity, and thus an underestimate of the distance). These two effects act in opposite directions, reducing the overall error. Jones *et al.* (2009) modelled the effect of extinction corrections for their 12 SNe and found that changes of 0.5 mag in assumed host galaxy extinction produced changes of less than 20% in the derived distance.

However, the EPM is highly dependent on model calculations for several crucial corrections, particularly the 'dilution factor' ζ and the relation between velocities measured from spectral lines and the photospheric expansion velocity. Since a limited number of model calculations are used by most studies, the potential for common systematic errors is clearly high. Jones *et al.* (2009) showed that the tendency of past EPM distances to be lower than Cepheid-based distances, as seen in the case of SN 1999em (Baron *et al.* 2004), is a consequence of adopting the Eastman *et al.* (1996) model atmospheres: their distances to SN 1999em using the Dessart and Hillier (2005) model are in agreement with the Cepheid distance.

5.2.2.2 *The Spectral-Fitted Expanding Atmosphere Method*

The systematic problems with the EPM can largely be traced back to the need to use model-atmosphere calculations to derive the dilution factor ζ. The difference in ζ between Eastman *et al.* (1996) and Dessart and Hillier (2005) is the primary cause of the large dispersion in H_0 found by Jones *et al.* (2009). The use of dilution factors can be avoided by fitting the observed spectra directly to synthetic spectra calculated using a stellar atmosphere computer code. This method was introduced by Baron *et al.* (1995), although they initially simply presented it as a variant of the EPM. It has since become known as the spectral-fitted expanding atmosphere method (Baron *et al.* 1996, 2004; Lentz *et al.* 2001; Mitchell *et al.* 2002).

In SEAM, SN spectra taken at a number of epochs are compared to synthetic spectra generated using the stellar atmosphere code PHOENIX (Hauschildt and Baron 1999). The absolute magnitude corresponding to the fitted spectrum is then calculated by convolving the theoretical luminosity with the filter response,

$$M_X = -2.5 \log \int_0^{\infty} S_X(\lambda) L_\lambda \, d\lambda + C_X, \tag{5.41}$$

where X is the passband, S_X the filter response, L_λ the luminosity per unit wavelength and C_X the zero point of the filter derived from standard stars. The distance modulus is calculated by comparing this absolute magnitude with the extinction-corrected apparent magnitude, as in Equation (1.1). In principle, only one photometric measurement is needed, although multiple passbands are clearly desirable to provide consistency checks and constraints on the extinction. The radius is found similarly as in the standard EPM, although in SEAM the velocity is determined by fitting the entire spectrum.

SEAM is, in principle, applicable to all types of SNe provided that good models exist to generate the synthetic spectra and that high-quality spectra are available for the target SN. The latter requirement is, unfortunately, rather stringent and currently restricts the method to relatively local SNe. Observations of SNe II-P in the Hubble flow at a distance of ~60 Mpc – or mid-plateau r-band magnitudes of ~17 – would require a lot of high-quality spectroscopic observing time (3–5 Å spectral resolution) on current 8–10 m class telescopes. This is why only a handful of distances have been calculated in the last 15 years using this method. Future very large (20–40 m-class) ground-based telescopes and the *James Webb Space Telescope* have the potential to extend this method to SNe out to cosmologically interesting distances, $z \sim 0.2$–0.3, where the mid-plateau r-band magnitudes would be of order 23 mag. This could provide a useful complement to the existing Cepheid-based distance ladder.

5.2.2.3 *The Empirical Expansion Velocity–Bolometric Luminosity Relation*

A much simpler, empirical relationship offering potentially higher resulting distance accuracy has become fashionable in recent years. Although SNe II-P show a wide range of luminosities at all epochs (~5 mag, but lower than characteristic of SNe Ia), during their plateau phase they exhibit highly correlated bolometric plateau luminosities (observed roughly during the middle of the plateau/nebular-phase transition, some 50 days after explosion) and envelope expansion velocities, i.e. the expansion velocities of their ejecta (see Figure 5.7).

Since the downward speed of the cooling and recombination wave travelling through the SN's photosphere is close to the outward velocity expansion, the photospheric radius changes only slowly during the plateau phase. Combined with a roughly constant effective temperature (which is of order the recombination temperature, 5000–7000 K), this leads to an approximately constant luminosity (and, hence, the appearance of a plateau). The expansion velocity can be derived from spectral measurements of the Fe II λ5169 Å or Hβ λ4861 Å lines, for instance. (The Fe II line takes a few weeks past explosion to fully develop.) The correlation between these two parameters reflects the physical notion that with increasing explosion energy, the kinetic and internal energies also increase. Or, in other words, the larger the velocity, the larger the SN's photosphere and, hence, the object is larger and more luminous.

Hamuy and Pinto (2002; see also Hamuy 2004a,b) used these characteristics to formulate a standard candle method for SNe II-P, provided that they are located in the smooth Hubble flow. This correlation allowed them to reduce the scatter in the Hubble diagram from ~1 mag to approximately 0.4 and 0.3 mag in the V and I filters, respectively, or ~15%

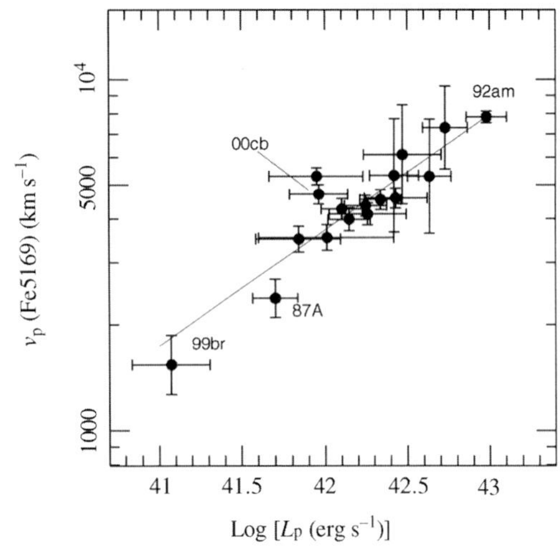

Figure 5.7 *Expansion velocities, v_p, from FeII λ 5169 Å versus bolometric luminosity, L_p, both measured in the middle of the plateau (day 50). Individual supernovae are indicated using the last two digits of the year in which they occurred and their sequence number (Hamuy and Pinto 2002).* (Reprinted from M. Hamuy and P. A. Pinto, *Astrophysical Journal*, **566**, Type II supernovae as standardized candles, L63–L65, Copyright 2001, with permission of the AAS and M. Hamuy.)

precision in distance estimates. There is potential for further accuracy improvements ($\lesssim$9% in distance) once the systematic uncertainties affecting the method are better understood. The latter include extinction corrections and deviations of local sample objects from the large-scale Hubble flow because of peculiar motions (e.g. Nugent *et al.* 2006; Poznanski *et al.* 2009). Extinction corrections usually rely on the assumption that all SNe II-P attain the same intrinsic colour at the end of the plateau phase. This is based on the underlying physics that the opacity in SNe II-P is dominated by electron scattering (a radiative cooling and recombination wave travelling through the hot, opaque envelope originally heated by the explosion), so all objects of this class should reach the same hydrogen recombination temperature at the end of the plateau phase (e.g. Grassberg *et al.* 1971; Eastman *et al.* 1996; but see Nugent *et al.* 2006 for refinements of the extinction correction and Olivares *et al.* 2010 for a more robust approach).

The immediate application of SNe II-P as distance indicators is for independent verification of the acceleration of the expansion of the Universe between $z \simeq 0.3$ and 0.5. Hamuy (2004b) provides a distance calibration that depends on the value of the Hubble constant, which was converted by Hendry *et al.* (2006) for use with their observations of SN 2004A in the V and I filters taken at a phase of 50 days (V_{50}, I_{50}),

$$D(V) = \mathrm{H}_0^{-1} 10^{\frac{1}{5}[V_{50} - A_V + 6.249(\pm 1.35)\log(v_{50}/5000) + 1.464(\pm 0.15)]} \tag{5.42}$$

and

$$D(I) = \mathrm{H}_0^{-1} 10^{\frac{1}{5}[I_{50} - A_I + 5.445(\pm 0.91)\log(v_{50}/5000) + 1.923(\pm 0.11)]}, \tag{5.43}$$

where v_{50} is the expansion velocity (in km s^{-1}) at the same time.

Alternatively, Olivares *et al.* (2010) suggest a simpler, linear relationship between their luminosity and expansion velocity data,

$$M_{B,-30} - 5\log(\mathrm{H}_0/70) = 3.50(\pm0.30)\log[v_{\mathrm{FeII},-30}/5000] - 16.01(\pm0.20), \quad (5.44)$$

$$M_{V,-30} - 5\log(\mathrm{H}_0/70) = 3.08(\pm0.25)\log[v_{\mathrm{FeII},-30}/5000] - 17.06(\pm0.14), \quad (5.45)$$

$$M_{I,-30} - 5\log(\mathrm{H}_0/70) = 2.62(\pm0.21)\log[v_{\mathrm{FeII},-30}/5000] - 17.61(\pm0.10), \quad (5.46)$$

where $v_{\mathrm{FeII},-30}$ is the expansion velocity determined on the basis of the FeII absorption line 30 days before the middle of the plateau phase (indicated by the subscript '−30'). With a reliable value of the Hubble constant and measurements of the expansion velocity and apparent magnitude of a given SN II-P, a distance modulus and, hence, a distance follows directly.

The method as originally proposed in Hamuy and Pinto (2002) suffered from a number of practical drawbacks. These included the fact that use of the end of the plateau phase to determine extinction, although well grounded physically, was inconvenient observationally: it required extended photometric follow-up of the SN over a period of several months, which is often impractical, especially for higher-z SNe. Second, as noted regarding its use in the EPM, the FeII λ5169 Å line is weak and can be difficult to measure in SNe at intermediate redshifts because of limited signal-to-noise ratios. (At higher redshifts, it gets even worse, because the line moves into a region of the near-IR spectrum which is highly populated by **telluric lines**: the **OH forest**.) Finally, it may not be practical to acquire a spectrum of the SN at the calibration time of 50 days post-maximum.

Nugent *et al.* (2006) presented a number of practical improvements to the method. They used $V - I$ colours at day 50 to estimate reddening instead of end-of-plateau colour: this means that the SN need not be followed past day 50. They studied the relation between the measured velocity from FeII λ5169 Å and Hβ. This allows use of the latter, stronger, line to determine the expansion velocity. They also studied the evolution of the FeII λ5169 Å velocity over time, presenting a power-law fit,

$$v_{50} = v(t)\left(\frac{t}{50}\right)^{0.464\pm0.017}, \quad (5.47)$$

where v_{50} and $v(t)$ are the velocities at day 50 and day t, respectively. This fit is valid for $7 \le t \le 75$ days, which offers much more flexibility in acquiring the spectrum necessary to measure the velocity. Nugent *et al.* (2006) fit the relation

$$M_I = -\alpha\log\left(\frac{v_{\mathrm{FeII}}}{5000\,\mathrm{km\,s^{-1}}}\right) - 1.36\left[(V-I)-(V-I)_0\right] + M_{I_0}, \quad (5.48)$$

where they adopted $(V - I)_0 = 0.53$ mag as the nominal, unreddened $V - I$ colour. Their fit to nearby SNe gave $\alpha = 6.69 \pm 0.50$ and $M_{I_0} = -17.49 \pm 0.08$ mag for $\mathrm{H}_0 = 70\,\mathrm{km\,s^{-1}\,Mpc^{-1}}$. The dependence on an assumed value for H_0 arises because this was used to infer the distances to the calibrating SNe. The inferred value of M_{I_0} is completely correlated with the adopted value of $(V - I)_0$, but this is unimportant for relative distances, provided that the reddening laws of different host galaxies are comparable (but see Section 6.1.1). Unfortunately, upon application of this method to a sample of high-z SNe II-P from the Supernova Legacy Survey, they found a significant difference between the best fits in the α–M_{I_0} plane, indicating a systematic difference between their low- and high-z samples. Given their limited sample (19 local SNe, of which only eight were in the Hubble flow, and only five in the high-z data set), they could not determine whether this difference was real or a consequence of low-number statistics. However, the intrinsic dispersion, estimated from the scatter about the Hubble law fit for the 24 SNe in the sample, was only 0.12 mag, indicating that the technique could potentially provide reliable distances.

Poznanski *et al.* (2009) increased the low-z sample by incorporating 19 additional SNe. They also modified the fit function used by Nugent *et al.* (2006) to

$$M_I = -\alpha \log\left(\frac{v_{\rm FeII}}{5000\ {\rm km\ s^{-1}}}\right) - R_I\left[(V-I)-(V-I)_0\right] + M_{I_0}, \tag{5.49}$$

introducing a free parameter, R_I, to allow for nonstandard extinction in the host galaxies. They also introduced cross correlation with template spectra to measure the expansion velocity, instead of relying on a single line. For a cleaned, combined sample of 40 SNe, they derived $\alpha = 4.6 \pm 0.7$, $R_V = 1.5 \pm 0.5$ and $M_{I_0} = -17.43 \pm 0.10$ mag, with a systematic error of 0.38 mag. The latter is driven by three outliers, which can be removed to give a dispersion of 0.22 mag in the Hubble diagram (see also Olivares *et al.* 2010 for similar accuracy). Note that the derived value of R_V is much smaller than the standard, Galactic value of 3.1, in agreement with several results from SNe Ia (see Section 5.2.1). The intrinsic scatter of approximately 0.2 mag corresponds to a scatter in cosmological distances of ~10% (see also Poznanski *et al.* 2010). Maguire *et al.* (2010) predict that a similar level of distance accuracy may be possible based on near-IR observations of SNe II-P. Note, however, that part of the remaining scatter may be systematic in nature, in the sense that different progenitor stellar masses and radii, ejecta masses, explosion energies and metallicities may contribute (e.g. Litvinova and Nadyozhin 1985; Kasen and Woosley 2009).

Recently, D'Andrea *et al.* (2010) applied the method to a sample of 15 SNe II-P obtained from the SDSS Supernova Search. Their sample was unexpectedly homogeneous for SNe II, with an intrinsic I-band dispersion of also only 0.22 mag, which they convincingly attribute to selection effects: the SDSS survey was optimized for SNe Ia and candidates were prioritized for spectroscopic follow-up in such a way that brighter objects were much more likely to have spectroscopic data. Like Nugent *et al.* (2006), they found that their two samples – in this case, the SDSS and Poznanski *et al.* (2009) samples – preferred significantly different fits in the $\alpha - M_{I_0}$ plane and also in the $\alpha - R_I$ plane for reasons that are not adequately understood.

It appears that this method, although promising, still suffers from uncontrolled systematic errors and requires further study. In addition, the method has, as yet, no absolute calibration: the published studies all normalize to an assumed value of H_0, which means that they could, in principle, be used for studies of dark energy, but cannot deliver absolute distances. EPM distances cannot be used as calibrators, because these are affected by serious model-dependent systematics. SEAM distances might be used, but as yet are very few in number. The *HST* Key Project found only three galaxies with both SNe II and Cepheid distances, and one of those is SN 1987A, which is not helpful for this method since it is not the explosion of a red supergiant. Hence, this aspect also requires further work.

5.2.2.4 Independent Distance Calibrations

A potentially more serious systematic effect was uncovered by Leonard *et al.* (2003), who obtained a Cepheid-based distance to NGC 1637, which also hosts SN 1999em, and compared it with SNe II-P-derived distances based on both the expanding photosphere and standard candle methods. They point out that their Cepheid distance to this galaxy is nearly 50% larger than earlier EPM distance estimates to SN 1999em. This provides the first direct comparison between these two primary distance-determining methods for a galaxy hosting a well-observed, spectroscopically and photometrically normal SN II-P. They perform extensive consistency checks, which show strong evidence to support the Cepheid distance scale, and conclude that the SN II-P distance scale may need revision. Additional calibrating galaxies are clearly desirable to resolve this worrisome discrepancy.

The 'expanding shock front' method was recently used successfully to obtain a geometric distance estimate to SN 1993J in M81 (Bartel *et al.* 2007; see Figure 5.8), the optically brightest SN in the northern hemisphere since SN 1954A. The underlying principles of the method are not unlike those used for application of the expansion parallax approach to planetary nebulae (see Section 3.7.1). The authors used the technique of Very Long Baseline Interferometry (see Section 2.1.2) to measure the radio shell's angular expansion velocities at high angular resolution over the period from 7 days until $\sim$9 yr after shock breakout (e.g. Bartel *et al.* 2002). The shell's expansion has been highly circularly symmetric (e.g. Bietenholz *et al.* 2001), thus facilitating a relatively straightforward analysis. They then compared the results with the linear expansion velocities of the ejected gas measured from the Doppler shifts of the Hα, Hβ, HeI, O[III] and NaI emission lines in the object's optical spectra. The resulting distance to M81, $d = 3.96 \pm 0.05$ (statistical) $\pm$ 0.29 (systematic) Mpc is – within the 1σ uncertainties – similar to earlier determinations by Freedman *et al.* (1994) and Huterer *et al.* (1995), both of which were based on Cepheid distance calibrations. The largest contribution to the systematic uncertainty is related to matching the locations of the optical and radio emission, followed by uncertainties in the degree of anisotropy of transverse and radial expansion, and changes of the distance estimate as a function of time (Bartel *et al.* 2007). The prospects of using this approach for other nearby SNe are slim, however, given the extensive observational campaigns required for derivation of similarly precise distances. Earlier attempts to determine the distance to SN 1979C in the Virgo cluster failed because of the much larger distance and the more limited amount and quality of the available observational material (Bartel 1985; Bartel *et al.* 1985; Bartel and Bietenholz 2003, 2005).

5.2.2.5 *Distance Estimates Based on the Ejected* ^{56}*Ni Mass*

A few alternative approaches to determining distances to SNe II-P have been proposed. Elmhamdi *et al.* (2003) attempted to standardize the luminosities of SNe II-P, based on the steepness of the V-band light curve slope at the inflection time, $t_{\rm i}$ (i.e. the transition from the plateau to the tail, which is driven by radioactive ^{56}Co decay) versus ejected ^{56}Ni mass. They showed that if mass, energy and mixing conditions vary by less than a factor of 1.4 among SNe II-P, then with an accuracy of better than 10%, Hα luminosity is proportional to ^{56}Ni mass between approximately 200 and 400 days after the explosion. The correlation between ^{56}Ni mass and the light curve decay rate at the inflection point is such that the steeper the decline at the inflection point, the lower is the ^{56}Ni mass,

$$\log M(^{56}\mathrm{Ni}) = -0.438 M_V(t_{\rm i} - 35) - 8.46, \tag{5.50}$$

so that a comparison with the SN's apparent magnitude will yield a distance estimate. The formal definition of $t_{\rm i}$ is the moment when the steepness of the light curve, $-\mathrm{d}M_V/\mathrm{d}t$, is maximal. It appears, therefore, that radiative diffusion times and ^{56}Ni masses are linked.

Although controversial, Nadyozhin (2003) also proposed to use the ejected ^{56}Ni mass, which he claims is sufficiently well correlated with the explosion energy, E (but see Smartt 2009; Smartt *et al.* 2009 for counterarguments), as the basis for distance determinations to

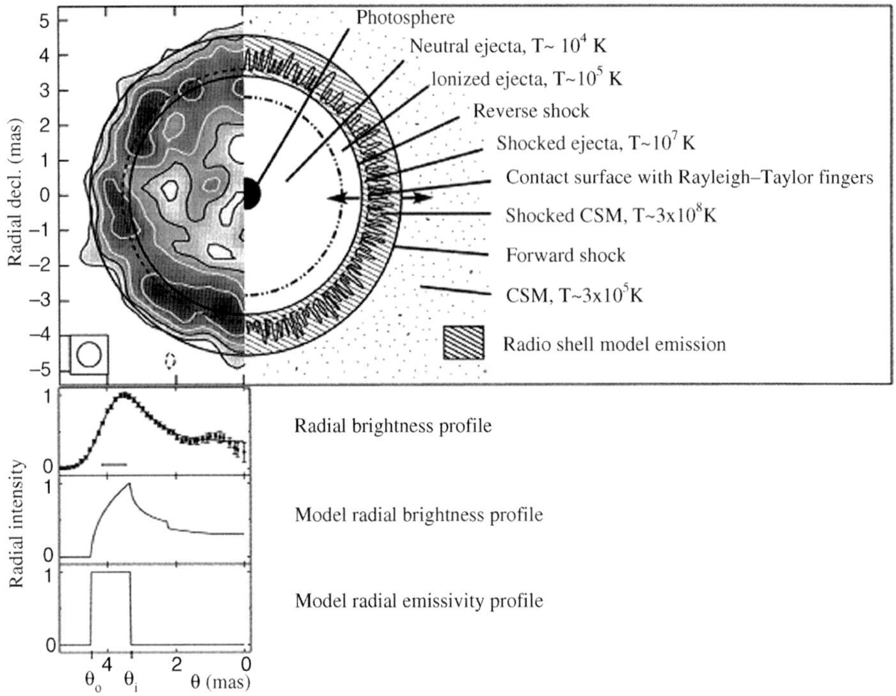

Figure 5.8 *(Top) Part of the composite Very Long Baseline Interferometry image of SN 1993J of Bartel et al. (2007). The outer circle shows the best-fitting outer radius, θ_o, of the shell model with absorption in the centre and indicates the expected location of the forward shock front, which is expanding into the circumstellar medium (CSM). The inner circle shows the inner radius, θ_i, of that model and indicates the expected location of the reverse shock front. The forward and reverse shock fronts travel in comoving opposite directions (arrows) from the contact discontinuity (dashed circle), where the ejecta hit the CSM. Fingers expanding into the shocked CSM caused by Rayleigh–Taylor instabilities at the contact surface are also shown. The smaller, dash-dotted circle indicates the boundary between the colder, neutral ejecta and the hotter, ionized ejecta. Broad-line emission is expected from the ionized ejecta up to the contact surface. (Second panel) Brightness profile of the composite image as a function of angular radius, θ. The resolution is 0.70 milli-arcseconds (mas). Also indicated is the corresponding radial profile of the projected best-fitting spherical shell model with uniform emissivity and an absorption disc in the centre, convolved to the resolution of 0.70 mas. (Third panel) Corresponding radial profile of the projected, unconvolved shell model with absorption. (Bottom) Radial profile of the volume emissivity of the shell model (without absorption). The shell model is limited by θ_o and θ_i.* (Reprinted from N. Bartel et *al.*, *Astrophysical Journal*, **668**, SN 1993J VLBI. IV. A geometric distance to M81 with the expanding shock front method, p. 924–940, Copyright 2007, with permission of the AAS and N. Bartel.)

SNe II-P. Hydrodynamical modelling of SNe II-P light curves predicts a correlation between three observable parameters – plateau duration, Δt, absolute magnitude and photospheric velocity in the middle of the plateau, v_{ph} – on the one hand and three physical parameters – E, the mass of the expelled envelope and the radius of the progenitor star – on the other.

By combining the exponential decay of the radioactive tails of SNe II-P light curves with their ejected ^{56}Ni masses, Nadyozhin (2003) proposes a relation of the form

$$\log d(\text{Mpc}) = -0.374 \log(\xi Q) + 0.0504(V - A_V) + 0.875 \log \Delta t + 1.17 \log v_{\text{ph}} - 2.482, \quad (5.51)$$

where Δt is given in days, v_{ph} in units of 1000 km s^{-1}, ξ is an adjustable parameter of order unity (a proportionality factor between E and the ejected ^{56}Ni mass), and

$$Q \equiv F_{41}(t) \exp\left(\frac{t}{111.3}\right), \quad (5.52)$$

where $F_{41}(t)$ is the bolometric tail luminosity (in units of 10^{41} erg s^{-1}) measured at time t (in days) and the e-folding time of radioactive ^{56}Co decay is 111.3 days. If the underlying assumptions are correct and not subject to significant scatter, the only free parameter in Equation (5.51) is ξ, so that this approach could potentially yield independent distance estimates if we can constrain ξ sufficiently well. Nadyozhin (2003) concludes that this parameter cannot be constrained more precisely than to $0.5 \lesssim \xi \lesssim 2$ at present, however, with a most likely value of $0.75 \leq \xi \leq 1.2$, depending on the distance calibration applied. He predicts that availability of additional SNe II-P at larger distances, particularly in the smooth Hubble flow, offers prospects for better constraints. Note, however, that SNe are 2–3 mag fainter in the tail than in their plateau phase, thus rendering observational confirmation troublesome.

Several large-scale sky surveys are currently in progress or in advanced planning stages, including the Palomar Transient Factory (Rau *et al.* 2009), the Panoramic Survey Telescope and Rapid Response System (Pan-STARRS; Kaiser *et al.* 2002), the Large Synoptic Survey Telescope (LSST; Tyson *et al.* 2003), the Visible and Infrared Survey Telescope (VISTA; Emerson *et al.* 2004), the VLT Survey Telescope (Capaccioli *et al.* 2003), the Dark Energy Survey (Castander 2007) and SkyMapper (Granlund *et al.* 2006). These offer prospects of discovering thousands of new SNe, hence the expectation is that our SN samples for cosmological distance measurements will soon be increased by several orders of magnitude.

5.2.3 A Link to Gamma-Ray Bursts as Standard Candles?

Gamma-ray bursts (GRBs), intense flashes of gamma rays lasting from milliseconds to minutes, are the most energetic events known in the Universe. Although they were first detected (Klebesadel *et al.* 1973) in 1967 by the US Vela 3 and 4 satellites as part of their classified missions to monitor compliance of the Soviet Union with the Nuclear Test Ban Treaty, it took until the 1991 launch of the Compton Gamma-Ray Observatory before their extragalactic nature was ascertained. This was achieved by observations with the Burst and Transient Source Explorer (BATSE) gamma-ray detector, which showed that the distribution of GRBs is isotropic and not associated with e.g. the Galactic plane or the Galactic Centre (Meegan *et al.* 1992; Piran 1992). Note, however, that some scientists continued to support a Galactic outer halo origin until the first reliable redshifts had been measured.

Although GRB light curves are extremely diverse and complex, they can be roughly divided into broadly overlapping ‘long–soft’ and ‘short–hard’ classes, where the labels ‘soft’ and ‘hard’ refer to their spectra, with hard representing highly energetic gamma rays. The boundary between long and short GRBs is usually set at 2 s. The former last, on average, 30 s from peak brightness until the end of their afterglow. They are mostly associated with highly focussed, core-collapse SN explosions in actively star-forming galaxies, hence suggesting massive progenitor stars (e.g. Woosley and Bloom 2006; see also Pontzen *et al.* 2010 for high-z GRBs). The majority of GRBs detected to date are of the long–soft variety; because of their long afterglows – compared to enhanced peak intensities lasting for several hundred milliseconds and much fainter afterglows for the short–hard bursts, which may be related to mergers of neutron stars (e.g. Nakar 2007) – they have also been studied most extensively.

Despite their wide variety, some features of the light curves of long–soft bursts have been proposed as possible distance indicators. Given that they can potentially be detected at any redshift (e.g. Inoue *et al.* 2004; Bromm and Loeb 2006; see also Schaefer 2007), this is of particular importance for distances beyond which other tracers are rendered inoperable. They could also be useful complementary cosmological probes to SNe Ia at high redshifts. Figure 5.9 shows the promise of using GRBs as cosmological standardizable candles out to large redshifts.

Cardone *et al.* (2009) provide a detailed overview of the state of the art in attempts to use GRBs as standard candles. Most of these are based on taking weighted averages of multiple empirical correlations between directly observable and distance-dependent parameters, (x, y). Very generally, correlations are of the form

$$\log y = k + 2\log d_{\rm L}(z) = a + b\log x, \tag{5.53}$$

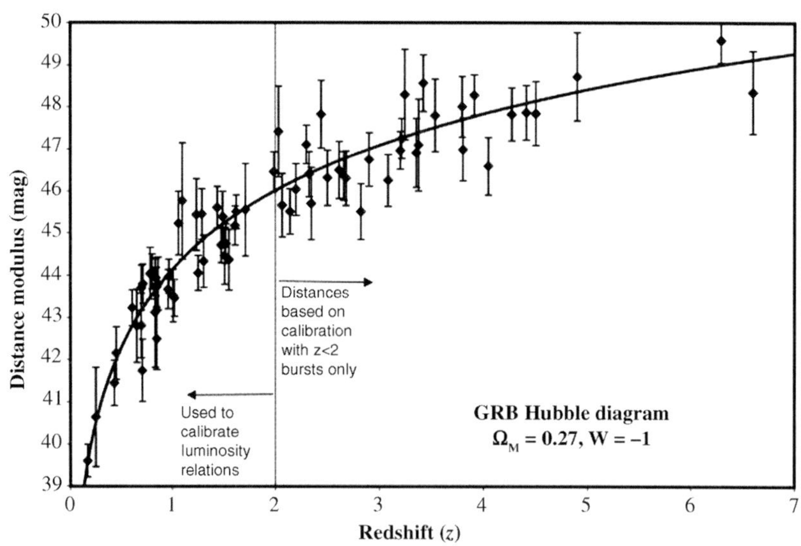

Figure 5.9 *GRB Hubble diagram calibrated only with $z < 2$ bursts (Schaefer 2007). The shape of the Hubble diagram is already accurately known for $z < 2$ from SNe (taken to be the concordance model displayed as a curve), so the 37 $z < 2$ bursts can be used to calibrate the luminosity relations in a largely cosmology-independent manner.* (Reprinted from B. Schaefer, *Astrophysical Journal*, **660**, The Hubble diagram to redshift >6 from 69 gamma-ray bursts, p. 16–46, Copyright 2007, with permission of the AAS and B. Schaefer.)

where x is a distance-independent property, $y = kd_L^2(z)$, k is a redshift-independent constant and the luminosity distance is given by Equation (5.14). If parameters (a, b) are known, as for a calibrated relationship, the distance modulus for an object at redshift z follows from

$$\mu(z) = 25 + 5 \log d_L(z) \tag{5.54}$$

$$= 25 + \frac{5}{2}(\log y - k) \tag{5.55}$$

$$= 25 + \frac{5}{2}(a + b \log x - k), \tag{5.56}$$

where all parameters are known.

Schaefer (2007) and Cardone *et al.* (2009) provide overviews of usable calibration relations. Most of these rely on the use of a distance-dependent quantity related to the isotropic absolute luminosity, L, the collimation-corrected energy, E_γ (i.e. corrected for jet geometry and measured either at the peak or during the afterglow) and observationally accessible GRB parameters. Using the nomenclature of Cardone *et al.* (2009), these relations are of the form

$$\log R = a \log Q + b, \tag{5.57}$$

where R is distance dependent and Q is not. They include

1. the relation proposed by Ghirlanda *et al.* (2004, 2006; however, see Butler *et al.* 2009 for serious objections). Here, $R = E_\gamma = 4\pi d_L^2(z) S_{\rm bolo} F_{\rm beam}(1+z)^{-1}$, where $S_{\rm bolo}$ is the **bolometric fluence** (i.e. the flux integrated over the burst duration) and $F_{\rm beam} = 1 - \cos\theta_{\rm jet}$ is the beaming factor, where $\theta_{\rm jet}$ is the jet's opening angle at the rest-frame time of the wavelength-independent break, T_a, in the afterglow light curve. This break is predicted by collimated GRB models (Rhoads 1997; Sari *et al.* 1999) and corresponds to the time when the light curve suddenly steepens. In this empirical correlation, $Q = E_p(1+z)/300$ keV, where E_p is the peak energy;

four luminosity-related correlations, i.e. $R = L = 4\pi d_L^2(z) P_{\rm bolo}$, where $P_{\rm bolo}$ is the bolometric flux, with

2. $Q = E_p(1+z)/300$ keV (Schaefer 2003; Yonetoku *et al.* 2004);
3. $Q = \tau_{\rm lag}(1+z)^{-1}/0.1$ s (Norris *et al.* 2000), where $\tau_{\rm lag}$ is the time offset between the arrival of low- and high-energy photons (initially based on the BATSE energy bands, i.e. 25–50, 50–100, 100–300 and $\geq$300 keV), which is usually between 0.01 and 0.5 s;
4. $Q = \tau_{\rm RT}(1+z)^{-1}/0.1$ s (Schaefer 2007), where $\tau_{\rm RT}$ is the rise time, i.e. the shortest time over which the light curve rises by half of the peak flux during the burst; and
5. $Q = V(1+z)/0.002$ (Fenimore and Ramirez-Ruiz 2000; Reichart *et al.* 2001; Schaefer 2007), where the variability V quantifies the smoothness of the light curve;

and a correlation involving the X-ray afterglow luminosity, L_X, with

6. $R = L_X(T_a)$ at time T_a and $Q = T_a/(1+z)$ (Dainotti *et al.* 2008; Ghisellini *et al.* 2008; with semi-theoretical basis provided by Yamazaki 2009; see also Cardone *et al.* 2009).

The potential use of GRBs as 'standardizable' candles is controversial. A number of authors have raised serious doubts as regards the validity of the empirical relations. Although they are very luminous, GRB detection is riddled with difficulties, in the sense that a range of biases may favour the detection of a particular class of GRBs. If so, these empirical correlations might not be universal or may even simply represent selection effects (e.g. Butler *et al.* 2007, 2009; Shahmoradi and Nemiroff 2009; see also Cardone *et al.* 2009). One serious problem is that for some of the relations the number of GRBs for which high-quality data are available to determine the required parameters is very small. A good example is the Ghirlanda relation, where one needs to identify the jet break. We now know that many breaks seen in GRB light curves are not achromatic, and so presumably are not jet breaks. The early data used to identify the Ghirlanda relation had to use many bursts for which even the existence of a break was open to doubt. In fact, for most of these relations one can only 'reliably' determine the necessary parameters for a small and possibly biased subset, typically of brighter events.

Butler *et al.* (2009) argue that a physically based correlation must exhibit significantly reduced scatter in the rest frame with respect to the observer and must not persist if the assumed redshifts are scattered. They argue that this requirement is not robustly met by some of the empirical correlations in common use. In addition, there is some evidence that high-z GRBs are significantly affected by gravitational lensing and magnification bias (e.g. Porciani *et al.* 2007; Wyithe *et al.* 2010). This must be properly corrected for to avoid large biases in their derived distances (see also Bolejko 2010 for complicating factors relating to the inhomogeneity of the matter distribution in the Universe).

Another common objection is that there tends to be some circularity in the use of GRBs for cosmology: one must first generally assume a given cosmology to determine rest-frame properties, and then use these relations to derive the cosmological parameters. Only if one had access to many GRBs at some common redshift, then the luminosity relations could be established (up to a zero point) without having to assume a specific cosmology. Note that this is less of a problem for SNe Ia, for which much larger samples are currently available, and at lower redshifts. Although the *Swift* satellite (Gehrels *et al.* 2004) has now accumulated a fairly large sample of GRBs, which help to provide better statistics, it has also generally shown that such luminosity relations exhibit more scatter than initially thought, and so in practice appear not yet ready to yield strong constraints on cosmological parameters.

5.3 Indirect Techniques to Measure H_0

Now that we are reaching distances well into the Hubble flow, the key to understanding the structure of the Universe on the largest scales is by getting a handle on the value of the Hubble constant, both instantaneously and as a function of time. Given the predominant degeneracies among the cosmological parameters, this is not an easy feat. Here, we discuss three of the most important methods that could potentially be used to constrain H_0, ideally in combination with additional constraints obtained independently.

5.3.1 Gravitational Lensing: Time Delays

Albert Einstein's (1915) theory of general relativity explains gravitation as a distortion of the structure of spacetime by matter. It predicts that light from distant sources is 'gravitationally lensed', i.e. that it is 'bent' around a massive foreground object, multiply imaged (depending on the exact geometry of the foreground mass distribution and the background object) and magnified. This idea was first published by Orest Chwolson in 1924 (see also Einstein 1936; Zwicky 1937). The effect was initially confirmed by Arthur Eddington's expedition (Dyson *et al.* 1920) during the solar eclipse of 1919. They used observations taken on the island of Príncipe in the southern Atlantic Ocean to show that star light that passed close to the Sun was slightly bent, so that these stars appeared marginally out of position with respect to a sample of reference stars. Although already predicted by Zwicky (1937) for individual galaxies, it took more than 40 years before Walsh *et al.* (1979) accidentally discovered that, on much larger scales, galaxy clusters also distort the propagation of light originating from background objects.

In general relativity, spacetime connects the past with the present. Light rays are affected by distortions of spacetime caused by massive objects located along their paths: they will both be deflected (i.e. forced to travel along a longer path, a geometric delay) and slow down in the presence of a strong gravitational field (a gravitational delay known as the **Shapiro time-delay effect**; Shapiro 1964). The latter only depends on well-known physical constants as well as on the inverse of the 2D Laplacian differential operator (∇) of the lensing system's mass density profile, $\Sigma(\vec{\theta})$ (where $\vec{\theta}$ is the image position),

$$\Delta t_{\rm grav}(\vec{\theta}) = (1 + z_{\rm d})\frac{8\pi G}{c^3}\nabla^{-2}\Sigma(\vec{\theta}), \tag{5.58}$$

where $z_{\rm d}$ is the redshift of the lens. This implies that the main uncertainties affecting the Shapiro time delay are associated with the projected 2D mass profile shape of the lens and its slope, particularly the *radial* slope (cf. Courbin 2003).

In essence, the gravitational potential of the lens produces an 'index of refraction', so that the principles of optical lensing can be applied to the gravitational lensing phenomenon. The deflection follows directly from **Fermat's principle** (see also Blandford and Narayan 1986), which requires that a given photon traces a path that is a minimum, a maximum or a saddle point of the travel time. The gravitational attraction can be thought of as either the movement of undisturbed objects in a curved background geometry or, alternatively, as the response of objects to a force in a flat geometry. The amount of deflection ($\theta_{\rm s}$, in radians) depends on the mass (M) of the foreground lens and the distance to the source, r,

$$\theta_{\rm s} = \frac{4GM}{rc^2}. \tag{5.59}$$

Refsdal (1964) first realized that one could, in principle, determine an independent value of H_0 by measuring arrival time differences between multiple images of the same source as a function of the angular separation of strongly lensed, time-variable sources (e.g. quasars, active galactic nuclei or SNe). The arrival time differences between multiply lensed images originating from the special geometry associated with the strong-lensing phenomenon are proportional to the inverse of the Hubble constant and depend only slightly ($\sim$10%: cf. Myers 1999; Freedman and Madore 2010) on cosmological parameters such as $\Omega_{\rm M}$ and Ω_Λ. By measuring the time delays and modelling the lens potential, one can derive the value for the angular diameter distance. If one can directly measure this time-delay distance, this implies that gravitational lensing can provide distance determinations that are completely independent of other secondary distance tracers (i.e. distance ladders).

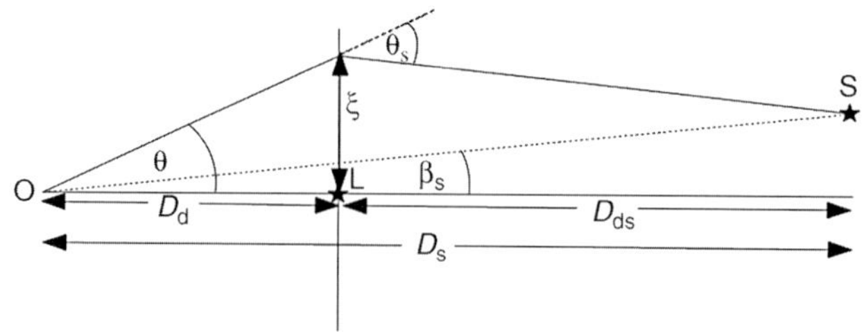

Figure 5.10 Schematic diagram of a gravitational lens, with the centre of the lens L at angular size distance D_d and the source S at angular size distance D_s. O: observer. Note that, despite the appearance in the diagram, for cosmological distances it is not in general true that $D_s = D_d + D_{ds}$.

In general, for strong lensing in a homogeneous, isotropic FLRW universe, the time delay of a lensed image at angular position $\vec{\theta}$ corresponding to source position $\vec{\beta}_s$ compared to the no-lensing case is (Suyu *et al.* 2010; see also Blandford and Narayan 1986; Courbin 2003)

$$\Delta t(\vec{\theta}, \vec{\beta}_s) = \frac{1}{c}\frac{D_d D_s}{D_{ds}}(1 + z_d)\phi(\vec{\theta}, \vec{\beta}_s), \tag{5.60}$$

where D_d, D_s and D_{ds} are the angular diameter distances between, respectively, observer and lens, observer and source, and lens and source (see Figure 5.10), while the **Fermat potential** $\phi(\vec{\theta}, \vec{\beta}_s)$ (the mathematical surface which determines the increase in travel time for a light ray passing through a gravitational field) is defined as a combination of the geometric path difference (deflection) caused by strong lensing and the lens potential, $\psi(\vec{\theta})$,

$$\phi(\vec{\theta}, \vec{\beta}_s) \equiv \left[\frac{(\vec{\theta} - \vec{\beta}_s)^2}{2} - \psi(\vec{\theta})\right], \tag{5.61}$$

which – in the thin-lens approximation – can be derived by integrating the 3D Newtonian gravitational potential, Φ_{3D}, along the line of sight, ℓ, (Weinberg 1972; Misner *et al.* 1973; Blandford and Narayan 1986)

$$\psi(\vec{\theta}) = \frac{D_{ds}}{D_s}\int_{\rm observer}^{\rm source}\frac{2\Phi_{3D}}{c^2}\frac{d\ell}{D_d}. \tag{5.62}$$

The deflection, $(\vec{\theta} - \vec{\beta}_s)$, is equivalent to the gradient of the 2D potential. It follows from the *lens equation*,

$$\vec{\theta} - \vec{\beta}_s = \vec{\nabla}_s\psi(\vec{\theta}). \tag{5.63}$$

Equation (5.60) can be simplified to

$$\Delta t(\vec{\theta}, \vec{\beta}_s) = \frac{D_{\Delta t}}{c}\phi(\vec{\theta}, \vec{\beta}_s) \tag{5.64}$$

$$\propto \frac{1}{H_0}\phi(\vec{\theta}, \vec{\beta}_s) \tag{5.65}$$

and the time-delay distance is

$$D_{\Delta t} \equiv \frac{D_d D_s}{D_{ds}}(1 + z_d). \tag{5.66}$$

This implies that we can use time-delay lens configurations to constrain H_0 by modelling $\psi(\vec{\theta})$ and $\vec{\beta}_s$ (cf. Suyu *et al.* 2010).

Although the underlying physics is well understood from first principles (see, for a review, Blandford and Narayan 1986), determining H_0 on this basis has proven extremely difficult because of many complicating effects, both intrinsic and observational (see, for reviews, Schechter 2005; Freedman and Madore 2010). From an observational perspective, few simple and easy-to-model gravitational lens systems have been identified, the amplitude of quasar variability is generally small (~5–15% rms at optical wavelengths, characterized by timescales of order a year; cf. Cristiani *et al.* 1996; Myers 1999; Vanden Berk *et al.* 2004) and often compounded by the uncorrelated effects of microlensing (cf. Myers 1999; Schechter 2005) and dust attenuation at optical wavelengths. As a consequence, accurately measuring time delays is anything but a *sinecure*. In addition, the geometry of gravitational lenses is not known *a priori*, particularly if they consist of galaxy groups or clusters, hence leading to a strong degeneracy between the mass distributions of the lensing systems and H_0 (e.g. Gorenstein *et al.* 1988; Wucknitz 2002; Suyu *et al.* 2006, 2009).

Moreover, this technique is significantly affected by the so-called 'mass sheet degeneracy'. This systematic effect usually dominates the error budget and implies that an arbitrary symmetric paraboloid, gradient sheet and constant can be added to the underlying potential without changing the predicted lensed image, although the relative Fermat potential does change (e.g. Falco *et al.* 1985; Courbin 2003; Kochanek *et al.* 2006; Suyu *et al.* 2009, 2010; and references therein). To overcome this degeneracy, independent information on the underlying mass distribution, i.e. of the lens itself and its 3D environment, is required (see Suyu *et al.* 2010 for a review). The distribution of velocity dispersion as a function of clustercentric radius can be used to constrain the latter, although obtaining velocity dispersions outside a galaxy cluster's central regions has proven difficult. Suyu *et al.* (2010) show that the effect of this mass sheet degeneracy on the value of the Hubble constant derived without correction for the environment is an underestimate by a factor of $(1 - \kappa_{ext})$, where κ_{ext} is the 'external convergence' parameter, which quantifies the contributions of the group environment to the lensing signal. Formally, it is the 2D differential Laplacian operator given by the divergence of the gradient of the 2D potential,

$$\kappa_{ext} = \frac{1}{2}\nabla_{\theta}^{2}\psi(\vec{\theta}), \tag{5.67}$$

so it is proportional to the surface mass density (cf. Kochanek 2002; Schechter 2005).

Courbin (2003) and Suyu *et al.* (2010) provide useful overviews of recent efforts to constrain H_0 on the basis of gravitational lens time delays. They show that the approach has benefited from significant improvements in accuracy and precision since early efforts (see, for improved time-delay measurements, Vuissoz *et al.* 2007, 2008; Paraficz and Hjorth 2009; see also Courbin 2003 for PG 1115+080). Current values for H_0 based on application of this technique range from approximately 50 to 85 km s^{-1} Mpc^{-1} (cf. Suyu *et al.* 2010), where most of the uncertainties originate from sometimes poorly constrained model assumptions, e.g. isothermal profiles, sometimes in the presence of external tidal fields, the central concentration degeneracy, environmental density distributions or multiple lenses (cf. Blandford and Narayan 1986; Schechter 2005; Suyu *et al.* 2006, 2009, 2010; Freedman and Madore 2010).

Adopting the currently favoured cosmological parameters (Komatsu *et al.* 2009, 2011), i.e. $\Omega_M = 0.3$, $\Omega_\Lambda = 0.7$ and $w = -1$, Suyu *et al.* (2010) improve the determination of H_0 by roughly a factor of two for the quadruple lens system B1608+656 (source and lens redshifts $z_s = 1.394$, $z_l = 0.630$: Fassnacht *et al.* 1996) based on a very detailed analysis of *HST* pixel-by-pixel photometry (see also Koopmans 2005) corrected for the effects of dust obscuration, low-resolution spectroscopy obtained with the Keck telescope on Hawai'i to constrain the system's dynamics, cosmological N-body simulations (ray tracing combined with number counts) and assuming a proper, extended source intensity distribution (allowing reconstruction of the underlying mass distribution in an iterative process, until the relative potential correction between any pair of images is $<0.1\%$), resulting in $H_0 = 70.6 \pm 3.1$ km s^{-1} Mpc^{-1} (see Figure 5.11). Combined with the *WMAP* 5-year results and assuming

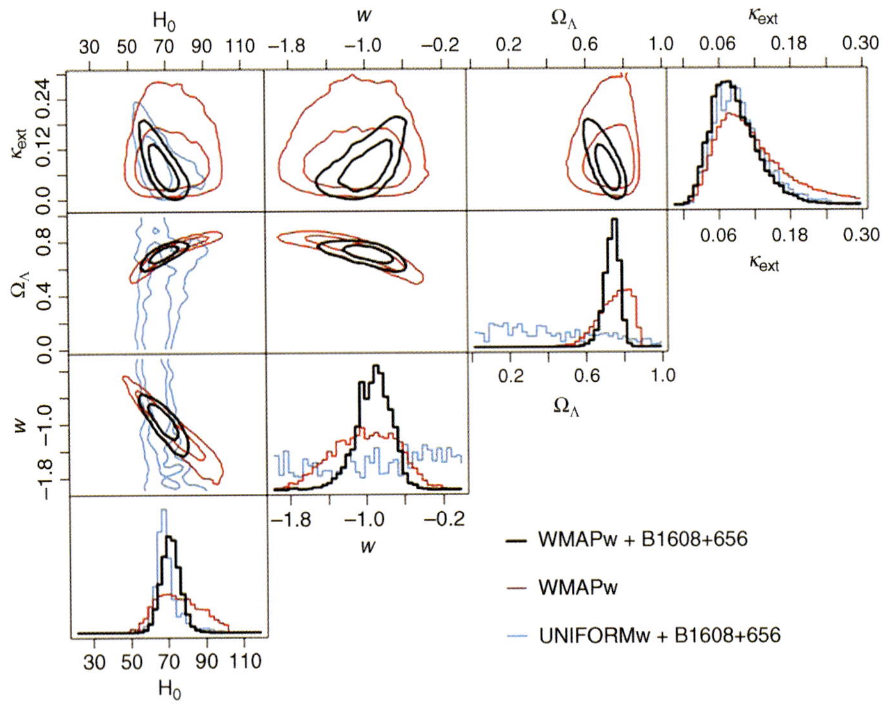

Figure 5.11 *B1608+656 probability density functions for H_0, Ω_Λ, w and κ_{ext} in a flat cosmological model (Suyu et al. 2010); contours are 68 and 95% confidence levels. The three sets of coloured contours correspond to three different prior/data set combinations. Blue: B1608+656 constraints, given a uniform w prior; red: prior provided by the WMAP 5-year data set alone; black: joint constraints from combining WMAP and B1608+656.* (Reprinted from S. Suyu et al., *Astrophysical Journal*, **711**, Dissecting the gravitational lens B1608+656. II. Precision measurements of the Hubble constant, spatial curvature, and the dark energy equation of state, p. 201–221, Copyright 2010, with permission of the AAS and S. Suyu.)

a flat geometry for the Universe, the resulting values are $H_0 = 69.7^{+4.9}_{-5.0}$ km s^{-1} Mpc^{-1} and $w = -0.94^{+0.17}_{-0.19}$ (68% confidence levels; Suyu *et al.* 2010), which is on a par with the results from BAOs (see Section 5.3.3), while the curvature of the Universe is constrained at the same level of accuracy as by Type Ia SNe.

Unfortunately, B1608+656 is the only four-image gravitational lens system with all three independent time delays between the images measured with errors of only a few percent (Fassnacht *et al.* 1999, 2002) and availability of an extended source surface brightness distribution for accurate lens modelling. The radio double lens JVAS0218+357 (e.g. Biggs *et al.* 2003; Wucknitz *et al.* 2004) and another radio lens, CLASS1608+656 (Fassnacht *et al.* 2002), may be suitable for a similar analysis. The former has a measured time-delay accuracy of better than 5% (Biggs *et al.* 1999), but its lensing system is a relatively low-mass spiral galaxy (York *et al.* 2005) that has proven difficult to model. The radio analysis yields $H_0 = 78 \pm 6$ km s^{-1} Mpc^{-1}. The latter object has been monitored sufficiently well to produce high-quality light curves, but the lens configuration is problematic. It appears that the lens consists of two interacting galaxies, which hampers straightforward modelling (cf. Koopmans *et al.* 2003; Schechter 2005).

It is likely that the LSST as well as Pan-STARRS will discover large numbers of time-delay lenses. Coe and Moustakas (2009) suggest that, in combination with a prior (i.e. a predefined estimate) on the probability density function from the *Planck* satellite, and if systematic effects such as environmental density variations – represented by the external convergence parameter $\kappa_{\rm ext}$ – are reduced, a sample of strong lenses observed with the LSST could reach subpercent-level precision in H_0 and constrain w to 3% or better.

5.3.2 The Sunyaev–Zel'dovich Effect

Inverse Compton scattering of low-energy CMB photons off a distribution of high-energy electrons in the ambient X-ray gas of rich galaxy clusters results in a small spectral distortion of the CMB radiation. In turn, this leads to a redistribution of ~1% of CMB photons passing through a cluster's central regions. Some of the latter move from the long-wavelength Rayleigh–Jeans tail of the Planck blackbody curve to the shorter wavelength Wien side, since they gain energy by roughly $k_{\rm B}T_{\rm e}/m_{\rm e}c^2$, where $k_{\rm B}$ is the Boltzmann constant, $T_{\rm e}$ the electron plasma temperature and $m_{\rm e}$ the electron mass. The impact of this redistribution is a small measured shift of order 1 mK in the peak of the Planck curve, which is referred to as the Sunyaev–Zel'dovich (SZ; Sunyaev and Zel'dovich 1969, 1972, 1980) effect. The SZ effect causes a decrease in CMB intensity at frequencies below 218 GHz and an increase at higher frequencies (cf. Carlstrom *et al.* 2002).

While the measured X-ray fluxes from galaxy clusters are distance dependent, the SZ decrement is not. It is a direct measurement of the intracluster medium's column density weighted by temperature. In other words, it represents the total pressure or, equivalently, the temperature-weighted mass integrated along the line of sight for hydrostatically supported gas (cf. Carlstrom *et al.* 2002). The SZ effect is due to predominantly CMB photons scattering off hot electrons that achieved large, potentially relativistic velocities (see Carlstrom *et al.* 2002 for a review) because of their high temperatures ($T_{\rm e} \sim 10^7$–10^8 K or $k_{\rm B}T_{\rm e} \sim 10$ keV: a thermal effect), combined with polarization (an effect on the order of micro-Kelvin-level temperature shifts; cf. Sazonov and Sunyaev 1999; Carlstrom *et al.* 2002) and scattering off electrons travelling with large bulk motions (a kinetic effect). The latter is a Doppler effect of secondary importance and sometimes referred to as the **Ostriker–Vishniac (1986) effect** (see, for reviews, Birkinshaw 1999; Carlstrom *et al.* 2002).[5]

Physically, this redistribution of photon energies and the shift in the brightness temperature peak of the blackbody curve, $\Delta T_{\rm SZE}$, is proportional to the electron density, $n_{\rm e}$,

$$\Delta T_{\rm SZE} = f_{(x,T_{\rm e})} T_{\rm CMB} \int_{\ell} \sigma_{\rm T} n_{\rm e} \frac{k_{\rm B} T_{\rm e}}{m_{\rm e} c^2} \, {\rm d}\ell \propto \int_{\ell} n_{\rm e} T_{\rm e} {\rm d}\ell = \int_{\theta} n_{\rm e} T_{\rm e} d_{\rm A} \, {\rm d}\theta, \tag{5.68}$$

where θ is the angular size, $f_{(x,T_{\rm e})}$ represents the frequency dependence of the SZ effect (see Carlstrom *et al.* 2002; Bonamente *et al.* 2006), $x = h\nu / k_{\rm B} T_{\rm CMB}$ ($T_{\rm CMB} = 2.728$ K; Fixsen *et al.* 1996), $\sigma_{\rm T}$ is the differential **Thomson scattering** cross section for charged particles, $m_{\rm e} c^2$ is the electron rest-mass energy and ℓ is again the path length along the line of sight. Note that equating the final two terms in Equation (5.68) is only allowed if the galaxy cluster is spherically symmetric. The angular size is related to the angular diameter distance derived from X-ray imaging and projection of a 3D model of the gas distribution under spherical symmetry,

$$d_{\rm A} = {\rm d}\ell / {\rm d}\theta \tag{5.69}$$

$$\propto \frac{\Delta T_{\rm SZE}^2 \Lambda}{S_{\rm X} T_{\rm e}^2} \tag{5.70}$$

(Bonamente *et al.* 2006), which in turn is inversely proportional to H_0. (See Equation (13) in Bonamente *et al.* (2006) for a properly normalized relation between $d_{\rm A}$ and H_0.) Λ is the X-ray cooling function of the gas, which describes the radiative cooling rate. It depends on the plasma temperature, which is usually on the order of the virial temperature at which the gas is in equilibrium with gravity (see Reese *et al.* 2000; and references therein), spatial temperature profile and energy, as well as its chemical composition. It consists of contributions from relativistic electron–ion and electron–electron thermal **Bremsstrahlung**, and recombination and two-photon (γ–γ) collisional processes.

The X-ray emission from rich galaxy clusters, $S_{\rm X}$, is caused by thermal Bremsstrahlung (i.e. 'braking' or **free–free radiation**, caused by accelerations of electrons due to other charged particles). It is proportional to the square of the electron density,

$$S_{\rm X} = \frac{1}{4\pi(1+z)^4} \int_{\ell} \Lambda n_{\rm e}^2 \, {\rm d}\ell \propto \int_{\theta} \Lambda n_{\rm e}^2 d_{\rm A} \, {\rm d}\theta, \tag{5.71}$$

where z is the galaxy cluster's redshift.

The factor of $(1+z)^4$ arises from the difference between luminosity distance and angular size distance. Surface brightness is defined as flux per unit solid angle, so the surface brightness of a cluster of galaxies of angular radius δ is

$$S_X = \frac{F_X}{\pi\delta^2} = \frac{L_X}{4\pi d_{\rm L}^2 \pi\delta^2}, \tag{5.72}$$

[5] For a typical line-of-sight peculiar cluster velocity of 300 km s^{-1} and a cluster temperature $T_{\rm e} = 8$ keV, the kinetic effect is 4% of the thermal SZ effect, which is thus associated with an uncertainty of 8% in cluster distance, because the angular diameter $d_{\rm A} \propto \Delta T_{\rm SZE}^2$ (Bonamente *et al.* 2006).

where F_X is the X-ray flux from, and L_X the X-ray luminosity of, the whole cluster. By definition, $\delta = r/d_A$, where r is the cluster's linear radius. Therefore,

$$S_X = \frac{L_X d_A^2}{4\pi d_L^2 \pi r^2} = \frac{S_{X0} d_A^2}{4\pi d_L^2}, \tag{5.73}$$

where S_{X0} is the X-ray luminosity of the cluster per unit area. Since $d_L = d_A (1+z)^2$, the factor $4\pi(1+z)^4$ appears in Equation (5.71).

By combining Equations (5.68) and (5.71) with a model of the intracluster plasma distribution, we can eliminate n_e and obtain the angular diameter distance, assuming that the cluster is roughly spherical (e.g. Birkinshaw 1999; Carlstrom *et al.* 2002; Reese *et al.* 2002; Bonamente *et al.* 2004, 2006; see also Wang and Fan 2006), and hence the Hubble constant. This distance independence of the SZ effect is very important, because it means that galaxy clusters at high redshifts can be detected just as easily as their nearby counterparts. In addition, the angular scale–redshift relation changes little for $0.3 < z < 2$, so that clusters between these redshifts have similar sizes on the sky.

The SZ approach to determining the Hubble constant can be applied well into the Hubble flow at cosmological distances (e.g. Silk and White 1978; Barbosa *et al.* 1996; Bonamente *et al.* 2006; and references therein). The distance range accessible through SZ effect measurements (in practice, however, up to $z \sim 1$) bridges that spanned by nearby objects and expansion values derived from CMB anisotropies (see Section 5.3.3) and SNe Ia measurements (see Section 5.2.1).

5.3.2.1 *Systematic Effects*

This method uses a combination of two distinct observations, neither of which are straightforward. The main drawbacks of the method's application, requiring careful analysis of the uncertainties (see Bonamente *et al.* 2006 for a detailed overview), result from potential substructure in a cluster's X-ray gas distribution, projection effects, the assumption that the gas is in hydrostatic equilibrium, model uncertainties in the gas and electron densities – including of extended steep-spectrum radio haloes, radio relics and the Galactic foreground hydrogen column density, although they contribute to the error budget at levels $\Delta d_A \leq 1.5\%$ each – X-ray background subtraction and calibration issues (cf. Reese *et al.* 2010) and potential point source contamination (cf. Carlstrom *et al.* 2002; Freedman and Madore 2010; Komatsu *et al.* 2011). Other effects, including the kinetic SZ effect contribution, have either already been discussed elsewhere in this chapter or contribute at negligible levels. Given that $\Lambda \propto T_e^{1/2}$ and $d_A \propto \Delta T_{SZE}^2$, distance determinations strongly depend on accurate measurements of the SZ effect decrement and the X-ray temperature.

Cluster Asphericity. Almost all analyses, apart from those that consider only one cluster, assume that clusters are spherically symmetric. This is certainly an approximation: neither the X-ray maps nor the SZ effect maps appear circular in projection. De Filippis *et al.* (2005) studied the effect of this assumption by comparing angular size distances obtained by previous authors (Mason *et al.* 2001; Reese *et al.* 2002), assuming spherical symmetry, with new distances obtained under the assumption that the clusters are oblate spheroids, by fitting the X-ray data to infer the oblateness. The average fractional change in inferred distance for the 25 clusters in their sample is 27%, somewhat larger than the expectation of

~15% quoted by Bonamente *et al.* (2006) on the basis of a simulation study by Sulkanen (1999), and the ~20% obtained by the simulation study of Ameglio *et al.* (2006). Since clusters are expected to be oriented randomly with respect to the line of sight, this error, while substantial for individual distances, should not bias attempts to determine H_0.

Clumpiness of the Intracluster Medium. All methods assume that the gas distribution is smooth. If it is significantly clumpy, the X-ray emission will be enhanced by a factor

$$C = \frac{\langle n_e^2 \rangle}{\langle n_e \rangle^2}, \tag{5.74}$$

while the SZ effect will be largely unaffected since it is linear in n_e. This will cause the true X-ray surface brightness to be higher than expected, leading to an underestimate of the distance.

The *Chandra X-ray Observatory*'s angular resolution is capable of identifying clumps in the X-ray emission. Since the emission is optically thin, there is no risk of clumps near the far edge of the cluster to be somehow 'hidden'. Analyses using *Chandra* data can, therefore, mask out the clumps, as is routinely done for foreground point sources. Bonamente *et al.* (2006) argue that the error from this source is negligible if such masking is carried out, although Reese *et al.* (2002), who did not attempt to mask clumps, state that the effect may amount to as much as ~20% in the absence of masking. This is a one-sided systematic: it can only decrease the estimated distance (e.g. Jeltema *et al.* 2005; Bonamente *et al.* 2006; LaRoque *et al.* 2006).

This systematic error is potentially redshift dependent, as high-z and, hence, younger clusters may be systematically less relaxed, and therefore clumpier, than nearby objects. Such an effect could be important when SZ effect distances are used to constrain cosmological models.

X-Ray and SZ Effect Calibrations. Vikhlinin *et al.* (2005) report a 7% systematic difference between temperatures extracted from *XMM–Newton* and *Chandra* data, suggesting a systematic error in calibration of ~5% for each experiment. This error arises from a combination of uncertainties in the photon energy calibration and the relative, energy-dependent calculation of effective area. Since d_A varies approximately as $T^{1.5}$, the implied systematic error on d_A is ~7.5%. There is also an absolute uncertainty on the effective area of the mirrors, which affects the measurement of S_X but not the temperature. For *Chandra*, this amounts to approximately 3–5%.

Reese *et al.* (2010) recently assessed the effects of X-ray calibration issues on the resulting Hubble constant for the Bonamente *et al.* (2006) sample, assuming isothermal cluster profiles (see below) for simplicity, and concluded that different temperature calibrations may have an effect of ~23% (spanning a range from −13 to +10%) on the value of H_0 at their most extreme, because $H_0 \propto T_e^2$.

The SZ effect calibration uncertainty depends on the experiment, but is typically a few percent (Udomprasert *et al.* 2004; Bonamente *et al.* 2006; Vanderlinde *et al.* 2010). Calibrations are usually based on planetary observations, although the South Pole Telescope normalizes to *WMAP*. Given that $d_A \propto \Delta T_{SZE}^2$, we deduce a 6–8% uncertainty on d_A.

Point Source Subtraction. Point sources can 'fill in' the SZ effect temperature decrement, resulting in an underestimate of $\Delta T_{\rm SZE}$. If they occur in a negative sidelobe of an interferometer, they can also enhance the decrement. This is, therefore, a point-to-point systematic error. Although it is systematic in the sense that it will not be reduced by repeated measurements of a particular cluster, it can be expected to average out over measurements of multiple clusters. The size of the effect will depend on the experimental set-up, the operating frequency (because the spectra of radio galaxies and dusty starbursts are different from that of the CMB, their relative importance will vary with frequency), and the exact technique used to perform the subtraction. Experiments operating at around 30 GHz estimate an uncertainty of $\sim$8% on individual cluster distances from this source. Experiments operating at 150 GHz should probably do better, but the analyses published so far (Menanteau *et al.* 2010; Vanderlinde *et al.* 2010) do not consider this in detail.

Because of the frequency dependence, observations at multiple frequencies would help to constrain this error source, and also the kinetic SZ effect and CMB anisotropies. However, almost all SZ effect experiments to date have concentrated on a single frequency, making it impossible to apply such a check. The multiwaveband observations expected from the *Planck* satellite will be very interesting in this context.

Choice of Profile Model. The isothermal β surface brightness model used in most early work does not produce good fits to the detailed X-ray profiles produced by *Chandra* data. The β model is a commonly used model to fit X-ray surface brightness profiles, where β refers to the power-law slope of the gas distribution profile (Cavaliere and Fusco-Femiano 1976; Mohr *et al.* 1999; LaRoque 2005; see also Bonamente *et al.* 2006). Several studies have been done to investigate the effect of different models (cf. Wang and Fan 2006).

One of the best SZ-based constraints of H_0 was recently obtained using Markov chain[6] **Monte Carlo analysis** of new X-ray observations of 38 galaxy clusters at redshifts $0.14 < z < 0.89$ obtained with the *Chandra X-ray Observatory* (Bonamente *et al.* 2006; see Figure 5.12). This provided significant improvements in precision and accuracy with respect to previous H_0 determinations based on individual or a few (up to 18) galaxy clusters (e.g. Myers *et al.* 1997; Mason *et al.* 2001; Reese *et al.* 2002; Jones *et al.* 2005; see for a review Bonamente *et al.* 2006). In combination with radio interferometric observations with the Berkeley–Illinois–Maryland Association (BIMA) array and Owens Valley Radio Observatory (OVRO), Bonamente *et al.* (2006) derived $H_0 = 76.9^{+3.9}_{-3.4}{}^{+10.0}_{-8.0}$ and $73.7^{+4.6}_{-3.8}{}^{+9.5}_{-7.6}$ km s^{-1} Mpc^{-1} (statistical and systematic uncertainties, respectively, at the 68% confidence level) assuming hydrostatic equilibrium and an isothermal β model, respectively. Changing their radial range (to $>$100 kpc) to avoid potential cool cluster cores (in case the radiative cooling time is less than the cluster's age) increased the resulting Hubble constant to $H_0 = 77.6^{+4.8}_{-4.3}{}^{+10.1}_{-8.2}$ km s^{-1} Mpc^{-1}, adopting the same convention regarding the uncertainties. The consistency between these values shows that H_0 determinations based on the SZ effect in X-rays are relatively insensitive to the model assumptions, indicating an overall systematic error of $<$5%, although the effect for individual clusters can be very

[6] The Markov chain (Gamerman and Lopes 1996) is a method of 'directing' a random walk in parameter space so that it will converge on the best fit. Once convergence is reached, probability distributions can be calculated for all variables involved in the fit.

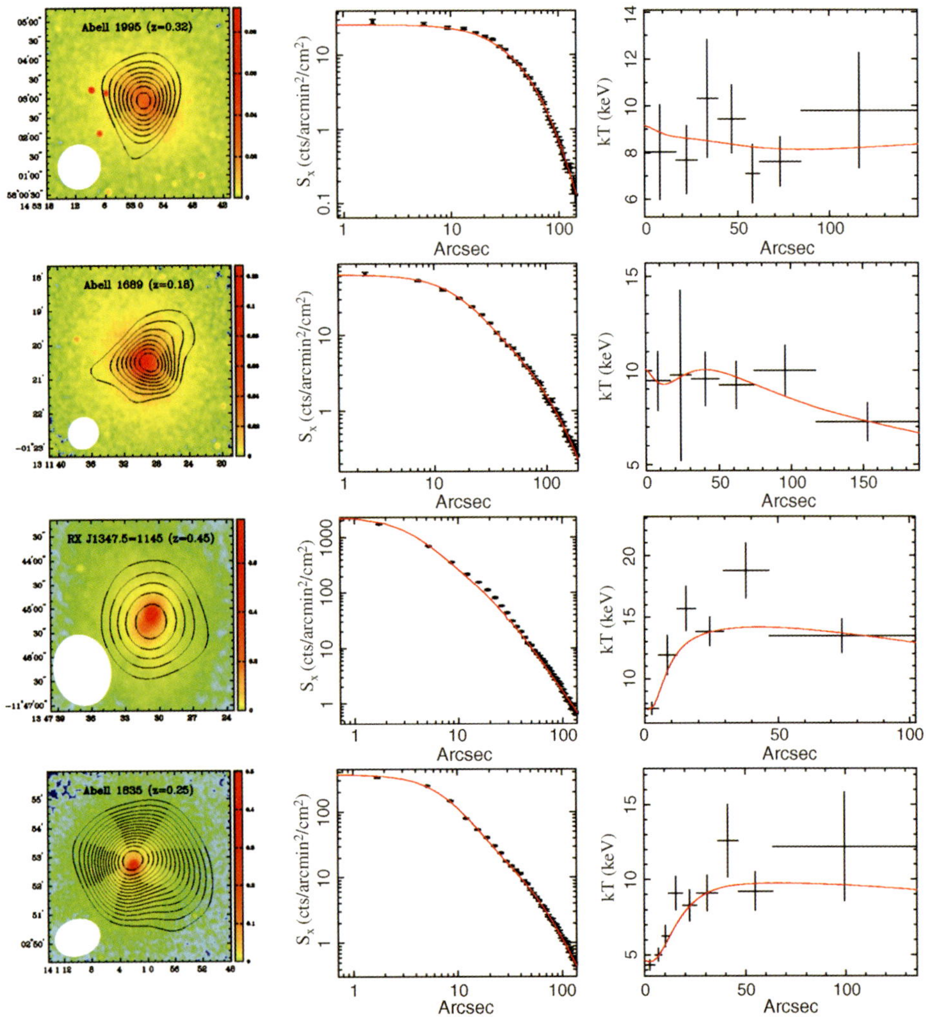

Figure 5.12 *(Left) Chandra images of the X-ray surface brightness in the 0.7–7 keV band in units of counts per pixel (197 pixels) for selected clusters (Bonamente et al. 2006). Overlaid are the SZ effect decrement contours, with contour levels (+ 1, −1, −2, −3, −4, . . .) times the rms noise in each image. The effective point spread functions of the SZ effect synthesized beams are shown in the bottom left-hand corners. (Middle) Radial profiles of the background-subtracted X-ray surface brightness. The solid lines are the best-fitting models. (Right) Radial profiles of the Chandra temperatures and best-fitting hydrostatic equilibrium models.* (Reprinted from M. Bonamente *et al.*, *Astrophysical Journal*, **647**, Determination of the cosmic distance scale from Sunyaev–Zel'dovich effect and Chandra X-ray measurements of high-redshift galaxy clusters, p. 25–54, Copyright 2006, with permission of the AAS and M. Bonamente.)

substantial: cluster MS1137.5+6625 has a corresponding estimated distance of $2.85^{+0.52}_{-0.63}$, $3.65^{+1.25}_{-0.97}$ and $5.07^{+1.96}_{-1.43}$ Gpc, respectively.

A simulation study by Ameglio *et al.* (2006), in contrast, expected a considerably larger bias, with isothermal models underestimating d_A by $\sim$20%. The authors acknowledge that this is an upper limit, partly because their simulation is known to overestimate the temperature gradient in the centres of clusters and partly because they have not accounted for the tendency of interferometric measurements to give the cluster centre higher weight than the outer regions. They also remark that the similar study by Hallman *et al.* (2006), although focussing on effects on cluster gas mass determinations rather than d_A, actually predicts – if translated into an effect on d_A – a bias in the opposite direction, overestimating d_A rather than underestimating it. This seems to be more in keeping with the results of Bonamente *et al.* (2006), where H_0 for the isothermal model is smaller than for the other two models, consistent with overestimated distances. It is also consistent with the compilation of Huchra (2010), where the values of H_0 derived from SZ effect measurements show a slight but definite tendency to increase over time, suggesting that more recent, and presumably more sophisticated, analyses yield smaller distances on average.

5.3.2.2 Prospects

Holanda *et al.* (2010) recently obtained $H_0 = 73.2^{+4.3}_{-3.7}$ and $71.4^{+4.4}_{-3.4}$ km s^{-1} Mpc^{-1} (1σ uncertainties) based on a sample of 25 angular diameter distances for galaxy clusters analyzed using the X-ray/SZ effect, based on geometry assumptions of a standard ΛCDM cosmology and a flat Universe with constant equation-of-state parameter, respectively. They combined their SZ analysis with results from BAOs and the 'shift parameter' related to dark energy which is apparent in the CMB signature (cf. Bond *et al.* 1997; Nesseris and Perivolaropoulos 2007), although the exact value of the Hubble constant does not depend sensitively on the assumed geometry or models of dark energy.

SZ measurements, although difficult in the early days because of the small but measurable shift in temperature, are now becoming routine at radio (single-dish and interferometric observations; centimetre wavelengths), millimetre and submillimetre wavelengths (see, for reviews, Carlstrom *et al.* 2002; Bonamente *et al.* 2006). The number of clusters with SZ-effect-derived angular size distances is now quite large: Reese (2004) collected 41 distances (although several clusters are represented multiple times), Udomprasert *et al.* (2004) list 24 clusters with SZ-effect data, but derive distances for only seven, De Filippis *et al.* (2005) has 25 and Bonamente *et al.* (2006) have 38. There is considerable overlap among these samples, but there must now be at least 50 clusters with angular diameter distances derived based on modern data. It is noticeable that the number of publications reporting on H_0 from SZ effect data has declined in recent years: it is now much more common to use the SZ effect to make estimates of the gas mass in clusters, rather than using it to define distances.

Nevertheless, current and planned galaxy cluster surveys, such as with the South Pole Telescope (e.g. Carlstrom *et al.* 2009; Andersson *et al.* 2010), the Atacama Cosmology Telescope (e.g. Fowler *et al.* 2007; Hincks *et al.* 2010; Marriage *et al.* 2010; Swetz *et al.* 2010; Vanderlinde *et al.* 2010; and references therein), APEX-SZ (the Atacama Pathfinder Experiment-SZ project; e.g. Schwan *et al.* 2003; Kneissl *et al.* 2008) and *Planck*, as well as the eROSITA survey (onboard the Russian Spectrum–Röntgen–Gamma satellite, which will be placed in an orbit at the second Lagrangian point, L2; e.g. Predehl *et al.* 2006), have the potential of detecting thousands of galaxy clusters at a large range of redshifts, hence increasing cluster sample sizes for potential combined X-ray and SZ-based H_0 determinations to unprecedented accuracy.

5.3.2.3 Galaxy Clusters as Potential 'Standard Candles'

From a physical point of view, this raises the question as to whether galaxy clusters are good 'standard candles' (e.g. Verde *et al.* 2002; Afshordi 2008). Haiman *et al.* (2001) already pointed out that the use of large galaxy cluster surveys for cosmological purposes will most likely be limited by the validity or otherwise of this assumption rather than by statistical uncertainties. If galaxy clusters are closely approximated by virialized, spherical haloes, they span a cluster 'fundamental plane' at every redshift range defined by their virial temperature, SZ decrement and angular size. Verde *et al.* (2002) showed that this fundamental plane and its redshift evolution are sensitive to both the clusters' internal evolution and the underlying cosmological parameters. Importantly, the cluster fundamental plane's properties can be used to constrain both. Deviations from the model assumptions (e.g. in the form of energy injection or feedback, stochastic scatter in the observables or deviations from virial equilibrium) create measurable deformations in the cluster fundamental plane. If the model assumptions hold, the cluster fundamental plane projection involving the size of the SZ decrement and virial temperature will be very narrow (Verde *et al.* 2002). This tightness can be used to quantify physical cluster properties, scatter in the observables and deviations from virial equilibrium and, hence, to test whether clusters are indeed suitable standard candles.

Alternatively, if clusters are good standard candles, deformations of the cluster fundamental plane can be used to reliably determine cosmological parameters at a comparable level to those resulting from CMB anisotropies and SZ measurements in galaxy cluster centres (Verde *et al.* 2002). However, Afshordi (2008) argued that one can release the model assumption of virial equilibrium and, hence, of hydrostatic equilibrium and self-similar pressure profiles among galaxy clusters to derive an alternative fundamental plane defined by cluster *mass* (M), half-mass radius (defined by SZ measurements: $R_{\rm SZ,2}$) and SZ flux or thermal energy, $Y_{\rm SZ}$,

$$M \propto (Y_{\rm SZ}/R_{\rm SZ,2})^{3/4}. \tag{5.75}$$

The best-fitting cluster fundamental plane (with 14% scatter) is

$$M_{200} = (7.8 \times 10^{14} {\rm M}_\odot h^{-1}) Y_{\rm SZ}^{0.75} \left(\frac{R_{\rm SZ,2}}{{\rm Mpc}^{-1} h} \right)^{-0.76}, \tag{5.76}$$

where M_{200} is the mass within a radius inside which the mean density of the cluster is 200 times the critical density of the Universe. Use of this alternative cluster fundamental plane leads to ~34% smaller statistically errors than simply using masses derived from a temperature or SZ flux–mass conversion (Afshordi 2008).

5.3.3 Anisotropies in the Cosmic Microwave Background

Peebles and Yu (1970) and Sunyaev and Zel'dovich (1970) predicted the presence of acoustic oscillations in the CMB. Both CMB and baryon acoustic oscillations are caused by the excitation of sound-like waves in the primordial, hot, dense plasma consisting of fluctuating density distributions of electrons and baryons, which – in the early Universe – were co-located with dark matter density fluctuations. Both are believed to have been nearly scale-invariant, adiabatic Gaussian perturbations, best described by a power law spectrum (e.g. Komatsu *et al.* 2011 and references therein). At overdensities, at the '**surface of last scattering**' ($z \sim 1089$, where matter and radiation decoupled shortly after the Big Bang), gravity counteracted the effects of the radiation pressure generated by the heat of photon–matter interactions in the same overdensities. In turn, this led to a pressure-induced, outward-expanding 'acoustic' wave, taking along with it the photons and baryonic matter. (The dark matter would stay behind because it only interacts gravitationally.)

Photons and baryons remained tightly coupled before the '**epoch of recombination**', when the expansion of the Universe exceeded the rate of Compton scattering. (The **Compton effect** refers to inelastic collisions of photons in matter, resulting in an energy decrease of X- and γ-ray photons.) At this time, when the early Universe had cooled down to ~3000 K some 400 000 years after the Big Bang, protons and electrons 'recombined' to form neutral hydrogen. As a consequence, photons and baryonic matter became 'decoupled' and the Universe became transparent to electromagnetic radiation (i.e. photons). This produced the anisotropic CMB radiation as we observe it today. Since photons rarely interact with neutral matter (i.e. neutral hydrogen and neutrons), they diffused away from their baryonic environment, because their mean free path essentially increased to the size of the Universe. The temperature fluctuations in current CMB maps are best represented by a set of **spherical harmonics**. The latter are solutions to **Laplace's equation**, $\nabla^2\phi = 0$, where ∇^2 is the Laplace operator and ϕ a scalar function, a differential equation describing the gravitational potential associated with a set of point masses. The **angular power spectrum** as a function of multipole anisotropy moment (l), or inverse angular scale (higher multipole moments correspond to smaller angular scales) – see Figure 5.13 – follows naturally from the tight coupling between baryons and photons before recombination in the early Universe.

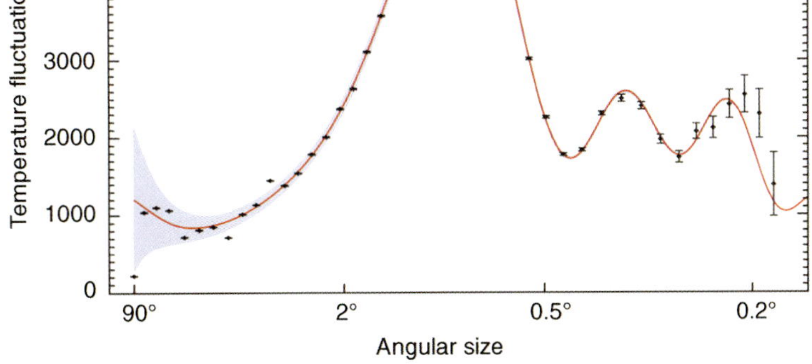

Figure 5.13 *Angular power spectrum of the fluctuations in the 5-year WMAP full-sky map. This shows the relative brightness of the temperature fluctuations in the map versus their size.* (Courtesy of the NASA/WMAP Science Team.)

The acoustic oscillations which occurred in the early Universe left a characteristic scale in the spatial distribution of baryonic matter. The baryonic matter was left behind at a fixed radius, determined by the epoch of recombination, which is often referred to as the '**sound horizon**'. Since the photon-induced radiation pressure had all but disappeared, the main force acting on the remaining baryons became gravitational. Over a Hubble time, this led to further growth of the initial, δ-function-like density perturbations, both at their origin and in a shell at the sound horizon, eventually resulting in characteristic clustering patterns in the present-day galaxy distribution. Although the size of the sound horizon can be calculated fairly easily from first principles in cosmology, the Universe has been expanding since its inception, and this expansion is apparently accelerating, thus leading to the notion of dark energy (cf. Section 5.1.3). BAO observations can help us understand the nature of dark energy and the accelerated expansion of the Universe by comparing the angle subtended by the sound horizon at redshift z, derived from galaxy-clustering observations, to the corresponding angle at the time of recombination, which is provided to high accuracy by measurements of the CMB. The CMB peak positions tightly constrain the **conformal** ('true, angle-preserving') **distance** to the surface of last scattering. Thus, BAO observations provide an independent standard ruler. The galaxy two-point correlation function exhibits a peak on scales of $100h^{-1}$ Mpc (Eisenstein *et al.* 2005), which is approximately the size of the linear comoving sound horizon at the epoch of recombination. This thus provides a standard ruler to measure the distance ratio between the surface of last scattering and any other redshift.

Although CMB power spectrum measurements can be obtained with very high statistical precision, a number of degeneracies affect the accuracy with which cosmological parameters can be determined (e.g. Efstathiou and Bond 1999). For the present discussion, the most important degeneracy is between H_0 and both Ω_Λ and w. Thus, CMB observations do not directly measure the local expansion rate of the Universe (i.e. H_0), but instead determine the conformal distance to the surface of last scattering and the matter/radiation (baryon/photon) ratio (as e.g. obtained from deuterium/hydrogen abundances in the spectra of distant quasars) through the amplitude of the early **integrated Sachs–Wolfe (1967) contribution** relative to the height of the first peak of the angular power spectrum.[7]

For a contemporary, standard ΛCDM cosmological model, this is enough information to predict the local expansion rate. For instance, Spergel *et al.* (2003) obtain $H_0 = 72 \pm 5$ km s^{-1} Mpc^{-1} (68% confidence) assuming a basic cosmological model for the geometry of the Universe consisting of a flat Universe containing radiation, baryons, cold dark matter and a cosmological constant, and a power-law power spectrum of adiabatic primordial fluctuations. Their value for the Hubble constant is remarkably close to that resulting from the *HST* Key Project, $H_0 = 72 \pm 3$ (statistical) $\pm$ 7 (systematic) km s^{-1} Mpc^{-1} (Freedman *et al.* 2001; see Section 4.1) in view of the large differences between both models in terms of observables, underlying physics and model assumptions. This striking match between both determinations of H_0 places strong constraints on the dark energy's equation-of-state parameter, w $(= -1)$, that is usually assumed for a ΛCDM Universe

[7] The integrated Sachs–Wolfe effect is caused by gravitational redshifting which originates from the time evolution of the gravitational potential: stronger gravitational potentials cause electromagnetic radiation to shift towards longer wavelengths. This occurred between the surface of last scattering and an observer on Earth. The 'early' integrated Sachs–Wolfe effect occurred immediately after the nonintegrated effect produced the primordial CMB, when photons travelled through a young Universe dominated by density fluctuations while its expansion was still significantly affected by the lingering radiation.

(cf. Spergel *et al.* 2003; Komatsu *et al.* 2011), because cosmological models which differ from a cosmological constant ($w \neq -1$) require smaller values of H_0. For instance, a model with $\Omega_M = 0.47$, $w = -0.5$ and $H_0 = 57$ km s^{-1} Mpc^{-1} yields a nearly identical angular power spectrum as the currently favoured ΛCDM model (Spergel *et al.* 2003).

In general, CMB measurements must be combined with other suitable observations to constrain any of the degenerate parameters. For instance, a maximum-likelihood analysis of the 7-year *WMAP* CMB data combined with BAO constraints and priors on the value of H_0 yields $H_0 = 70.4 \pm 2.5$ km s^{-1} Mpc^{-1} (Komatsu *et al.* 2011; Larson *et al.* 2010), while the older, 5-year *WMAP* observations in combination with both SNe Ia and BAO constraints result in $H_0 = 70.5 \pm 1.3$ km s^{-1} Mpc^{-1} (Komatsu *et al.* 2009). Assuming a ΛCDM Universe, combining BAO measurements from the SDSS Data Release 7 (DR7) and the Two-Degree-Field (2dF) Galaxy Redshift Survey with constraints on the baryon and CDM densities from *WMAP-5* and SNe Ia yields $H_0 = 68 \pm 2$ km s^{-1} Mpc^{-1} (Freedman and Madore 2010).

5.3.4 The Drive for Improved Accuracy

The measurement of cosmological distances is challenging. It is tempting to look at the most recent *WMAP* results (e.g. Komatsu *et al.* 2011) and conclude that the determination of cosmological distances is an unnecessary luxury. However, this is not the case. The *WMAP* fits are very sensitive to assumptions made about the cosmological model: while the 'headline' value of H_0 appears highly precise, it relies on the assumption of a flat geometry and the presence of a cosmological constant, and when the former constraint is relaxed the fitted value moves to $H_0 = 53^{+15}_{-13}$ km s^{-1} Mpc^{-1}: the central value has changed considerably, and the precision is much reduced. Therefore, direct measurements of H_0 are still essential to constrain fits and reduce correlations.

In addition, although the remaining statistical uncertainties in the value of the Hubble constant have been reduced to unprecedentedly low levels of <5%, the quest for errors at <2 or even 1% levels continues. The field has now reached the era of precision cosmology, potentially allowing scrutiny of fundamental questions that were thus far impossible to address. These include prediction of the critical density of the Universe, ρ_0, constraining the dark energy's equation of state and determining the neutrino mass (e.g. Dolgov 1996; Crotty *et al.* 2004; Hannestad 2006; see also Freedman and Madore 2010). As regards the latter, one of the most significant limitations to determining the neutrino mass from the CMB power spectrum is caused by its strong degeneracy with H_0 (Komatsu *et al.* 2009). Improved estimates of H_0 to an accuracy of 2–3%, combined with CMB data from the *Planck* satellite, will improve the precision of neutrino mass measurements by an order of magnitude (Freedman and Madore 2010).

On the assumption of a flat CDM cosmological model (e.g. Komatsu *et al.* 2009, 2011), the fluctuations in the CMB observed by *WMAP* imply a similar value and slightly reduced uncertainty of H_0 compared to the *HST* Key Project. However, if the underlying flatness assumption is abandoned, *WMAP* data do not place strong constraints on the Hubble constant, because they principally constrain the actual matter density, $\Omega_M h^2$. Clearly, therefore, improved independent and more accurate determinations of H_0 will further constrain Ω_M. In this context, Olling (2007) points out that the current best value of the Hubble constant is the result of using the *WMAP* observations in combination with constraints from galaxy redshift surveys and **Lyman-α forest** observations, which both probe large-scale structure, SNe Ia in the distant Universe, the SZ effect and gravitational lensing. (The CMB data alone are insufficient to provide strong constraints on the value of H_0, given its strong degeneracy with the dark energy's equation of state, w; Albrecht *et al.* 2006.) However, the underlying physics associated with these independent methods may be more complex than that of the CMB and/or 'geometric' determinations of the Hubble constant, thus potentially leading to serious biases in the derived cosmological parameters (e.g. Seljak *et al.* 2003; Efstathiou 2005). Olling (2007) therefore proposes to consider using H_0 as an independent constraint on the matter–energy density in the Universe rather than as a free parameter that is yet to be determined.

The discovery of accelerated expansion of the Universe, and the associated need for dark energy (Riess *et al.* 1998; Perlmutter *et al.* 1999; see, for reviews, Filippenko 2005; Frieman *et al.* 2008) has highlighted the need for measurements of H_0 with improved precision. Riess *et al.* (2009a) argue that possibly the most fundamental question which new observations may constrain with higher accuracy is whether dark energy is static (i.e. a cosmological constant) or dynamical (i.e. an inflation-like scalar field), or indeed whether Einstein's theory of general relativity can accommodate the presence of this large, negative pressure comfortably.

Current constraints imply that the Universe is close to flat, in which case $\Omega_{tot} = 0.996$ ($\pm 1.6\%$; Olling 2007). The resulting value for $w = -0.95$ ($\pm 12\%$). Its value decreases to $w = -0.98$ ($\pm 7.2\%$) if the flatness assumption is abandoned, which is comfortably within the uncertainty range of the equivalent determination based on the *WMAP* 7-year data (Komatsu *et al.* 2011), $w = -1.10 \pm 0.14$ (68% confidence interval, not including data from high-z SNe Ia). Nevertheless, better constraints on the value of the equation of state, at the percent level and as a function of redshift (cf. Hu 2005), are required to assess whether w might be (or have been) evolving, which requires both better CMB data (one of the aims of *Planck*) and much smaller errors on H_0.

The most promising approach to reducing the systematic uncertainties associated with the present-day expansion rate of the Universe is by trying to achieve improved *local* calibrations of primary distance indicators and their derivatives, including the zero point of the Cepheid period–luminosity relation, at both optical and – potentially with much higher accuracy (i.e. reduced scatter) – at IR wavelengths, for instance through trigonometric parallax measurements of a carefully selected sample of Cepheids by *Gaia* (see Section 2.1.2). Riess *et al.* (2009a) and Freedman and Madore (2010) predict that improvements in the Cepheid distance scale will be achievable to a level of 3–4% or better, based on mid-IR observations with the *James Webb Space Telescope*. In addition, improved ($\sim$2%) calibration of the SNe Ia distance scale by cross-correlation with Cepheid-based distances to SNe Ia host galaxies will allow us to tie local samples (i.e. a locally determined Hubble constant) to the intermediate- and high-z Universe. This thus also provides an independent check on the accuracy and any remaining systematic effects. Prospects for further improvements will come from tying the extragalactic distance scale to more galaxies with well-determined, geometric maser distances, improvements in the calibration of Tully–Fisher distances – particularly at IR wavelengths – as well as better modelling and higher sensitivity of SZ clusters, the CMB power spectrum (measured by *Planck* at higher angular resolution than available to date) and BAOs at a range of redshifts (see Freedman and Madore 2010 for an overview; see also Olling 2007).

Bibliography

Adams DN 1979 *The Hitchhiker's Guide to the Galaxy* (Introduction). Pan Books (McMillan Publ.).

Afshordi N 2008 Fundamental Plane of Sunyaev–Zel'dovich clusters. *Astrophys. J.* **686**, 201–205.

Albrecht A, Bernstein G, Cahn R, Freedman WL, Hewitt J, Hu W, Huth J, Kamionkowski M, Kolb EW, Knox L, Mather JC, Staggs S and Suntzeff NB 2006 *Report of the Dark Energy Task Force*. Unpublished (astro-ph/0609591).

Ameglio S, Borgani S, Diaferio A and Dolag K 2006 Angular-diameter distance estimates from the Sunyaev–Zel'dovich effect in hydrodynamical cluster simulations. *Mon. Not. R. Astron. Soc.* **369**, 1459–1468.

Andersson K, Benson BA, Ade PAR, Aird KA, Armstrong B, Bautz M, Bleem LE, Brodwin M, Carlstrom JE, Chang CL, Crawford TM, Crites AT, de Haan T, Desai S, Dobbs MA, Dudley JP, Foley RJ, Forman WR, Garmire G, George EM, Gladders MD, Halverson NW, High FW, Holder GP, Holzapfel WL, Hrubes JD, Jones C, Joy M, Keisler R, Knox L, Lee AT, Leitch EM, Lueker M, Marrone DP, McMahon JJ, Mehl J, Meyer SS, Mohr JJ, Montroy TE, Murray SS, Padin S, Plagge T, Pryke C, Reichardt CL, Rest A, Ruel J, Ruhl JE, Schaffer KK, Shaw L, Shirokoff E, Song J, Spieler HG, Stalder B, Staniszewski Z, Stark AA, Stubbs CW, Vanderlinde K, Vieira JD, Vikhlinin A, Williamson R, Yang Y and Zahn O 2010 X-ray properties of the first SZE-selected galaxy cluster sample from the South Pole Telescope. *Astrophys. J*, submitted (arXiv:1006.3068).

Baade W and Zwicky F 1934 On supernovae. *Publ. Nat'l Acad. Sci. USA*, **20**, 254–259.

Barbosa D, Bartlett JG, Blanchard A and Oukbir J 1996 The Sunyaev–Zel'dovich effect and the value of Ω_0. *Astron. Astrophys.* **314**, 13–17.

Baron E, Hauschildt PH, Branch D, Austin S, Garnavich P, Ann HB, Wagner RM, Filippenko AV, Matheson T and Liebert J 1995 Non-LTE spectral analysis and model constraints on SN 1993J. *Astrophys. J.* **441**, 170–181.

Baron E, Hauschildt PH, Nugent P and Branch D 1996 Non-local thermodynamic equilibrium effects in modelling of supernovae near maximum light. *Mon. Not. R. Astron. Soc.* **293**, 287–315.

Baron E, Nugent PE, Branch D and Hauschildt PH 2004 Type IIP supernovae as cosmological probes: a spectral-fitting expanding atmosphere model distance to SN 1999em. *Astrophys. J.* **616**, L91–L94.

Barris BJ, Tonry JL, Blondin S, Challis P, Chornock R, Clocchiatti A, Filippenko AV, Garnavich P, Holland ST, Jha S, Kirshner RP, Krisciunas K, Leibundgut B, Li W, Matheson T, Miknaitis G, Riess AG, Schmidt BP, Smith RC, Sollerman J, Spyromilio J, Stubbs CW, Suntzeff NB, Aussel H, Chambers KC, Connelley MS, Donovan D, Henry JP, Kaiser N, Liu MC, Martín EL and Wainscoat RJ 2004 Twenty-three high-redshift supernovae from the Institute for Astronomy Deep Survey: doubling the supernova sample at $z > 0.7$. *Astrophys. J.* **602**, 571–594.

Bartel N 1985 Angular diameter determinations of radio supernovae and the distance scale. In *Supernovae as Distance Indicators* (ed. Bartel N), pp. 107–122. Springer.

Bartel N and Bietenholz MF 2003 SN 1979C VLBI: 22 years of almost free expansion. *Astrophys. J.* **591**, 301–315.

Bartel N and Bietenholz MF 2005 Supernova distances with the expanding shock front method. In *Future Directions in High Resolution Astronomy: The 10th Anniversary of the VLBA* (eds Romney J and Reid M), pp. 293–297. Astronomical Society of the Pacific.

Bartel N, Rogers AEE, Shapiro II, Gorenstein MV, Gwinn CR, Marcaide JM and Weiler KW 1985 Hubble's constant determined using very-long baseline interferometry of a supernova. *Nature* **318**, 25–30.

Bartel N, Bietenholz MF, Rupen MP, Beasley AJ, Graham DA, Altunin VI, Venturi T, Umana G, Cannon WH and Conway JE 2002 SN 1993J VLBI. II. Related changes of the deceleration, flux density decay, and spectrum. *Astrophys. J.* **581**, 404–426.

Bartel N, Bietenholz MF, Rupen MP and Dwarkadas VV 2007 SN 1993J VLBI. IV. A geometric distance to M81 with the expanding shock front method. *Astrophys. J.* **668**, 924–940.

Basset BA and Kunz M 2004 Cosmic distance-duality as a probe of exotic physics and acceleration. *Phys. Rev. D* **69**, 101305.

Berlind AA, Frieman J, Weinberg DH, Blanton MR, Warren MS, Abazajian K, Scranton R, Hogg DW, Scoccimarro R, Bahcall NA, Brinkmann J, Gott III JR, Kleinman SJ, Krzesinski J, Lee BC, Miller

CJ, Nitta A, Schneider DP, Tucker DL and Zehavi I 2006 Percolation galaxy groups and clusters in the SDSS redshift survey: identification, catalogs, and the multiplicity function. *Astrophys. J. Suppl. Ser.* **167**, 1–25.

Bersten MC and Hamuy M 2009 Bolometric light curves for 33 Type II plateau supernovae. *Astrophys. J.* **701**, 200–208.

Bietenholz MF, Bartel N and Rupen MP 2001 SN 1993J VLBI. I. The center of the explosion and a limit on anisotropic expansion. *Astrophys. J.* **557**, 770–781.

Biggs AD, Browne IWA, Helbig P, Koopmans LVE, Wilkinson PN and Perley RA 1999 Time delay for the gravitational lens system B0218+357. *Mon. Not. R. Astron. Soc.* **304**, 349–358.

Biggs AD, Wucknitz O, Porcas RW, Browne IWA, Jackson NJ, Mao S and Wilkinson PN 2003 Global 8.4-GHz VLBI observations of JVAS B0218+357. *Mon. Not. R. Astron. Soc.* **338**, 599–608.

Birkinshaw M 1999 The Sunyaev–Zel'dovich effect. *Phys. Rep.* **310**, 97–195.

Blandford R and Narayan R 1986 Fermat's principle, caustics, and the classification of gravitational lens images. *Astrophys. J.* **310**, 568–582.

Blondin S, Dessart L, Leibundgut B, Branch D, Höflich P, Tonry JL, Matheson T, Foley RJ, Chornock R, Filippenko AV, Sollerman J, Spyromilio J, Kirshner RP, Wood-Vasey WM, Clocchiatti A, Aguilera C, Barris B, Becker AC, Challis P, Covarrubias R, Davis TM, Garnavich P, Hicken M, Jha S, Krisciunas K, Li W, Miceli A, Miknaitis G, Pignata G, Prieto JL, Rest A, Riess AG, Salvo ME, Schmidt BP, Smith RC, Stubbs CW and Suntzeff NB 2006 Using line profiles to test the fraternity of Type Ia supernovae at high and low redshifts. *Astron. J.* **131**, 1648–1666.

Blondin S, Davis TM, Krisciunas K, Schmidt BP, Sollerman J, Wood-Vasey WM, Becker AC, Challis P, Clocchiatti A, Damke G, Filippenko AV, Foley RJ, Garnavich PM, Jha SW, Kirshner RP, Leibundgut B, Li W, Matheson T, Miknaitis G, Narayan G, Pignata G, Rest A, Riess AG, Silverman JM, Smith RC, Spyromilio J, Stritzinger M, Stubbs CW, Suntzeff NB, Tonry JL, Tucker BE and Zenteno A 2008 Time dilation in Type Ia supernova spectra at high redshift. *Astrophys. J.* **682**, 724–736.

Bolejko K 2010 Weak lensing and the Dyer–Roeder approximation. *Mon. Not. R. Astron. Soc.* **412**, 1937–1942.

Bonamente M, Joy MK, Carlstrom JE, Reese ED and LaRoque SJ 2004 Markov chain Monte Carlo joint analysis of Chandra X-ray imaging spectroscopy and Sunyaev–Zel'dovich effect data. *Astrophys. J.* **614**, 56–63.

Bonamente M, Joy MK, LaRoque SJ, Carlstrom JE, Reese ED and Dawson KS 2006 Determination of the cosmic distance scale from Sunyaev–Zel'dovich effect and Chandra X-ray measurements of high-redshift galaxy clusters. *Astrophys. J.* **647**, 25–54.

Bond JR, Efstathiou G and Tegmark M 1997 Forecasting cosmic parameter errors from microwave background anisotropy experiments. *Mon. Not. R. Astron. Soc.* **291**, L33–L41.

Bonometto SA, Iovino A, Guzzo L, Giovanelli R and Haynes M 1993 Correlation functions from the Perseus–Pisces redshift survey. *Astrophys. J.* **419**, 451–458.

Botticella MT, Riello M, Cappellaro E, Benetti S, Altavilla G, Pastorello A, Turatto M, Greggio L, Patat F, Valenti S, Zampieri L, Harutyunyan A, Pignata G and Taubenberger S 2008 Supernova rates from the Southern inTermediate Redshift ESO Supernova Search (STRESS). *Astron. Astrophys.* **479**, 49–66.

Bremer MN, Silk J, Davies LJM and Lehnert MD 2010 A redshift survey towards the cosmic microwave background cold spot. *Mon. Not. R. Astron. Soc.* **404**, L69–L73.

Bromm V and Loeb A 2006 High-redshift gamma-ray bursts from Population III progenitors. *Astrophys. J.* **642**, 382–388.

Burrows A 2000 Supernova explosions in the Universe. *Nature* **403**, 727–733.

Burstein D 1990 Large-scale motions in the Universe: a review. *Rep. Prog. Phys.* **53**, 421–481.

Butler NR, Kocevski D, Bloom JS and Curtis JL 2007 A complete catalog of Swift gamma-ray burst spectra and durations: demise of a physical origin for pre-Swift high-energy correlations. *Astrophys. J.* **671**, 656–677.

Butler NR, Kocevski D and Bloom JS 2009 Generalized tests for selection effects in gamma-ray burst high-energy correlations. *Astrophys. J.* **694**, 76–83.

Capaccioli M, Mancini D and Sedmak G 2003 VST: The VLT Survey Telescope. *Mem. Soc. Astron. It.* **74**, 450–451.

Cappellaro E, Evans R and Turatto M 1999 A new determination of supernova rates and a comparison with indicators for galactic star formation. *Astron. Astrophys.* **351**, 459–466.

Cardelli JA, Clayton GC and Mathis JS 1989 The relationship between infrared, optical, and ultraviolet extinction. *Astrophys. J.* **345**, 245–256.

Cardone VF, Capozziello S and Dainotti MG 2009 An updated gamma-ray bursts Hubble diagram. *Mon. Not. R. Astron. Soc.* **400**, 775–790.

Carlstrom JE, Holder GP and Reese ED 2002 Cosmology with the Sunyaev–Zel'dovich effect. *Annu. Rev. Astron. Astrophys.* **40**, 643–680.

Carlstrom JE, Ade PAR, Aird KA, Benson BA, Bleem LE, Busetti S, Chang CL, Chauvin E, Cho H-M, Crawford TM, Crites AT, Dobbs MA, Halverson NW, Heimsath S, Holzapfel WL, Hrubes JD, Joy M, Keisler R, Lanting TM, Lee AT, Leitch EM, Leong J, Lu W, Lueker M, McMahon JJ, Mehl J, Meyer SS, Mohr JJ, Montroy TE, Padin S, Plagge T, Pryke C, Ruhl JE, Schaffer KK, Schwan D, Shirokoff E, Spieler HG, Staniszewski Z, Stark AA, Vieira K and Vanderlinde JD 2009 The 10 meter South Pole Telescope. *Publ. Astron. Soc. Pac.*, submitted (arXiv:0907.4445).

Castander FJ and the Dark Energy Survey Collaboration 2007 Cosmology with the largest scale structures: probing dark energy. In *Proc. Highlights of Spanish Astrophysics IV* (eds Figueras F, Girart JM, Hernanz M and Jordi C), pp. 193–200. Springer.

Cavaliere A and Fusco-Femiano R 1976 X-rays from hot plasma in clusters of galaxies. *Astron. Astrophys.* **49**, 137–144.

Chandrasekhar S 1931 The maximum mass of ideal white dwarfs. *Astrophys. J.* **74**, 81–82.

Chwolson O 1924 Über eine mögliche Form fiktiver Doppelsterne. *Astron. Nachr.* **221**, 329–330.

Coe D and Moustakas LA 2009 Cosmological constraints from gravitational lens time delays. *Astrophys. J.* **706**, 45–59.

Coil AL, Matheson T, Filippenko AV, Leonard DC, Tonry J, Riess AG, Challis P, Clocchiatti A, Garnavich PM, Hogan CJ, Jha S, Kirshner RP, Leibundgut B, Phillips MM, Schmidt BP, Schommer RA, Smith RC, Soderberg AM, Spyromilio J, Stubbs C, Suntzeff NB and Woudt P 2000 Optical spectra of Type Ia supernovae at $z = 0.46$ and $z = 1.2$. *Astrophys. J.* **544**, L111–L114.

Colless M, Dalton G, Maddox S, Sutherland W, Norberg P, Cole S, Bland-Hawthorn J, Bridges T, Cannon R, Collins C, Couch W, Cross N, Deeley K, De Propris R, Driver SP, Efstathiou G, Ellis RS, Frenk CS, Glazebrook K, Jackson C, Lahav O, Lewis I, Lumsden S, Madgwick D, Peacock JA, Peterson BA, Price I, Seaborne M and Taylor K 2001 The 2dF Galaxy Redshift Survey: spectra and redshifts. *Mon. Not. R. Astron. Soc.* **328**, 1039–1063.

Conley A, Carlberg RG, Guy J, Howell DA, Jha S, Riess AG and Sullivan M 2007 Is there evidence for a Hubble bubble? The nature of Type Ia supernova colors and dust in external galaxies. *Astrophys. J.* **664**, L13–L16.

Contreras C, Hamuy M, Phillips MM, Folatelli G, Suntzeff NB, Persson SE, Stritzinger M, Boldt L, González S, Krzeminski W, Morrell N, Roth M, Salgado F, José Maureira M, Burns CR, Freedman WL, Madore BF, Murphy D, Wyatt P, Li W and Filippenko AV 2010 The Carnegie Supernova Project: first photometry data release of low-redshift Type Ia supernovae. *Astron. J.* **139**, 519–539.

Cooray A and Caldwell RR 2006 Large-scale bulk motions complicate the Hubble diagram. *Phys. Rev. D* **73**, 103002.

Courbin F 2003 Quasar lensing: the observer's point of view. In *Gravitational Lensing: A Unique Tool for Cosmology* (eds Valls-Gabaud D and Kneib J-P). Unpublished (astro-ph/0304497).

Cristiani S, Trentini S, La Franca F, Aretxaga I, Andreani P, Vio R and Gemmo A 1996 The optical variability of QSOs. *Astron. Astrophys.* **306**, 395–407.

Crotty P, Lesgourgues J and Pastor S 2004 Current cosmological bounds on neutrino masses and relativistic relics. *Phys. Rev. D* **69**, 123007.

Cruz M, Cayón L, Martínez-González E, Vielva P and Jin J 2007 The non-Gaussian cold spot in the 3 year Wilkinson Microwave Anisotropy Probe data. *Astrophys. J.* **655**, 11–20.

Dainotti MG, Cardone VF and Capozziello S 2008 A time–luminosity correlation for γ-ray bursts in the X-rays. *Mon. Not. R. Astron. Soc.* **391**, L79–L83.

D'Andrea CB, Sako M, Dilday B, Frieman JA, Holtzman J, Kessler R, Konishi K, Schneider DP, Sollerman J, Wheeler JC, Yasuda N, Cinabro D, Jha S, Nichol RC, Lampeitl H, Smith M, Atlee DW, Bassett B, Castander FJ, Goobar A, Miquel R, Nordin J, Östman L, Prieto JL, Quimby R,

Riess AG and Stritzinger M 2010 Type II-P supernovae from the SDSS-II Supernova Survey and the standardized candle method. *Astrophys. J.* **708**, 661–674.
Davis M and Peebles PJE 1983 Evidence for local anisotropy of the Hubble flow. *Annu. Rev. Astron. Astrophys.* **21**, 109–130.
De Bernardis F, Giusarma E and Melchiorri A 2006 Constraints on dark energy and distance duality from Sunyaev–Zel'dovich effect and Chandra X-ray measurements. *Int'l J. Mod. Phys. D* **15**, 759–766.
De Filippis E, Sereno M, Bautz MW and Longo G 2005 Measuring the three-dimensional structure of galaxy clusters. I. Application to a sample of 25 clusters. *Astrophys. J.* **625**, 108–120.
Dessart L and Hillier DJ 2005 Distance determinations using type II supernovae and the expanding photosphere method. *Astron. Astrophys.* **439**, 671–685.
Dessart L and Hillier DJ 2006 Quantitative spectroscopic analysis of and distance to SN1999em. *Astron. Astrophys.* **447**, 691–707.
Dessart L, Blondin S, Brown PJ, Hicken M, Hillier DJ, Holland ST, Immler S, Kirshner RP, Milne P, Modjaz M and Roming PWA 2008 Using quantitative spectroscopic analysis to determine the properties and distances of Type II plateau supernovae: SN 2005cs and SN 2006bp. *Astrophys. J.* **675**, 644–669.
Dolgov AD 1996 Neutrinos in cosmology. *Nucl. Phys. B Proc. Suppl.* **48**, 5–12.
Dressler A 1984 Internal kinematics of galaxies in clusters. I. Velocity dispersions for elliptical galaxies in Coma and Virgo. *Astrophys. J.* **281**, 512–524.
Dressler A, Faber SM, Burstein D, Davies RL, Lynden-Bell D, Terlevich RJ, Wegner G 1987 Spectroscopy and photometry of elliptical galaxies. A large-scale streaming motion in the local Universe. *Astrophys. J.* **313**, L37–L42.
Dyson FW, Eddington AS and Davidson CR 1920 A determination of the deflection of light by the sun's gravitational field from observations made at the solar eclipse of May 29, 1919. *Phil. Trans. R. Soc. A* **220**, 291–333.
Eastman RG and Kirshner RP 1989 Model atmospheres for SN 1987A and the distance to the Large Magellanic Cloud. *Astrophys. J.* **347**, 771–793.
Eastman RG, Schmidt BP and Kirshner R 1996 The atmospheres of Type II supernovae and the expanding photosphere method. *Astrophys. J.* **466**, 911–937.
Efstathiou G 2005 Effects of destriping errors on estimates of the CMB power spectrum. *Mon. Not. R. Astron. Soc.* **356**, 1549–1558.
Efstathiou G and Bond JR 1999 Cosmic confusion: degeneracies among cosmological parameters derived from measurements of microwave background anisotropies. *Mon. Not. R. Astron. Soc.* **304**, 75–97.
Einstein A 1915 Die Feldgleichungen der Gravitation (The field equations of gravitation). Königlich Preussische Akademie der Wissenschaften, pp. 844–847.
Einstein A 1936 Lenslike action of a star by the deviation of light in the gravitational field. *Science* **84**, 506–507.
Eisenstein DJ, Zehavi I, Hogg DW, Scoccimarro R, Blanton MR, Nichol RC, Scranton R, Seo H-J, Tegmark M, Zheng Z, Anderson SF, Annis J, Bahcall N, Brinkmann J, Burles S, Castander FJ, Connolly A, Csabai I, Doi M, Fukugita M, Frieman JA, Glazebrook K, Gunn JE, Hendry JS, Hennessy G, Ivezić Ž, Kent S, Knapp GR, Lin H, Loh Y-S, Lupton RH, Margon B, McKay TA, Meiksin A, Munn JA, Pope A, Richmond MW, Schlegel D, Schneider DP, Shimasaku K, Stoughton C, Strauss MA, SubbaRao, M, Szalay AS, Szapudi I, Tucker DL, Yanny B and York DG 2005 Detection of the baryonl acoustic peak in the large-scale correlation function of SDSS luminous red galaxies. *Astrophys. J.* **633**, 560–574.
Elias JH, Matthews K, Neugebauer G and Persson SE 1985 Type I supernovae in the infrared and their use as distance indicators. *Astrophys. J.* **296**, 379–389.
Ellis GFR 2007 On the definition of distance in general relativity: I. M. H. Etherington *(Philosophical Magazine Ser. 7, vol. 15, 761 (1933)). Gen. Rel. Grav.* **39**, 1047–1042.
Elmhamdi A, Chugai NN and Danziger IJ 2003 Light curves and Hα luminosities as indicators of ^{56}Ni mass in type II-P supernovae. *Astron. Astrophys.* **404**, 1077–1086.
Emerson JP, Sutherland WJ, McPherson AM, Craig SC, Dalton GB and Ward AK 2004 The Visible & Infrared Survey Telescope for Astronomy. *ESO Messenger* **117**, 27–32.

Etherington IMH 1933 On the definition of distance in general relativity. *Phil. Mag.* **15**, 761–773. (Reprinted in 2007 *Gen. Rel. Grav.* **39**, 1055–1067.)

Falco EE, Gorenstein MV and Shapiro II 1985 On model-dependent bounds on H_0 from gravitational images. Application to Q0957+561A,B. *Astrophys. J.* **289**, L1–L4.

Fassnacht CD, Womble DS, Neugebauer G, Browne IWA, Readhead ACS, Matthews K and Pearson TJ 1996 1608+656: a gravitationally lensed post-starburst radio galaxy. *Astrophys. J.* **460**, L103–L106.

Fassnacht CD, Pearson TJ, Readhead ACS, Browne IWA, Koopmans LVE, Myers ST and Wilkinson PN 1999 A determination of H_0 with the CLASS gravitational lens B1608+656. I. Time delay measurements with the VLA. *Astrophys. J.* **527**, 498–512.

Fassnacht CD, Xanthopoulos E, Koopmans LVE and Rusin D 2002 A determination of H_0 with the CLASS gravitational lens B1608+656. III. A significant improvement in the precision of the time delay measurements. *Astrophys. J.* **581**, 823–835.

Fenimore EE and Ramirez-Ruiz E 2000 Redshifts for 220 BATSE gamma-ray bursts determined by variability and the cosmological consequences. Unpublished (astro-ph/0004176).

Filippenko AV 1997 Optical spectra of supernovae. *Annu. Rev. Astron. Astrophys.* **35**, 309–355.

Filippenko AV 2005 Type Ia supernovae and cosmology. In *White Dwarfs: Cosmological and Galactic Probes* (eds Sion E, Vennes S and Shipman H), *Astrophys. Space Sci. Libr.* **332**, 97–133.

Fixsen DJ, Cheng ES, Gales JM, Mather JC, Shafer RA and Wright EL 1996 The cosmic microwave background spectrum from the full COBE FIRAS data set. *Astrophys. J.* **473**, 576–587.

Foley RJ and Kasen D 2011 Measuring ejecta velocity improves Type Ia supernova distances. *Astrophys. J.* **729**, 55.

Foley RJ, Filippenko AV, Leonard DC, Riess AG, Nugent P and Perlmutter S 2005 A definitive measurement of time dilation in the spectral evolution of the moderate-redshift Type Ia supernova 1997ex. *Astrophys. J.* **626**, L11–L14.

Foley RJ, Filippenko AV, Aguilera C, Becker AC, Blondin S, Challis P, Clocchiatti A, Covarrubias R, Davis TM, Garnavich PM, Jha SW, Kirshner RP, Krisciunas K, Leibundgut B, Li W, Matheson T, Miceli A, Miknaitis G, Pignata G, Rest A, Riess AG, Schmidt BP, Smith RC, Sollerman J, Spyromilio J, Stubbs CW, Suntzeff NB, Tonry JL, Wood-Vasey WM and Zenteno A 2008 Constraining cosmic evolution of Type Ia supernovae. *Astrophys. J.* **684**, 68–87.

Fowler JW, Niemack MD, Dicker SR, Aboobaker AM, Ade PAR, Battistelli ES, Devlin MJ, Fisher RP, Halpern M, Hargrave PC, Hincks AD, Kaul M, Klein J, Lau JM, Limon M, Marriage TA, Mauskopf PD, Page L, Staggs ST, Swetz DS, Switzer ER, Thornton RJ and Tucker CE 2007 Optical design of the Atacama Cosmology Telescope and the Millimeter Bolometric Array Camera. *Appl. Opt.* **46**, 3444–3454.

Fraley GS 1968 Supernovae explosions induced by pair-production instability. *Astrophys. Space Sci.* **2**, 96–114.

Freedman WL and Madore BF 2010 The Hubble constant. *Annu. Rev. Astron. Astrophys.* **48**, 673–710.

Freedman WL, Hughes SM, Madore BF, Mould JR, Lee MG, Stetson P, Kennicutt RC, Turner A, Ferrarese L, Ford H, Graham JA, Hill R, Hoessel JG, Huchra J and Illingworth GD 1994 The Hubble Space Telescope extragalactic distance scale Key Project. I. The discovery of Cepheids and a new distance to M81. *Astrophys. J.* **427**, 628–655.

Freedman WL, Madore BF, Gibson BK, Ferrarese L, Kelson DD, Sakai S, Mould JR, Kennicutt Jr RC, Ford HC, Graham JA, Huchra JP, Hughes SMG, Illingworth GD, Macri LM and Stetson PB 2001 Final results from the Hubble Space Telescope Key Project to measure the Hubble constant. *Astrophys. J.* **553**, 47–72.

Freedman WL, Burns CR, Phillips MM, Wyatt P, Persson SE, Madore BF, Contreras C, Folatelli G, Gonzalez ES, Hamuy M, Hsiao E, Kelson DD, Morrell N, Murphy DC, Roth M, Stritzinger M, Sturch L, Suntzeff NB, Astier P, Balland C, Bassett B, Boldt L, Carlberg RG, Conley AJ, Frieman JA, Garnavich PM, Guy J, Hardin D, Howell DA, Kessler R, Lampeitl H, Marriner J, Pain R, Perrett K, Regnault N, Riess AG, Sako M, Schneider DP, Sullivan M and Wood-Vasey M 2009 The Carnegie Supernova Project: first near-infrared Hubble diagram to $z \sim 0.7$. *Astrophys. J.* **704**, 1036–1058.

Friedmann A 1922 Über die Krümming des Raumes. *Zeitschr. der Phys.* **10**, 377–386.

Friedmann A 1924 Über die Möglichkeit einer Welt mit konstanter negativer Krümmung des Raumes. *Z. Phys.* **21**, 326–332.
Frieman JA, Turner MS and Huterer D 2008 Dark energy and the accelerating Universe. *Annu. Rev. Astron. Astrophys.* **46**, 385–432.
Gal-Yam A, Mazzali P, Ofek EO, Nugent PE, Kulkarni SR, Kasliwal MM, Quimby RM, Filippenko AV, Cenko SB, Chornock R, Waldman R, Kasen D, Sullivan M, Beshore EC, Drake AJ, Thomas RC, Bloom JS, Poznanski D, Miller AA, Foley RJ, Silverman JM, Arcavi I, Ellis RS, Deng J 2009 Supernova 2007bi as a pair-instability explosion. *Nature* **462**, 624–627.
Gamerman D and Lopes HF 1996 *Markov Chain Monte Carlo: Stochastic Simulation for Bayesian Inference*, 2nd edn. Chapman and Hall/CRC Texts in Statistical Science. Boca Raton, FL, Taylor and Francis Group.
Gamezo VN, Khokhlov AM, Oran ES, Chtchelkanova AY and Rosenberg RO 2003 Thermonuclear supernovae: simulations of the deflagration stage and their implications. *Science* **299**, 77–81.
Gehrels N 2010 The Joint Dark Energy Mission (JDEM) Omega. Unpublished (arXiv:1008.4936).
Gehrels N, Chincarini G, Giommi P, Mason KO, Nousek JA, Wells AA, White NE, Barthelmy SD, Burrows DN, Cominsky LR, Hurley KC, Marshall FE, Mészáros P, Roming PWA, Angelini L, Barbier LM, Belloni T, Campana S, Caraveo PA, Chester MM, Citterio O, Cline TL, Cropper MS, Cummings JR, Dean AJ, Feigelson ED, Fenimore EE, Frail DA, Fruchter AS, Garmire GP, Gendreau K, Ghisellini G, Greiner J, Hill JE, Hunsberger SD, Krimm HA, Kulkarni SR, Kumar P, Lebrun F, Lloyd-Ronning NM, Markwardt CB, Mattson BJ, Mushotzky RF, Norris JP, Osborne J, Paczyński B, Palmer DM, Park H.-S, Parsons AM, Paul J, Rees MJ, Reynolds CS, Rhoads JE, Sasseen TP, Schaefer BE, Short AT, Smale AP, Smith IA, Stella L, Tagliaferri G, Takahashi T, Tashiro M, Townsley LK, Tueller J, Turner MJL, Vietri M, Voges W, Ward MJ, Willingale R, Zerbi FM and Zhang WW 2004 The Swift gamma-ray burst mission. *Astrophys. J.* **611**, 1005–1020.
Geller MJ and Huchra JP 1989 The Center for Astrophysics Redshift Survey. Recent results. In *Large Scale Structure and Motions in the Universe* (ed. Mezzetti M), *Astrophys. Space Sci. Libr.* **15**, 3–18.
Ghirlanda G, Ghisellini G, Lazzati D and Firmani C 2004 Gamma-ray bursts: new rulers to measure the Universe. *Astrophys. J.* **613**, L13–L16.
Ghirlanda G, Ghisellini G and Firmani C 2006 Gamma-ray bursts as standard candles to constrain the cosmological parameters. *New J. Phys.* **8**, 123.
Ghisellini G, Ghirlanda G, Tavecchio F, Fraternali F and Pareschi G 2008 Ultra-high energy cosmic rays, spiral galaxies and magnetars. *Mon. Not. R. Astron. Soc.* **390**, L88–L92.
Gilfanov M and Bogdán Á 2010 An upper limit on the contribution of accreting white dwarfs to the Type Ia supernova rate. *Nature* **463**, 924–925.
Goldhaber G, Groom DE, Kim A, Aldering G, Astier P, Conley A, Deustua SE, Ellis R, Fabbro S, Fruchter AS, Goobar A, Hook I, Irwin M, Kim M, Knop RA, Lidman C, McMahon R, Nugent PE, Pain R, Panagia N, Pennypacker CR, Perlmutter S, Ruiz-Lapuente P, Schaefer B, Walton NA and York T 2001 Timescale stretch parameterization of Type Ia supernova *B*-band light curves. *Astrophys. J.* **558**, 359–368.
Gorenstein MV, Cohen NL, Shapiro II, Rogers AEE, Bonometti RJ, Falco EE, Bartel N and Marcaide JM 1988 VLBI observations of the gravitational lens system 0957+561. Structure and relative magnification of the A and B images. *Astrophys. J.* **334**, 42–58.
Gott III JR, Jurić M, Schlegel D, Hoyle F, Vogeley M, Tegmark M, Bahcall N and Brinkmann J 2005 A map of the Universe. *Astrophys. J.* **624**, 463–484.
Granett BR, Szapudi I and Neyrinck MC 2010 Galaxy counts on the cosmic microwave background cold spot. *Astrophys. J.* **714**, 825–833.
Granlund A, Conroy PG, Keller SC, Oates AP, Schmidt B, Waterson MF, Kowald E and Dawson MI 2006 A large-format imager for the SkyMapper Survey Telescope. *Proc. SPIE* **6269**, 626927.
Grassberg EK, Imshennik VS and Nadyozhin DK 1971 On the theory of the light curves of supernovae. *Astrophys. Space Sci.* **10**, 28–51.
Guy J, Astier P, Baumont S, Hardin D, Pain R, Regnault N, Basa S, Carlberg RG, Conley A, Fabbro S, Fouchez D, Hook IM, Howell DA, Perrett K, Pritchet CJ, Rich J, Sullivan M, Antilogus P, Aubourg E, Bazin G, Bronder J, Filiol M, Palanque-Delabrouille N, Ripoche P and Ruhlmann-

Kleider V 2007 SALT2: using distant supernovae to improve the use of Type Ia supernovae as distance indicators. *Astron. Astrophys.* **466**, 11–21.

Haiman Z, Mohr JJ and Holder GP 2001 Constraints on cosmological parameters from future galaxy cluster surveys. *Astrophys. J.* **553**, 545–561.

Hallman EJ, Motl PM, Burns JO and Norman ML 2006 Challenges for Precision cosmology with X-ray and Sunyaev–Zel'dovich effect gas mass measurements of galaxy clusters. *Astrophys. J.* **648**, 852–867.

Hamuy M, Phillips MM, Suntzeff NB, Schommer RA, Maza J, Antezan AR, Wischnjewsky M, Valladares G, Muena C, Gonzales LE, Aviles R, Wells LA, Smith RC, Navarrete M, Covarrubias R, Williger GM, Walker AR, Layden AC, Elias JH, Baldwin JA, Hernandez M, Tirado H, Ugarte P, Elston R, Saavedra N, Barrientos F, Costa E, Lira P, Ruiz M T, Anguita C, Gomez X, Ortiz P, della Valle M, Danziger J, Storm J, Kim Y-C, Bailyn C, Rubenstein EP, Tucker D, Cersosimo S, Mendez RA, Siciliano L, Sherry W, Chaboyer B, Koopmann RA, Geisler D, Sarajedini A, Dey A, Tyson N, Rich RM, Gal R, Lamontagne R, Caldwell N, Guhathakurta P, Phillips AC, Szkody P, Prosser C, Ho LC, McMahan R, Baggley G, Cheng K-P, Havlen R, Wakamatsu K, Janes K, Malkan M, Baganoff F, Seitzer P, Shara M, Sturch C, Hesser J, Hartig ANP, Hughes J, Welch D, Williams TB, Ferguson H, Francis PJ, French L, Bolte M, Roth J, Odewahn S, Howell S and Krzeminski W 1996 *BVRI* light curves for 29 Type Ia supernovae. *Astron. J.* **112**, 2408–2437.

Hamuy MA 2001 Type II supernovae as distance indicators. PhD thesis, The University of Arizona, USA.

Hamuy MA 2004a Observed and physical properties of Type II plateau supernovae. In *Cosmic Explosions in Three Dimensions: Asymmetries in Supernovae and Gamma-Ray Bursts* (eds Höflich P, Kumar P and Wheeler JC), pp. 43–49. Cambridge University Press.

Hamuy MA 2004b The latest version of the standardized candle method for Type II supernovae. In *Measuring and Modeling the Universe* (ed. Freedman WL), online proceedings. *Carnegie Obs.* (astro-ph/0301281).

Hamuy MA and Pinto PA 2002 Type II supernovae as standardized candles. *Astrophys. J.* **566**, L63–L65.

Hannestad S 2006 Neutrinos in cosmology. *Prog. Part. Nucl. Phys.* **57**, 309–323.

Hauschildt PH and Baron E 1999 Numerical solution of the expanding stellar atmosphere problem. *J. Comput. Appl. Math.* **109**, 41–63.

Haynes MP and Giovanelli R 1986 The connection between Pisces–Perseus and the Local Supercluster. *Astrophys. J.* **306**, L55–L59.

Heger A and Woosley SE 2002 The nucleosynthetic signature of Population III. *Astrophys. J.* **567**, 532–543.

Hendry MA, Smartt SJ, Crockett RM, Maund JR, Gal-Yam A, Moon D-S, Cenko SB, Fox DW, Kudritzki RP, Benn CR and Østensen R 2006 SN 2004A: another Type II-P supernova with a red supergiant progenitor. *Mon. Not. R. Astron. Soc.* **369**, 1303–1320.

Hershkowitz S, Linder E and Wagoner RV 1986a Spectral flux from low-density photospheres. Numerical results. *Astrophys. J.* **301**, 220–229.

Hershkowitz S, Linder E and Wagoner RV 1986b Spectral flux from low-density photospheres. Approximate results. *Astrophys. J.* **303**, 800–809.

Hicken M, Garnavich PM, Prieto JL, Blondin S, DePoy DL, Kirshner RP and Parrent J 2007 The luminous and carbon-rich supernova 2006gz: a double degenerate merger? *Astrophys. J.* **669**, L17–L20.

Hicken M, Wood-Vasey WM, Blondin S, Challis P, Jha S, Kelly PL, Rest A and Kirshner RP 2009 Improved dark energy constraints from $\sim$100 new CfA supernova Type Ia light curves. *Astrophys. J.* **700**, 1097–1140.

Hillebrandt W and Niemeyer JC 2000 Type Ia supernova explosion models. *Annu. Rev. Astron. Astrophys.* **38**, 191–230.

Hincks AD, Acquaviva V, Ade P, Aguirre P, Amiri M, Appel JW, Barrientos LF, Battistelli ES, Bond JR, Brown B, Burger B, Chervenak J, Das S, Devlin MJ, Dicker, S, Doriese WB, Dunkley J, Dünner R Essinger-Hileman T, Fisher RP, Fowler JW, Hajian A, Halpern M, Hasselfield M, Hernández-Monteagudo C, Hilton GC, Hilton M, Hlozek R, Huffenberger K, Hughes D, Hughes JP, Infante L, Irwin KD, Jimenez R, Juin JB, Kaul M, Klein J, Kosowsky A, Lau JM, Limon M, Lin Y-T,

Lupton RH, Marriage T, Marsden D, Martocci K, Mauskopf P, Menanteau F, Moodley K, Moseley H, Netterfield CB, Niemack MD, Nolta MR, Page LA, Parker L, Partridge B, Quintana H, Reid B, Sehgal N, Sievers J, Spergel DN, Staggs ST, Stryzak O, Swetz D, Switzer E, Thornton R, Trac H, Tucker C, Verde L, Warne R, Wilson G, Wollack E and Zhao Y 2010 The Atacama Cosmology Telescope (ACT): beam profiles and first SZ cluster maps. *Astrophys. J. Suppl. Ser.* **191**, 423–438.

Holanda RFL, Lima JAS and Ribeiro MB 2010 Testing the distance-duality relation with galaxy clusters and Type Ia supernovae. *Astrophys. J.* **722**, L233–L237.

Homeier NL 2005 The effect of Type Ibc contamination in cosmological supernova samples. *Astrophys. J.* **620**, 12–20.

Hook IM, Howell DA, Aldering G, Amanullah R, Burns MS, Conley A, Deustua SE, Ellis R, Fabbro S, Fadeyev V, Folatelli G, Garavini G, Gibbons R, Goldhaber G, Goobar A, Groom DE, Kim AG, Knop RA, Kowalski M, Lidman C, Nobili S, Nugent PE, Pain R, Pennypacker CR, Perlmutter S, Ruiz-Lapuente P, Sainton G, Schaefer BE, Smith E, Spadafora AL, Stanishev V, Thomas RC, Walton NA, Wang L and Wood-Vasey WM 2005 Spectra of high-redshift Type Ia supernovae and a comparison with their low-redshift counterparts. *Astron. J.* **130**, 2788–2803.

Howell DA, Sullivan M, Perrett K, Bronder TJ, Hook IM, Astier P, Aubourg E, Balam D, Basa S, Carlberg RG, Fabbro S, Fouchez D, Guy J, Lafoux H, Neill JD, Pain R, Palanque-Delabrouille N, Pritchet CJ, Regnault N, Rich J, Taillet R, Knop R, McMahon RG, Perlmutter S and Walton NA 2005 Gemini spectroscopy of supernovae from the Supernova Legacy Survey: improving high-redshift supernova selection and classification. *Astrophys. J.* **634**, 1190–1201.

Howell DA, Sullivan M, Nugent PE, Ellis RS, Conley AJ, Le Borgne D, Carlberg RG, Guy J, Balam D, Basa S, Fouchez D, Hook IM, Hsiao EY, Neill JD, Pain R, Perrett KM and Pritchet CJ 2006 The Type Ia supernova SNLS-03D3bb from a super-Chandrasekhar-mass white dwarf star. *Nature* **443**, 308–311.

Hsiao EY, Conley A, Howell DA, Sullivan M, Pritchet CJ, Carlberg RG, Nugent PE and Phillips MM 2007 *K*-corrections and spectral templates of Type Ia supernovae. *Astrophys. J.* **663**, 1187–1200.

Hu W 2005 Dark energy probes in light of the CMB. In *Observing Dark Energy* (eds Wolff SC and Lauer TD), pp. 215–234. Astronomical Society of the Pacific.

Hubble E 1929a A relation between distance and radial velocity among extra-galactic nebulae. *Publ. Nat'l Acad. Sci. USA* **15**, 168–173.

Hubble EP 1929b A spiral nebula as a stellar system. Messier 31. *Astrophys. J.* **69**, 103–158.

Huchra J 2010 Estimates of the Hubble constant. https://www.cfa.harvard.edu/~huchra/hubble.plot.dat. Accessed 20 December 2010.

Hudson MJ, Smith RJ, Lucey JR and Branchini E 2004 Streaming motions of galaxy clusters within 12,000 km s^{-1}. V. The peculiar velocity field. *Mon. Not. R. Astron. Soc.* **352**, 61–75.

Hui L and Greene PB 2006 Correlated fluctuations in luminosity distance and the importance of peculiar motion in supernova surveys. *Phys. Rev. D* **73**, 123526.

Huterer D, Sasselov DD and Schechter PL 1995 Distances to nearby galaxies: combining fragmentary data using four different methods. *Astron. J.* **110**, 2705–2714.

Inoue AK, Yamazaki R and Nakamura T 2004 Near-infrared colors of gamma-ray burst afterglows and cosmic reionization history. *Astrophys. J.* **601**, 644–653.

Inoue KT and Silk J 2006 Local voids as the origin of large-angle cosmic microwave background anomalies. I. *Astrophys. J.* **648**, 23–30.

Inoue KT and Silk J 2007 Local voids as the origin of large-angle cosmic microwave background anomalies: the effect of a cosmological constant. *Astrophys. J.* **664**, 650–659.

Inoue KT, Sakai N and Tomita K 2010 Evidence of quasi-linear super-structures in the cosmic microwave background and galaxy distribution. *Astrophys. J.* **724**, 12–25.

Jeltema TE, Canizares CR, Bautz MW and Buote DA 2005 The evolution of structure in X-ray clusters of galaxies. *Astrophys. J.* **624**, 606–629.

Jha S, Riess AG and Kirshner RP 2007 Improved distances to Type Ia supernovae with multicolor light-curve shapes: MLCS2k2. *Astrophys. J.* **659**, 122–148.

Jones ME, Edge AC, Grainge K, Grainger WF, Kneissl R, Pooley GG, Saunders R, Miyoshi SJ, Tsuruta T, Yamashita K, Tawara Y, Furuzawa A, Harada A and Hatsukade I 2005 H_0 from an orientation-unbiased sample of Sunyaev–Zel'dovich and X-ray clusters. *Mon. Not. R. Astron. Soc.* **357**, 518–526

Jones MI, Hamuy M, Lira P, Maza J, Clocchiatti A, Phillips M, Morrell N, Roth M, Suntzeff NB, Matheson T, Filippenko AV, Foley RJ and Leonard DC 2009 Distance determination to 12 Type II supernovae using the expanding photosphere method. *Astrophys. J.* **696**, 1176–1194.

Kaiser N 1987 Clustering in real space and in redshift space. *Mon. Not. R. Astron. Soc.* **227**, 1–21.

Kaiser N, Aussel H, Burke BE, Boesgaard H, Chambers K, Chun MR, Heasley JN, Hodapp K-W, Hunt B, Jedicke R, Jewitt D, Kudritzki R, Luppino GA, Maberry M, Magnier E, Monet DG, Onaka PM, Pickles AJ, Rhoads PHH, Simon T, Szalay A, Szapudi I, Tholen DJ, Tonry JL, Waterson M and Wick J 2002 Pan-STARRS: a Large Synoptic Survey Telescope array. *Proc. SPIE* **4836**, 154–164.

Kasen D and Woosley SE 2009 Type II supernovae: model light curves and standard candle relationships. *Astrophys. J.* **703**, 2205–2216.

Kessler R, Becker AC, Cinabro D, Vanderplas J, Frieman JA, Marriner J, Davis TM, Dilday B, Holtzman J, Jha SW, Lampeitl H, Sako M, Smith M, Zheng C, Nichol RC, Bassett B, Bender R, Depoy DL, Doi M, Elson E, Filippenko AV, Foley RJ, Garnavich PM, Hopp U, Ihara Y, Ketzeback W, Kollatschny W, Konishi K, Marshall JL, McMillan RJ, Miknaitis G, Morokuma T, Mörtsell E, Pan K, Prieto JL, Richmond MW, Riess AG, Romani R, Schneider DP, Sollerman J, Takanashi N, Tokita K, van der Heyden K, Wheeler JC, Yasuda N and York D 2009 First-year Sloan Digital Sky Survey-II supernova results: Hubble diagram and cosmological parameters. *Astrophys. J. Suppl. Ser.* **85**, 32–84.

Khokhlov A, Müller E and Höflich P 1993 Light curves of Type Ia supernova models with different explosion mechanisms. *Astron. Astrophys.* **270**, 223–248.

Kirshner RP and Kwan J 1974 Distances to extragalactic supernovae. *Astrophys. J.* **193**, 27–36.

Klebesadel RW, Strong IB and Olson RA 1973 Observations of gamma-ray bursts of cosmic origin. *Astrophys. J.* **182**, L85–L88.

Kneissl R and the APEX-SZ collaboration 2008 APEX-SZ: a Sunyaev–Zel'dovich galaxy cluster survey. In *Einstein's Relativistic Astrophysics Legacy and Cosmology* (eds Aschenbach B, Burwitz V, Hasinger G and Leibundgut B), *ESO Astrophys. Symp. Ser.*, pp. 331–333.

Knop RA, Aldering G, Amanullah R, Astier P, Blanc G, Burns MS, Conley A, Deustua SE, Doi M, Ellis R, Fabbro S, Folatelli G, Fruchter AS, Garavini G, Garmond S, Garton K, Gibbons R, Goldhaber G, Goobar A, Groom DE, Hardin D, Hook I, Howell DA, Kim AG, Lee BC, Lidman C, Mendez J, Nobili S, Nugent PE, Pain R, Panagia N, Pennypacker CR, Perlmutter S, Quimby R, Raux J, Regnault N, Ruiz-Lapuente P, Sainton G, Schaefer B, Schahmaneche K, Smith E, Spadafora AL, Stanishev V, Sullivan M, Walton NA, Wang L, Wood-Vasey WM, Yasuda N and the Supernova Cosmology Project 2003 New constraints on Ω_M, Ω_Λ, and w from an independent set of 11 high-redshift supernovae observed with the Hubble Space Telescope. *Astrophys. J.* **598**, 102–137.

Kochanek CS 2002 What do gravitational lens time delays measure? *Astrophys. J.* **578**, 25–32.

Kocevski DD and Ebeling H 2006 On the origin of the Local Group's peculiar velocity. *Astrophys. J.* **645**, 1043–1053.

Kochanek CS, Morgan ND, Falco EE, McLeod BA, Winn JN, Dembicky J and Ketzeback B 2006 The time delays of gravitational lens HE 0435−1223: an early-type galaxy with a rising rotation curve. *Astrophys. J.* **640**, 47–61.

Komatsu E, Dunkley J, Nolta MR, Bennett CL, Gold B, Hinshaw G, Jarosik N, Larson D, Limon M, Page L, Spergel DN, Halpern M, Hill RS, Kogut A, Meyer SS, Tucker GS, Weiland JL, Wollack E and Wright EL 2009 Five-year Wilkinson Microwave Anisotropy Probe observations: cosmological interpretation. *Astrophys. J. Suppl. Ser.* **180**, 330–376.

Komatsu E, Smith KM, Dunkley J, Bennett CL, Gold B, Hinshaw G, Jarosik N, Larson D, Nolta MR, Page L, Spergel DN, Halpern M, Hill RS, Kogut A, Limon M, Meyer SS, Odegard N, Tucker GS, Weiland JL, Wollack E and Wright EL 2011 Seven-year Wilkinson Microwave Anisotropy Probe (WMAP) observations: cosmological interpretation. *Astrophys. J. Suppl. Ser.* **192**, 18.

Koopmans LVE 2005 Gravitational imaging of cold dark matter substructures. *Mon. Not. R. Astron. Soc.* **363**, 1136–1144.

Koopmans LVE, Treu T, Fassnacht CD, Blandford RD and Surpi G 2003 The Hubble constant from the gravitational lens B1608+656. *Astrophys. J.* **599**, 70–85.

Kowal CT 1968 Absolute magnitudes of supernovae. *Astron. J.* **73**, 1021–1024.

LaRoque SJ 2005 Constraints on cluster structure and cosmology from X-ray and Sunyaev–Zel'dovich effect properties of galaxy clusters. PhD thesis, The University of Chicago, IL, USA.
LaRoque SJ, Bonamente M, Carlstrom JE, Joy MK, Nagai D, Reese ED and Dawson KS 2006 X-ray and Sunyaev–Zel'dovich effect measurements of the gas mass fraction in galaxy clusters. *Astrophys. J.* **652**, 917–936.
Larson D, Dunkley J, Hinshaw G, Komatsu E, Nolta MR, Bennett CL, Gold B, Halpern M, Hill RS, Jarosik N, Kogut A, Limon M, Meyer SS, Odegard N, Page L, Smith KM, Spergel DN, Tucker GS, Weiland JL, Wollack E and Wright EL 2010 Seven-year Wilkinson Microwave Anisotropy Probe (WMAP) observations: power spectra and WMAP-derived parameters. *Astrophys. J. Suppl. Ser.* **192**, 16.
Lavaux G, Tully RB, Mohayaee R and Colombi S 2010 Cosmic flow from Two Micron All-Sky Redshift Survey: the origin of cosmic microwave background dipole and implications for ΛCDM cosmology. *Astrophys. J.* **709**, 483–498.
Leaman J, Li W, Chornock R and Filippenko AV 2011 Nearby supernova rates from the Lick Observatory Supernova Search. I. The methods and database. *Mon. Not. R. Astron. Soc.* **412**, 1419–1440.
Leibundgut B 2001 Cosmological implications from observations of Type Ia supernovae. *Annu. Rev. Astron. Astrophys.* **39**, 67–98.
Lemaître G 1927 Un Univers homogène de masse constante et de rayon croissant rendant compte de la vitesse radiale des nébuleuses extra-galactiques. *Ann. Soc. Sci. Bruxelles* **A47**, 49–59.
Lemaître G 1931 The beginning of the world from the point of view of quantum theory. *Nature* **127**, 706–709.
Lemaître G 1933 L'Univers en expansion. *Ann. Soc. Sci. Bruxelles* **A53**, 51–85.
Lentz EJ, Baron E, Branch D and Hauschildt PH 2001 Non-LTE synthetic spectral fits to the Type Ia supernova 1994D in NGC 4526. *Astrophys. J.* **557**, 266–278.
Leonard DC, Filippenko AV, Ardila DR and Brotherton MS 2001 Is it round? Spectropolarimetry of the Type II-P supernova 1999em. *Astrophys. J.* **553**, 861–885.
Leonard DC, Kanbur SM, Ngeow CC and Tanvir NR 2003 The Cepheid distance to NGC 1637: a direct test of the expanding photosphere method distance to SN 1999em. *Astrophys. J.* **594**, 247–278.
Lidman C, Howell DA, Folatelli G, Garavini G, Nobili S, Aldering G, Amanullah R, Antilogus P, Astier P, Blanc G, Burns MS, Conley A, Deustua SE, Doi M, Ellis R, Fabbro S, Fadeyev V, Gibbons R, Goldhaber G, Goobar A, Groom DE, Hook I, Kashikawa N, Kim AG, Knop RA, Lee BC, Mendez J, Morokuma T, Motohara K, Nugent PE, Pain R, Perlmutter S, Prasad V, Quimby R, Raux J, Regnault N, Ruiz-Lapuente P, Sainton G, Schaefer BE, Schahmaneche K, Smith E, Spadafora AL, Stanishev V, Walton NA, Wang L, Wood-Vasey WM, Yasuda N and the Supernova Cosmology Project 2005 Spectroscopic confirmation of high-redshift supernovae with the ESO VLT. *Astron. Astrophys.* **430**, 843–851.
Linder EV 2008 Mapping the cosmological expansion. *Rep. Prog. Phys.* **71**, 056901.
Litvinova IY and Nadyozhin DK 1985 Determination of integrated parameters for Type II supernovae. *Sov. Astron. Lett.* **11**, 145–147.
Lucey J, Radburn-Smith D and Hudson M 2005 Beta, local SN Ia data and the Great Attractor. In *Nearby Large-Scale Structures and the Zone of Avoidance* (eds Fairall KP and Woudt PA), pp. 21–26. Astronomical Society of the Pacific.
Lynden-Bell D, Faber SM, Burstein D, Davies RL, Dressler A, Terlevich RJ and Wegner G 1988 Spectroscopy and photometry of elliptical galaxies. V. Galaxy streaming toward the new supergalactic center. *Astrophys. J.* **326**, 19–49.
Maguire K, Kotak R, Smartt SJ, Pastorello A, Hamuy M and Bufano F 2010 Type II-P supernovae as standardized candles: improvements using near-infrared data. *Mon. Not. R. Astron. Soc.* **403**, L11–L15.
Mandelbaum R, Seljak U, Hirata CM, Bardelli S, Bolzonella M, Bongiorno A, Carollo M, Contini T, Cunha CE, Garilli B, Iovino A, Kampczyk P, Kneib J-P, Knobel C, Koo DC, Lamareille F, Le Fèvre O, Le Borgne J-F, Lilly SJ, Maier C, Mainieri V, Mignoli M, Newman JA, Oesch PA, Perez-Montero E, Ricciardelli E, Scodeggio M, Silverman J and Tasca L 2008 Precision photometric redshift calibration for galaxy–galaxy weak lensing. *Mon. Not. R. Astron. Soc.* **386**, 781–806.
Marriage TA, Acquaviva V, Ade PAR, Aguirre P, Amiri M, Appel JW, Barrientos LF, Battistelli ES, Bond JR, Brown B, Burger B, Chervenak J, Das S, Devlin MJ, Dicker SR, Doriese WB, Dunkley

J, Dunner R, Essinger-Hileman T, Fisher RP, Fowler JW, Hajian A, Halpern M, Hasselfield M, Hernández-Monteagudo C, Hilton GC, Hilton M, Hincks AD, Hlozek R, Huffenberger KM, Hughes DH, Hughes JP, Infante L, Irwin KD, Juin JB, Kaul M, Klein J, Kosowsky A, Lau JM, Limon M, Lin Y-T, Lupton RH, Marsden D, Martocci K, Mauskopf P, Menanteau F, Moodley K, Moseley H, Netterfield CB, Niemack MD, Nolta MR, Page LA, Parker L, Partridge B, Quintana H, Reese ED, Reid B, Sehgal N, Sherwin BD, Sievers J, Spergel DN, Staggs ST, Swetz DS, Switzer ER, Thornton R, Trac H, Tucker C, Warne R, Wilson G, Wollack E and Zhao Y 2011 The Atacama Cosmology Telescope: Sunyaev–Zel'dovich selected galaxy clusters at 148 GHz in the 2008 survey. *Astrophys. J.* 731, 100.

Mason BS, Myers ST and Readhead ACS 2001 A measurement of H_0 from the Sunyaev–Zel'dovich effect. *Astrophys. J.* **555**, L11–L15.

Matheson T, Blondin S, Foley RJ, Chornock R, Filippenko AV, Leibundgut B, Smith RC, Sollerman J, Spyromilio J, Kirshner RP, Clocchiatti A, Aguilera C, Barris B, Becker AC, Challis P, Covarrubias R, Garnavich P, Hicken M, Jha S, Krisciunas K, Li W, Miceli A, Miknaitis G, Prieto JL, Rest A, Riess AG, Salvo ME, Schmidt BP, Stubbs CW, Suntzeff NB and Tonry JL 2005 Spectroscopy of high-redshift supernovae from the ESSENCE project: the first 2 years. *Astron. J.* **129**, 2352–2375.

Meegan CA, Fishman GJ, Wilson RB, Horack JM, Brock MN, Paciesas WS, Pendleton GN and Kouveliotou C 1992 Spatial distribution of gamma-ray bursts observed by BATSE. *Nature* **355**, 143–145.

Menanteau F, González J, Juin J-B, Marriage TA, Reese ED, Acquaviva V, Aguirre P, Appel JW, Baker AJ, Barrientos LF, Battistelli ES, Bond JR, Das S, Deshpande AJ, Devlin MJ, Dicker S, Dunkley J, Dünner R, Essinger-Hileman T, Fowler JW, Hajian A, Halpern M, Hasselfield M, Hernández-Monteagudo C, Hilton M, Hincks AD, Hlozek R, Huffenberger KM, Hughes JP, Infante L, Irwin KD, Klein J, Kosowsky A, Lin Y-T, Marsden D, Moodley K, Niemack MD, Nolta MR, Page LA, Parker L, Partridge B, Sehgal N, Sievers J, Spergel DN, Staggs ST, Swetz D, Switzer E, Thornton R, Trac H, Warne R and Wollack E 2010 The Atacama Cosmology Telescope: physical properties and purity of a galaxy cluster sample selected via the Sunyaev–Zel'dovich effect. *Astrophys. J.* **723**, 1523–1541.

Minkowski R 1941 Spectra of supernovae. *Publ. Astron. Soc. Pac.* **53**, 224–225.

Misner CW, Thorne KS and Wheeler JA 1973 *Gravitation*. WH Freeman and Co.

Mitchell RC, Baron E, Branch D, Hauschildt PH, Nugent PE, Lundqvist P, Blinnikov S and Pun CSJ 2002 Detailed spectroscopic analysis of SN 1987A: the distance to the Large Magellanic Cloud using the spectral-fitting expanding atmosphere method. *Astrophys. J.* **574**, 293–305.

Mohr JJ, Mathiesen B and Evrard AE 1999 Properties of the intracluster medium in an ensemble of nearby galaxy clusters. *Astrophys. J.* **517**, 627–649.

Mould JR, Huchra JP, Freedman WL, Kennicutt Jr RC Ferrarese L, Ford HC, Gibson BK, Graham JA, Hughes SMG, Illingworth GD, Kelson DD, Macri LM, Madore BF, Sakai S, Sebo KM, Silbermann NA and Stetson PB 2000 The Hubble Space Telescope Key Project on the extragalactic distance scale. XXVIII. Combining the constraints on the Hubble constant. *Astrophys. J.* **529**, 786–794.

Myers ST 1999 Scaling the Universe: gravitational lenses and the Hubble constant. *Publ. Nat'l Acad. Sci. USA* **96**. 4236–4239.

Myers ST, Baker JE, Readhead ACS, Leitch EM and Herbig T 1997 Measurements of the Sunyaev–Zel'dovich effect in the nearby clusters A478, A2142, and A2256. *Astrophys. J.* **485**, 1–21.

Nadyozhin DK 2003 Explosion energies, nickel masses and distances of Type II plateau supernovae. *Mon. Not. R. Astron. Soc.* **346**, 97–104.

Nakar E 2007 Short-hard gamma-ray bursts. *Phys. Rep.* **442**, 166–236.

Nesseris S and Perivolaropoulos L 2007 Tension and systematics in the Gold06 SN Ia data set. *J. Cosmol. Astropart. Phys.* **02**, 025.

Norris JP, Marani GF and Bonnell JT 2000 Connection between energy-dependent lags and peak luminosity in gamma-ray bursts. *Astrophys. J.* **534**, 248–257.

Nugent P, Kim A and Perlmutter S 2002 *K*-corrections and extinction corrections for Type Ia supernovae. *Publ. Astron. Soc. Pac.* **114**, 803–819.

Nugent P, Sullivan M, Ellis R, Gal-Yam A, Leonard DC, Howell DA, Astier P, Carlberg RG, Conley A, Fabbro S, Fouchez D, Neill JD, Pain R, Perrett K, Pritchet CJ and Regnault N 2006 Toward a cosmological Hubble diagram for Type II-P supernovae. *Astrophys. J.* **645**, 841–850.

Olivares EF, Hamuy M, Pignata G, Maza J, Bersten M, Phillips MM, Suntzeff NB, Filippenko AV, Morrel NI, Kirshner RP and Matheson T 2010 The standardized candle method for Type II plateau supernovae. *Astrophys. J.* **715**, 833–853.

Olling RP 2007 Accurate extragalactic distances and dark energy: anchoring the distance scale with rotational parallaxes. *Mon. Not. R. Astron. Soc.* **378**, 1385–1399.

Ostriker JP and Vishniac ET 1986 Generation of microwave background fluctuations from nonlinear perturbations at the era of galaxy formation. *Astrophys. J.* **306**, L51–L54.

Paraficz D and Hjorth J 2009 Gravitational lenses as cosmic rulers: Ω_M, Ω_Λ from time delays and velocity dispersions. *Astron. Astrophys.* **507**, L49–L52.

Peebles PJE and Yu JT 1970 Primeval adiabatic perturbation in an expanding Universe. *Astrophys. J.* **162**, 815–836.

Perlmutter S 2003 Supernovae, dark energy, and the accelerating Universe. *Phys. Today* **56**, 53–60.

Perlmutter S, Gabi S, Goldhaber G, Goobar A, Groom DE, Hook IM, Kim AG, Kim MY, Lee JC, Pain R, Pennypacker CR, Small IA, Ellis RS, McMahon RG, Boyle BJ, Bunclark PS, Carter D, Irwin MJ, Glazebrook K, Newberg HJM, Filippenko AV, Matheson T, Dopita M, Couch WJ and the Supernova Cosmology Project 1997 Measurements of the cosmological parameters Ω and Λ from the first seven supernovae at $z \geq 0.35$. *Astrophys. J.* **483**, 565–581.

Perlmutter S, Aldering G, Goldhaber G, Knop RA, Nugent P, Castro PG, Deustua S, Fabbro S, Goobar A, Groom DE, Hook IM, Kim AG, Kim MY, Lee JC, Nunes NJ, Pain R, Pennypacker CR, Quimby R, Lidman C, Ellis RS, Irwin M, McMahon RG, Ruiz-Lapuente P, Walton N, Schaefer B, Boyle BJ, Filippenko AV, Matheson T, Fruchter AS, Panagia N, Newberg HJM, Couch WJ and the Supernova Cosmology Project 1999 Measurements of Ω and Λ from 42 high-redshift supernovae. *Astrophys. J.* **517**, 565–586.

Phillips MM 1993 The absolute magnitudes of Type Ia supernovae. *Astrophys. J.* **413**, L105–L108.

Phillips MM, Lira P, Suntzeff NB, Schommer RA, Hamuy M and Maza J 1999 The reddening-free decline rate versus luminosity relationship for Type Ia supernovae. *Astron. J.* **118**, 1766–1776.

Piran T 1992 The implications of the Compton (GRO) observations for cosmological gamma-ray bursts. *Astrophys. J.* **389**, L45–L48.

Pirsig RM 1974 *Zen and the Art of Motorcycle Maintenance: An Inquiry into Values*. William Morrow & Co.

Pontzen A, Deason A, Governato F, Pettini M, Wadsley J, Quinn T, Brooks A, Bellovary J and Fynbo JPU 2010 The nature of HI absorbers in gamma-ray burst afterglows: clues from hydrodynamic simulations. *Mon. Not. R. Astron. Soc.* **402**, 1523–1535.

Porciani C, Viel M and Lilly SJ 2007 Strong MgII systems in quasar and gamma-ray burst spectra. *Astrophys. J.* **659**, 218–224.

Poznanski D, Butler N, Filippenko AV, Ganeshalingam M, Li W, Bloom JS, Chornock R, Foley RJ, Nugent PE, Silverman JM, Cenko SB, Gates EL, Leonard DC, Miller AA, Modjaz M, Serduke FJD, Smith N, Swift BJ and Wong DS 2009 Improved standardization of Type II-P supernovae: application to an expanded sample. *Astrophys. J.* **694**, 1067–1079.

Poznanski D, Nugent PE and Filippenko AV 2010 Type II-P Supernovae as standard candles: the SDSS-II sample revisited. *Astrophys. J.* **721**, 956–959.

Predehl P, Hasinger G, Böhringer H, Briel U, Brunner H, Churazov E, Freyberg M, Friedrich P, Kendziorra E, Lutz D, Meidinger N, Pavlinsky M, Pfeffermann E, Santangelo A, Schmitt J, Schuecker P, Schwope A, Steinmetz M, Strüder L Sunyaev R and Wilms J 2006 eROSITA. *Proc. SPIE* **6266**, 62660P.

Pskovskii YuP 1984 Photometric classification and basic parameters of type I supernovae. *Sov. Astron.* **28**, 658–664.

Ramella M, Geller MJ and Huchra JP 1992 The distribution of galaxies within the 'Great Wall'. *Astrophys. J.* **384**, 396–403.

Rau A, Kulkarni SR, Law NM, Bloom JS, Ciardi D, Djorgovski GS, Fox DB, Gal-Yam A, Grillmair CC, Kasliwal MM, Nugent PE, Ofek EO, Quimby RM, Reach WT, Shara M, Bildsten L, Cenko SB, Drake AJ, Filippenko AV, Helfand DJ, Helou G, Howell DA, Poznanski D and Sullivan M 2009 Exploring the optical transient sky with the Palomar Transient Factory. *Publ. Astron. Soc. Pac.* **121**, 1334–1351.

Reese ED 2004 Measuring the Hubble constant with the Sunyaev–Zel'dovich effect. In *Measuring and Modeling the Universe* (ed. Freedman WL), *Carnegie Obs. Astrophys. Ser.*, pp. 138–158. Cambridge University Press.

Reese ED, Mohr JJ, Carlstrom JE, Joy M, Grego L, Holder GP, Holzapfel WL, Hughes JP, Patel SK and Donahue M 2000 Sunyaev–Zel'dovich effect-derived distances to the high-redshift clusters MS 0451.6−0305 and CL 0016+16. *Astrophys. J.* **533**, 38–49.

Reese ED, Carlstrom JE, Joy M, Mohr JJ, Grego L and Holzapfel WL 2002 Determining the cosmic distance scale from interferometric measurements of the Sunyaev–Zel'dovich effect. *Astrophys. J.* **581**, 53–85.

Reese ED, Kawahara H, Kitayama T, Ota N, Sasaki S and Suto Y 2010 Impact of Chandra calibration uncertainties on galaxy cluster temperatures: application to the Hubble constant. *Astrophys. J.* **721**, 653–669.

Refsdal S 1964 On the possibility of determining Hubble's parameter and the masses of galaxies from the gravitational lens effect. *Mon. Not. R. Astron. Soc.* **128**, 307–310.

Reichart DE, Lamb DQ, Fenimore EE, Ramirez-Ruiz E, Cline TL and Hurley K 2001 A possible Cepheid-like luminosity estimator for the long gamma-ray bursts. *Astrophys. J.* **552**, 57–71.

Rhoads JE 1997 How to tell a jet from a balloon: a proposed test for beaming in gamma-ray bursts. *Astrophys. J.* **487**, L1–L4.

Riess AG, Press WH and Kirshner RP 1996 A precise distance indicator: Type Ia supernova multicolor light-curve shapes. *Astrophys. J.* **473**, 88–109.

Riess AG, Filippenko AV, Challis P, Clocchiatti A, Diercks A, Garnavich PM, Gilliland RL, Hogan CJ, Jha S, Kirshner RP, Leibundgut B, Phillips MM, Reiss D, Schmidt BP, Schommer RA, Smith RC, Spyromilio J, Stubbs C, Suntzeff NB and Tonry J 1998 Observational evidence from supernovae for an accelerating Universe and a cosmological constant. *Astron. J.* **116**, 1009–1038.

Riess AG, Macri L, Casertano S, Sosey M, Lampeitl H, Ferguson HC, Filippenko AV, Jha SW, Li W, Chornock R and Sarkar D 2009a A redetermination of the Hubble constant with the Hubble Space Telescope from a differential distance ladder. *Astrophys. J.* **699**, 539–563.

Riess AG, Macri L, Li W, Lampeitl H, Casertano S, Ferguson HC, Filippenko AV, Jha SW, Chornock R, Greenhill L, Mutchler M, Ganeshalingham M and Hicken M 2009b Cepheid calibrations of modern Type Ia supernovae: implications for the Hubble constant. *Astrophys. J. Suppl. Ser.* **183**, 109–141.

Robertson HP 1935 Kinematics and world-structure. *Astrophys. J.* **82**, 284–301.

Robertson HP 1936a Kinematics and world-structure. II. *Astrophys. J.* **83**, 187–201.

Robertson HP 1936b Kinematics and world-structure. III. *Astrophys. J.* **83**, 257–271.

Sachs RK and Wolfe AM 1967 Perturbations of a cosmological model and angular variations of the microwave background. *Astrophys. J.* **147**, 73–90.

Saha A, Sandage A, Tammann GA, Dolphin AE, Christensen J, Panagia N and Macchetto FD 2001 Cepheid calibration of the peak brightness of Type Ia supernovae. XI. SN 1998aq in NGC 3982. *Astrophys. J.* **562**, 314–336.

Sandage A, Tammann GA, Saha A, Reindl B, Macchetto FD and Panagia N 2006 The Hubble constant: a summary of the Hubble Space Telescope program for the luminosity calibration of Type Ia supernovae by means of Cepheids. *Astrophys. J.* **653**, 843–860.

Sari R, Piran T and Halpern JP 1999 Jets in gamma-ray bursts. *Astrophys. J.* **519**, L17–L20.

Sazonov SY and Sunyaev RA 1999 Microwave polarization in the direction of galaxy clusters induced by the CMB quadrupole anisotropy. *Mon. Not. R. Astron. Soc.* **310**, 765–772.

Schaefer BE 2003 Explaining the gamma-ray burst $E_{\rm peak}$ distribution. *Astrophys. J.* **583**, L71–L74.

Schaefer BE 2007 The Hubble diagram to redshift >6 from 69 gamma-ray bursts. *Astrophys. J.* **660**, 16–46.

Schechter PL 2005 The Hubble constant from gravitational lens time delays. In *Gravitational Lensing Impact on Cosmology* (eds Mellier Y and Meylan G), *Proc. Int'l Astron. Union. Symp.* **225**, 281–296.

Schmidt BP, Kirshner RP and Eastman RG 1992 Expanding photospheres of Type II supernovae and the extragalactic distance scale. *Astrophys. J.* **395**, 366–386.

Schmidt BP, Kirshner RP, Eastman RG, Phillips MM, Suntzeff NB, Hamuy M, Maza J and Aviles R 1994 The distances to five Type II supernovae using the expanding photosphere method, and the value of H_0. *Astrophys. J.* **432**, 42–48.

Schwan D, Bertoldi F, Cho S, Dobbs M, Guesten R, Halverson NW, Holzapfel WL, Kreysa E, Lanting TM, Lee AT, Lueker M, Mehl J, Menten K, Muders D, Myers M, Plagge T, Raccanelli A, Schilke P, Richards PL, Spieler H and White M 2003 APEX-SZ: a Sunyaev–Zel'dovich galaxy cluster survey. *New Astron. Rev.* **47**, 933–937.

Seljak U, McDonald P and Makarov A 2003 Cosmological constraints from the cosmic microwave background and Lyman α forest revisited. *Mon. Not. R. Astron. Soc.* **342**, L79–L84.

Shahmoradi A and Nemiroff R 2009 How real detector thresholds create false standard candles. In *Gamma-Ray Bursts* (eds Meegan C, Kouveliotou C and Gehrels N), *AIP Conf. Proc.* **1133**, 425–427.

Shapiro II 1964 Fourth test of general relativity. *Phys. Rev. Lett.* **13**, 789–791.

Shapley H 1919 On the existence of external galaxies. *Publ. Astron. Soc. Pac.* **31**, 261–268.

Silk J and White SDM 1978 The determination of Q_0 using X-ray and microwave observations of galaxy clusters. *Astrophys. J.* **226**, L103–L106.

Smartt SJ 2009 Progenitors of core-collapse supernovae. *Annu. Rev. Astron. Astrophys.* **47**, 63–106.

Smartt SJ, Eldridge JJ, Crockett RM and Maund JR 2009 The death of massive stars. I. Observational constraints on the progenitors of Type II-P supernovae. *Mon. Not. R. Astron. Soc.* **395**, 1409–1437.

Spergel DN, Verde L, Peiris HV, Komatsu E, Nolta MR, Bennett CL, Halpern M, Hinshaw G, Jarosik N, Kogut A, Limon M, Meyer SS, Page L, Tucker GS, Weiland JL, Wollack E and Wright EL 2003 First-year Wilkinson Microwave Anisotropy Probe (WMAP) observations: determination of cosmological parameters. *Astrophys. J. Suppl. Ser.* **148**, 175–194.

Sulkanen ME 1999 Galaxy cluster shapes and systematic errors in H_0 measured by the Sunyaev–Zel'dovich effect. *Astrophys. J.* **522**, 59–65.

Sunyaev RA and Zel'dovich YaB 1969 The interaction of matter and radiation in a hot-model Universe. *Astrophys. Space Sci.* **4**, 301–316.

Sunyaev RA and Zel'dovich YaB 1970 Small-scale fluctuations of relic radiation. *Astrophys. Space Sci.* **7**, 3–19.

Sunyaev RA and Zel'dovich YaB 1972 The observations of relic radiation as a test of the nature of X-ray radiation from the clusters of galaxies. *Comm. Astrophys. Space Phys.* **4**, 173–178.

Sunyaev RA and Zel'dovich YaB 1980 Microwave background radiation as a probe of the contemporary structure and history of the Universe. *Annu. Rev. Astron. Astrophys.* **18**, 537–560.

Suyu SH, Marshall PJ, Hobson MP and Blandford RD 2006 A Bayesian analysis of regularized source inversions in gravitational lensing. *Mon. Not. R. Astron. Soc.* **371**, 983–998.

Suyu SH, Marshall PJ, Blandford RD, Fassnacht CD, Koopmans LVE, McKean JP and Treu T 2009 Dissecting the gravitational lens B1608+656. I. Lens potential reconstruction. *Astrophys. J.* **691**, 277–298.

Suyu SH, Marshall PJ, Auger MW, Hilbert S, Blandford RD, Koopmans LVE, Fassnacht CD and Treu T 2010 Dissecting the gravitational lens B1608+656. II. Precision measurements of the Hubble constant, spatial curvature, and the dark energy equation of state. *Astrophys. J.* **711**, 201–221.

Swetz DS, Ade PAR, Amiri M, Appel JW, Battistelli ES, Burger B, Chervenak J, Devlin MJ, Dicker SR, Doriese WB, Dünner R, Essinger-Hileman T, Fisher RP, Fowler JW, Halpern M, Hasselfield M, Hilton GC, Hincks AD, Irwin KD, Jarosik N, Kaul M, Klein J, Lau JM, Limon M, Marriage TA, Marsden D, Martocci K, Mauskopf P, Moseley H, Netterfield CB, Niemack MD, Nolta, MR, Page LA, Parker L, Staggs ST, Stryzak O, Switzer ER, Thornton R, Tucker C, Wollack E and Zhao Y 2010 The Atacama Cosmology Telescope: the receiver and instrumentation. *Astrophys. J. Suppl. Ser.*, submitted (arXiv:1007.0290).

Tammann GA, Sandage A and Reindl B 2008 Comparison of distances from RR Lyrae stars, the tip of the red giant branch, and classical Cepheids. *Astrophys. J.* **679**, 52–70.

Tanaka M, Kawabata KS, Yamanaka M, Maeda K, Hattori T, Aoki K, Nomoto K, Iye M, Sasaki T, Mazzali PA and Pian E 2010 Spectropolarimetry of the extremely luminous Type Ia supernova 2009dc: nearly spherical explosion of super-Chandrasekhar mass white dwarf. *Astrophys. J.* **714**, 1209–1216.

Tonry JL, Blakeslee JP, Ajhar EA and Dressler A 2000 The surface brightness fluctuation survey of galaxy distances. II. Local and large-scale flows. *Astrophys. J.* **530**, 625–651.

Tonry JL, Schmidt BP, Barris B, Candia P, Challis P, Clocchiatti A, Coil AL, Filippenko AV, Garnavich P, Hogan C, Holland ST, Jha S, Kirshner RP, Krisciunas K, Leibundgut B, Li W, Matheson T, Phillips MM, Riess AG, Schommer R, Smith RC, Sollerman J, Spyromilio J, Stubbs CW and Suntzeff NB 2003 Cosmological results from high-z supernovae. *Astrophys. J.* **594**, 1–24.

Tully RB 1986 Alignment of clusters and galaxies on scales up to $0.1c$. *Astrophys. J.* **303**, 25–38.

Tyson JA, Wittman DM, Hennawi JF and Spergel DN 2003 LSST: a complementary probe of dark energy. *Nucl. Phys. B Proc. Suppl.* **124**, 21–29.

Udomprasert PS, Mason BS, Readhead ACS and Pearson TJ 2004 An unbiased measurement of H_0 through Cosmic Background Imager observations of the Sunyaev–Zel'dovich effect in nearby galaxy clusters. *Astrophys. J.* **615**, 63–81.

Utrobin VP 2007 An optimal hydrodynamic model for the normal Type IIP supernova 1999em. *Astron. Astrophys.* **461**, 233–251.

Uzan JP, Aghanim N and Mellier Y 2004 Distance duality relation from X-ray and Sunyaev–Zel'dovich observations of clusters. *Phys. Rev. D* **70**, 083533.

Vanden Berk DE, Wilhite BC, Kron RG, Anderson SF, Brunner RJ, Hall PB, Ivezić Ž, Richards GT, Schneider DP, York DG, Brinkmann JV, Lamb DQ, Nichol RC and Schlegel DJ 2004 The ensemble photometric variability of ~25,000 quasars in the Sloan Digital Sky Survey. *Astrophys. J.* **601**, 692–714.

Vanderlinde K, Crawford TM, de Haan T, Dudley JP, Shaw L, Ade PAR, Aird KA, Benson BA, Bleem LE, Brodwin M, Carlstrom JE, Chang CL, Crites AT, Desai S, Dobbs MA, Foley RJ, George EM, Gladders MD, Hall NR, Halverson NW, High FW, Holder GP, Holzapfel WL, Hrubes JD, Joy M, Keisler R, Knox L, Lee AT, Leitch EM, Loehr A, Lueker M, Marrone DP, McMahon JJ, Mehl J, Meyer SS, Mohr JJ, Montroy TE, Ngeow C-C, Padin S, Plagge T, Pryke C, Reichardt CL, Rest A, Ruel J, Ruhl JE, Schaffer KK, Shirokoff E, Song J, Spieler HG, Stalder B, Staniszewski Z, Stark AA, Stubbs CW, van Engelen A, Vieira JD, Williamson R, Yang Y, Zahn O and Zenteno A 2010 Galaxy clusters selected with the Sunyaev–Zel'dovich effect from 2008 South Pole Telescope observations. *Astrophys. J.* **722**, 1180–1196.

van Kerkwijk MH, Chang P and Justham S 2010 Sub-Chandrasekhar white dwarf mergers as the progenitors of Type Ia supernovae. *Astrophys. J.* **722**, L157–L161.

Verde L, Haiman Z and Spergel DN 2002 Are clusters standard candles? Galaxy cluster scaling relations with the Sunyaev–Zel'dovich effect. *Astrophys. J.* **581**, 5–19.

Vielva P, Martínez-González E, Barreiro RB, Sanz JL and Cayón L 2004 Detection of non-Gaussianity in the Wilkinson Microwave Anisotropy Probe first-year data using spherical wavelets. *Astrophys. J.* **609**, 22–34.

Vikhlinin A, Markevitch M, Murray SS, Jones C, Forman W and Van Speybroeck L 2005 Chandra temperature profiles for a sample of nearby relaxed galaxy clusters. *Astrophys. J.* **628**, 655–672.

Vuissoz C, Courbin F, Sluse D, Meylan G, Ibrahimov M, Asfandiyarov I, Stoops E, Eigenbrod A, Le Guillou L, van Winckel H and Magain P 2007 COSMOGRAIL: the COSmological MOnitoring of GRAvItational Lenses. V. The time delay in SDSS J1650+4251. *Astron. Astrophys.* **464**, 845–851.

Vuissoz C, Courbin F, Sluse D, Meylan G, Chantry V, Eulaers E, Morgan C, Eyler ME, Kochanek CS, Coles J, Saha P, Magain P and Falco EE 2008 COSMOGRAIL: the COSmological MOnitoring of GRAvItational Lenses. VII. Time delays and the Hubble constant from WFI J2033−4723. *Astron. Astrophys.* **488**, 481–490.

Walker AG 1937 On Milne's theory of world-structure. *Proc. London Math. Soc.* **42**, 90–127.

Walsh D, Carswell RF and Weymann RJ 1979 0957+561A,B: twin quasistellar objects or gravitational lens? *Nature* **279**, 381–384.

Wang Y-G and Fan Z-H 2006 Systematic errors in the determination of the Hubble constant due to the asphericity and nonisothermality of clusters of galaxies. *Astrophys. J.* **643**, 630–640.

Weinberg S 1972 *Gravitation and Cosmology: Principles and Applications of the General Theory of Relativity*. Wiley-VCH.

Wheeler JC and Benetti S 2000 Supernovae. In *Allen's Astrophysical Quantities* (ed. Cox AN), 4th edn. American Institute of Physics Press/Springer.

Wheeler JC and Harkness RP 1990 Type I supernovae. *Rep. Prog. Phys.* **53**, 1467–1557.

Whelan J and Iben Jr I 1973 Binaries and supernovae of Type I. *Astrophys. J.* **186**, 1007–1014.

Wilson OC 1939 Possible applications of supernovae to the study of the nebular red shifts. *Astrophys. J.* **90**, 634–636.

Wood-Vasey WM, Miknaitis G, Stubbs CW, Jha S, Riess AG, Garnavich PM, Kirshner RP, Aguilera C, Becker AC, Blackman JW, Blondin S, Challis P, Clocchiatti A, Conley A, Covarrubias R, Davis TM, Filippenko AV, Foley RJ, Garg A, Hicken M, Krisciunas K, Leibundgut B, Li W, Matheson T, Miceli A, Narayan G, Pignata G, Prieto JL, Rest A, Salvo ME, Schmidt BP, Smith RC, Sollerman J, Spyromilio J, Tonry JL, Suntzeff NB and Zenteno A. 2007 Observational constraints on the nature of dark energy: first cosmological results from the ESSENCE supernova survey. *Astrophys. J.* **666**, 694–715.

Wood-Vasey WM, Friedman AS, Bloom JS, Hicken M, Modjaz M, Kirshner RP, Starr DL, Blake CH, Falco EE, Szentgyorgyi AH, Challis P, Blondin S, Mandel KS and Rest A 2008 Type Ia supernovae are good standard candles in the near infrared: evidence from PAIRITEL. *Astrophys. J.* **689**, 377–390.

Woosley SE and Bloom JS 2006 The supernova gamma-ray burst connection. *Annu. Rev. Astron. Astrophys.* **44**, 507–556.

Woudt PA, Kraan-Korteweg RC, Lucey J, Fairall AP and Moore SAW 2008 The Norma cluster (ACO 3627). I. A dynamical analysis of the most massive cluster in the Great Attractor. *Mon. Not. R. Astron. Soc.* **383**, 445–457.

Wright EL 2004 Theoretical Overview of Cosmic Microwave Background Anisotropy. In *Measuring and Modeling the Universe* (ed Freedman WL), *Carnegie Obs. Astrophys. Ser.*, pp. 291–308. Cambridge University Press.

Wucknitz O 2002 Degeneracies and scaling relations in general power-law models for gravitational lenses. *Mon. Not. R. Astron. Soc.* **332**, 951–961.

Wucknitz O, Biggs AD and Browne IWA 2004 Models for the lens and source of B0218+357: a LENSCLEAN approach to determine H_0. *Mon. Not. R. Astron. Soc.* **349**, 14–30.

Wyithe S, Oh SP and Pindor B 2010 A gravitational lensing explanation for the excess of strong MgII absorbers in GRB afterglow spectra. *Mon. Not. R. Astron. Soc.*, submitted (arXiv:1004.2081).

Yamazaki R 2009 Prior emission model for X-ray plateau phase of gamma-ray burst afterglows. *Astrophys. J.* **690**, L118–L121.

Yonetoku D, Murakami T, Nakamura T, Yamazaki R, Inoue AK and Ioka K 2004 Gamma-ray burst formation rate inferred from the spectral peak energy–peak luminosity relation. *Astrophys. J.* **609**, 935–951.

York T, Jackson N, Browne IWA, Wucknitz O and Skelton JE 2005 The Hubble constant from the gravitational lens CLASS B0218+357 using the Advanced Camera for Surveys. *Mon. Not. R. Astron. Soc.* **357**, 124–134.

Young DR, Smartt SJ, Valenti S, Pastorello A, Benetti S, Benn CR, Bersier D, Botticella MT, Corradi RLM, Harutyunyan AH, Hrudkova M, Hunter I, Mattila S, de Mooij EJW, Navasardyan H, Snellen IAG, Tanvir NR and Zampieri L 2010 Two Type Ic supernovae in low-metallicity, dwarf galaxies: diversity of explosions. *Astron. Astrophys.* **512**, A70.

Zehavi I, Riess AG, Kirshner RP and Dekel A 1998 A local Hubble bubble from Type Ia supernovae? *Astrophys. J.* **503**, 483–491.

Zwicky F 1937 On the masses of nebulae and of clusters of nebulae. *Astrophys. J.* **86**, 217–246.

6

Systematic Uncertainties and Common Pitfalls

If your experiment needs statistics, then you ought to have done a better experiment.
– Lord Ernest Rutherford (unsourced), First Baron Rutherford of Nelson (1871–1937),
New Zealand-born British physicist and 1908 Nobel laureate (chemistry)

Uncertainties in distance estimates can translate into equal or larger fractional errors in derived properties such as masses, linear sizes, dynamical timescales, star formation rates and ages. Accurate distance determinations are, therefore, essential both for 3D mapping of velocity fields and matter distributions and for comparisons of the physical properties of objects of any size and mass, and at any distance.

Although we have already highlighted throughout this book where the key remaining uncertainties reside for all of the techniques of distance determination discussed, in this chapter we first address the most common biases and pitfalls specifically associated with estimation of cosmic distances. In particular, we will highlight the issues involved in extinction mapping and the choice of extinction law, followed by a technical discussion of two of the most common types of statistical sample selection effects and measurement biases, the Lutz–Kelker and Malmquist biases. We will then proceed to explore one of the most long-standing systematic disagreements in this field, that of the high versus low estimates of the Hubble constant that affected scientific discourse throughout the twentieth century. Note that a similarly long-standing systematic disagreement affected distance determinations to the Large Magellanic Cloud (LMC), which we briefly discussed in Section 1.2 (see also e.g. de Vaucouleurs 1993). We conclude the chapter by focussing on the current state of play in constraining distance uncertainties on cosmic scales, linking 'local' uncertainties to the expansion rate of the Universe at high redshifts.

An Introduction to Distance Measurement in Astronomy, First Edition. Richard de Grijs.

6.1 Common Biases

6.1.1 Extinction: Spatial Distribution and Wavelength Dependence

The effects of extinction, a combination of absorption and scattering by dust and gas, are among the most significant in the context of systematic uncertainties feeding through into distance determinations. This notion goes back at least as far as Trumpler's (1930) realization that the distribution of the known open clusters at that time was not defined by the same centre as that of Shapley's (1918a,b) centroid of the Galactic globular clusters (see Section 1.1.1). Since the effects of extinction or 'attenuation' correspond to the addition of a free parameter, A_λ (where λ denotes the relevant wavelength), to the distance modulus, they render observational comparisons with 'standard candles' or other calibrating sources intrinsically uncertain. As detailed in Equation (1.1) this immediately affects the main-sequence fitting technique (see Section 3.2.1). Additional examples of techniques potentially affected by extinction include the use of red clump stars (see Section 3.2.2; Paczyński and Stanek 1998; Stanek and Garnavich 1998), novae (see Section 3.6; Shafter 1997) and supernovae (Type Ia: Section 5.2.1; Type II-P: Section 5.2.2; Nugent *et al.* 2006; Jones *et al.* 2009; Poznanski *et al.* 2009). At greater distances, on average, extinction affects the planetary nebulae luminosity function (see Section 4.4) as a standard candle and the Tully–Fisher (see Section 4.5) and colour–magnitude relations (see Section 4.7) as diagnostic tools. This is, hence, driving development of 'reddening-free' approaches: see e.g. the Wesenheit relations for period–luminosity analysis (Section 3.5) and the equivalent reddening-free approach taken by the Carnegie Supernova Project, e.g. Equation (5.32) in Section 5.2.1.

In addition to adding a variable term to the distance modulus, the corresponding wavelength dependence (i.e. the *extinction/attenuation law*; see below) causes differential effects on the individual magnitude measurements comprising an object's colour. Since extinction effects decrease towards longer wavelengths, which is related to the typical dust grain size (dust grains are of similar size to the wavelength of blue light, which is, hence, most easily absorbed), and because colours are defined as a magnitude at a shorter wavelength minus that at a longer wavelength, this gives rise to a reddening of the colour, i.e. a positive *colour excess*. The latter is defined as the difference between an object's observed and intrinsic colours. For instance, the $(B - V)$ colour excess is defined as

$$E(B - V) \equiv (m_B - m_V) - (m_{B,0} - m_{V,0}) = A_B - A_V, \tag{6.1}$$

where the subscript '0' indicates intrinsic, extinction-corrected parameters and $A_\lambda \equiv m_\lambda - m_{\lambda,0}$.

The amount of extinction and reddening along a given line of sight depends on the density of dust in that direction, i.e. the total *optical depth*. The ratio of the extinction at a wavelength λ to that at a reference wavelength, often the V passband[1] (with a central wavelength, $\lambda_c \approx 550$ nm), is called an *extinction law* or extinction curve. One needs to be careful in adopting the most appropriate extinction law for one's object of interest, since 'the' Galactic extinction law differs systematically from that in the Magellanic Clouds

[1] Note that Binney and Merrifield (1998) point out that 'the' extinction curve is universal redward of the R band ($\lambda_c \approx 720$ nm) but not in the V band and at bluer wavelengths. This implies that, strictly speaking, extinction and reddening values should be normalized redward of the V band.

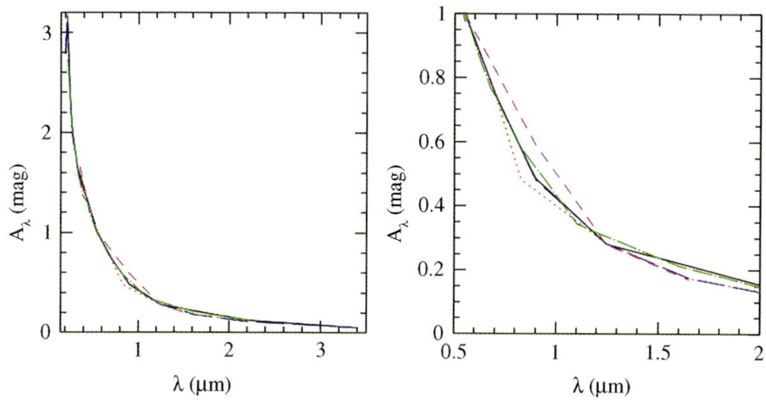

Figure 6.1 *Comparison of commonly used extinction curves (de Grijs et al. 2005); the right-hand panel is a zoomed-in version of the left-hand panel. Galactic extinction laws: solid lines, Savage and Mathis (1979); dotted lines, Rieke and Lebofsky (1985); short-dashed lines, Voshchinnikov and Il'in (1987); and long-dashed lines, Fitzpatrick (1999). Starburst galaxy extinction law: dot/long-dashed lines, Calzetti et al. (1994).*

(e.g. Prévot *et al.* 1984; Fitzpatrick 1986; Gordon and Clayton 1998; Misselt *et al.* 1999; Dobashi *et al.* 2009; Bot *et al.* 2010) and in starburst galaxies (e.g. Calzetti *et al.* 1994; Gordon *et al.* 1997), and even as a function of sight line (Cardelli *et al.* 1989; Mathis and Cardelli 1992; Valencic *et al.* 2004; Gordon *et al.* 2009; see also e.g. Jansen *et al.* 1994 for a comparison of extinction laws among nearby, highly inclined spiral galaxies). This is probably related to variations in dust grain properties (e.g. Weingartner and Draine 2001; see also the reviews in Calzetti 2001 and Piovan *et al.* 2006a,b).

In the left-hand panel of Figure 6.1, we show a number of commonly used Galactic extinction laws in relation to each other over the wavelength range from the ultraviolet to the near-infrared (near-IR) regimes, normalized at an extinction of 1 mag in the V band (de Grijs *et al.* 2005). In the right-hand panel, we zoom in to display the differences among the individual extinction laws from $\lambda = 0.5$ to 2.0 μm. From a comparison of the individual extinction curves in the right-hand panel, it is clear that the differences are generally $\lesssim 0.05$ mag at wavelengths longward of 1 μm and shortward of $\sim$0.8 μm, with the exception of the Voshchinnikov and Il'in (1987) extinction law. In the intermediate wavelength range, the differences are mainly driven by the Rieke and Lebofsky (1985) Galactic extinction law on the one hand and the Voshchinnikov and Il'in (1987) curve on the other. Nevertheless, representative differences from the mean generally do not exceed 0.1 mag, even at these wavelengths, and are often significantly smaller. The peak of the extinction curve occurs at approximately 73 nm (not shown) in the far-ultraviolet regime. Shorter-wavelength radiation (X-rays) passes through the dust grains (but see Section 7.3), while wavelengths in the far-IR regime and beyond refract around the grains. The significantly decreased effects of extinction at IR wavelengths, combined with the often smaller scatter of physical properties and the increased current availability of IR observing facilities, is driving research efforts – including in the field of distance determination, e.g. in relation to period–luminosity relations for which extinction corrections are often still ambiguous (see Section 3.5) – from the classical, optical regime to these longer wavelengths.

Classical extinction laws are defined by the slope of the extinction curve in the V band, i.e.

$$R_V \equiv \frac{A_V}{A_B - A_V} = \frac{A_V}{E(B-V)}, \tag{6.2}$$

where $R_V \simeq 3.1$ for the Milky Way for lines of sight that do not include dense dust clouds (e.g. Schultz and Wiemer 1975; Sneden *et al.* 1978; see also Cardelli *et al.* 1989 for an analysis of the likely uncertainties and applicable range). Thus, by adopting a suitable value for R_V, one can use Equation (6.2) to determine A_V from measurements of $E(B-V)$ (or any other combination of filters and colours).

Although this approach might work to first order, it assumes that (i) one knows the relevant intrinsic magnitude of the object of interest and (ii) the dust is distributed in a 'foreground screen' geometry. Both assumptions introduce additional and often significant uncertainties. Where relevant, we have discussed uncertainties associated with the former throughout this book. As regards the second assumption, choosing the most appropriate extinction law requires a detailed knowledge of the geometry of the mixture of dust and stars, including e.g. circumstellar dust or dust associated predominantly with spiral arms (see e.g. Tempel *et al.* 2011 for a dust model of M31), and the relevant filling factor, allowing for patchy versus smooth distributions of the dust component; a foreground screen is often an oversimplification. For the same optical depth, a uniform mixture of dust and stars causes less extinction than the classical foreground screen model because part of the extinction lies behind the source. These effects are, particularly in external galaxies, often compounded by the unknown effects caused by population changes, also known as the 'age–extinction(–metallicity) degeneracy' (see e.g. Figure 10 in Bruzual and Charlot 2003 for an illustrative example of the age–metallicity degeneracy). Finally, one has to consider the possibility that even if the extinction component acts as an obscuring layer in front of the object of interest, it may not represent a *uniform* layer. In this case, referred to as *differential extinction* – i.e. variable extinction across the face of the object – one may introduce significant uncertainties by applying a uniform foreground screen model (see, for examples, Cohen 2006; Bastian *et al.* 2007).

For completeness, we note that when correcting extragalactic flux or magnitude measurements for the effects of extinction, one should take into account both internal extinction associated with the object of interest and foreground extinction along the line of sight caused by dust in the Milky Way. The latter can, in fact, be suitably represented by a true 'foreground screen' and has been well characterized (see e.g. Schlegel *et al.* 1998; see also Marshall *et al.* 2006 for a 3D extinction model of the Milky Way).

6.1.2 Parallaxes: Lutz–Kelker Bias

Lutz and Kelker (1973) first discussed theoretically how trigonometric parallax measurements are affected by measurement bias (see Smith 2003 for an in-depth assessment of the original analysis), building on earlier work by Wallerstein (1967) and West (1969), and a similar approach by Ljunggren and Oja (1966). (See Trumpler and Weaver (1953) for early observational arguments of *sample bias* and Sandage and Saha (2002) for an historical overview.) Consider a uniformly distributed sample of objects in space. For a survey of fixed solid angle, the number of objects per distance unit increases as d^2 (where d is the distance), so that the number of objects per unit of parallax is not constant. This biases the statistical likelihood towards finding objects at larger distances, where more volume is sampled and, hence, more sources are located. This 'Lutz–Kelker' bias – sometimes referred to as the 'Lutz–Kelker–Hanson' bias after Hanson (1979), who included the power-law distribution of proper motions of the parent population – causes measured parallaxes to be higher, on average, than their real values. In turn, this causes underestimates of derived distances and, hence, of luminosities if the latter are based on measurements of apparent magnitudes.

This systematic effect for samples of objects limited by a minimum-parallax criterion and, importantly, uniformly distributed in space is caused by observational uncertainties. These cause objects which are in reality located outside the adopted lower limit to scatter into the sample's selection criterion and *vice versa*. Since there are more objects just outside than just inside the selection boundary, more objects will be scattered into than out of the sample, so that a systematic bias is introduced. Lutz and Kelker (1973) asserted that there is nothing special about the adopted lower parallax limit, and that the bias is not introduced by adoption of any parallax limit (it is, instead, caused by observational uncertainties and by the fact that the number density of homogeneously distributed stars increases for smaller parallax angles), so that this effect applies to parallax measurements of any sample of objects.

However, Smith (2003) pointed out that there is, properly speaking, no universal Lutz–Kelker bias of *individual parallaxes*. He argues that there is a bias for objects that are members of samples, but this is different from the Lutz–Kelker bias, yet it often has the same form and is given the same name. The overall bias for samples selected according to relative parallax *error* (not parallax as such) is sometimes referred to as a Lutz–Kelker bias (e.g. Koen 1992; Arenou and Lori 2002), although it is almost the same as the sample bias discussed by Trumpler and Weaver (1953). An example of this type of bias for *Hipparcos* parallaxes is discussed in Oudmaijer *et al.* (1998) and, although various authors have since dismissed their result as an artefact, Smith (2003) points out that it is, in fact, real and in essence similar to the Trumpler and Weaver (1953) bias (see also Butkevich *et al.* 2005a). Nevertheless, Lutz–Kelker corrections can often be used to reduce the latter bias. However, these corrections applied to isolated objects, independent of sample properties, are incomplete refinements of absolute magnitude derivations calculated directly from the parallax, not a correction for bias (Smith 2003).

Corrections for the Lutz–Kelker bias applied to carefully selected samples of objects have become relevant in the context of the *Hipparcos* and anticipated *Gaia* precision parallax measurements (see Section 2.1.2) because this statistical underestimate of the stellar distance depends on the precision of the parallax measurement. Lutz and Kelker (1973) analytically derived the probability distribution, $p(\varpi|\varpi_0)$, of the actual parallax, ϖ,

$$p(\varpi|\varpi_0) \propto \left(\frac{\varpi_0}{\varpi}\right)^4 \exp\left[-\frac{(\varpi-\varpi_0)^2}{2\sigma^2}\right], \tag{6.3}$$

given the measured parallax, ϖ_0, and its standard deviation, σ. This is the probability density that ϖ of a given star is found within the range $[\varpi, \varpi + \mathrm{d}\varpi]$ given ϖ_0, which can be rephrased using **Bayes' theorem** as

$$p(\varpi|\varpi_0) = \frac{p(\varpi_0|\varpi)p(\varpi)}{p(\varpi_0)}. \tag{6.4}$$

Here, $p(\varpi_0)$ is the 'prior' probability of the measurement, i.e. the probability distribution which represents the uncertainty before the data have been taken into account. Because we want to know $p(\varpi|\varpi_0)$ as a function of ϖ at fixed ϖ_0, the prior probability of $p(\varpi_0)$ is not important, so we can set it to a constant value. If our parallax measurements are dominated by Gaussian noise,

$$p(\varpi_0|\varpi) = \frac{1}{\sqrt{2\pi}\sigma} \exp\left[-\frac{1}{2}\left(\frac{\varpi - \varpi_0}{\sigma}\right)^2\right]. \tag{6.5}$$

The term $p(\varpi)$ in Equation (6.4), the prior probability distribution of the parallax, is determined by both the sampled volume and the intrinsic luminosity function of the objects considered (see e.g. Binney and Merrifield 1998 for stars; Verbiest *et al.* 2010 for pulsars).

For our 3D homogeneously distributed sample of objects, the prior on the sampled volume follows immediately: $p(d) \propto \rho d^2$, where ρ is the object density. This corresponds to the parallax probability distribution effectively determined by Lutz and Kelker (1973), our 'volumetric prior',

$$p_d(\varpi) = \left|\frac{\partial d}{\partial \varpi}\right| p(d) \propto \varpi^{-4}. \tag{6.6}$$

This strong tendency of the probability $p(\varpi|\varpi_0)$ to yield small values of ϖ is the original Lutz–Kelker bias. Small values of ϖ are associated with bright absolute magnitudes, so that the Lutz–Kelker bias also depends on the luminosity function of the objects of interest, which was first included by Lutz (1979).

Following Binney and Merrifield (1998) and Verbiest *et al.* (2010), we can expand the prior information by taking advantage of our knowledge of the intrinsic luminosity functions of stars and pulsars, respectively. For the present purposes, we will adopt a power-law luminosity function for reasons of simplicity, $p(L) \propto L^\alpha$, where L is the intrinsic luminosity and α the power-law slope (but see Binney and Merrifield 1998 and below). Translating this luminosity probability distribution into its equivalent parallax function, we get the 'luminosity prior' (Verbiest *et al.* 2010),

$$p_L(\varpi) = \left|\frac{\partial L}{\partial \varpi}\right| p(L) \propto \varpi^{-2\alpha-3}. \tag{6.7}$$

We can now obtain the prior probability distribution of the parallax by combining Equations (6.6) and (6.7),

$$p(\varpi) = p_d(\varpi)p_L(\varpi) \propto \varpi^{-2\alpha-7}, \tag{6.8}$$

which, combined with Equations (6.4) – Bayes' theorem – and (6.5), leads to the probability distribution of the true parallax,

$$p(\varpi|\varpi_0) \propto \varpi^{-2\alpha-7} \exp\left[-\frac{1}{2}\left(\frac{\varpi - \varpi_0}{\sigma}\right)^2\right]. \tag{6.9}$$

For its maximum, requiring a zero derivative – $\partial p(\varpi|\varpi_0)/\partial\varpi = 0$, thus resulting in a *local* maximum – we get

$$(-2\alpha - 7)\varpi^{-2\alpha-8} \exp\left[-\frac{1}{2}\left(\frac{\varpi - \varpi_0}{\sigma}\right)^2\right] + \varpi^{-2\alpha-7}\frac{\varpi - \varpi_0}{\sigma^2} \exp\left[-\frac{1}{2}\left(\frac{\varpi - \varpi_0}{\sigma}\right)^2\right] = 0, \quad (6.10)$$

so that by rearranging this expression, the Lutz–Kelker bias in its current form can be expressed formally as

$$\varepsilon_\varpi \equiv \frac{\varpi}{\sigma} = \frac{1}{2}\left[\frac{\varpi_0}{\sigma} \pm \sqrt{\frac{\varpi_0^2}{\sigma^2} - 4(2\alpha + 7)}\right], \quad (6.11)$$

although in practice only inclusion of the positive additive term is adopted. The bias is, thus, a function of both the significance of the parallax measurements (and not of just the parallax itself) – i.e. as increasingly precise parallax measurements are published, their values will converge towards decreasing bias – and the slope of the luminosity function, with a critical slope at $\alpha = -3.5$, where the volumetric and luminosity priors – Equations (6.6) and (6.7) – cancel each other out exactly. For shallower slopes, as usually found for both stellar and pulsar samples (for the latter see e.g. Lorimer *et al.* 2006; $\alpha \approx -1.7$), the measured parallax values are likely overestimated.

Unfortunately, derivation of the Lutz–Kelker bias contains a degree of circularity, which was already realized by Lutz (1979, 1983): to correct for this bias, we need to know the shape and properties of the relevant luminosity function, which is, however, the quantity we generally want to determine when we encounter the Lutz–Kelker bias. Lutz (1983) stated that he preferred not to assume a luminosity function at all and rejected any special assumptions, in the hope to retain some of the generality of the corrections. Standard Lutz–Kelker corrections (e.g. Lutz and Kelker 1975; Smith 1987a,b,c,d, 2003; Koen 1992) become significant for observed parallaxes with $\varepsilon_\varpi \geq 0.05$ and diverge for $\varepsilon_\varpi \geq 0.175$ (Maíz Apellániz 2005; see Figure 6.2). For stars observed with *Gaia*, unobscured G and M dwarfs are expected to have $\varepsilon_\varpi = 0.05$ at $d \approx 2$ kpc and ~500 pc, respectively (see Reid 1997 for an application to *Hipparcos* parallaxes of subdwarfs). Given *Gaia*'s expected operational distance range (see Figure 2.4 and Section 2.1.2), it is clear that Lutz–Kelker corrections must be applied.

The two main assumptions underlying this analytical derivation are that (i) the objects of interest are distributed homogeneously in space and (ii) their luminosity function is a simple power law. Both assumptions are invalid in general, however. Binney and Merrifield (1998) argue – following Houk *et al.* (1997)'s results on the distribution of main-sequence stars with good *Hipparcos* parallaxes – that the stellar luminosity functions for a given spectral type are better approximated by Gaussian distributions (see also Turon Lacarrieu and Crézé 1977; Lutz 1979; Oudmaijer *et al.* 1998), while Faucher-Giguère and Kaspi (2006) showed that the intrinsic pulsar luminosity function is also better approximated by a lognormal distribution.

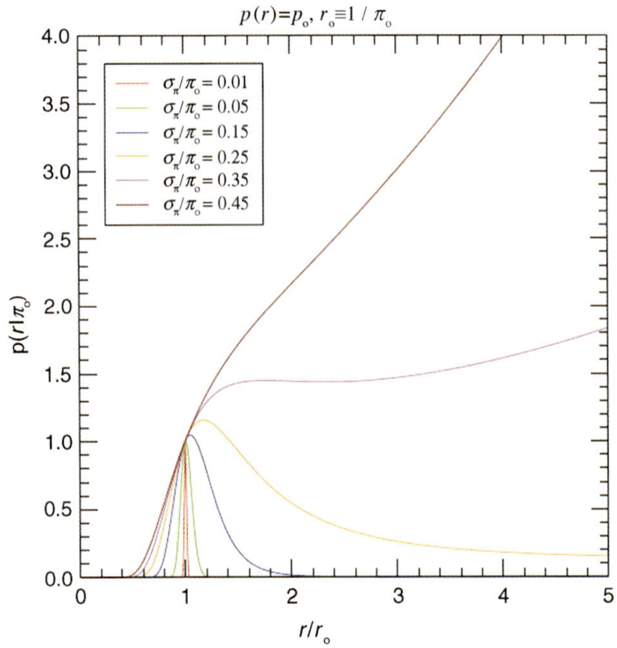

Figure 6.2 *Real distance probability distributions for Gaussian uncertainties and an underlying constant spatial distribution assuming different values of the Lutz–Kelker bias,* $\varepsilon_{\varpi} \equiv \sigma_{\pi}/\pi_0$ (Reprinted from J. Maíz-Apellániz, *ESA Special Publications*, **576**, In The Three-Dimensional Universe with Gaia (eds. Turon, C., O'Flaherty, K. S. and Perryman, M. A. C.), Self-consistent distance determinations for Lutz–Kelker-limited samples, p. 179–182, Copyright 2005, with permission of ESA and J. Maíz-Apellániz.).

Although the assumption of a roughly homogeneous volumetric distribution of stars for which trigonometric parallaxes can be measured with current technology is not disastrous to first order (see also Maíz-Apellániz 2005 for an assessment of the effects of a more realistic density distribution), pulsars are strongly concentrated towards the Galactic plane and, hence, not homogeneously distributed in space. To correct for the latter inconsistency, Verbiest *et al.* (2010) performed detailed Monte Carlo simulations of the Gaussian measurement uncertainty and both priors independently to evaluate the Lutz–Kelker bias. They conclude that a comparison of historic pulsar parallax values to the most recent measurements confirms that Lutz–Kelker-type bias effects are present in the relevant observations, but correction is complicated by systematic measurement errors and underestimation of parallax uncertainties.

Thus, it is important to be aware, quantitatively, of the effects of Lutz–Kelker-type biases on parallax measurements of given samples of objects. On the other hand, if one has a set of parallax measurements, it may be possible to avoid the Lutz–Kelker bias mostly or entirely by taking a direct approach to distance determination. An assumed model describing the distance distribution of one's sample objects can be converted into a predicted parallax distribution, which in turn could be fitted to the parallax data in an iterative approach.

This could eventually lead to a robust 3D model distribution and its associated distance estimates, even if the individual parallax measurements are not extremely accurate. This is because if we work in the parallax domain, we do not have to take into account the complicated higher-order propagation of the uncertainties in the distance domain. (Note that intrinsic cosmic variation implies that the model-predicted parameters are not uniquely determined but have an intrinsic dispersion. The latter will act in a similar way as the formal parallax errors and depend on the relative amplitude of the dispersion. However, this situation resembles a 'reverse Lutz–Kelker' effect, for which there is no distance dependence.) Future modelling efforts of *Gaia* data will likely adopt such an approach (see also Butkevich *et al.* 2005a,b; van Leeuwen 2007, his Chapters 3.3 and 5). As in all fields of astrophysics dealing with statistical approaches, sample selection criteria feed through into the final analysis, so that one needs to apply careful corrections to arrive at statistically sound results.

6.1.3 Malmquist Bias

Any astronomical observation is characterized by a minimum flux density (i.e. it is flux or magnitude limited), F, below which we cannot detect any object with a given signal-to-noise ratio. Since $F \propto L/d^2$, higher luminosity objects can be detected to greater distances than their intrinsically fainter counterparts. This implies that the relative numbers of intrinsically bright and faint objects in an observational sample may not be representative of their relative numbers per unit volume. Instead, luminous objects are likely overrepresented and the average luminosity of the sample objects increases with distance and, consequently, with respect to the average luminosity of the unbiased sample of the parent population. In honour of the Swedish astronomer Gunnar Malmquist (1893–1982), this type of observational selection effect is referred to as the *Malmquist bias* (Malmquist 1922, 1936). Malmquist (1922) considered three types of samples, including stars of and brighter than a given apparent magnitude, and stars between two limits of apparent magnitude. He obtained analytical expressions for the bias in the mean absolute magnitude for an arbitrary spatial distribution and a Gaussian luminosity function, which he later generalized to include absorption effects (Malmquist 1936). Butkevich *et al.* (2005b) provide a more detailed historical overview of efforts to correct for this bias.

Following Binney and Merrifield (1998) and Butkevich *et al.* (2005b), the number of stars, N, in a magnitude-limited sample with absolute and apparent magnitudes in the ranges $[M, M + \mathrm{d}M]$ and $[m, m + \mathrm{d}m]$ is given by

$$\frac{\mathrm{d}^2 N}{\mathrm{d}m\mathrm{d}M} = \Phi(M)\frac{\mathrm{d}n}{\mathrm{d}d}\left(\frac{\partial d}{\partial m}\right)_M, \tag{6.12}$$

where $d(m, M)$ is the distance to a star characterized by apparent and absolute magnitudes m and M, respectively, $\Phi(M)$ the stellar luminosity function and $\mathrm{d}n$ the number of stars contained within the region of sky observed at distances $(d, d + \mathrm{d}d)$. For observations covering a solid angle ω, the volume occupied by this partial shell is $\omega d^2\,\mathrm{d}d$, so that

$$\frac{\mathrm{d}n}{\mathrm{d}d} = \omega d^2 \nu(d) \tag{6.13}$$

and $\nu(d)$ is the stellar number density. We are interested in the number of stars per unit magnitude, $A(m) \equiv \mathrm{d}N/\mathrm{d}m$,

$$A(m) = \int_{-\infty}^{\infty} \mathrm{d}M \frac{\mathrm{d}^2 N}{\mathrm{d}m\, \mathrm{d}M} = \omega \int_{-\infty}^{\infty} \mathrm{d}M\, \Phi(M) \left(\frac{\partial d}{\partial m}\right)_M d^2 \nu(d). \tag{6.14}$$

Using the definition of the distance modulus, $m - M = 5 \log d - 5$ (where d is expressed in parsecs), we can now change the integration variable from M to d, so that

$$A(m) = \omega \int_0^{\infty} \mathrm{d}d\, \Phi(M) d^2 \nu(d), \tag{6.15}$$

because $(\partial M/\partial d)_M = -(\partial m/\partial d)_M$ (cf. the definition of the distance modulus). The mean absolute magnitude of the stars in the sample with apparent magnitude m is then given by

$$\langle M \rangle_m = \frac{\int_0^{\infty} \mathrm{d}d\, M\Phi(M) d^2 \nu(d)}{\int_0^{\infty} \mathrm{d}d\, \Phi(M) d^2 \nu(d)} \tag{6.16}$$

and we see that the denominator of Equation (6.16) is equivalent to $A(m)$: see Equation (6.15). In addition, M and m are linked through the distance modulus with $(\partial M/\partial m)_d = 1$. Therefore, if we differentiate Equation (6.15) with respect to m, we get

$$\frac{\mathrm{d}A}{\mathrm{d}m} = \omega \int_0^{\infty} \mathrm{d}d \frac{\mathrm{d}\Phi}{\mathrm{d}M} d^2 \nu(d). \tag{6.17}$$

From Equation (6.16), it now follows that

$$\frac{1}{A}\frac{\mathrm{d}A}{\mathrm{d}m} = \left\langle \frac{1}{\Phi}\frac{\mathrm{d}\Phi}{\mathrm{d}M} \right\rangle_m \tag{6.18}$$

and

$$\frac{1}{A}\frac{\mathrm{d}^2 A}{\mathrm{d}m^2} = \left\langle \frac{1}{\Phi}\frac{\mathrm{d}^2\Phi}{\mathrm{d}M^2} \right\rangle_m . \tag{6.19}$$

Binney and Merrifield (1998) provide further insight into the underlying cause of the Malmquist bias by adopting a Gaussian distribution for the form of $\Phi(M)$ (see also Butkevich *et al.* 2005b; and Jaschek and Gómez 1985 for a generalized, non-Gaussian luminosity function). We will follow this example here, too:

$$\Phi(M) = \frac{1}{\sqrt{2\pi\sigma^2}} \exp\left[-\frac{(M - M_0)^2}{2\sigma^2}\right], \tag{6.20}$$

where the local stellar density is unity, M_0 is the mean absolute magnitude and σ is the dispersion of a *volume-limited* stellar sample. With this functional form for the stellar luminosity function, we obtain

$$\frac{1}{A}\frac{\mathrm{d}A}{\mathrm{d}m} = -\left\langle \frac{M - M_0}{\sigma^2} \right\rangle_m , \tag{6.21}$$

or

$$\langle M \rangle_m - M_0 = -\sigma^2 \frac{\mathrm{d} \ln A}{\mathrm{d}m}. \tag{6.22}$$

In most magnitude-limited samples, $\mathrm{d}A/\mathrm{d}m > 0$, so that Equation (6.22) implies that the observed sample of objects will, on average, have a higher luminosity than the mean luminosity of the parent population.

Smith (1987a) also considered the case of a lognormal (Gaussian) luminosity function, Equation (6.20), although in the context of the Lutz–Kelker bias discussed in the previous section. In this case, the prior probability distribution function for the true parallax is given by

$$p(\varpi_0|m) = p_0 \exp\left[-\frac{(M - M^*)^2}{2\sigma^2}\right], \tag{6.23}$$

where p_0 is a normalization constant and $M^* = M_0 - 1.84\sigma^2$ is the most likely value of the absolute magnitude for an individual object. The latter equality immediately links the Lutz–Kelker bias to the Malmquist effect for an assumed infinite, uniform space distribution of the parent population which is contained in the equation's second term along with a contribution related to the transformation from M to ϖ. Luri *et al.* (1993, 1996) and Butkevich *et al.* (2005b) discuss implications resulting from a proper treatment of the actual stellar distribution (i.e. a density anisotropy) in the Milky Way.

The probability distribution function for the parallax now becomes

$$p(\varpi_0|\varpi, m) = p'_0 \exp\left[-\frac{(\varpi_0 - \varpi)^2}{2\sigma_\varpi^2} - \frac{(M - M^*)^2}{2\sigma^2}\right], \tag{6.24}$$

where p'_0 includes the relevant normalization (Smith 1987a). The most probable value for the parallax of an *individual* object, ϖ^*, can be derived from

$$\frac{\varpi_0 - \varpi}{\sigma_\varpi^2} + \frac{2.17}{\varpi_0}\frac{M - M^*}{\sigma^2} = 0 \tag{6.25}$$

or (see Smith 1987a)

$$\varpi_0^2 - \varpi\,\varpi_0 + \frac{2.17}{\sigma^2}\sigma_\varpi^2(M - M_0) + 4\sigma_\varpi^2 = 0. \tag{6.26}$$

Thus, this implies that there is an apparent link between Lutz–Kelker- and Malmquist-type biases. The corrections for each of these biases individually work in opposite directions, although the total correction remains positive (Smith 1987a). The exact value depends on the intrinsic variations in both σ_ϖ and σ.

We have thus far only discussed the *integral* Malmquist bias, without addressing any distance dependence explicitly (see e.g. Butkevich *et al.* 2005a for an application to a sample of *Hipparcos* K0V-type stars). In extragalactic astronomy, the importance of an explicit treatment of the Malmquist bias's distance dependence goes back to Teerikorpi (1975) and was significantly expanded by Sandage (1994). Sandage (1995) and Teerikorpi (1997) provided comprehensive reviews of the current state of our understanding of the effects of

Malmquist bias, while Teerikorpi (1998) generalized the classical bias in Euclidean space to Friedmann's cosmological models.

In the ideal case when observations are not affected by interstellar extinction, an absolute magnitude is uniquely tied to an apparent magnitude and a distance. This implies that we can average over any of these variables while keeping the other two fixed. The result, referred to as *differential* bias, could be either *magnitude* dependent (when the mean is calculated for stars of a fixed apparent magnitude) or *distance* dependent (when the averaging is done over stars located at the same true distance). Teerikorpi (1993, 1997) suggested to call these differential manifestations Malmquist biases of the first and second kind, respectively.

Butkevich *et al.* (2005b) show that the functional form of the Malmquist bias for the magnitude- and distance-dependent varieties is similar to that of the integral Malmquist bias in Equation (6.22), i.e.

$$\langle \Delta M_m \rangle = -\sigma^2 \frac{\mathrm{d} \ln a(m)}{\mathrm{d}m} \tag{6.27}$$

and

$$\langle \Delta M_r \rangle = -\sigma^2 \frac{\mathrm{d} \ln \psi(M_{\mathrm{lim}})}{\mathrm{d}M_{\mathrm{lim}}}, \tag{6.28}$$

where $\langle M_m \rangle$ and $\langle M_r \rangle$ are the mean absolute magnitudes of stars of a given apparent magnitude and at a fixed distance, respectively, and $a(m)$ and $\psi(M_{\mathrm{lim}})$ are the distributions of apparent magnitudes, m, and the fraction of stars brighter than M_{lim} at a fixed distance, respectively (see Butkevich *et al.* 2005b for the full derivation and a comprehensive discussion of its effects on sample selection). Of these, Equation (6.28) is the classical Malmquist bias (Malmquist 1922), while the integral version of Equation (6.22), $\langle \Delta M_{\mathrm{int}} \rangle$, was derived subsequently (Malmquist 1936). In the unrealistically simplified case when the spatial distribution of stars is homogeneous, $\langle \Delta M_m \rangle = \langle \Delta M_{\mathrm{int}} \rangle = -1.382\sigma^2$.

Butkevich *et al.* (2005b) conclude that at large distances, where the bias dominates, any magnitude-limited sample consists only of stars with apparent magnitudes close to the selection limit. This causes the spectroscopic distance or parallax to approach a constant limit as the true distance increases. In turn, this illustrates how ignoring the Malmquist bias can lead to serious misinterpretations in statistical studies dealing with the distance dependence of physical quantities.

Malmquist-type biases are of importance in nearly every astronomical subfield. The obvious solution to deal with this problem is to use a sample that is not magnitude limited, e.g. one that is volume or diameter limited, or selected based on magnitude-independent properties such as proper motions; see Sandage *et al.* (2006) for Type Ia supernovae or Poznanski *et al.* (2010), who discuss the distance-independent sample selection criteria of Type II-P supernovae employed by D'Andrea *et al.* (2010) versus those of Nugent *et al.* (2006) and Poznanski *et al.* (2009). In extragalactic astronomy, the region unaffected by Malmquist bias is often referred to as the '**unbiased plateau**', which can be easily visualized using the so-called **Spaenhauer (1978) diagram** (e.g. Sandage 1994; Teerikorpi 1997; Butkevich *et al.* 2005b; see Figure 6.3). This is a 2D representation of the absolute magnitude/distance space applicable to a given set of observations. Spaenhauer's (1978) diagrams reveal how the average absolute magnitude of a standard candle changes with increasing kinematic distance when imposed magnitude limits cut away progressively larger

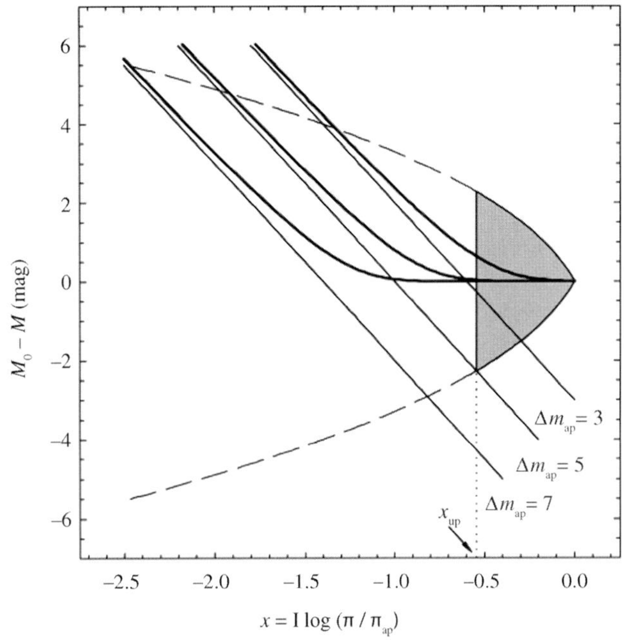

Figure 6.3 *Spaenhauer diagrams for $\sigma = 1$ mag and three different apparent magnitude limits, $\Delta\, m_{ap} = 3$, 5 and 7 mag (Butkevich et al. 2005b). The heavy lines are the Malmquist bias curves, while magnitude limits, m_{lim}, are shown as straight lines. The unbiased, volume-limited region is contained within the shaded area. The boundary of the unbiased plateau, x_{up}, is shown by the arrow. The bias curve shifts leftwards, preserving its form and extending the unbiased region as fainter objects are included in a sample.*

parts of the objects' luminosity function (shown by how the straight lines in Figure 6.3 affect the shape of the observable grey area).

6.2 High Versus Low Values of the Hubble Constant: Science or Philosophy?

For a large fraction of the twentieth century, a heated, polarized and often bitter, personal debate between two camps pervaded the science underpinning the extragalactic distance scale. The opposing views were spearheaded by Allan Sandage (1926–2010) of the Carnegie Observatories (Mt Wilson and Palomar Observatories) in California (USA) on the one hand and Gérard de Vaucouleurs (1918–1995) of the University of Texas at Austin (USA) on the other. It is instructive to see how this debate grew and eventually came close to a consensus because this gives insights into how systematic effects or even viewpoints may dominate scientific discussions. Therefore, here we briefly recount the establishment of the Hubble constant as a major cosmological parameter for distance determination beyond the nearest galaxies and subsequently focus on the development of the ensuing debate (see also the historical overview by Trimble 1996).

Based on pioneering spectral analysis, Vesto Melvin Slipher (1875–1969) of Lowell Observatory (Arizona, USA) noticed that a few ‘Fraunhofer groups’ (absorption lines) in the “dark-line spectra” (his designation) of the faint spiral nebulae which were known at that time were shifted with respect to their rest frame, i.e. nonmoving wavelengths, particularly towards the red end of the spectrum (e.g. Slipher 1915). For instance, among the earliest radial velocity measurements of external galaxies was Slipher’s (1913) ‘rounded off’ mean radial velocity determination to the Andromeda Nebula of $v_r = -300$ km s^{-1}, which he estimated to be well within the accuracy of the observations, hence validating this rounding. (He was correct: the statistical mean and 1σ uncertainty of his measurements were $v_r = -297 \pm 9$ km s^{-1}.) This discovery ultimately led Edwin Hubble (1889–1953), based at Mt Wilson Observatory, to propose a linear relationship between galaxy redshifts or recessional velocities and their distances, d, from the local standard of rest (Hubble 1929),

$$v_r = H_0 d, \tag{6.29}$$

where the proportionality constant has since been known as the Hubble constant. In fact, Hubble himself used ‘K’ as proportionality constant. Its current designation, ‘Hubble constant’, was first used as such by Bondi (1952), while Robertson (1955) was the first to have used the symbol H (see Trimble 1996 for historical notes).

The first values for the Hubble constant may have been derived by Lemaître (1927) and Robertson (1928), in fact, who used Slipher’s radial velocity measurements and Hubble’s mean absolute magnitude for ‘nebulae’ to obtain a proportionality constant of 526 and 460 km s^{-1} Mpc^{-1}, respectively. However, Hubble deserves the credit for discovering the cosmic expansion because he started a systematic programme to measure distances to nearby galaxies. In addition, his discovery of Cepheid variable stars in M31 (Hubble 1925; although Humason claimed to have spotted them earlier: see Christenson 1995; Trimble 1996) and his actual ‘Hubble diagram’ (Hubble 1929) were instrumental in convincing the astronomical community of the reality of the relationship. Hubble’s (1929) original determination and that of Hubble and Humason (1931) and other determinations around the same time were all approximately 500 km s^{-1} Mpc^{-1}, with quoted uncertainties ‘in the area of 10%’.

This high value posed an immediate problem, at least in the context of our current understanding of the evolution of the Universe: for a Hubble constant of 500 km s^{-1} Mpc^{-1}, the corresponding expansion age of the Universe was only 2 Gyr, while in the 1930s geologists had already deduced that the age of the Earth was approximately 3 Gyr, based on radioactive dating techniques (see also Trimble 1996). In addition, such a large Hubble constant implied that the Milky Way was significantly larger than any other nearby galaxy, with the possible exception of the Andromeda Nebula. Although the Dutch astronomer Jan Hendrik Oort (1931) derived a revised estimate of $H_0 = 290^{+290}_{-145}$ km s^{-1} Mpc^{-1}, it took until Baade’s (1956) revision of the distance scale to nearby galaxies (first announced at the International Astronomical Union’s General Assembly in Rome, Italy; Baade 1952) before the astronomical community accepted a reduction in H_0 by approximately a factor of 2 (see also Behr 1951). Baade’s revision to the underlying period–luminosity calibration was inspired by his recognition that Hubble had mistaken Population II W Virginis stars (see Section 3.5.4) for Population I Cepheids (see Section 3.5.2).

This first major revision to the Hubble constant was followed by further revisions in the late 1950s. In a classic paper, Humason *et al.* (1956) determined $H_0 = 180$ km s^{-1} Mpc^{-1}. This was subsequently updated to 75 km s^{-1} Mpc^{-1} by Sandage (1958), Hubble's successor at the Carnegie Observatories, although he recognized that the uncertainty could still be a factor of 2. Sandage realized that at distances beyond the nearest galaxies where individual bright Cepheids could be resolved, Hubble had mistaken HII regions for individual stellar standard candles. He may also have mistaken star clusters for bright stars. In addition, he relied on redshifts in the spectra of the brightest stars in his sample galaxies and, at even greater distances, on galaxy spectra as a whole.

By the early to mid-1970s, estimates by the team of Sandage and Gustav Tammann of the University of Basle (Switzerland) were firmly stuck at a value near $H_0 =$ 55 km s^{-1} Mpc^{-1} (Sandage and Tammann 1974a,b,c,d, 1975a,b; see Figure 6.4). Their estimates remained near this value for the next few decades. Nevertheless, at a 1976 meeting of the International Astronomical Union, de Vaucouleurs challenged the accuracy of the Sandage and Tammann work (e.g. de Vaucouleurs 1976, 1977), citing a dozen "blunders" in their papers, including incorrect handling of extinction effects and unwarranted scaling between large, luminous spiral galaxies and their fainter counterparts. This signalled the beginning of a great controversy in this area that persists until today, although it has become less polarized over the years (see also the compilation of Huchra 2010). During the tenure of Sandage and de Vaucouleurs, this rivalry was often bitter and represented a true clash

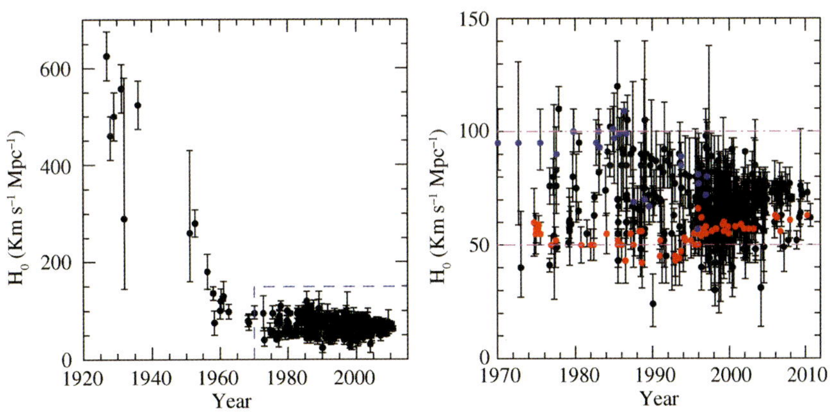

Figure 6.4 *Evolution of determinations of H_0. The first six data points in the left-hand panel are from Lemaître (1927), Robertson (1928), Hubble (1929), Hubble and Humason (1931), Oort (1931) and Hubble (1936), followed by the corrected calibrations by Behr (1951) and Baade (1956). Note that we left out a significantly discrepant data point derived by Eddington (1935), $H_0 = 865 \pm 100$ km s^{-1} Mpc^{-1}. In the right-hand panel (a section indicated in the left-hand panel by the dashed box), the fiducial lines at 50 and 100 km s^{-1} Mpc^{-1} are meant to guide the eye. Red points: determinations by Sandage, Tammann and collaborators. Blue points: determinations by de Vaucouleurs, van den Bergh and co-workers. (Based on data collected by Huchra 2010.)*

of personalities. The results of de Vaucouleurs, Sidney van den Bergh of the Dominion Astrophysical Observatory in Victoria (Canada) and their collaborators suggested a higher value, closer to $H_0 = 100$ km s^{-1} Mpc^{-1} (e.g. van den Bergh 1960a,b, 1975, 1994; de Vaucouleurs and Peters 1981; de Vaucouleurs 1993; and references therein). The fundamental implications of these differing points of view are that de Vaucouleurs was, in fact, arguing for a younger, more rapidly expanding Universe than Sandage's team.

Although this direct challenge to the eminent successor to Edwin Hubble was at first met with alarm and disbelief, de Vaucouleurs continued to advertise his camp's results at astronomical conferences. He constructed an elaborate distance scale which he based on both the commonly used standard candles and new reference objects, including bright star clusters and ring-like structures within certain galaxies (e.g. de Vaucouleurs and Buta 1980a,b; Buta and de Vaucouleurs 1982, 1983). He made sure to verify and cross-check the results from different methods, and eventually averaged them all. In addition, he accused Sandage's camp of making unwarranted assumptions about the smoothness of the Universe. In fact, he argued that the Milky Way is part of a local (super)cluster of galaxies (e.g. de Vaucouleurs and Bollinger 1979), which causes a slowdown of the local cosmic expansion, i.e. he predicted the effects of what we now know as 'peculiar motions' (see Section 5.1.2).

While de Vaucouleurs gained significant support and momentum,[2] the main problem with his proposed value of $H_0 \simeq 100$ km s^{-1} Mpc^{-1} was that it implied an age for the Universe of 10 Gyr. The issue was that careful studies at that time had led to ages for the oldest globular clusters of $\sim$17 Gyr, although de Vaucouleurs speculated that the models used to derive these ages might be wrong. He was correct, to some extent: Chaboyer *et al.* (1996) used the 18 oldest Galactic globular clusters and up-to-date stellar evolutionary models and their uncertainties to determine the age of the Universe at $>$ 12.07 Gyr with 95% confidence. Chaboyer (1998) subsequently derived a best fitting age of 14.6 ± 1.7 Gyr for the oldest Galactic globular clusters, while Gratton *et al.* (1997) and Reid (1997) found ages in the range of 8.5–13.3 Gyr, with a most likely age of 12.1 Gyr, and 11–13 Gyr, respectively. Finally, Chaboyer *et al.* (1998) revised their mean age of the oldest globular clusters to 11.5 ± 1.3 Gyr.

In fact, since Sandage's value for the Hubble constant had decreased over the years, de Vaucouleurs implied that the final drop – from 100 to 50 km s^{-1} Mpc^{-1} – had been caused by a desire to accommodate the globular cluster results. Indignantly, Sandage denied this accusation. For several decades, estimates of the Hubble constant clustered around 50–55 and 80–90 km s^{-1} Mpc^{-1} (e.g. de Vaucouleurs 1993; see Figure 6.4). This polarized situation is a telltale sign of 'publication bias' – results that match one's expectations are more likely to get published than significantly discrepant values – which is a type of sociological behaviour (human nature!) that affects many fields in science, technology and medicine (see e.g. Scargle 2000; Ioannidis 2005; Schonemann and Scargle 2008; see also Vaughan 2005). We already saw a similar, recent example affecting distance measurements to the LMC in Section 1.2 (cf. Schaefer 2008).

The large, systematic discrepancy in the 'best' value of the Hubble constant was clearly unsatisfactory. Eventually, this led to the conception of the ambitious, aptly named *Hubble*

[2] Tully and Fisher 1977 derived $H_0 = 80 \pm 8$ km s^{-1} Mpc^{-1} based on their newly developed luminosity calibration, although Aaronson *et al.* 1979 reduced this value to 61 ± 4 km s^{-1} Mpc^{-1} using a slightly modified Tully–Fisher relation at IR wavelengths (see also Sandage and Tammann 1976 and Section 4.5).

Space Telescope (HST) project. Constraining the Hubble constant to better than 10% was one of the original goals driving its development (see for the final results of the *HST* Key Project on the Extragalactic Distance Scale, Freedman *et al.* 2001). However, even today, significant tension – at the 2–3σ level – continues to affect determinations of H_0 based on different approaches (e.g. Riess *et al.* 2009), even using similar kinds of data but different techniques and assumptions (e.g. Freedman *et al.* 2001 versus Sandage *et al.* 2006).

In summary, Hubble's first measurements of H_0 and, thus, the current expansion rate of the Universe, seemed straightforward to refine in principle. After all, the only observables required were the velocities and distances of one's sample galaxies. However, unambiguous determination of the Hubble constant has proven more troublesome than anticipated. This field has now made some progress towards reaching closure, although the remaining disparities are still too large for comfort.

The uncertainties in the Hubble constant are, at the present time, still dominated by uncertainties in the Cepheid period–luminosity calibration, at least locally. In addition, uncertainties in the strength of local peculiar motions are of secondary importance. Calibration of the Cepheid distance scale has seen much progress, although it still has not reached full convergence (see e.g. Tammann *et al.* 2008). Even in the post-*Hipparcos* era, the continuing cycle of revision and counter-revision shows no sign of abating. Nevertheless, it seems unlikely that we will see another change in the zero point of the period–luminosity relation of the size introduced by Baade (1956), of order 1.5 mag, but revisions of ~0.1–0.2 mag continue to appear in the literature. An important issue of contention in this context is the use of a possibly not universal period–luminosity versus a period–luminosity–colour relationship: what is the true nature of the colour terms? Multiwavelength and, particularly, IR observational campaigns might eventually decouple the combined effects of extinction, metallicity and temperature (see also Tammann *et al.* 2008). Second, the 'universality' of the period–luminosity behaviour has been called into question, particularly in terms of differences between the Cepheid variables in the Large Magellanic Cloud compared to the Milky Way (e.g. Feast and Catchpole 1997; Madore and Freedman 1998; Reid 1999; Groenewegen and Oudmaijer 2000; Di Benedetto 2002; Sandage and Tammann 2006, 2008; Fouqué *et al.* 2007; Tammann *et al.* 2008; Bono *et al.* 2010). In addition, the use of many diagnostic methods simultaneously has the potential to achieve significantly reduced systematic uncertainties (e.g. Section 5.3.4).

To prevent readers from jumping to conclusions too quickly, we conclude with a word of caution: Fernie (1969) quoted Hubble's comments on the Cepheid distance scale at a lecture he gave in 1935, 'Further revision is expected to be of minor importance'. This was before Baade's first significant correction.

Bibliography

Aaronson M, Huchra J and Mould J 1979 The infrared luminosity/HI velocity–width relation and its application to the distance scale. *Astrophys. J.* **229**, 1–13.

Arenou F and Luri X 2002 Statistical effects from Hipparcos astrometry. *Highlights in Astronomy* (ed. Rickman H), vol. **12**, pp. 661–664. Astronomical Society of the Pacific.

Baade W 1952 Extragalactic nebulae. Report to IAU Commission 28. *Int'l Astron. Union Trans.* **8**, 397–399.

Baade W 1956 The period–luminosity relation of the Cepheids. *Publ. Astron. Soc. Pac.* **68**, 5–16.

Bastian N, Konstantopoulos I, Smith LJ, Trancho G, Westmoquette MS and Gallagher JS 2007 A detailed study of the enigmatic cluster M82F. *Mon. Not. R. Astron. Soc.* **379**, 1333–1342.

Behr A 1951 Zur Entfernungsskala der extragalaktischen Nebel. *Astron. Nachr.* **279**, 97–104.

Binney J and Merrifield MR 1998 *Galactic Astronomy*. Princeton University Press.

Bondi H 1952 *Cosmology*. Cambridge University Press.

Bono G, Caputo F, Marconi M and Musella I 2010 Insights into the Cepheid distance scale. *Astrophys. J.* **715**, 277–291.

Bot C, Ysard N, Paradis D, Bernard JP, Lagache G, Israel FP and Wall WF 2010 *Astron. Astrophys.* **523**, A20.

Bruzual A G and Charlot S 2003 Stellar population synthesis at the resolution of 2003. *Mon. Not. R. Astron. Soc.* **344**, 1000–1028.

Buta R and de Vaucouleurs G 1982 Inner ring structures in galaxies as distance indicators. II. Calibration of inner ring diameters as quaternary indicators. *Astrophys. J. Suppl. Ser.* **48**, 219–237.

Buta R and de Vaucouleurs G 1983 Inner ring structures in galaxies as distance indicators. III. Distances to 453 spiral and lenticular galaxies. *Astrophys. J. Suppl. Ser.* **51**, 149–169.

Butkevich AG, Berdyugin AV and Teerikorpi P 2005a The absolute magnitude of K0V stars from Hipparcos data using an analytical treatment of the Malmquist bias. *Astron. Astrophys.* **435**, 949–954.

Butkevich AG, Berdyugin AV and Teerikorpi P 2005b Statistical biases in stellar astronomy: the Malmquist bias revisited. *Mon. Not. R. Astron. Soc.* **362**, 321–330.

Calzetti D 2001 The dust opacity of star-forming galaxies. *Publ. Astron. Soc. Pac.* **113**, 1449–1485.

Calzetti D, Kinney AL and Storchi-Bergmann T 1994 Dust extinction of the stellar continua in starburst galaxies: the ultraviolet and optical extinction law. *Astrophys. J.* **429**, 582–601.

Cardelli JA, Clayton GC and Mathis JS 1989 The relationship between infrared, optical, and ultraviolet extinction. *Astrophys. J.* **345**, 245–256.

Chaboyer B 1998 The age of the Universe. *Phys. Rep.* **307**, 23–30.

Chaboyer B, Demarque P, Kernan PJ and Krauss LM 1996 A lower limit on the age of the Universe. *Science* **271**, 957–961.

Chaboyer B, Demarque P, Kernan PJ and Krauss LM 1998 The age of globular clusters in light of Hipparcos: resolving the age problem? *Astrophys. J.* **494**, 96–110.

Christenson GE 1995 *Edwin Hubble: Mariner of the Nebulae*. New York, Farrar, Straus, Giroux.

Cohen JG 2006 The not so extraordinary globular cluster 037–B327 in M31. *Astrophys. J.* **653**, L21–L23.

D'Andrea CB, Sako M, Dilday B, Frieman JA, Holtzman J, Kessler R, Konishi K, Schneider DP, Sollerman J, Wheeler JC, Yasuda N, Cinabro D, Jha S, Nichol RC, Lampeitl H, Smith M, Atlee DW, Bassett B, Castander FJ, Goobar A, Miquel R, Nordin J, Östman L, Prieto JL, Quimby R, Riess AG and Stritzinger M 2010 Type II-P supernovae from the SDSS-II Supernova Survey and the standardized candle method. *Astrophys. J.* **708**, 661–674.

de Grijs R, Anders P, Lamers HJGLM, Bastian N, Fritze-von Alvensleben U, Parmentier G, Sharina ME and Yi S 2005 Systematic uncertainties in the analysis of star cluster parameters based on broad-band imaging observations. *Mon. Not. R. Astron. Soc.* **359**, 874–894.

de Vaucouleurs G 1976 Supergalactic studies. V. The supergalactic anisotropy of the redshift–magnitude relation derived from nearby groups and Sc galaxies. *Astrophys. J.* **205**, 13–28.

de Vaucouleurs G 1977 Distances of the Virgo, Fornax and Hydra clusters of galaxies and the local value of the Hubble ratio. *Nature* **266**, 126–129.

de Vaucouleurs G 1993 The extragalactic distance scale. VIII. A comparison of distance scales. *Astrophys. J.* **415**, 10–32.

de Vaucouleurs G and Bollinger G 1979 The extragalactic distance scale. VII. The velocity–distance relations in different directions and the Hubble ratio within and without the local supercluster. *Astrophys. J.* **233**, 433–452.

de Vaucouleurs G and Buta R 1980a Diameters of nuclei, lenses, and inner and outer rings in 532 galaxies. *Astron. J.* **85**, 637–648.

de Vaucouleurs G and Buta R 1980b Inner ring structures in galaxies as distance indicators. I. Dimensionless systematics of inner rings. *Astrophys. J. Suppl. Ser.* **44**, 451–479.

de Vaucouleurs G and Peters WL 1981 Hubble ratio and solar motion from 200 spiral galaxies having distances derived from the luminosity index. *Astrophys. J.* **248**, 395–407.

Di Benedetto GP 2002 On the absolute calibration of the Cepheid distance scale using Hipparcos parallaxes. *Astron. J.* **124**, 1213–1220.

Dobashi K, Bernard J-P, Kawamura A, Egusa F, Hughes A, Paradis D, Bot C and Reach WT 2009 Extinction map of the Small Magellanic Cloud based on the Sirius and 6X 2MASS point source catalogs. *Astron. J.* **137**, 5099–5109.

Eddington AS 1935 The speed of recession of the galaxies. *Mon. Not. R. Astron. Soc.* **95**, 636–638.

Faucher-Giguère C-A and Kaspi VM 2006 Birth and evolution of isolated radio pulsars. *Astrophys. J.* **643**, 332–355.

Feast MW and Catchpole RM 1997 The Cepheid period–luminosity zero-point from Hipparcos trigonometrical parallaxes. *Mon. Not. R. Astron. Soc.* **286**, L1–L5.

Fernie JD 1969 The period–luminosity relation: a historical review. *Publ. Astron. Soc. Pac.* **81**, 707–731.

Fitzpatrick EL 1986 An average interstellar extinction curve for the Large Magellanic Cloud. *Astron. J.* **92**, 1068–1073.

Fitzpatrick EL 1999 Correcting for the effects of interstellar extinction. *Publ. Astron. Soc. Pac.* **111**, 63–75.

Fouqué P, Arriagada P, Storm J, Barnes TG, Nardetto N, Mérand A, Kervella P, Gieren W, Bersier D, Benedict GF and McArthur BE 2007 A new calibration of Galactic Cepheid period–luminosity relations from *B* to *K* bands, and a comparison to LMC relations. *Astron. Astrophys.* **476**, 73–81.

Freedman WL, Madore BF, Gibson BK, Ferrarese L, Kelson DD, Sakai S, Mould JR, Kennicutt Jr RC, Ford HC, Graham JA, Huchra JP, Hughes SMG, Illingworth GD, Macri LM and Stetson PB 2001 Final results from the Hubble Space Telescope Key Project to measure the Hubble constant. *Astrophys. J.* **553**, 47–72.

Gordon KD and Clayton GC 1998 Starburst-like dust extinction in the Small Magellanic Cloud. *Astrophys. J.* **500**, 816–824.

Gordon KD, Calzetti D and Witt AN 1997 Dust in starburst galaxies. *Astrophys. J.* **487**, 625–635.

Gordon KD, Cartledge S and Clayton GC 2009 FUSE measurements of far-ultraviolet extinction. III. The dependence on $R(V)$ and discrete feature limits from 75 Galactic sightlines. *Astrophys. J.* **705**, 1320–1335.

Gratton RG, Fusi Pecci F, Carretta E, Clementini G, Corsi CE and Lattanzi M 1997 Ages of globular clusters from Hipparcos parallaxes of local subdwarfs. *Astrophys. J.* **491**, 749–771.

Groenewegen MAT and Oudmaijer RD 2000 Multi-colour *PL*-relations of Cepheids in the Hipparcos catalogue and the distance to the LMC. *Astron. Astrophys.* **356**, 849–872.

Hanson RB 1979 A practical method to improve luminosity calibrations from trigonometric parallaxes. *Mon. Not. R. Astron. Soc.* **186**, 875–896.

Houk N, Swift CM, Murray CA, Penston MJ and Binney JJ 1997 The properties of main-sequence stars from Hipparcos data. In *Proc. ESA Symp. 'Hipparcos – Venice '97'* (ed. Perryman MAC), *ESA Spec. Publ.* **402**, 279–282.

Hubble EP 1925 Cepheids in spiral nebulae. *Pop. Astron.* **33**, 252–255.

Hubble E 1929 A relation between distance and radial velocity among extra-galactic nebulae. *Publ. Nat'l Acad. Sci. USA* **15**, 168–173.

Hubble EP 1936 *The Realm of the Nebulae*. Yale University Press.

Hubble E and Humason ML 1931 The velocity–distance relation among extra-galactic nebulae. *Astrophys. J.* **74**, 43–80.

Huchra J 2010 Estimates of the Hubble constant. https://www.cfa.harvard.edu/~huchra/hubble.plot.dat. Accessed 20 December 2010.

Humason ML, Mayall NU and Sandage AR 1956 Redshifts and magnitudes of extragalactic nebulae. *Astron. J.* **61**, 97–162.

Ioannidis JPD 2005 Why most published research fundings are false. *PLoS Med.* **2** (8), e124.

Jansen RA, Knapen JH, Beckman JE, Peletier RF and Hes R 1994 Measurements of dust extinction in highly inclined spiral galaxies. *Mon. Not. R. Astron. Soc.* **270**, 373–389.

Jaschek C and Gómez AE 1985 The Malmquist correction. *Astron. Astrophys.* **146**, 387–388.

Jones MI, Hamuy M, Lira P, Maza J, Clocchiatti A, Phillips M, Morrell N, Roth M, Suntzeff NB, Matheson T, Filippenko AV, Foley RJ and Leonard DC 2009 Distance determination to 12 Type II supernovae using the expanding photosphere method. *Astrophys. J.* **696**, 1176–1194.
Koen C 1992 Confidence intervals for the Lutz–Kelker correction. *Mon. Not. R. Astron. Soc.* **256**, 65–68.
Lemaître G 1927 Un Univers homogène de masse constante et de rayon croissant rendant compte de la vitesse radiale des nébuleuses extra-galactiques. *Ann. Soc. Sci. Bruxelles* **A47**, 49–59.
Ljunggren B and Oja T 1966 Absolute magnitudes of late-type giant stars. *Ark. Astron.* **3**, 501–529.
Lorimer DR, Faulkner AJ, Lyne AG, Manchester RN, Kramer M, McLaughlin MA, Hobbs G, Possenti A, Stairs IH, Camilo F, Burgay M, D'Amico N, Corongiu A and Crawford F 2006 The Parkes Multibeam Pulsar Survey. VI. Discovery and timing of 142 pulsars and a Galactic population analysis. *Mon. Not. R. Astron. Soc.* **372**, 777–800.
Luri X, Mennessier MO, Torra J and Figueras F 1993 A new approach to the Malmquist bias. *Astron. Astrophys.* **267**, 305–307.
Luri X, Mennessier MO, Torra J and Figueras F 1996 A new maximum likelihood method for luminosity calibrations. *Astron. Astrophys. Suppl. Ser.* **117**, 405–415.
Lutz TE 1979 On the use of trigonometric parallaxes for the calibration of luminosity systems. II. *Mon. Not. R. Astron. Soc.* **189**, 273–278.
Lutz TE 1983 The calibration of absolute magnitudes from trigonometric parallaxes: sampling. In *The Nearby Stars and the Stellar Luminosity Function* (eds Philip AGD and Upgren AR), *Int'l Astron. Union Colloq.* **76**, 41–50.
Lutz TE and Kelker DH 1973 On the use of trigonometric parallaxes for the calibration of luminosity systems: theory. *Publ. Astron. Soc. Pac.* **85**, 573–578.
Lutz TE and Kelker DH 1975 Calibration of the Wilson–Bappu effect using trigonometric parallaxes. *Publ. Astron. Soc. Pac.* **87**, 617–624.
Madore BF and Freedman WL 1998 Hipparcos parallaxes and the Cepheid distance scale. *Astrophys. J.* **492**, 110–115.
Maíz-Apellániz J 2005 Self-consistent distance determinations for Lutz–Kelker-limited samples. In *The Three-Dimensional Universe with Gaia* (eds Turon C, O'Flaherty KS and Perryman MAC), *ESA Spec. Publ.* **576**, 179–182.
Malmquist KG 1922 On some relations in stellar statistics. *Lund Medd. Ser.* **I 100**, 1–52.
Malmquist KG 1936 The effect of an absorption of light in space upon some relations in stellar statistics. *Stockholms Obs. Medd.*, No. 26.
Marshall DJ, Robin AC, Reylé C, Schultheis M and Picaud S 2006 Modelling the Galactic interstellar extinction distribution in three dimensions. *Astron. Astrophys.* **453**, 635–651.
Mathis JS and Cardelli JA 1992 Deviations of interstellar extinctions from the mean *R*-dependent extinction law. *Astrophys. J.* **398**, 610–620.
Misselt KA, Clayton GC and Gordon KD 1999 Reanalysis of the ultraviolet extinction from interstellar dust in the Large Magellanic Cloud. *Astrophys. J.* **515**, 128–139.
Nugent P, Sullivan M, Ellis R, Gal-Yam A, Leonard DC, Howell DA, Astier P, Carlberg RG, Conley A, Fabbro S, Fouchez D, Neill JD, Pain R, Perrett K, Pritchet CJ and Regnault N 2006 Toward a cosmological Hubble diagram for Type II-P supernovae. *Astrophys. J.* **645**, 841–850.
Oort JH 1931 Some problems concerning the distribution of luminosities and peculiar velocities of extragalactic nebulae. *Bull. Astron. Inst. Neth.* **6**, 155–159.
Oudmaijer RD, Groenewegen MAT and Schrijver H 1998 The Lutz–Kelker bias in trigonometric parallaxes. *Mon. Not. R. Astron. Soc.* **294**, L41–L46.
Paczyński B and Stanek KZ 1998 Galactocentric distance with the Optical Gravitational Lensing Experiment and Hipparcos red clump stars. *Astrophys. J.* **494**, L219–L222.
Piovan L, Tantalo R and Chiosi C 2006a Modelling galaxy spectra in presence of interstellar dust. I. The model of interstellar medium and the library of dusty single stellar populations. *Mon. Not. R. Astron. Soc.* **366**, 923–944.
Piovan L, Tantalo R and Chiosi C 2006b Modelling galaxy spectra in presence of interstellar dust. II. From the ultraviolet to the far-infrared. *Mon. Not. R. Astron. Soc.* **370**, 1454–1478.
Poznanski D, Butler N, Filippenko AV, Ganeshalingam M, Li W, Bloom JS, Chornock R, Foley RJ, Nugent PE, Silverman JM, Cenko SB, Gates EL, Leonard DC, Miller AA, Modjaz M, Serduke FJD,

Smith N, Swift BJ and Wong DS 2009 Improved standardization of Type II-P supernovae: application to an expanded sample. *Astrophys. J.* **694**, 1067–1079.

Poznanski D, Nugent PE and Filippenko AV 2010 Type II-P Supernovae as standard candles: the SDSS-II sample revisited. *Astrophys. J.* **721**, 956–959.

Prévot ML, Lequeux J, Maurice E, Prévot L and Rocca-Volmerange B 1984 The typical interstellar extinction in the Small Magellanic Cloud. *Astron. Astrophys.* **132**, 389–392.

Reid IN 1997 Younger and brighter: new distances to globular clusters based on Hipparcos parallax measurements of local subdwarfs. *Astron. J.* **114**, 161–179.

Reid IN 1999 The HR diagram and the Galactic distance scale after Hipparcos. *Annu. Rev. Astron. Astrophys.* **37**, 191–237.

Rieke GH and Lebofsky MJ 1985 The interstellar extinction law from 1 to 13 microns. *Astrophys. J.* **288**, 618–621.

Riess AG, Macri L, Casertano S, Sosey M, Lampeitl H, Ferguson HC, Filippenko AV, Jha SW, Li W, Chornock R and Sarkar D 2009 A redetermination of the Hubble constant with the Hubble Space Telescope from a differential distance ladder. *Astrophys. J.* **699**, 539–563.

Robertson HP 1928 On relativistic cosmology. *Phil. Mag.* **5**, 835–848.

Robertson HP 1955 The theoretical aspects of the nebular redshift. *Publ. Astron. Soc. Pac.* **67**, 82–98.

Sandage A 1958 Current problems in the extragalactic distance scale. *Astrophys. J.* **127**, 513–526.

Sandage A 1994 Bias properties of extragalactic distance indicators. I. The Hubble constant does not increase outward. *Astrophys. J.* **430**, 1–12.

Sandage A 1995 Practical cosmology: inventing the past. In *The Deep Universe* (eds Binggeli B and Buser R), *Saas–Fee Advanced Course* (Swiss Society for Astrophysics and Astronomy) **23**, 1–232. Springer.

Sandage A and Saha A 2002 Bias properties of extragalactic distance indicators. XI. Methods to correct for observational selection bias for RR Lyrae absolute magnitudes from trigonometric parallaxes expected from the Full-sky Astrometric Mapping Explorer satellite. *Astron. J.* **123**, 2047–2069.

Sandage A and Tammann GA 1974a Steps toward the Hubble constant. Calibration of the linear sizes of extra-galactic HII regions. *Astrophys. J.* **190**, 525–538.

Sandage A and Tammann GA 1974b Steps toward the Hubble constant. II. The brightest stars in late-type spiral galaxies. *Astrophys. J.* **191**, 603–621.

Sandage A and Tammann GA 1974c Steps toward the Hubble constant. III. The distance and stellar content of the M101 group of galaxies. *Astrophys. J.* **194**, 223–243.

Sandage A and Tammann GA 1974d Steps toward the Hubble constant. IV. Distances to 39 galaxies in the general field leading to a calibration of the galaxy luminosity classes and a first hint of the value of H_0. *Astrophys. J.* **194**, 559–568.

Sandage A and Tammann GA 1975a Steps toward the Hubble constant. V. The Hubble constant from nearby galaxies and the regularity of the local velocity field. *Astrophys. J.* **196**, 313–328.

Sandage A and Tammann GA 1975b Steps toward the Hubble constant. VI. The Hubble constant determined from redshifts and magnitudes of remote Sc I galaxies: the value of q_0. *Astrophys. J.* **197**, 265–280.

Sandage A and Tammann GA 1976 Steps toward the Hubble constant. VII. Distances to NGC 2403, M101, and the Virgo cluster using 21 centimeter line widths compared with optical methods: the global value of H_0. *Astrophys. J.* **210**, 7–24.

Sandage A and Tammann GA 2006 Absolute magnitude calibrations of Population I and II Cepheids and other pulsating variables in the instability strip of the Hertzsprung–Russell diagram. *Annu. Rev. Astron. Astrophys.* **44**, 93–140.

Sandage A and Tammann GA 2008 Temperature differences in the Cepheid instability strip require differences in the period–luminosity relation in slope and zero point. *Astrophys. J.* **686**, 779–784.

Sandage A, Tammann GA, Saha A, Reindl B, Macchetto FD and Panagia N 2006 The Hubble constant: a summary of the Hubble Space Telescope program for the luminosity calibration of Type Ia supernovae by means of Cepheids. *Astrophys. J.* **653**, 843–860.

Savage BD and Mathis JS 1979 Observed properties of interstellar dust. *Annu. Rev. Astron. Astrophys.* **17**, 73–111.

Scargle JD 2000 The 'file-drawer' problem in scientific inference. *J. Sci. Expl.* **14**, 91–106.

Schaefer BE 2008 A problem with the clustering of recent measures of the distance to the Large Magellanic Cloud. *Astron. J.* **135**, 112–119.
Schlegel DJ, Finkbeiner DP and Davis M 1998 Maps of dust infrared emission for use in estimation of reddening and cosmic microwave background radiation foregrounds. *Astrophys. J.* **500**, 525–553.
Schonemann PH and Scargle JD 2008 A generalized publication bias model. *Chin. J. Psychol.* **50**, 21–29.
Schultz GV and Wiemer W 1975 Interstellar reddening and IR-excess of O and B stars. *Astron. Astrophys.* **43**, 133–139.
Shafter AW 1997 On the nova rate in the Galaxy. *Astrophys. J.* **487**, 226–236.
Shapley H 1918a Studies based on the colors and magnitudes in stellar clusters. Seventh paper: The distances, distribution in space, and dimensions of 69 globular clusters. *Contrib. Mount Wilson Obs.* **152**, 1–28.
Shapley H 1918b Studies based on the colors and magnitudes in stellar clusters. VII. The distances, distribution in space, and dimensions of 69 globular clusters. *Astrophys. J.* **48**, 154–181.
Slipher VM 1913 The radial velocity of the Andromeda Nebula. *Lowell Obs. Bull.* **1**, 56–57.
Slipher VM 1915 Spectrographic observations of nebulae. *Pop. Astron.* **23**, 21–24.
Smith Jr H 1987a The calibration problem. I. Estimation of mean absolute magnitude using trigonometric parallaxes. *Astron. Astrophys.* **171**, 336–341.
Smith Jr H 1987b The calibration problem. II. Trigonometric parallaxes selected according to proper-motion and the problem of statistical parallaxes. *Astron. Astrophys.* **171**, 342–347.
Smith Jr H 1987c The calibration problem. III. First-order solution for mean absolute magnitude and dispersion. *Astron. Astrophys.* **181**, 391–393.
Smith Jr H 1987d The calibration problem. IV. The Lutz–Kelker correction. *Astron. Astrophys.* **188**, 233–238.
Smith Jr H 2003 Is there really a Lutz–Kelker bias? Reconsidering calibration with trigonometric parallaxes. *Mon. Not. R. Astron. Soc.* **338**, 891–902.
Sneden C, Gehrz RD, Hackwell JA, York DG and Snow TP 1978 Infrared colors and the diffuse interstellar bands. *Astrophys. J.* **223**, 168–179.
Spaenhauer AM 1978 A systematic comparison of four methods to derive stellar space densities. *Astron. Astrophys.* **65**, 313–321.
Stanek KZ and Garnavich PM 1998 Distance to M31 with the Hubble Space Telescope and Hipparcos red clump stars. *Astrophys. J.* **503**, L131–L134.
Tammann GA, Sandage A and Reindl B 2008 Comparison of distances from RR Lyrae stars, the tip of the red giant branch, and classical Cepheids. *Astrophys. J.* **679**, 52–70.
Teerikorpi P 1975 On the effect of the luminosity selection on the redshift–distance relationship. *Astron. Astrophys.* **45**, 117–124.
Teerikorpi P 1993 On general Malmquist corrections to direct and inverse Tully–Fisher distance moduli. *Astron. Astrophys.* **280**, 443–450.
Teerikorpi P 1997 Observational selection bias affecting the determination of the extragalactic distance scale. *Annu. Rev. Astron. Astrophys.* **35**, 101–136.
Teerikorpi P 1998 Cosmological Malmquist bias in the Hubble diagram at high redshifts. *Astron. Astrophys.* **339**, 647–657.
Tempel E, Tuvikene T, Tamm A and Tenjes P 2011 SDSS surface photometry of M31 with absorption corrections. *Astron. Astrophys.* **526**, A155.
Trimble V 1996 H_0: the incredible shrinking constant 1925–1975. *Publ. Astron. Soc. Pac.* **108**, 1073–1082.
Trumpler RJ 1930 Preliminary results on the distances, dimensions and space distribution of open star clusters. *Lick Obs. Bull.* **XIV**, 154–188.
Trumpler RJ and Weaver HF 1953 *Statistical Astronomy*. Dover Publications.
Tully RB and Fisher JR 1977 A new method of determining distances to galaxies. *Astron. Astrophys.* **54**, 661–673.
Turon Lacarrieu C and Crézé M 1977 On the statistical use of trigonometric parallaxes. *Astron. Astrophys.* **56**, 273–281.
Valencic LA, Clayton GC and Gordon KD 2004 Ultraviolet extinction properties in the Milky Way. *Astrophys. J.* **616**, 912–924.

van den Bergh S 1960a The extragalactic distance scale. *Z. Astrophys.* **49**, 198–200.
van den Bergh S 1960b The extragalactic distance scale. *J. R. Astron. Soc. Can.* **54**, 49–58.
van den Bergh S 1975 The extragalactic distance scale. In *Galaxies and the Universe* (eds Sandage A, Sandage M and Kristian J), *Stars Stell. Syst.* **9**, 509–540.
van den Bergh S 1994 The Hubble parameter revisited. *Publ. Astron. Soc. Pac.* **106**, 1113–1119.
van Leeuwen F 2007 Hipparcos, the new reduction of the raw data. *Astrophys. Space Sci. Libr.* **350**. doi: 10.1007/978-1-4020-6342-8_1.
Vaughan S 2005 A simple test for periodic signals in red noise. *Astron. Astrophys.* **431**, 391–403.
Verbiest JPW, Lorimer DR and McLaughlin MA 2010 Lutz–Kelker bias in pulsar parallax measurements. *Mon. Not. R. Astron. Soc.* **405**, 564–572.
Voshchinnikov NV and Il'in VB 1987 An improved method for determining interstellar extinction from color excesses. *Sov. Astron. Lett.* **13**, 157–160.
Wallerstein G 1967 On the use of small parallaxes for calibration purposes. *Publ. Astron. Soc. Pac.* **79**, 317–321.
Weingartner JC and Draine BT 2001 Dust grain-size distributions and extinction in the Milky Way, Large Magellanic Cloud, and Small Magellanic Cloud. *Astrophys. J.* **548**, 296–309.
West RM 1969 On the calibration of $M_V(K)$ for giants by means of trigonometric parallaxes. *Astron. Astrophys.* **3**, 1–4.

7

Promises and Prospects

It is a capital mistake to theorise before one has data. Insensibly one begins to twist facts to suit theories instead of theories to suit facts.
– Sir Arthur Conan Doyle (1859–1930), British physician and novelist, in the words of his fictitious character Sherlock Holmes

No one wants to learn from mistakes, but we cannot learn enough from successes to go beyond the state of the art.
– Henry Petroski (born 1942), American engineer

7.1 The Way Forward: Where Are Significant Gains Achievable?

It will have become clear to even the casual reader that significant progress has been achieved in recent times in establishing an increasingly firm and robust distance ladder, where possible based on well-understood physics. Nevertheless, uncertainties – both systematic and statistical – persist, even for the nearest and presumably best understood rungs of the distance ladder, resulting from different observational or technical approaches, as well as from our incomplete theoretical understanding of relevant physical aspects. An example of such lingering systematic uncertainties and the associated controversy, related to the role of the Pleiades open cluster as a crucial nearby rung of the cosmic distance ladder, is discussed at length in Section 7.2.

Reconciliation of these systematic differences and uncertainties may require further advances in theoretical research, e.g. in terms of a more detailed and improved understanding of the late stages of stellar evolution, stellar atmospheric and pulsation physics, horizontal branch morphologies, including the so-called '**second parameter effect**', and mass-loss processes, among others, as a function of stellar mass. Throughout this book, we have attempted to indicate those areas where theoretical advances are needed to clarify the

An Introduction to Distance Measurement in Astronomy, First Edition. Richard de Grijs.

physical processes at work at a range of physical scales and improve the resulting distance accuracies. Other desirable theoretical improvements, which are not necessarily distance intrinsic, are related to larger-scale diagnostics, such as the Galactic free-electron distribution (see Section 3.8) or Galactic as well as extragalactic extinction mapping and the shape of the relevant reddening laws (cf. Section 6.1.1).

From an observational perspective, the future looks bright across the entire observable wavelength range. Although much current focus is on designing ever larger telescopes, the astronomical community must carefully consider whether the field is best served by having access to the next generation of these extremely large telescopes at optical/near-infrared (near-IR) wavelengths and the Square Kilometre Array in the radio domain or if significant progress can still be made with dedicated 2–4 m-class optical telescopes and upgraded current-generation radio interferometers. Clearly, although they will have small fields of view, larger optical and near-IR telescopes have larger light-collecting areas and will, thus, be able to apply current techniques to objects at greater distances: think of e.g. eclipsing binary analysis potentially at Virgo cluster distances, monitoring Cepheid variables spanning a reasonable period distribution in Coma cluster galaxies and RR Lyrae variables in both spirals and ellipticals in the Virgo cluster, thus providing an independent calibration of Type Ia supernovae (SNe Ia) distances and finally linking the different stellar population tracers (cf. the planetary nebulae luminosity function in Section 4.4).

On the other hand, one only has to consider the tremendous success of surveys with small telescopes, such as the Optical Gravitational Lensing Experiment (OGLE) and the Sloan Digital Sky Survey (SDSS), to realize that smaller, dedicated telescopes still have an important role to play in the overall context of astrophysical distance measurement (see also the efforts by the Stellar Oscillations Network Group to establish a worldwide network of small telescopes for asteroseismology, which will also have an impact on securing a more robust cosmic distance scale; Grundahl *et al.* 2006). After all, in many cases currently unresolved questions benefit from being allocated significant amounts of observing time rather than access to the deep Universe. In this context, the European Southern Observatory's *VISTA* telescope (Emerson *et al.* 2004) will likely play an important role in e.g. achieving firmer zero points for period–luminosity relations at near-IR wavelengths by surveying the Magellanic Clouds as well as the Galactic Centre region and the inner disc through the *VISTA* near-IR YJK_s survey of the Magellanic System (VMC; Cioni *et al.* 2008, 2011) and the *VISTA* Variables in the Vía Láctea (VVV; Minniti *et al.* 2010) public surveys, respectively.

Looking beyond the immediate future, many new ground-based observatories and space-based missions are currently in the design, construction or early operations phases, at wavelengths across the electromagnetic spectrum, from the very high-frequency X-rays (e.g. in the context of improving Sunyaev–Zel'dovich effect measurements; see Section 5.3.2) to low-frequency radio waves (e.g. Section 7.5). Where appropriate, we have highlighted expected progress throughout this book and direct the reader to refer to the relevant chapters for more details. Nevertheless, we point out once again one of the key forthcoming space-based missions of relevance to the field of astrophysical distance measurement. The Milky Way's structure will be characterized to unprecedented levels of accuracy within a few years of the launch of *Gaia* (see Section 2.1.2). Prior to this, in August 2011, a Japanese-led consortium is set to launch its *nano-JASMINE* mission (e.g. Kobayashi *et al.* 2010), the precursor and technical demonstrator of the more ambitious *JASMINE* satellite (Japan Astrometry Satellite Mission for Infrared Exploration; Gouda *et al.* 2005, 2008).

With a mirror size of only 5 cm but a field of view of 0.5×0.5 deg^2, its main scientific aim is to perform IR ($0.6 - 1.0$ μm) astrometry of nearby bright stars with an accuracy of 1 milli-arcsecond (mas), i.e. at the same level as the *Hipparcos* measurements. During its lifetime of >2 years, *nano-JASMINE* will determine highly accurate distances to more than 8000 stars based on annual parallax measurements.

Somewhat further afield, the *James Webb Space Telescope (JWST)* will give us an unprecedentedly high-resolution mid-IR view of the Universe, promising e.g. significant reduction in the uncertainties in mid-IR Cepheid period–luminosity relations (e.g. Madore *et al.* 2009a,b; see also Section 3.5.2) and red giant branch bump validation as a distance indicator (e.g. Valenti *et al.* 2004), among others. Observations at IR wavelengths hold significant promise in relation to improved or alternative methods of distance determination. For example, Barrau *et al.* (2008) suggested that the absorption of very high-energy ($10^{11} - 10^{13}$ eV) γ rays from distant active galactic nuclei by the **cosmic** *infrared* **background** (CIB, spanning wavelengths from a few microns to a few millimetres; the absorption is caused by electron–positron pair production) could be used to obtain a lower limit on the Hubble constant. They suggest that since the spectral distribution of the CIB is well characterized (cf. Dole *et al.* 2006), it is possible to recover the intrinsic spectrum of a given active galactic nucleus as a function of the integrated CIB density along the line of sight. Given that the absorption is proportional to $\exp(1/H_0)$, a firm lower limit on H_0 can be derived. This is based on the idea that for larger intrinsic distances and, hence, higher values of H_0, the number of CIB photons along the line of sight will be larger. For sufficiently low values of H_0, the shape of the recovered spectrum will become unacceptable and, hence, a robust lower limit can be obtained. Based on conservative hypotheses, Barrau *et al.* (2008) thus obtained $H_0 > 74$ km s^{-1} Mpc^{-1} (68% confidence level). This method could be a viable and independent alternative to other current techniques, particularly in view of new contributions to our knowledge of the CIB between 60 and 110 μm by the *Herschel* mission (e.g. Franceschini *et al.* 2006) and between 5 and 60 μm by the *JWST* (e.g. Windhorst *et al.* 2006).

Finally, novel methods for distance determination are proposed on a regular basis, although they do not always measure up to their promises upon closer inspection or application in practice. Particularly at greater distances, most currently favoured distance tracers are based on physical properties of the objects of interest (i.e. resulting in luminosity distances). Elvis and Karovska (2002) proposed a novel method to determine *geometric* distances to radio-quiet quasars using the size of their **broad-emission-line regions** (Peterson 1993, 2001) as 'standard ruler' in what essentially corresponds to a **quasar parallax**, i.e. an inverted trigonometric parallax approach (see Section 2.1.1; see also Komberg 2004 for radio-loud quasars and radio galaxies featuring jets). An equivalent approach yielded a distance to SN 1987A with an accuracy of 6% (Panagia *et al.* 1991; Binney and Merrifield 1998). The linear size of these broad-emission-line regions, which are caused by photo-ionization of gas by a highly luminous source with a very broad spectrum, thus producing a wide range of ionization levels, is known from light travel time measurements (Blandford and McKee 1982; Peterson 1993, 2001; Netzer and Peterson 1997) – also known as **reverberation mapping** – while their angular size can be measured using interferometry. Despite significant challenges, Elvis and Karovska (2002) confidently predict that interferometers with 0.01 mas resolution will soon be able to plausibly measure the size of the emission-line regions of $z = 2$ quasars to sufficient accuracy to constrain the prevailing cosmological parameters.

In the remainder of this chapter, we first revisit the distance controversy associated with one of the lowest rungs of the distance ladder, the Pleiades open cluster. Sections 7.3, 7.4 and 7.5 discuss, at some level of detail, the prospects and promises of using (i) scattered X-ray haloes to determine distances to nearby galaxies, (ii) **gravitational wave** sources as 'standard sirens' and (iii) the redshifted 21 cm HI line as a diagnostic tool to map the 3D distribution of neutral hydrogen at high redshifts. Finally, in Section 7.6, we provide an updated version of the distance ladder, in essence summarizing the entire book in a single figure.

7.2 The Pleiades Distance Controversy

The Pleiades open cluster is a crucial rung of the local distance ladder, whose calibration affects many fundamental aspects of stellar astrophysics. In fact, this open cluster has been used for calibration purposes ever since the very first 'Hertzsprung–Russell' diagram was published (Rosenberg 1910). However, the original *Hipparcos* parallaxes (Mermilliod *et al.* 1997; van Leeuwen and Hansen-Ruiz 1997; Robichon *et al.* 1999; van Leeuwen 1999), as well as the recalibrated astrometry (van Leeuwen 2007a,b; see Section 2.1.2), yielded distances to the individual member stars and the open cluster as a whole that were systematically lower than those resulting from previous ground-based distance determinations. The latter were predominantly based on the main-sequence fitting technique (see Section 3.2.1) because prior to the successful *Hipparcos* mission stellar parallaxes at the distance of the Pleiades were too small to be measurable reliably with contemporary instrumentation. The 1997 *Hipparcos*-based parallaxes implied a distance of 118 ± 4 pc (i.e. a distance modulus of $m - M = 5.37 \pm 0.06$ mag or $d = 390 \pm 10$ lightyears; Mermilliod *et al.* 1997; van Leeuwen *et al.* 1997), compared to a long succession of prior research, advocating a distance in the range of 130–135 pc (420–440 lightyears; $m - M = 5.60 \pm 0.04$; e.g. Pinsonneault *et al.* 1998; An *et al.* 2007; and references therein).

Doubt was initially cast on the original *Hipparcos* analysis, which required advanced mathematical techniques to solve simultaneously for the positions, motions and distances of 118 000 stars. Pan *et al.* (2004) analysed the orbital parameters of the bright binary star Atlas (HR1178/HD23850), a member of the Pleiades' Seven Sisters, using long-baseline optical/IR interferometry with the Palomar Testbed Interferometer, and reported a firm lower limit of $d = 127$ pc (414 lightyears), with a most likely distance range of $133 < d < 137$ pc (or 440 ± 7 lightyears). Zwahlen *et al.* (2004) confirmed this distance estimate for Atlas, reporting $d = 132 \pm 4$ pc based on interferometric astrometry combined with radial velocity information (a completely independent approach). Subsequently, Soderblom *et al.* (2005) derived relative trigonometric parallaxes and proper motions for three members of the Pleiades using the *Hubble Space Telescope (HST)*'s Fine Guidance Sensor, which they attempted to tie to the absolute astrometric reference frame based on the observed colours, spectral types and luminosity classes of a subset of reference-frame stars. They reported an absolute parallax of $\varpi_{\rm abs} = 7.43 \pm 0.17$ (systematic) ± 0.20 (statistical) mas (but see van Leeuwen 2009 for a critical assessment of these uncertainty estimates), corresponding to $d = 134.6 \pm 3.1$ pc ($m - M = 5.65 \pm 0.05$ mag). In addition, combination of this result with that of Pan *et al.* (2004) and an independent eclipsing-binary-based distance (Munari *et al.* 2004: $d = 132 \pm 2$ pc, *modulo* a somewhat uncertain reddening correction) leads to $d = 133.5 \pm 1.2$ pc ($m - M = 5.63 \pm 0.02$ mag). Note that many of these distances were derived for individual stars and are, therefore, not necessarily representative of the cluster as a whole, particularly in the presence of a significant dispersion along the line of sight.

Soderblom *et al.* (2005) validate their distance determination by pointing out the good agreement with recent estimates from main-sequence fitting (see also Table 3 in van Leeuwen 2009; note that these methods are not strictly independent), i.e. $d = 132 \pm 4$ pc (Stello and Nissen 2001), 132 ± 2 pc (Pinsonneault *et al.* 1998) and 131 ± 7 pc (Gatewood *et al.* 2000), and an independent estimate of $d = 131 \pm 24$ pc (Narayanan and Gould 1999) based on the gradient in the radial velocities in the direction of the cluster's proper motion, although the large associated uncertainty makes this 'agreement' rather meaningless. Makarov (2002) reanalysed *Hipparcos* data, taking into account the relevant magnitude-dependent astrometric noise characteristics (van Leeuwen and Fantino 2003), to derive $d = 129 \pm 3$ pc (but note that this result and the proposed correction are refuted by van Leeuwen 2005, 2009).

Although the *Hipparcos* recalibration reduced the discrepancy slightly, with an updated distance of 122 ± 2 pc (399 ± 6 lightyears), the difference remains too large for comfort: the variation in distance modulus implied is approximately 0.2–0.3 mag (Pinsonneault *et al.* 1998), while the difference in parallax required is of order 1 mas, but note that the absolute uncertainty in *Hipparcos* parallaxes is only 0.1 mas (Arenou *et al.* 1995; Lindegren 1995). However, the *Hipparcos* team (e.g. van Leeuwen 2009 and references therein) points out that these independent distance determinations are based on small numbers of stars covering small fields of view, while van Leeuwen *et al.* (2007a,b) included parallax measurements of a larger sample of 53 Pleiades members spread across 9–10 degrees on the sky (see e.g. Figure 7.1, which also includes an indication of the expected parallax for a distance of 132 pc). In addition, *HST*'s parallaxes are relative – they were derived differentially against background stars within a small field – while the *Hipparcos* parallaxes are absolute (i.e. the outcome of a global solution over the whole sky).

The controversy has, thus, not been fully resolved, and all methods applied to date are affected by their own unique sets of uncertainties (see e.g. Valls-Gabaud 2007). From the perspective of the underlying physics, to account for the *Hipparcos* distance, stellar models would require changes in physics – e.g. in the Pleiades' characteristic metallicity or helium abundance, or the mixing length (see, for a review, Valls-Gabaud 2007) – or input parameters, such as an age differential between local and Pleiades member stars or an unusual spatial distribution, i.e. depth effects (cf. Stello and Nissen 2001; but see Soderblom *et al.* 2005), that are too radical to be reasonable. The independent approach of main-sequence fitting compares the Pleiades main sequence with a mean sequence constructed from nearby stars characterized by large parallaxes. It is often argued that a small correction for evolution is necessary – the Pleiades cluster is approximately 100 Myr old (Pinsonneault *et al.* 1998), while the nearby field stars are typically as old as the Sun – but the cluster stars appear to have essentially the same metallicity as the Sun (e.g. Boesgaard and Friel 1990), although perhaps not the same surface gravity (e.g. van Leeuwen 2000). On the other hand, most models applied to resolve the Pleiades controversy include many assumptions and simplifications in terms of stellar structure (rotation, convection, magnetic fields), stellar evolution and stellar atmospheres, which may well dominate or negate the need for the proposed small evolutionary correction.

Given the crucial role of the Pleiades as a distance anchor, it is entirely unsatisfactory that the current distance controversy remains. Valls-Gabaud (2007) suggests to use a physically

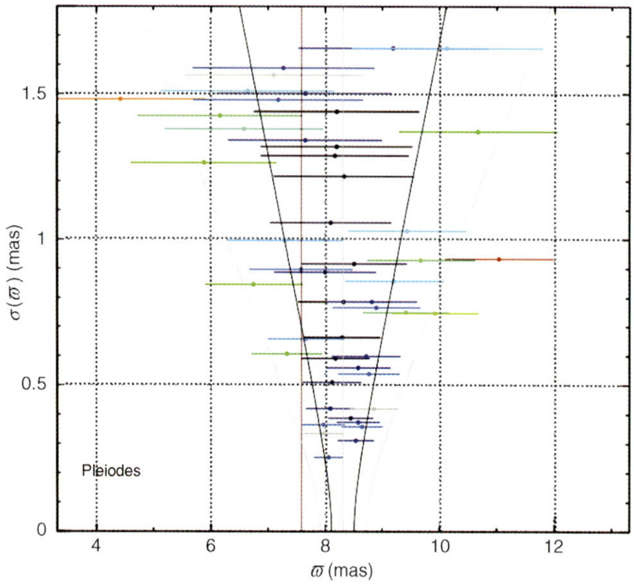

Figure 7.1 *Distribution of individual stellar Pleiades parallaxes (ϖ) versus their formal errors, $\sigma(\varpi)$, based on Hipparcos data. The central vertical (grey) line represents the mean parallax, the black and grey lines on either side show the 1 and 2σ uncertainty levels, respectively, including the internal distance dispersion. The red vertical line corresponds to a distance of 132 pc (van Leeuwen 2009).* (Reprinted from F. van Leeuwen, *Astronomy and Astrophysics*, **497**, Parallaxes and proper motions for 20 open clusters as based on the new Hipparcos catalogue, p. 209–242, Copyright 2009, with permission of ESO.)

independent means of distance determination by focusing on eclipsing binary stars, in particular the star HD 23642, of which 'professional' observations go back as far as Galileo (1610)! Modern analyses (e.g. Giannuzzi 1995; Lastennet and Valls-Gabaud 2002; Munari *et al.* 2004; Southworth *et al.* 2005) yield distances in the same range as the main-sequence fitting technique, i.e. Southworth *et al.* (2005) derived $d = 139 \pm 3$ pc for reasonable estimates of the binary components' effective temperatures, compared to a *Hipparcos* distance of 111 ± 12 pc for this star.

The US National Radio Astronomy Observatory recently approved an ambitious 'Key Science Project' with the dual aims to solve this controversy prior to the release of the relevant *Gaia* data sets (foreseen for 2015) *and* determine the distance to the Pleiades to an accuracy of better than 1% using its Very Long Baseline Array (VLBA; see Section 2.1.2). The relative astrometric accuracy of $\sim$10 mas achievable with the VLBA surpasses the specifications of the *Gaia* satellite for most stars in its catalogue. In 2010, the VLBA began a long-term programme to determine the complete 3D structure of the Milky Way by measuring parallaxes with $\lesssim$ 10 mas accuracy to some 400 high-mass star-forming regions, aimed at measuring the fundamental Galactic parameters with 1% accuracy. The combination of this VLBA Key Science Project with the high-precision astrometric accuracy offered by *Gaia* promises to eventually resolve the Pleiades distance controversy and, hence, provide a highly robust first rung of the distance ladder.

7.3 X-Ray Scattering Haloes

When X-rays interact with interstellar dust grains, the signal is both absorbed and scattered, particularly over small angles (a unique property of X-rays because of their high energies[1]). As a consequence, X-ray sources observed behind large dust columns are surrounded by haloes of faint and diffuse radiation, analogous to traffic lights on a foggy night. Overbeck (1965) predicted this effect (see also Slysh 1969; Hayakawa 1970, 1973; Martin 1970) and first discussed its use as a powerful diagnostic tool, provided that the dust grain size distribution and its scattering properties are accurately known (cf. Mathis and Lee 1991; Draine 2003; Draine and Tan 2003; Vaughan *et al.* 2004, 2006), but it took another 20 years until the first detection of an X-ray halo with the *Einstein X-ray Observatory* (Catura 1983; Rolf 1983; see also Draine and Tan 2003). Predehl and Schmitt (1995) analysed an extensive data set of *ROSAT (Röntgen Satellite)* observations and found a strong correlation between the X-ray halo strength or, equivalently, the scattering optical depth, the X-ray absorption and the visual extinction for X-ray sources with known optical counterparts (see also Mauche and Gorenstein 1986). In turn, these results enable us to disentangle interstellar and local matter because both extinction and X-ray absorption are produced by the total column density between source and observer (representing the 'local' environment), but the scattering is only due to dust on large scales (the interstellar medium).

Trümper and Schönfelder (1973) already suggested that observations of X-ray haloes associated with variable X-ray sources could be used to obtain precise geometric distances, although distance estimates depend somewhat on the assumed density distribution of dust (see, for details, Draine and Tan 2003). The underlying physics is based on the fact that scattered light must travel along a slightly longer path than direct, unscattered light. Any intensity variations of the source will, therefore, be seen somewhat delayed in the halo. Following Predehl *et al.* (2000; see also Draine and Bond 2004 and Ling *et al.* 2009 for detailed mathematical treatments), we point out that the time delay Δt (in seconds) of a single-scattered photon with respect to an unscattered photon is given by (see Figure 7.2)

$$\Delta t = \frac{d}{2c}\frac{x\alpha^2}{1-x} = 1.15d\frac{x\alpha^2}{1-x}, \tag{7.1}$$

where d is the actual distance (in kpc), c the speed of light in a vacuum, x the fractional distance from the observer where the scattering occurs (i.e. $x \ll 1$ is the ratio of the distance between the X-ray source and the scattering medium to that between the source and the observer) and α the observed angle (in arcseconds). Alternatively, Draine and Bond (2004) derive

$$\Delta t \sim 140 d\alpha^2 \text{ days}, \tag{7.2}$$

for d and α expressed in Mpc and units of 100 arcsec, respectively. (For a detailed mathematical treatment of both single and multiple scattering, see Draine and Tan 2003.) The halo radiation is delayed and smeared out, and originates from scattering along the entire line of sight. This thus produces different time delays for different distances and annuli,

[1] This approach cannot be used with optical telescopes because visible light has lower energy and is scattered through much larger angles. Note, however, that Xu *et al.* (1995) discuss using visible light scattered by dust (i.e. a light echo; cf. Section 3.7.2) to study the 3D structure of the interstellar medium in front of SN 1987A in the Large Magellanic Cloud.

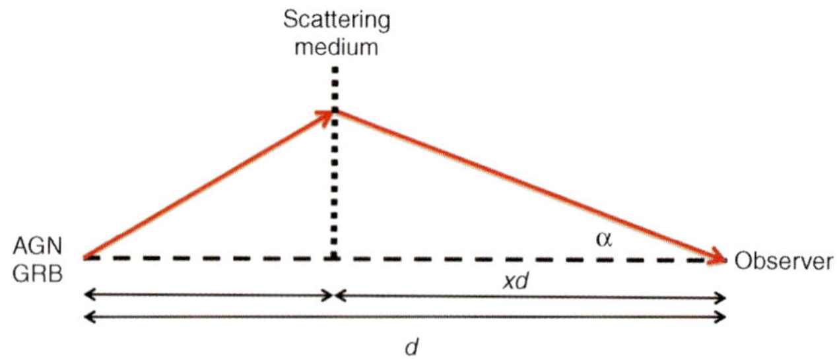

Figure 7.2 *Geometry for single scattering of X-rays by dust. AGN: active galactic nucleus. GRB: gamma-ray burst.*

although most scattered light originates from dust grains close to the observer. If we now plot the dust distance distribution d, in the ideal case we obtain a single peak centred on the real dust distance, as well as a width representing the depth of the dust distribution, superimposed on a background distribution (see Figure 7.3).

Although the basic idea is simple, measuring this effect in practice is difficult because the amplitude and timescale of the intensity variations, combined with the distance to the source, must match the observational angular resolution (e.g. Molnar and Mauche 1986; Kitamoto *et al.* 1989). Because the scattered radiation is both delayed and smeared out, the intensity variations are reduced. This damping is small for the inner halo regions, while the variability can no longer be observed at large distances in the halo because the delay of the radiation scattered off interstellar dust becomes comparable to or longer than the timescales of the intrinsic intensity variations. Thus, the distance to the source can, in principle, be determined based on observations of time delays and their variations throughout X-ray

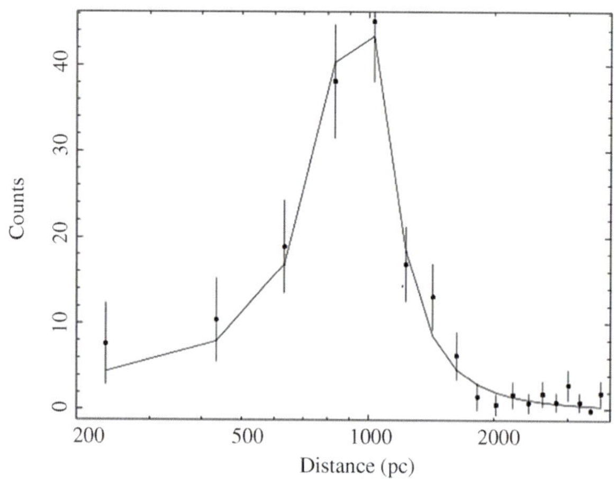

Figure 7.3 *Background-subtracted dust distance distribution of GRB 061019 (Vianello et al. 2007).* (Reprinted from G. Vianello *et al.*, *Astronomy and Astrophysics*, **473**, Dust-scattered X-ray halos around two Swift gamma-ray bursts: GRB 061019 and GRB 0701292007, p. 423–427, Copyright 2007, with permission of ESO.)

haloes (e.g. Draine and Bond 2004). The first successful application of this technique was achieved using high-resolution X-ray observations of the eclipsing binary system Cygnus X-3 with the *Chandra X-ray Observatory* (Predehl *et al.* 2000).

These authors started from the simplifying assumption that the scattering medium is located exactly halfway to the source, which resulted in a first-guess distance estimate of 10 kpc. They then used the system's light curves as a function of energy, combined with Monte Carlo simulations of simulated scattering events – assumed to be distributed uniformly around the source – which they then matched to their observations to yield a final distance estimate of 9^{+4}_{-2} kpc (see also Hu *et al.* 2003 for an independent attempt using the same data), where the level of uncertainty was determined by their limited temporal coverage. This approach has since been used to determine the distance to the bright X-ray source Nova V1974 Cygni 1992 (Draine and Tan 2003), the low-mass X-ray binary 4U 1624−490 (Xiang *et al.* 2007), the high-mass X-ray binary Cygnus X-1 (Ling *et al.* 2009) and – using gamma-ray burst emission observed in X-rays (Vianello *et al.* 2007; see also Shao and Dai 2007 and Shao *et al.* 2008 based on scattered emission close to the gamma-ray bursts themselves) – to the Galactic molecular cloud [KOY98] 66 (Kawamura *et al.* 1998), as well as the dust distribution and properties along a number of specific sight lines to gamma-ray bursts (e.g. Vaughan *et al.* 2004, 2006; Romano *et al.* 2006; Tiengo and Mereghetti 2006; Vianello *et al.* 2007). Distances to galaxies projected in front of bright gamma-ray bursts could, in principle, be determined with accuracies of better than 1% (Draine and Bond 2004), although the time-variable geometry of the X-ray halo would show an expanding ring structure rather than a static halo. The X-ray counts in this ring scale as $1/d$ (Draine and Bond 2004), so that closer galaxies would be better for distance determination. Specifically, for typical dust distances $d_{\rm dust}$ (hundreds of parsecs), these expanding rings have radii (θ) of order a few arcminutes after several thousand seconds post-burst, i.e. (Vianello *et al.* 2007)

$$\theta\ [\text{arcsec}] = \sqrt{\frac{827}{d_{\rm dust}\ [\text{pc}]}t\ [\text{s}]}. \tag{7.3}$$

Application of the scattering X-ray halo method to extragalactic objects, particularly through repeated X-ray imaging with the *Chandra X-ray Observatory* of time-varying active galactic nuclei or quasars located behind a foreground galaxy, which would act as origin of the X-ray halo, has been discussed by a number of authors (e.g. Rudak and Mészáros 1991; Klose 1996; Draine and Bond 2004), but not yet demonstrated in practice. This particular geometry would allow an independent determination of the distance to the foreground galaxy without significant systematic uncertainties. Draine and Bond (2004) explore the use of the **BL Lac** object 5C 3.76 to obtain a distance to M31 to an unprecedented absolute accuracy of ~1% (4%) in 4 (2) months of on-source observing time. Although they also performed simulations to assess the potential of distance determinations to the Large and Small Magellanic Clouds and M81, the lack of bright background sources behind these galaxies renders M31 the most favourable object.

Note that for application to extragalactic objects, one needs to take into account that the Galactic halo contains dust that acts as foreground extinction and also contributes to the scattered halo. In the direction of M31, it contributes $A_V \approx 0.3$ mag, so that the Galactic foreground signal caused by scattering off dust at a distance d will result in a time delay of $\sim 3600(d/300\ \text{pc})(\theta/100'')^2$ s, where θ is the angular distance from the source position. Draine and Bond (2004) argue that this timescale is short compared to the much longer time delays expected for active galactic nucleus or quasar variability, so that the Galactic foreground contribution is expected to predominantly affect the Poissonian noise characteristics.

Thus, the technique of distance determination through careful analysis of scattering X-ray haloes is promising yet thus far unproven. It is anticipated that dedicated observation campaigns with the *Chandra X-ray Observatory* have the potential to yield percent-level distance accuracy to M31, and potentially to other nearby galaxies if sufficiently bright background X-ray sources can be identified.

7.4 Standard Sirens: Listening to Gravitational Waves

At intermediate to high redshifts, SNe Ia provide very useful standard candles, with peak brightnesses that can be calibrated to 15% accuracy (see Section 5.2.1) and, hence, yield reliable luminosity distances. Unfortunately, however, we do not yet have a full and robust theoretical understanding of the physical basis of these explosive events. This is particularly worrisome given that any evolutionary effects in terms of their brightnesses would introduce unknown systematic errors (e.g. Drell *et al.* 2000; but see Section 5.2.1 for counterarguments).

To avoid these potential systematic effects, a number of authors have suggested a completely independent distance tracer which covers a similar distance range and extends to the cosmic horizon (e.g. Schutz 1986; Haehnelt 1998; Hughes 2002, 2003; Holz and Hughes 2005; Arun *et al.* 2009). **Inspiralling of massive binary black holes** (BBHs) – black holes which are widely separated and slowly spiral together – will generate gravitational waves (GWs, i.e. fluctuations in the curvature of spacetime, propagating outwards from their source in the form of waves), the back-reaction of which is responsible for this slow 'inspiralling'. This is understood and can be modelled very well using 'post-Newtonian' terms (i.e. an approximation to a general relativistic formulation; see e.g. Blanchet *et al.* 2002), particularly during the early inspiral phase. From inspiral GWs one can derive a binary's luminosity distance (but not its redshift, since this is degenerate with the binary's evolution[2]), as well as its position on the sky, its orientation, and information pertinent to mass and spin combinations (see Arun *et al.* 2004; Blanchet *et al.* 2004). This information can be gleaned from GW observations, independent of assumptions about the masses and orbital parameters of the binary system's components.

GWs are a natural consequence of Einstein's theory of general relativity (Einstein 1916). Einstein derived his 'quadrupole formula' (in its modern version corrected by a factor of 2; cf. Hughes 2003 and references therein), predicting the rate at which radiation caused by variations in an object's gravitational (mass) quadrupole moment propagates. In this context, a system's mass monopole and dipole represent its total mass energy and centre of mass, respectively. Neither change as a function of time, so they do not generate radiation. The mass quadrupole, on the other hand, may be time variable and contribute to GW generation. The quadrupole moment, $Q \sim$ (source mass)(source size)^2, is in essence a measure of how far from spherically symmetric the radiating source is.

[2] The timescales, t, of a binary's orbital evolution are determined by its component masses, m_i, as $t \sim Gm_i/c^3$, where G is the Newtonian gravitational constant. These timescales and – as a consequence, the binary masses – redshift, so that a binary with masses (m_1, m_2) at redshift z is indistinguishable from a binary system at $z = 0$ with masses $[(1 + z)m_1, (1 + z)m_2]$ (cf. Holz and Hughes 2005).

This can be understood by considering the analogy with electromagnetic radiation. A single, stationary charge (i.e. a monopole) cannot radiate because this would violate charge conservation. However, a single *oscillating* charge will radiate in the form of a dipole and higher-order radiation field. In general, the dipole moment produces by far the strongest radiation field, but it requires the centre of charge to be subject to acceleration with respect to its centre of mass, e.g. when the charges on a spinning rod have opposite signs at either end. An oscillating *mass* will also have a time-varying dipole moment. However, the only way in which it can oscillate is by oscillation of another mass with equal but opposite momentum because of conservation of either momentum or angular momentum. The time-varying dipole moment of this second component will exactly offset that of the first, so that no radiation will be generated, given that the centre of the 'gravitational charge' coincides with the system's centre of mass. Gravitational quadrupole radiation is similar to the configuration in which both ends of the spinning rod are given the same charge. In this case, the only feature of the system that changes is the extent and the shape of the charge distribution in space, so that little radiation is produced.

Returning now to GWs, these are oscillations of spacetime itself caused by the acceleration of mass, rather than of electric and magnetic fields that propagate through spacetime. The clear review by Hughes (2003) provides a useful analogy to understand this difference. Formally, the oscillating contribution to the curvature of spacetime induced by GWs varies on length scales $\lambda/2\pi$ (where λ is the GW wavelength), i.e. on much shorter scales than those over which all other important curvatures vary. Thus, GWs are more similar to propagating waves on the ocean's surface compared to the curvature of the Earth. They only weakly interact with matter – i.e. they will be affected by negligible absorption on their way to an observer on Earth, which also makes their detection difficult, however – and arise from the bulk dynamics of a dense source of mass energy (such as massive BBHs), so that they directly probe their source's dynamical state.

Since their wavelengths are generally comparable to or larger than the size of their source, GWs cannot be used to image these objects. Instead, GWs are characterized by two polarizations (named '+' and '×', referring to the axial orientation of their tidal squeezing and stretching behaviour; see Figure 7.4) which make them analogous to propagating aural

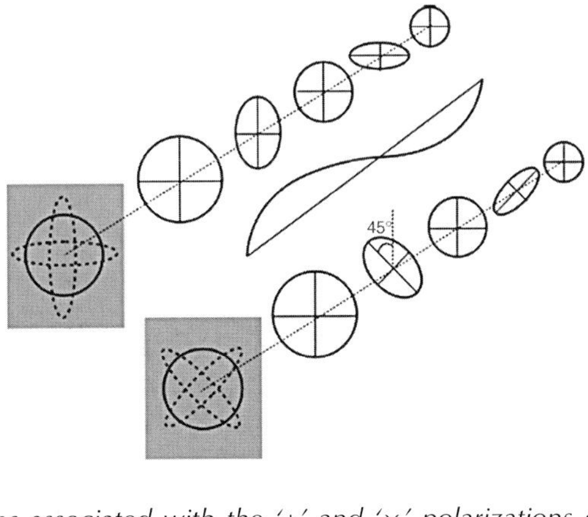

Figure 7.4 *Distortions associated with the '+' and '×' polarizations of gravitational waves (Kokkotas 2002).* (Reprinted from K. D. Kokkotas, *Encyclopedia of Physical Science and Technology, 7,* Gravitational wave physics, p. 67–85, Copyright 2002, with permission of Elsevier.)

(sound) waves. Therefore, GW events associated with inspiralling BBHs are referred to as *standard sirens*. They have the potential to determine luminosity distances, d_L, with typical accuracies of $\Delta d_L/d_L \sim 1$–10% (Hughes 2002; Holz and Hughes 2005; Arun *et al.* 2009). Distance estimates can be significantly improved, in theory to better than 0.5–1% in many cases (cf. Schutz 1986, 2002; Holz and Hughes 2005; Arun *et al.* 2009), if an electromagnetic counterpart (an astronomical object, an afterglow or even a precursor caused by gas infall; cf. Armitage and Natarajan 2002; Milosavljević and Phinney 2005; Lang and Hughes 2008) can be identified. It may also be possible to use the *distribution* of binary systems or their host galaxies for cosmological purposes, even to the extent that H_0 can potentially be constrained to ~1% levels, so that we do not need to identify individual objects (Chernoff and Finn 1993; Finn 1996; Wang and Turner 1997; MacLeod and Hogan 2008).

However, predominantly weak gravitational lensing of the GW signal along the line of sight will deteriorate distance accuracies, particularly for objects at $z \gtrsim 1.5$, to similar levels as currently achievable for SNe Ia, but with different systematics (cf. Holz and Hughes 2005; Kocsis *et al.* 2006). For a ΛCDM Universe (i.e. dominated by cold dark matter and a cosmological constant, Λ), the expected mean error rate caused by weak gravitational lensing is $\langle \Delta d_L/d_L \rangle \simeq 0.005$, with a standard deviation of $\langle (\Delta d_L/d_L)^2 \rangle \simeq 0.05$, reduction of which on a case-by-case basis is difficult (Dalal *et al.* 2003; Gunnarsson *et al.* 2006; but see Jönsson *et al.* 2007, Hilbert *et al.* 2010, Shapiro *et al.* 2010 and Shang and Haiman 2011 for recent, promising advances; see also Simon *et al.* 2009 for a tomographic reconstruction). This 5–10% effect is significantly greater than the intrinsic distance error associated with the use of GWs as standard sirens (Holz and Hughes 2005; Holz and Linder 2005). In addition, GW detectors are, in essence, 'all-sky' detectors with poor spatial resolution and, hence, their ability to localize the source of any GW detection is limited. As Hughes (2002) points out, this further strengthens the analogy of GW detection as 'listening to the Universe', given that human hearing is similarly poorly capable of locating the origin of sound waves, yet we usually have high-resolution vision. This completes the analogy with electromagnetic-wave detection.

The tidal squeezing and stretching induced by passing GWs is most apparent if these forces act on objects (test masses) with much smaller sizes than the GW wavelength (see Hughes 2003 for a review). This can, in principle, be determined by measuring the time-varying separation between test masses which are otherwise entirely isolated from other perturbing effects, using interferometric detectors such as e.g. the *(Advanced) Laser Interferometer Gravitational Wave Observatory (LIGO)* or the future *Laser Interferometer Space Antenna (LISA)* (e.g. Danzmann 1998; Holz and Hughes 2005; Arun *et al.* 2009). For reasons of energy conservation, the so-called **wave strain** h – a measure of the fractional size distortion caused by a passing GW – must fall off with distance as $1/r$ (Hughes 2003),

$$h \sim \frac{G}{c^4}\frac{\ddot{Q}}{r}, \tag{7.4}$$

where $\ddot{Q}$ is the second derivative of the quadrupole moment,

$$\ddot{Q} \sim 2Mv^2 \simeq 4E_{\rm kin}^{\rm ns}. \tag{7.5}$$

Here, M is the system's mass, v its internal velocity and $E_{\rm kin}^{\rm ns}$ the nonspherical part of its internal kinetic energy. Thus, it becomes apparent that objects which have strong nonspherical dynamics are good candidates for GW generation. In general, the sensitivity levels of h that must be reached for a positive source detection are of order 10^{-21}–10^{-22}. This implies, in practice, that for every kilometre length of an interferometer arm, we must be able to detect tidal distortions of better than 10^{-16} cm (for feasibility, see Weiss 1972 and the review by Hughes 2003).

The 'natural' frequency of GWs originating from a compact source is given by

$$f_{\rm GW}(M, R) = \frac{1}{2\pi}\frac{{\rm d}\Phi(t)}{{\rm d}t} \sim \frac{1}{2\pi}\sqrt{\frac{GM}{R^3}}, \tag{7.6}$$

where the phase $\Phi(t)$ depends on the masses – particularly the **chirp mass** (see below) – and spins of the binary components (e.g. Poisson and Will 1995) and R is the scale over which the source's dynamics vary, e.g. an object's actual size or the binary separation. Because the **Schwarzschild radius** – i.e. the radius at which, if all the mass were compressed inside of this distance, the escape speed would equal the speed of light – $R \sim 2GM/c^2$, which is a lower limit for most sources, the hard upper bound to the GW frequency is

$$f_{\rm GW}(M) < \frac{1}{4\sqrt{2}\pi}\frac{c^3}{GM} \simeq 10^4 {\rm Hz}\left(\frac{{\rm M}_\odot}{M}\right). \tag{7.7}$$

LISA will be sensitive to BBH GWs at frequencies $10^{-5} \lesssim f_{\rm GW} \lesssim 0.1$ Hz, i.e. to binaries with total masses $m_1 + m_2 \sim 10^3$–10^6 M$_\odot$ (Danzmann *et al.* 1998) located at $z \lesssim 5$–10 (Hughes 2002; Vecchio 2004). Observations of such objects will be most useful for efforts to constrain the cosmological parameters, including the evolution of the Hubble constant as a function of redshift (e.g. Nishizawa *et al.* 2010, who propose to use the method of Bonvin *et al.* 2006a,b to 10^5–10^6 inspiralling neutron-star binaries to achieve percent-level accuracies). One can derive the luminosity distances of BBHs by determining the quadrupole moment harmonics of both polarizations as

$$h_+ = \frac{2\mathcal{M}_z^{5/3}[\pi f_{\rm GW}(t)]^{2/3}}{d_{\rm L}}\left[1 + (\hat{L}\cdot\hat{n})^2\right]\cos[\Phi(t)], \tag{7.8}$$

$$h_\times = \frac{4\mathcal{M}_z^{5/3}[\pi f_{\rm GW}(t)]^{2/3}(\hat{L}\cdot\hat{n})}{d_{\rm L}}\sin[\Phi(t)]. \tag{7.9}$$

Here, $\mathcal{M}_z = (1+z)(m_1 m_2)^{3/5}/(m_1+m_2)^{1/5}$ is the binary's redshifted 'chirp mass', which defines the rate at which the inspiralling occurs, thus determining the 'chirp' of the orbital frequency. The object's sky position is defined by the unit vector $\hat{n}$, which points from the centre of the barycentre frame (centred on the solar system) to the binary system. Finally, $\hat{L}$ defines the binary's orientation, since it points along the direction of the system's orbital angular momentum. From Equations (7.8) and (7.9) it follows that we can determine $d_{\rm L}$ by fixing these angles.

Holz and Hughes (2005) show that the best *LISA*-based distance determinations can be obtained for binaries with total redshifted masses of $(1+z)(m_1+m_2) \simeq$ several $\times\ 10^5$ M$_\odot$ (see their tables 1–4). This is driven by the reduced signal-to-noise ratios of low-mass binary systems – which will also be closer to their final merger when their signals become strong enough for reliable measurements – on the one hand and the poor sensitivity at low frequencies on the other. The latter will limit *LISA*'s sensitivity to binary systems with $(1+z)(m_1+m_2) \lesssim$ a few $\times\ 10^6\ M_\odot$.

7.5 Three-Dimensional Mapping of Redshifted Neutral Hydrogen

A promising approach to measuring the prevailing cosmological parameters to high accuracy is provided by means of **21 cm tomography** (e.g. Mao *et al.* 2008), i.e. 3D neutral hydrogen mapping of the Universe using the redshifted 21 cm emission line (see Figure 7.5). Although this signal has not yet been detected, it can potentially tightly constrain the nature of dark matter, dark energy, the early Universe and the end of the cosmic '**dark ages**', the period

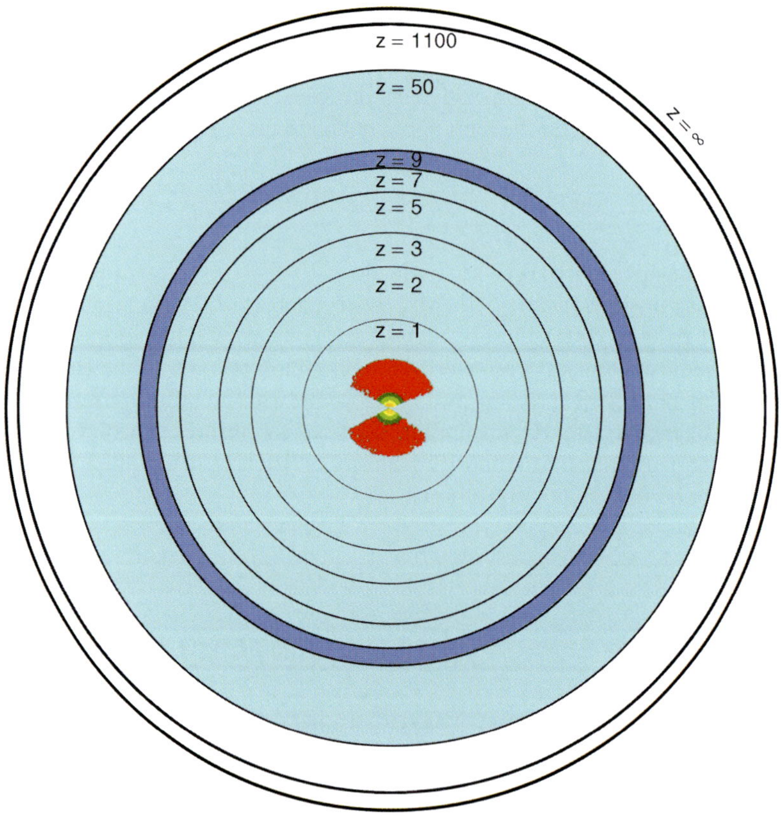

Figure 7.5 *The technique of 21 cm tomography can potentially map most of our observable Universe (light blue), while the cosmic microwave background probes mainly a thin shell at $z \sim 1100$ and current large-scale structure surveys (here represented by the* SDSS *and its luminous red galaxies) map only small volumes near the centre. The most convenient region for initial 21 cm tomography efforts is $6 \lesssim z \lesssim 9$ (dark blue).* (Reprinted from Y. Mao *et al.*, *Physics Review D*, **78**, How accurately can 21cm tomography constrain cosmology?, 023529, Copyright 2008, with permission of the APS and M. Tegmark.)

between recombination of electrons with hydrogen and helium nuclei and the epoch of recombination. It can also help us understand when precisely reionization occurred, so that we can potentially trace the transition from a neutral to the current, ionized Universe and possibly see the first stars as well.

In addition, direct 3D imaging of the high-z intergalactic medium will enable deduction of the structure of the 'cosmic web' with unprecedented accuracy. Mapping the redshifted 21 cm line – produced by spin flips in neutral hydrogen atoms, which is strictly forbidden by quantum-mechanical laws but happens either spontaneously on occasion or may be collisionally induced – offers the best approach to achieving this. Note that this **'hyperfine' transition** can be observed either in emission or absorption against the cosmic microwave background (CMB)'s blackbody spectrum (i.e. we do not require bright background objects), depending on whether the spin temperature is higher or lower than the CMB temperature, respectively. In essence, this implies that the measured CMB intensity will increase or decrease depending on the object's temperature. The HI spin temperature is likely lower than the CMB temperature only *before* or at the beginning of the epoch of recombination, so for all practical purposes we can assume that we are dealing with an emission line, with the proviso that future, higher sensitivity instruments may also be able to detect the 21 cm signal in absorption at these higher redshifts.

The alternatives, i.e. using observations of the CMB or quasar absorption lines – such as the Lyman-α forest, CIV absorption systems or future observations of the 21 cm forest, i.e. neutral hydrogen absorption against high-z radio sources – are less attractive, despite recent new technological advances enabling improved probing of the interstellar medium using the *Hubble Space Telescope*'s Cosmic Origins Spectrograph.[3] This is so, because the CMB maps the (2D) surface of last scattering at a given time (i.e. it is a 2D snapshot), while quasar absorption spectra provide only highly directional ($\sim$1D), pointed observations of the intergalactic medium along a given line of sight. The reason that HI observations allow mapping in three rather than two dimensions is that the redshift of the 21 cm line provides the radial coordinate along the line of sight. This signal can be observed from the dark ages (Shapiro *et al.* 2006; Lewis and Challinor 2007) – before any stars had formed – through the epoch of recombination and even to the present time. However, the redshifted 21 cm line is difficult to access at $z \gtrsim 6$ because of increasing synchrotron foreground contributions and the presence of less neutral hydrogen at those higher redshifts, corresponding to lower frequencies. Note that this technique does not, therefore, provide an immediate distance measurement, but it allows detailed 3D mapping of the largest accessible volumes in the Universe.

Highly redshifted 21 cm emission is expected to be dominated by large holes caused by HII regions blown by quasars (e.g. Tozzi *et al.* 2000; Barkana and Loeb 2001; Wyithe and Loeb 2004; Kohler *et al.* 2005; Wyithe *et al.* 2005; Fan *et al.* 2006) and galaxy clusters (e.g. Furlanetto *et al.* 2004; Santos *et al.* 2008). Although these large-scale features have not yet been detected, the prospects of their discovery are exciting, since they may well be the first signatures of the epoch of recombination which will be detected by the next generation of redshifted 21 cm experiments (cf. Furlanetto *et al.* 2006), including the Square Kilometre Array (SKA).[4] Their successful detection depends on a sufficient level of contrast between

[3] see e.g. `http://www.spacetelescope.org/images/heic09101/`.
[4] `http:///www.skatelescope.org`

neutral and ionized regions. A rule of thumb for the required (brightness-)temperature contrast is (cf. Furlanetto *et al.* 2006)

$$\Delta T_{\mathrm{b}} \approx 22 x_{\mathrm{HI}}(1+\delta)\sqrt{\frac{1+z}{7.5}}\ \mathrm{mK}, \tag{7.10}$$

where x_{HI} is the fraction of the intergalactic medium in neutral hydrogen, $(1+\delta)$ is the fractional baryon overdensity with respect to the mean (i.e. the hydrogen mass density fluctuation) and we assume that the 21 cm spin temperature (which is equivalent to the excitation temperature) $T_{\mathrm{S}} \gg T_{\gamma}$ (i.e. the CMB temperature at the relevant redshift). Note, however, that if the intergalactic medium is partially ionized by either X-rays or incomplete recombination in fossil HII regions, the contrast is reduced accordingly. Formally, the brightness temperature is the difference between the total observed flux,

$$T_{\mathrm{b,tot}} = T_{\gamma}(z)\exp(-\tau_{21}) + T_{\mathrm{S}}\left[1-\exp(-\tau_{21})\right], \tag{7.11}$$

where τ_{21} is the (redshifted) 21 cm optical depth (Field 1959; see also Santos *et al.* 2008), and the CMB temperature, *modulo* a redshift correction,

$$\Delta T_{\mathrm{b}}(\nu) = \frac{\left(T_{\mathrm{b,tot}} - T_{\gamma}(z)\right)\tau_{21}}{(1+z)} \tag{7.12}$$

and ν is the observational frequency, so that (Santos *et al.* 2008)

$$T_{\mathrm{b}}(\nu) \sim (1+\delta)x_{\mathrm{HI}}\left(1-\frac{1+z}{H(z)}\frac{\partial v}{\partial r}\right)\left(\frac{T_{\mathrm{S}}-T_{\gamma}}{T_{\mathrm{S}}}\right)\left(\frac{1+z}{10}\right)^{1/2}\ \mathrm{mK}, \tag{7.13}$$

where $\partial v/\partial r$ is the peculiar velocity gradient, which is a measure of the gravitational potential (see Section 5.1.2).

Furlanetto *et al.* (2006) speculate that if the mean HI fraction in the intergalactic medium is significantly high, $\langle x\rangle_{\mathrm{HI}} > 0.2$, at $z \sim 6.0$–6.5, large and continuously growing HII bubbles should be detectable with next-generation experiments such as the Low Frequency Array (LOFAR)[5] centred on the Netherlands, the 21 Centimeter Array (21CMA: see Figure 7.6)[6] in western China, the Murchison Widefield Array (MWA)[7] in Australia and the SKA, among others (see Furlanetto *et al.* 2006 for a review and technical details).

Finally, we have thus far assumed that we know the correct underlying cosmological model. Use of an incorrect cosmological model would generate apparent errors in the scaling of angular compared to line-of-sight sizes, which depend on the angular diameter distance and H_0, respectively. In turn, this would introduce an artificial anisotropy even in intrinsically isotropic galaxy distributions, referred to as the **Alcock–Paczyński (1979) effect**. Although this can be used to measure cosmological parameters, this has proven difficult in practice (Hui *et al.* 1999; Eriksen *et al.* 2005; see Furlanetto *et al.* 2006 for a review of recent efforts in this area). The main problems are the shallow depth of galaxy redshift surveys (but see Hu and Haiman 2003) versus the sparse coverage of deeper quasar surveys (e.g. Matsubara and Szalay 2002). Because the 21 cm signal covers the entire

[5] http://www.lofar.org.
[6] http://21cma.bao.ac.cn.
[7] http://www.haystack.mit.edu/ast/arrays/mwa/.

Figure 7.6 *The 21 Centimeter Array (21CMA) in the Xinjiang province of western China, consisting of approximately 10 000 dipole antennas.*

sky and, therefore, does not suffer from sparseness problems, using 21 cm tomography may allow a definitive detection of the Alcock–Paczyński effect (e.g. Scott and Rees 1990; Nusser 2005; Barkana 2006).

7.6 The Present-Day Distance Ladder

Remarkable and significant progress as regards the accuracy and robustness of cosmic distances at any scale has been made in the past few decades. The launch of the *HST* in the early 1990s proved a pivotal event in reducing the uncertainties in the Hubble constant, predominantly through carefully calibrated Cepheid-based extragalactic distances. The *Wilkinson Microwave Anisotropy Probe* has allowed determination of both the prevailing cosmological parameters and the Hubble constant at high redshift to unprecedented accuracy and precision, provided that the cosmological-model-dependent assumptions at the basis of these results retain their validity as ever more precise and larger-scale measurements are becoming available. Lower rungs of the distance ladder have also seen (at least partial) convergence of their absolute levels through cross-calibration with independent methods of distance determination. Nevertheless, establishing a fully robust distance ladder remains a lofty goal and may, in fact, be but an unreachable dream, given the significant uncertainties affecting many of the contributing methods, even the most robust techniques (cf. the Pleiades controversy discussed earlier in this chapter).

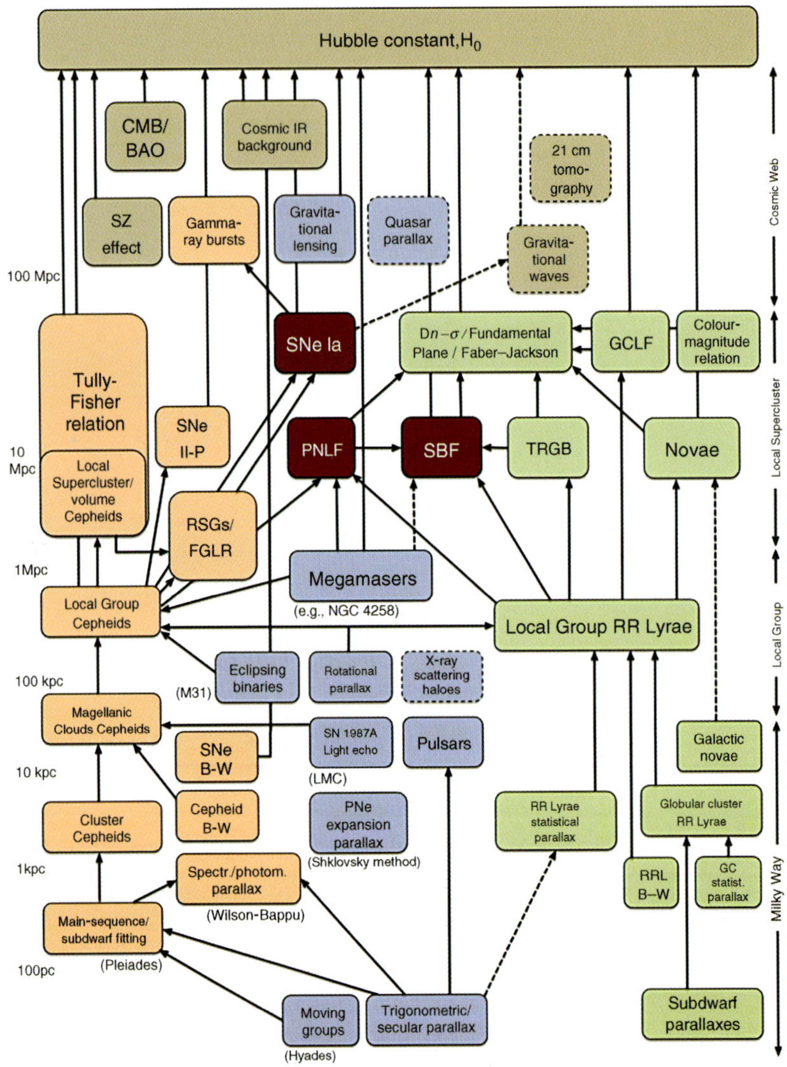

Figure 7.7 Updated, present-day distance ladder, based on an original idea by Ciardullo (2006). Light orange: methods of distance determination associated with active star formation ('Population I', intermediate- and high-mass stars). Light green: distance tracers associated with 'Population II' objects/low-mass stars. Blue: geometric methods. Red: supernovae (SNe) Ia, the planetary nebulae (PNe) luminosity function (PNLF) and surface brightness fluctuations (SBF) are applicable for use with both Populations I and II. Light brown: methods of distance or H_0 determination which are not immediately linked to a specific stellar population. Dashed boxes: proposed methods. Solid, dashed arrows: reasonably robust, poorly established calibrations. B–W: Baade–Wesselink. RRL: RR Lyrae. RSGs/FGLR: red supergiants/flux-weighted gravity–luminosity relationship. TRGB: tip of the red giant branch. GCLF: globular cluster (GC) luminosity function. SZ: Sunyaev–Zel'dovich. CMB/BAO: cosmic microwave background/baryon acoustic oscillations. Colour–magnitude relation: refers to galactic colours and magnitudes (Chapter 4.7).

We have tried to give an objective and comprehensive account of the current state of play in this very dynamic field. New and promising methods have appeared next to their well-established counterparts. This is, therefore, an opportune time to update the graphical representation of the distance ladder originally proposed by Ciardullo (2006). Figure 7.7 summarizes the applicability, distance range, mutual dependences and robustness of the most common methods of distance determination discussed in this book.

Bibliography

Alcock C and Paczyński B 1979 An evolution free test for non-zero cosmological constant *Nature* **281**, 358–359.

An D, Terndrup DM, Pinsonneault MH, Paulson DB, Hanson RB and Stauffer JR 2007 The distances to open clusters from main-sequence fitting. III. Improved accuracy with empirically calibrated isochrones. *Astrophys. J.* **655**, 233–260.

Arenou F Lindegren L, Froeschle M, Gomez AE, Turon C, Perryman MAC and Wielen R 1995 Zero-point and external errors of Hipparcos parallaxes. *Astron. Astrophys.* **304**, 52–60.

Armitage PJ and Natarajan P 2002 Accretion during the merger of supermassive black holes. *Astrophys. J.* **567**, L9–L12.

Arun KG, Blanchet L, Iyer BR and Qusailah MSS 2004 The 2.5PN gravitational wave polarizations from inspiralling compact binaries in circular orbits. *Class. Quant. Grav.* **21**, 3771–3801.

Arun KG, Babak S, Berti E, Cornish N, Cutler C, Gair J, Hughes SA, Iyer BR, Lang RN, Mandel I, Porter EK, Sathyaprakash BS, Sinha S, Sintes AM, Trias M, Van Den Broeck C and Volonteri M 2009 Massive black-hole binary inspirals: results from the LISA parameter estimation taskforce. *Class. Quant. Grav.* **26**, 094027.

Barkana R 2006 Separating out the Alcock–Paczyński effect on 21 cm fluctuations. *Mon. Not. R. Astron. Soc.* **372**, 259–264.

Barkana R and Loeb A 2001 In the beginning: the first sources of light and the reionization of the Universe. *Phys. Rep.* **349**, 125–238.

Barrau A, Gorecki A and Grain J 2008 An original constraint on the Hubble constant: $h > 0.74$. *Mon. Not. R. Astron. Soc.* **389**, 919–924.

Binney J and Merrifield MR 1998 *Galactic Astronomy.* Princeton University Press.

Blanchet L, Faye G, Iyer BR and Joguet B 2002 Gravitational-wave inspiral of compact binary systems to 7/2 post-Newtonian order. *Phys. Rev. D* **65**, 061501.

Blanchet L, Damour T, Esposito-Farèse G and Iyer BR 2004 Gravitational radiation from inspiralling compact binaries completed at the third post-Newtonian order. *Phys. Rev. Lett.* **93**, 091101.

Blandford RD and McKee CF 1982 Reverberation mapping of the emission line regions of Seyfert galaxies and quasars. *Astrophys. J.* **255**, 419–439.

Boesgaard AM and Friel ED 1990 Chemical composition of open clusters. I. Fe/H from high-resolution spectroscopy. *Astrophys. J.* **351**, 467–479.

Bonvin C, Durrer R and Gasparini MA 2006a Fluctuations of the luminosity distance. *Phys. Rev. D* **73**, 023523.

Bonvin C, Durrer R and Kunz M 2006b Dipole of the luminosity distance: a direct measure of $H(z)$. *Phys. Rev. Lett.* **96**, 191302.

Catura RC 1983 Evidence for X-ray scattering by interstellar dust. *Astrophys. J.* **275**, 645–651.

Chernoff DF and Finn LS 1993 Gravitational radiation, inspiraling binaries, and cosmology. *Astrophys. J.* **411**, L5–L8.

Ciardullo R 2006 The planetary nebula luminosity function. In *Planetary Nebulae Beyond the Milky Way* (eds Stanghellini L, Walsh JR and Douglas N), *ESO Astrophys. Symp.*, pp. 79–90.

Cioni M-RL, Bekki K, Clementini G, de Blok WJG, Emerson JP, Evans CJ, de Grijs R, Gibson BK, Girardi L, Groenewegen MAT, Ivanov VD, Leisy P, Marconi M, Mastropietro C, Moore B, Naylor T, Oliveira JM, Ripepi V, van Loon JT, Wilkinson MI and Wood PR 2008 The Magellanic Clouds as a template for the study of stellar populations and galaxy interactions. *Publ. Astron. Soc. Aust.* **25**, 121–128.

Cioni M-RL, Clementini G, Girardi L, Guandalini R, Gullieuszik M, Miszalski B, Moretti M-I, Ripepi V, Rubele S, Bagheri G., Bekki K., Cross N, de Blok WJG, de Grijs R, Emerson JP, Evans CJ, Gonzales-Solares E, Groenewegen MAT, Irwin M, Ivanov VD, Kerber L, Lewis J, Marconi M, Marquette J-B, Mastropietro C, Moore B, Napiwotzki R, Naylor T, Oliveira JM, Read M, Sutorius E, van Loon JT, Wilkinson MI and Wood PR 2011 The VMC Survey. I. Strategy and first data. *Astron. Astrophys.* 527, A116.

Dalal N, Holz DE, Chen X and Frieman JA 2003 Corrective lenses for high-redshift supernovae. *Astrophys. J.* **585**, L11–L14.

Danzmann K 1998 Laser-interferometric gravitational wave detectors: on the ground and in deep space. In *Relativistic Astrophysics* (eds Riffert H, Ruder H, Nollert H-P and Hehl FW), pp. 48–65. Vieweg Verlag.

Dole H, Lagache G, Puget J-L, Caputi KI, Fernández-Conde N, Le Floc'h E, Papovich C, Pérez-González PG, Rieke GH and Blaylock M 2006 The cosmic infrared background resolved by Spitzer. Contributions of mid-infrared galaxies to the far-infrared background. *Astron. Astrophys.* **451**, 417–429.

Doyle AC 1887 *A Study in Scarlet*, Part 1, Chapter 3. Ward Lock & Co.

Draine BT 2003 Interstellar dust grains. *Annu. Rev. Astron. Astrophys.* **41**, 241–289.

Draine BT and Bond NA 2004 Direct extragalactic distance determination using X-ray scattering. *Astrophys. J.* **617**, 987–1003.

Draine BT and Tan JC 2003 The scattered X-ray halo around Nova Cygni 1992: testing a model for interstellar dust. *Astrophys. J.* **594**, 347–362.

Drell PS, Loredo TJ and Wasserman I 2000 Type Ia supernovae, evolution, and the cosmological constant. *Astrophys. J.* **530**, 593–617.

Einstein A 1916 Die Grundlage der allgemeinen Relativitätstheorie. *Ann. Phys.* **354**, 769–822.

Elvis M and Karovska M 2002 Quasar parallax: method for determining direct geometrical distances to quasars. *Astrophys. J.* **581**, L67–L70.

Emerson JP, Sutherland WJ, McPherson AM, Craig SC, Dalton GB and Ward AK 2004 The Visible & Infrared Survey Telescope for Astronomy. *ESO Messenger* **117**, 27–32.

Eriksen KA, Marble AR, Impey CD, Bai L and Petry CE 2005 The Alcock–Paczyński test for the Lyα forest: first results from Magellan and the MMT. In *Observing Dark Energy* (eds Wolff SC and Lauer TR), *Astron. Soc. Pac. Conf. Ser.* **339**, 172–176.

Fan X, Strauss MA, Becker RH, White RL, Gunn JE, Knapp GR, Richards GT, Schneider DP, Brinkmann J and Fukugita M 2006 Constraining the evolution of the ionizing background and the Epoch of Reionization with $z \sim 6$ quasars. II. A sample of 19 quasars. *Astron. J.* **132**, 117–136.

Field GB 1959 An attempt to observe neutral hydrogen between the galaxies. *Astrophys. J.* **129**, 525–535.

Finn LS 1996 Binary inspiral, gravitational radiation, and cosmology. *Phys. Rev. D* **53**, 2878–2894.

Franceschini A, Vaccari M, Berta S, Rodighiero G and Lonsdale C 2006 Galaxy evolution through infrared surveys: from Spitzer to Herschel. In *Studying Galaxy Evolution with Spitzer and Herschel* (eds Charmandaris V, Rigopoulou D and Kylafis N) (astro-ph/0610897).

Furlanetto SR, Zaldarriaga M and Hernquist L 2004 The growth of HII regions during reionization. *Astrophys. J.* **613**, 1–15.

Furlanetto SR, Oh SP and Briggs FH 2006 Cosmology at low frequencies: the 21 cm transition and the high-redshift Universe. *Phys. Rep.* **433**, 181–301.

Galileo G 1610 *Sidereus Nuncius* (*Sidereal Messenger*), Venice.

Gatewood G, de Jonge JK and Han I 2000 The Pleiades, map-based trigonometric parallaxes of open clusters. V. *Astrophys. J.* **533**, 938–943.

Giannuzzi MA 1995 The spectroscopic binary HD 23642 and the distance of the Pleiades. *Astron. Astrophys.* **293**, 360–362.

Gouda N, Yano T, Yamada Y, Kobayashi Y, Tsujimoto T and the JASMINE Working Group 2005 Japan Astrometry Satellite Mission for Infrared Exploration (JASMINE). In *Proc. The Three-Dimensional Universe with Gaia* (eds Turon C, O'Flaherty KS and Perryman MAC), *ESA Spec. Publ.* **576**, 77–80.

Gouda N, Kobayashi Y, Yamada Y, Yano T and the JASMINE Working Group 2008 Infrared space astrometry project JASMINE. In *A Giant Step: from Milli- to Micro-arcsecond Astrometry* (eds Jin WJ, Platais I and Perryman MAC), *Proc. Int'l Astron. Union* **248**, 248–251.

Grundahl F, Kjeldsen H, Frandsen S, Andersen M, Bedding T, Arentoft T and Christensen-Dalsgaard J 2006 SONG: Stellar Oscillations Network Group. A global network of small telescopes for asteroseismology and planet searches. *Mem. Soc. Astron. It.* **77**, 458–459.

Gunnarsson C, Dahlén T, Goobar A, Jönsson J and Mörtsell E 2006 Corrections for gravitational lensing of supernovae: better than average? *Astrophys. J.* **640**, 417–427.

Haehnelt MG 1998 Supermassive black holes as sources for LISA. In *Second Int'l Laser Interferometer Space Antenna (LISA) Symp. on the Detection and Observation of Gravitational Waves in Space* (ed. Folkner WM), *Astron. Inst. Phys. Conf. Proc.* **456**, 45–49.

Hayakawa S 1970 Scattering of cosmic X-rays by interstellar dust grains. *Prog. Theor. Phys.* **43**, 1224–1230.

Hayakawa S 1973 Cosmic X-rays and interstellar dust. In *Interstellar Dust and Related Topics* (eds Greenberg JM and van de Hulst HC), *Proc. Int'l Astron. Union Symp.* **52**, 283–296.

Hilbert S, Gair JR and King LJ 2010 Reducing distance errors for standard candles and standard sirens with weak-lensing shear and flexion maps. *Mon. Not. R. Astron. Soc.* **412**, 1023–1037.

Holz DE and Hughes SA 2005 Using gravitational-wave standard sirens. *Astrophys. J.* **629**, 15–22.

Holz DE and Linder EV 2005 Safety in numbers: gravitational lensing degradation of the luminosity distance–redshift relation. *Astrophys. J.* **631**, 678–688.

Hu J, Zhang SN and Li TP 2003 Distance determination of variable Galactic sources. In *Stellar Astrophysics: A Tribute to Helmut A. Abt, Proc. 6th Pac. Rim Conf.* (eds Cheng KS, Leung KC and Li TP). *Astrophys. Space Sci. Libr.* **298**, 55–59.

Hu W and Haiman Z 2003 Redshifting rings of power. *Phys. Rev. D* **68**, 063004.

Hughes SA 2002 Tuning gravitational-wave detector networks to measure compact binary mergers. *Phys. Rev. D* **66**, 102001.

Hughes SA 2003 Listening to the Universe with gravitational-wave astronomy. *Ann. Phys.* **303**, 142–178.

Hui L, Stebbins A and Burles S 1999 A geometrical test of the cosmological energy contents using the Lyα forest. *Astrophys. J.* **511**, L5–L8.

Jönsson J, Goobar A and Mörtsell E 2007 Tuning gravitationally lensed standard sirens. *Astrophys. J.* **658**, 52–59.

Kawamura A, Onishi T, Yonekura Y, Dobashi K, Mizuno A, Ogawa H and Fukui Y 1998 A ^{13}CO survey of molecular clouds in Gemini and Auriga. *Astrophys. J. Suppl. Ser.* **117**, 387–425.

Kitamoto S, Miyamoto S and Yamamoto T 1989 Scattered X-ray halo and transient dips of Cygnus X-1. *Publ. Astron. Soc. Jpn* **41**, 81–95.

Klose S 1996 Dust-scattered X-ray halos and the quasar redshift controversy. *Astrophys. J.* **473**, 806–809.

Kobayashi Y, Yano T, Gouda N, Niwa Y, Murooka J, Yamada Y, Sako N and Nakasuka S 2010 Nano-JASMINE: current status and data output. *Proc. SPIE* **7731**, 77313Z.

Kocsis B, Frei Z, Haiman Z and Menou K 2006 Finding the electromagnetic counterparts of cosmological standard sirens. *Astrophys. J.* **637**, 27–37.

Kohler K, Gnedin NY, Miralda-Escudé J and Shaver PA 2005 Redshifted 21 cm emission from the pre-reionization era. II. HII regions around individual quasars. *Astrophys. J.* **633**, 552–559.

Kokkotas KD 2002 Gravitational wave physics. In *Encyclopedia of Physical Science and Technology* (ed. Meyers RA), 3rd edn, vol. 7, pp. 67–85. Academic Press.

Komberg BV 2004 A method for estimating the distances of extragalactic radio sources with jets using VLBI. *Astron. Rep.* **48**, 699–704.

Lang RN and Hughes SA 2008 Localizing coalescing massive black hole binaries with gravitational waves. *Astrophys. J.* **677**, 1184–1200.

Lastennet E and Valls-Gabaud D 2002 Detached double-lined eclipsing binaries as critical tests of stellar evolution. Age and metallicity determinations from the HR diagram. *Astron. Astrophys.* **396**, 551–580.

Lewis A and Challinor A 2007 21 cm angular-power spectrum from the dark ages. *Phys. Rev. D* **76**, 083005.

Lindegren L 1995 Estimating the external accuracy of Hipparcos parallaxes by deconvolution. *Astron. Astrophys.* **304**, 61–68.

Ling Z, Zhang SN and Tang S 2009 Determining the distance of Cyg X-3 with its X-ray dust scattering halo. *Astrophys. J.* **695**, 1111–1120.

MacLeod CL and Hogan CJ 2008 Precision of Hubble constant derived using black hole binary absolute distances and statistical redshift information. *Phys. Rev. D* **77**, 043512.

Madore BF, Rigby J, Freedman WL, Persson SE, Sturch L and Mager V 2009a The Cepheid period–luminosity relation (the Leavitt law) at mid-infrared wavelengths. III. Cepheids in NGC 6822. *Astrophys. J.* **693**, 936–939.

Madore BF, Freedman WL, Rigby J, Persson SE, Sturch L and Mager V 2009b The Cepheid period–luminosity relation (the Leavitt law) at mid-infrared wavelengths. II. Second-epoch LMC data. *Astrophys. J.* **695**, 988–995.

Makarov VV 2002 Computing the parallax of the Pleiades from the Hipparcos intermediate astrometry data: an alternative approach. *Astron. J.* **124**, 3299–3304.

Mao Y, Tegmark M, McQuinn M, Zaldarriaga M and Zahn O 2008 How accurately can 21 cm tomography constrain cosmology? *Phys. Rev. D* **78**, 023529.

Martin PG 1970 On the interaction of cosmic X-rays with interstellar grains. *Mon. Not. R. Astron. Soc.* , **149**, 221–235.

Mathis JS and Lee C-W 1991 X-ray halos as diagnostics of interstellar grains. *Astrophys. J.* **376**, 490–499.

Matsubara T and Szalay AS 2002 Cosmological parameters from redshift-space correlations. *Astrophys. J.* **574** 1–8.

Mauche CW and Gorenstein P 1986 Measurements of X-ray scattering from interstellar grains. *Astrophys. J.* **302**, 371–387.

Mermilliod J-C, Turon C, Robichon N, Arenou F and Lebreton Y 1997 The distance of the Pleiades and nearby clusters. In *Proc. ESA Symp. 'Hipparcos – Venice '97'* (ed. Perryman MAC), *ESA Spec. Publ.* **402**, 643–650.

Milosavljević M and Phinney ES 2005 The afterglow of massive black hole coalescence. *Astrophys. J.* **622**, L93–L96.

Minniti D, Lucas PW, Emerson JP, Saito RK, Hempel M, Pietrukowicz P, Ahumada AV, Alonso MV, Alonso-Garcia J, Arias JI, Bandyopadhyay RM, Barbá RH, Barbuy B, Bedin LR, Bica E, Borissova J, Bronfman L, Carraro G, Catelan M, Clariá JJ, Cross N, de Grijs R, Dékány I, Drew JE, Fariña C, Feinstein C, Fernández Lajús E, Gamen RC, Geisler D, Gieren W, Goldman B, Gonzalez OA, Gunthardt G, Gurovich S, Hambly NC, Irwin MJ, Ivanov VD, Jordán A, Kerins E, Kinemuchi K, Kurtev R, López-Corredoira M, Maccarone T, Masetti N, Merlo D, Messineo M, Mirabel IF, Monaco L, Morelli L, Padilla N, Palma T, Parisi MC, Pignata G, Rejkuba M, Roman-Lopes A, Sale SE, Schreiber MR, Schröder AC, Smith M, Sodré Jr L, Soto M, Tamura M, Tappert C, Thompson MA, Toledo I, Zoccali M and Pietrzyński G 2010 VISTA Variables in the Vía Láctea (VVV): the public ESO near-IR variability survey of the Milky Way. *New Astron.* **15**, 433–443.

Molnar LA and Mauche CW 1986 Effects of the X-ray scattering halo on the observational properties of Cygnus X-3. *Astrophys. J.* **310**, 343–353.

Munari U, Dallaporta S, Siviero A, Soubiran C, Fiorucci M and Girard P 2004 The distance to the Pleiades from orbital solution of the double-lined eclipsing binary HD 23642. *Astron. Astrophys.* **418**, L31–L34.

Narayanan VK and Gould A 1999 Correlated errors in Hipparcos parallaxes toward the Pleiades and the Hyades. *Astrophys. J.* **523**, 328–339.

Netzer H and Peterson BM 1997 Reverberation mapping and the physics of active galactic nuclei. In *Astronomical Time Series* (eds Maoz D, Sternberg A and Leibowitz EM), pp. 85–108. Kluwer.

Nishizawa A, Taruya A and Saito S 2011 Tracing the redshift evolution of Hubble parameter with gravitational-wave standard sirens. *Phys. Rev. D.* **83**, 084045.

Nusser A 2005 The Alcock–Paczyński test in redshifted 21 cm maps. *Mon. Not. R. Astron. Soc.* **364**, 743–750.

Overbeck JW 1965 Small-angle scattering of celestial X-rays by interstellar grains. *Astrophys. J.* **141**, 864–866.

Pan X, Shao M and Kulkarni SR 2004 A distance of 133–137 parsecs to the Pleiades star cluster. *Nature* **427**, 326–328.

Panagia N, Gilmozzi R, Macchetto FD, Adorf H-M and Kirshner RP 1991 Properties of the SN 1987A circumstellar ring and the distance to the Large Magellanic Cloud. *Astrophys. J.* **380**, L23–L26.

Peterson BM 1993 Reverberation mapping of active galactic nuclei. *Publ. Astron. Soc. Pac.* **105**, 247–268.

Peterson BM 2001 Multiwavelength monitoring of active galactic nuclei. In *Probing the Physics of Active Galactic Nuclei* (eds Peterson BM, Pogge RW and Polidan RS), *Astron. Soc. Pac. Conf. Proc.* **224**, 1–10.

Petroski H 2006 *Success Through Failure: The Paradox of Design*. Princeton University Press.

Pinsonneault MH, Stauffer J, Soderblom DR, King JR and Hanson RB 1998 The problem of Hipparcos distances to open clusters. I. Constraints from multicolor main-sequence fitting. *Astrophys. J.* **504**, 170–191.

Poisson E and Will CM 1995 Gravitational waves from inspiraling compact binaries: parameter estimation using second-post-Newtonian waveforms. *Phys. Rev. D* **52**, 848–855.

Predehl P and Schmitt JHMM 1995 X-raying the interstellar medium: ROSAT observations of dust scattering halos. *Astron. Astrophys.* **293**, 889–905.

Predehl P, Burwitz V, Paerels F and Trümper J 2000 Chandra measurement of the geometrical distance to Cyg X-3 using its X-ray scattering halo. *Astron. Astrophys.* **357**, L25–L28.

Robichon N, Arenou F, Mermilliod J-C and Turon C 1999 Open clusters with Hipparcos. I. Mean astrometric parameters. *Astron. Astrophys.* **345**, 471–484.

Rolf DP 1983 Evidence for the detection of X-ray scattering from interstellar dust grains. *Nature* **302**, 46–48.

Romano P, Guidorzi C, Moretti A and Page KL 2006 GRB 061019: dust-scattered X-ray halo detected in Swift/XRT. *GRB Coord. Netw. Circ. Serv.* **5737**, 1.

Rosenberg H 1910 Über den Zusammenhang von Helligkeit und Spektraltypus in den Plejaden. *Astron. Nachr.* **186**, 71–78.

Rudak B and Mészáros P 1991 Dust from early galaxies and the X-ray background radiation. *Astrophys. J.* **371**, 29–35.

Santos MG, Amblard A, Pritchard J, Trac H, Cen R and Cooray A 2008 Cosmic reionization and the 21 cm signal: comparison between an analytical model and a simulation. *Astrophys. J.* **689**, 1–16.

Schutz BF 1986 Determining the Hubble constant from gravitational wave observations. *Nature* **323**, 310–311.

Schutz BF 2002 Lighthouses of gravitational wave astronomy. In *Lighthouses of the Universe: The Most Luminous Celestial Objects and Their Use for Cosmology* (eds Gilfanov M, Sunyaev R and Churazov E), pp. 207–224. Springer.

Scott D and Rees MJ 1990 The 21 cm line at high redshift: a diagnostic for the origin of large scale structure. *Mon. Not. R. Astron. Soc.* **247**, 510–516.

Shang C and Haiman Z 2011 Cosmology with standard sirens: the importance of the shape of the lensing magnification distribution. *Mon. Not. R. Astron. Soc.* **411**, 9–22.

Shao L and Dai ZG 2007 Behavior of X-ray dust scattering and implications for X-ray afterglows of gamma-ray bursts. *Astrophys. J.* **660**, 1319–1325.

Shao L, Dai ZG and Mirabal N 2008 Echo emission from dust scattering and X-ray afterglows of gamma-ray bursts. *Astrophys. J.* **675**, 507–518.

Shapiro C, Bacon DJ, Hendry M and Hoyle B 2010 Delensing gravitational wave standard sirens with shear and flexion maps. *Mon. Not. R. Astron. Soc.* **404**, 858–866.

Shapiro PR, Ahn K, Alvarez MA, Iliev IT, Martel H and Ryu D 2006 The 21 cm background from the cosmic dark ages: minihalos and the intergalactic medium before reionization. *Astrophys. J.* **646**, 681–690.

Simon P, Taylor AN and Hartlap J 2009 Unfolding the matter distribution using three-dimensional weak gravitational lensing. *Mon. Not. R. Astron. Soc.* **399**, 48–68.

Slysh VI 1969 Scattering of pulsar X-ray radiation by interstellar dust particles. *Nature* **224**, 159–160.

Soderblom DR, Nelan E, Benedict GF, McArthur B, Ramirez I, Spiesman W and Jones BF 2005 Confirmation of errors in Hipparcos parallaxes from Hubble Space Telescope Fine Guidance Sensor astrometry of the Pleiades. *Astron. J.* **129**, 1616–1624.

Southworth J, Maxted PFL and Smalley B 2005 Eclipsing binaries as standard candles. HD 23642 and the distance to the Pleiades. *Astron. Astrophys.* **429**, 645–655.

Stello D and Nissen PE 2001 The problem of the Pleiades distance. Constraints from Strömgren photometry of nearby field stars. *Astron. Astrophys.* **374**, 105–115.

Tiengo A and Mereghetti S 2006 Dust-scattered X-ray halos around gamma-ray bursts: GRB 031203 revisited and the new case of GRB 050713A. *Astron. Astrophys.* **449**, 203–209.

Tozzi P, Madau P, Meiksin A and Rees MJ 2000 Radio signatures of HI at high redshift: mapping the end of the 'Dark Ages'. *Astrophys. J.* **528**, 597–606.

Trümper J and Schönfelder V 1973 Distance determination of variable X-ray sources. *Astron. Astrophys.* **25**, 445–450.

Valenti E, Ferraro FR and Origlia L 2004 Red giant branch in near-infrared colour–magnitude diagrams. I. Calibration of photometric indices. *Mon. Not. R. Astron. Soc.* **351**, 1204–1214.

Valls-Gabaud D 2007 The distance to the Pleiades revisited. In *Binary Stars as Critical Tools and Tests in Contemporary Astrophysics* (eds Hartkopf WI, Guinan EF and Harmanec P), *Proc. Int'l Astron. Union Symp.* **240**, 281–289.

van Leeuwen F 1999 Hipparcos distance calibrations for 9 open clusters. *Astron. Astrophys.* **341**, L71–L74.

van Leeuwen F 2000 Parallaxes for open clusters using the Hipparcos intermediates astrometric data. In *Stellar Clusters and Associations: Convection, Rotation, and Dynamos* (eds Pallavicini R, Micela G and Sciortino S), *Astron. Soc. Pac. Conf. Ser.* **198**, 85–94.

van Leeuwen F 2005 Rights and wrongs of the Hipparcos data. A critical quality assessment of the Hipparcos catalogue. *Astron. Astrophys.* **439**, 805–822.

van Leeuwen F 2007a Hipparcos, the new reduction of the raw data. *Astrophys. Space Sci. Libr.* **350**.

van Leeuwen F 2007b Validation of the new Hipparcos reduction. *Astron. Astrophys.* **474**, 653–664.

van Leeuwen F 2009 Parallaxes and proper motions for 20 open clusters as based on the new Hipparcos catalogue. *Astron. Astrophys.* **497**, 209–242.

van Leeuwen F and Fantino E 2003 Dynamic modelling of the Hipparcos attitude. *Space Sci. Rev.* **108**, 537–576.

van Leeuwen F and Hansen-Ruiz CS 1997 The parallax of the Pleiades cluster. In *Proc. ESA Symp. 'Hipparcos – Venice '97'* (ed. Perryman MAC), *ESA Spec. Publ.* **402**, 689–692.

van Leeuwen F, Feast MW, Whitelock PA and Yudin B 1997 First results from Hipparcos trigonometrical parallaxes of Mira-type variables. *Mon. Not. R. Astron. Soc.* **287**, 955–960.

Vaughan S, Willingale R, O'Brien PT, Osborne JP, Reeves JN, Levan AJ, Watson MG, Tedds JA, Watson D, Santos-Lleó M, Rodríguez-Pascual PM and Schartel N 2004 The discovery of an evolving dust-scattered X-ray halo around GRB 031203. *Astrophys. J.* **603**, L5–L8.

Vaughan S, Willingale R, Romano P, Osborne JP, Goad MR, Beardmore AP, Burrows DN, Campana S, Chincarini G, Covino S, Moretti A, O'Brien PT, Page KL, Supper MA and Tagliaferri G 2006 The dust-scattered X-ray halo around Swift GRB 050724. *Astrophys. J.* **639**, 323–330.

Vecchio A 2004 LISA observations of rapidly spinning massive black hole binary systems. *Phys. Rev. D* **70**, 042001.

Vianello G, Tiengo A and Mereghetti S 2007 Dust-scattered X-ray halos around two Swift gamma-ray bursts: GRB 061019 and GRB 070129. *Astron. Astrophys.* **473**, 423–427.

Wang Y and Turner EL 1997 Cosmological constant and advanced gravitational wave detectors. *Phys. Rev. D* **56**, 724–729.

Weiss R 1972 Electromagetically coupled broadband gravitational antenna. *Q. Prog. Rep.* (Research Laboratory of Electronics, Massachusetts Institute of Technology), LIGO-P720002-01-R (available at http://docuserv.ligo.caltech.edu/docs/public/P/P720002-01/P720002-01.pdf).

Windhorst RA, Cohen SH, Jansen RA, Conselice C and Yan H 2006 How JWST can measure first light, reionization and galaxy assembly. *New Astron. Rev.* **50**, 113–120.

Wyithe JSB and Loeb A 2004 A characteristic size of $\sim$10 Mpc for the ionized bubbles at the end of cosmic reionization. *Nature* **432**, 194–196.
Wyithe JSB, Loeb A and Barnes DG 2005 Prospects for redshifted 21 cm observations of quasar HII regions. *Astrophys. J.* **634**, 715–727.
Xiang J, Lee JC and Nowak MA 2007 Using the X-ray dust scattering halo of 4U 1624-490 to determine distance and dust distributions. *Astrophys. J.* **660**, 1309–1318.
Xu J, Crotts APS and Kunkel WE 1995 A three-dimensional study using light echoes of the structure of the interstellar medium in front of SN 1987A. *Astrophys. J.* **451**, 806–815.
Zwahlen N, North P, Debernardi Y, Eyer L, Galland F, Groenewegen MAT and Hummel CA 2004 A purely geometric distance to the binary star Atlas, a member of the Pleiades. *Astron. Astrophys.* **425**, L45–L48.

Glossary

(Keywords in **boldface text** refer to the relevant entries elsewhere in this glossary.)

Absolute parallax **Trigonometric parallax** measured with respect to a fixed reference frame (e.g. the International Celestial Reference Frame, provided by extragalactic radio sources).

AGB dredge up Process operating in asymptotic giant branch (AGB) stars in which material from the stellar core region is mixed into their outer layers because of **thermal pulses**, thus changing the stellar composition.

AGB manqué stars Stars which, after core helium exhaustion on the horizontal branch, evolve to higher temperatures at much lower luminosities, without going through an asymptotic giant branch phase. They are often classified as 'sdO' types, i.e. helium-rich subdwarf O stars.

AI Velorum See δ **Scuti**.

Alcock–Paczyński effect Introduction of an artificial anisotropy in galaxy distributions, even in intrinsically isotropic distributions, caused by assuming the incorrect underlying cosmological model.

Angular diameter distance A **standard ruler** method relating an object's angular and linear sizes.

Angular power spectrum Often associated with the **cosmic microwave background (CMB) radiation**, where it refers to the amplitude of regular variations in the CMB temperature as a function of angular frequency.

Anomalous Cepheids Anomalous Cepheids occupy parts of parameter space also covered by other types of variables, but they are not easily explained on the basis of differences in stellar evolution. Their periods, 0.8–2 days, overlap with those of **RR Lyrae** stars, yet the shapes of their light curves are distinctly different from the **BL Herculis** or XX Virginis-type stars in the same period range: they are of lower amplitude and more symmetrical than typical **RR Lyrae** (and XX Virginis)-like light curves, although they are brighter.

Baade–Wesselink method Method to determine distances to **Cepheid** variables. It depends on the fact that stellar pulsation involves physical expansions and contractions, leading to both brightness changes (yielding a proxy for angular diameter changes) and a periodic variation in the star's radial velocity.

Baize–Romaní algorithm Method to separate the system mass and **dynamical parallax** of binary systems consisting of main-sequence stars, without having access to spectroscopic orbital parameters (e.g. radial velocities). It assumes that the component masses follow a generic mass–luminosity relation, which can be combined with a **bolometric**

An Introduction to Distance Measurement in Astronomy, First Edition. Richard de Grijs.

correction and extinction-corrected magnitudes to solve for the system mass and **dynamical parallax** simultaneously.

Barycentre Centre of mass; equilibrium position between two objects.

Baryon acoustic oscillations Overdensity or clustering of baryonic matter at certain length scales caused by acoustic waves which propagated in the early Universe.

Baryons Composite particles consisting of three quarks, including protons and neutrons.

Bayes' theorem Bayes' theorem expresses the conditional or 'posterior' probability of a given hypothesis (i.e. its probability after data has been obtained) in terms of its 'prior probability' as well as the prior probability of the data and the conditional probability of the data given the hypothesis. The theorem implies that data have a stronger confirming effect if they were more unlikely before being observed.

BL Herculis stars Short-period **Population II Cepheid** stars.

BL Lac(ertae) object Active galactic nucleus with an optical spectrum dominated by a featureless, nonthermal continuum (i.e. $F \propto \nu^{-\alpha}$, where F and ν are the flux and frequency, respectively) and characterized by rapid, large-amplitude intensity variations and significant optical polarization (named after the class prototype, BL Lacertae).

Blažhko effect Amplitude or phase modulations on timescales of typically tens to hundreds of days, occurring in 20–30% of fundamental-mode pulsating **RR Lyrae** stars.

Bolometric Refers to an object's radiation output across the entire electromagnetic spectrum.

Bolometric fluence Time-integrated bolometric flux. For **gamma-ray bursts**, the bolometric fluence is the flux integrated over the burst duration.

Bremsstrahlung 'Braking' or free–free radiation caused by accelerations of electrons due to other charged particles.

Broad-emission-line region Region close to the black hole in an active galactic nucleus or quasar characterized by emission-line widths of usually 500–1000 km s^{-1}, with broader wings of up to several thousand km s^{-1} on permitted lines.

Cataclysmic variables Stellar binary systems consisting of a white dwarf primary and a mass-transferring 'donor' secondary star which irregularly significantly increase in brightness (hence their original name **novae**), then drop down back to a quiescent state.

Cepheids The most massive class of variable stars, with masses ≥ 5 $M_\odot$, and located at the top of the classical **instability strip**.

Chandrasekhar mass limit The maximum nonrotating mass (~ 1.38 $M_\odot$) which can be supported against gravitational collapse by **electron degeneracy pressure**.

Chirp mass A binary black hole's redshifted mass, which defines the rate at which **inspiralling** occurs (thus determining the 'chirp' of the orbital frequency).

CNO cycle Carbon–nitrogen–oxygen or Bethe–Weizsäcker cycle: a set of fusion reactions by which stars convert hydrogen to helium, and the dominant source of energy in stars more massive than approximately 1.3 $M_\odot$.

Colour excess Difference between an object's observed and intrinsic colours.

Comoving distance (coordinates, time) Comoving quantities correct for the expansion of the Universe, so that they do not change as a function of time. Comoving distance is by definition equal to the **proper distance** at the present time.

Compton effect Inelastic collisions of photons in matter, resulting in an energy decrease of X- and γ-ray photons.

Conformal distance True, angle-preserving distance.

Cosmic infrared background Faint background radiation at infrared wavelengths released by the processes of unresolved star, galaxy and structure formation that have occurred since the decoupling of matter and radiation following the Big Bang.

Cosmic microwave background (CMB) radiation Faint background glow originating from the **surface of last scattering**, believed to be a relic of the Big Bang and characterized by a blackbody temperature of approximately 2.7 K.

Cosmic web Large-scale structure of the Universe, consisting of galaxy clusters and superclusters interspersed with regions referred to as 'voids'.

Cosmological constant Energy density of the vacuum.

Cosmological density parameter (Ω_M) Sum of the mass–energy densities, ρ, in matter, defined as the fraction of ρ in the Universe to the critical value, ρ_0, that will asymptotically bring the expansion of the Universe to a halt at an infinite time in the future.

Cosmological redshift (z) Scale of the Universe at a given distance with respect to current scales: because of the large-scale Hubble expansion, the observed wavelength, λ, of light from a distant object appears to be stretched to an observed wavelength $\lambda(1 + z)$.

CY Aquarius Prototype of a δ **Scuti** subclass of high-velocity, metal-poor stars, jointly with **SX Phoenicis** and **DY Pegasus**.

Dark ages The period between recombination of electrons with hydrogen and helium nuclei and the **epoch of recombination**.

Dark energy Type of energy which is characterized by a large, negative pressure.

δ Scuti stars Prototype (with **AI Velorum**) pulsating variables in the lower part of the **instability strip** (i.e. on the main sequence or subgiant branch), with (ultra)short periods of $\sim$80 min to a few hours and small amplitudes ($\Delta V < 0.3$ mag); also known as **dwarf Cepheids**.

Dispersion measure Delay of **pulsar** pulse time of arrival between two different frequencies caused by dispersive signal propagation through the partially ionized interstellar medium.

Distance duality relation Etherington's (1933) reciprocity relation, linking an object's luminosity distance (d_L) to its angular size distance (d_A), $d_L = (1 + z)^2 d_A$, where z represents the object's redshift.

Distance modulus Difference between apparent and absolute magnitude (m, M) of an astronomical object, which – in the absence of or after correction for extinction (denoted by the subscript '0') – is a measure of the object's distance, d: $(m - M)_0 = 5 \log d - 5$, where d is expressed in pc.

$D_n - \sigma$ relation 'Edge-on' projection of the **Fundamental Plane** of elliptical galaxies, where D_n is the diameter within which the **effective surface brightness** $I_{\rm eff} = 20.75\mu_B$ (in the B filter) and σ is the stellar velocity dispersion.

Double degenerate In the double-degenerate model, two carbon–oxygen binary white dwarfs merge and create a **super-Chandrasekhar white dwarf**, in which carbon fusion is ignited.

Downsizing Phenomenon in present-day galaxies where the stars in more massive galaxies tend to have formed earlier and over a shorter timescale than those in their lower-mass counterparts.

Dwarf Cepheids See δ **Scuti** stars.

Dynamical parallax Parallax angle for a visual binary star derived from the masses of its two components and the size and period of their orbit.

DY Pegasus See **CY Aquarius**.

Eclipsing binary star Binary star in which the orbital plane of both components is aligned with the observer's line of sight, resulting in mutual eclipses.

Eddington luminosity Distance from a star or black hole where the inward gravitational force balances the outward continuum radiation pressure, assuming **hydrostatic equilibrium** and spherical symmetry; also known as 'Eddington limit'.

Effective radius Radius containing 50% of the light or mass of an object.

Effective surface brightness Surface brightness within the **effective radius**.

Einstein's field equations Set of 10 equations in Einstein's theory of **general relativity** which describe the fundamental interaction of gravitation as a result of the curvature of **spacetime** by matter and energy.

Electron degeneracy pressure Pressure caused by compression of electron-degenerate matter (as in white dwarfs), which in turn increases the electrons' kinetic energy.

Electron–positron pair production Creation of an electron and its antiparticle, a positron, usually from conversion of a high-energy photon ($\gamma \rightarrow e^+ + e^-$) through Einstein's well-known equation $E = mc^2$ (*modulo* a relativistic factor), where E, m and c are the photon energy, the equivalent mass and the speed of light in a vacuum, respectively.

Epoch of recombination Transition to the period when the expansion of the Universe exceeded the rate of **Compton scattering** and protons and electrons 'recombined' to form neutral hydrogen. As a consequence, photons and baryonic matter became 'decoupled' and the Universe became transparent to electromagnetic radiation.

Equation of state (w) In cosmology, this refers to the ratio of the pressure (P) and density (ρ) of the **dark energy**, $w = P/(\rho c^2)$, where c is the speed of light in a vacuum.

Expanding cluster parallax See **moving groups method**.

Expanding photosphere method **Baade–Wesselink**-type technique relating the expansion velocity of a supernova's photosphere with the object's increasing (angular) radius.

Expansion parallax This technique measures the angular expansion of a given planetary nebula on the plane of the sky and compares the resulting angular velocity to the radial expansion along the line of sight. Assuming isotropic expansion, a distance can be derived.

Extinction A combination of absorption and scattering by dust (and gas), giving rise to redder observed colours. *Differential* extinction refers to variable extinction across the face of an object.

Extinction (attenuation) law Ratio of the extinction at a wavelength λ to that at a reference wavelength, often the V passband, in which case it is denoted as R_V. Also defined as the ratio of total to selective absorption as a function of wavelength.

Faber–Jackson relation Proportionality between the luminosity, L, of early-type galaxies and their central velocity dispersion, σ_0, as $L \propto \sigma_0^4$.

Fermat potential Mathematical surface which determines the increase in travel time for a light ray passing through a gravitational field.

Fermat's principle 'Principle of least time', which requires that a given photon traces a path that is a minimum, a maximum or a saddle point of the travel time.

Finger-of-God effect Elongation of galaxy clusters in redshift space caused by the gravitationally driven **peculiar velocities** of their member galaxies.

Flux-weighted gravity–luminosity relation Proportionality between the flux-weighted gravity ($g_F \equiv g/T_{\rm eff}^4$, where $\log g$ and $T_{\rm eff}$ are the stellar surface gravity and effective temperature, respectively) and absolute bolometric magnitude, $M_{\rm bol}$, of (O-,) B- and A-type blue supergiants.

Free–free radiation See **Bremsstrahlung**.

Friedmann equation Solution of **Einstein's field equations** of **general relativity** for the homogeneously and isotropically expanding **Friedmann–Lemaître–Robertson–Walker (FLRW)** Universe.

Friedmann–Lemaître–Robertson–Walker (FLRW) metric Exact solution of **Einstein's field equations** of **general relativity** for a homogeneously and isotropically expanding Universe. Also called the Standard Model of cosmology.

Fundamental Plane Three-dimensional plane defined by the **effective radius**, (central) velocity dispersion and **effective surface brightness** of elliptical galaxies.

Gaia European Space Agency Cornerstone mission (launch foreseen for 2012) which will use high-accuracy astrometry to create a highly accurate map of the 3D distribution (positions and space motions) of approximately 10^9 stars in the Milky Way and their basic physical properties (e.g. luminosity and chemical composition).

Gamma-ray burst Intense flashes of gamma rays lasting from milliseconds to minutes, the most energetic events known in the Universe.

General relativity Einstein's generalization of the theory of special relativity and Newton's law of universal gravitation, combining space and time into a single continuum.

Globular cluster luminosity function Number of globular clusters as a function of magnitude (or luminosity) for a given galaxy, characterized by a near-universal peak magnitude.

Gravitational wave Fluctuation in the curvature of **spacetime**, propagating outwards from its source in the form of a wave.

Great Attractor Gravitational overdensity at the centre of the Hydra–Centaurus supercluster of galaxies.

Hayashi track Protostellar excursion in the **Hertzsprung–Russell diagram** after the protostellar cloud has reached approximate **hydrostatic equilibrium**.

Heisenberg uncertainty principle Quantum-mechanical inequality that prevents one from knowing both position and momentum to arbitrarily high precision.

Helium flash Runaway fusion of helium in a degenerate state, where it is supported against gravity by quantum-mechanical rather than thermal pressure.

Henyey track Path traced by >0.5 $M_\odot$ **pre-main-sequence stars** in the **Hertzsprung–Russell diagram** following the **Hayashi track**.

Hertzsprung–Russell diagram Graph defining the relationship between stellar luminosities (or absolute magnitudes) and spectral types (or, equivalently, spectral classifications, temperatures or colours).

Hipparcos Pioneering European Space Agency space astrometry mission (1989–1993) which recorded the positions of more than 100 000 stars with high and more than 10^6 stars with lesser precision.

H II regions Low-density ionized gas cloud featuring current and recent active star formation.

Hubble bubble Postulated local void characterized by an increased outflow of galaxies of $\sim$5% in the range $0.01 < z < 0.023$.

Hubble constant (H_0) Expansion rate of the Universe at the present time, relating a galaxy's galactocentric recessional velocity, v, to its distance, d, through $v = H_0 d$. It is usually expressed in units of km s^{-1} Mpc^{-1}.

Hubble diagram Diagram relating (galactic) distances to their radial velocities or, alternatively, providing a magnitude–redshift relation.

Hubble flow Large-scale expansion of the Universe following Hubble's law (see **Hubble constant**).

Hubble law See **Hubble constant**.

***Hubble (Space Telescope)* Key Project** Dedicated *Hubble Space Telescope*-based project to determine the **Hubble constant** with an accuracy of $\pm 10\%$ using systematic observations of **Cepheid** variable stars in several carefully selected galaxies.

Hubble time (H_0^{-1}) Inverse of the **Hubble constant**, to first order corresponding to the current age of the Universe.

Hydrostatic equilibrium Balance reached when gravitational compression in a stellar interior is equal to an outward pressure-gradient force (thermal pressure).

Hyperfine transition Atomic transition between split energy levels caused by a change in electron spin or an electron's orbital angular momentum.

Initial mass function Distribution of stellar masses at the time of star formation.

Instability strip Region in the **Hertzsprung–Russell diagram** in which pulsating variable stars occur.

Inspiraling binary black holes Black holes which are widely separated and slowly spiral together because of the back-reaction caused by **gravitational wave** emission.

Integrated Sachs–Wolfe effect Effect caused by gravitational redshifting: stronger gravitational potentials cause electromagnetic radiation to shift towards longer wavelengths.

Inverse Compton scattering Process where photons gain energy through collisions with electrons.

Isochrone Snapshot of a theoretical **simple stellar population** at a given time.

***K* correction** Redshift-dependent correction of an object's magnitude to its rest-frame equivalent.

κ mechanism **Cepheid** pulsation mechanism as heat engine, based on the opacity of helium. The ionized gas in the star's outer layers is opaque at the faintest part of the pulsation cycle. This leads to a temperature increase because of stellar radiation, hence leading to expansion. The expanding outer atmosphere enables the star to cool, and thus become less ionized and hence less opaque, which in turn allows radiation to escape. Gravitational attraction by the star's interior mass causes the process to stop and reverse, eventually leading to cyclic behaviour.

κ space Alternative three-dimensional 'fundamental plane' defining the properties of elliptical galaxies, with 'edge-on' projections driven by physical parameters.

Kaiser effect Effect caused by coherent, infalling motions of galaxies during cluster assembly, resulting in an apparent flattening of the cluster structure in redshift space (along the line of sight).

Laplace equation Differential equation describing the gravitational potential associated with a set of point masses, $\nabla^2 \phi = 0$, where ∇^2 is the Laplace operator and ϕ a scalar function.

Large Synoptic Survey Telescope (LSST) Planned 8.4 m ground-based telescope (based at Cerro Pachón in Chile) equipped with a 3200 megapixel camera that will produce an unprecedented wide-field astronomical survey of the southern sky.

Λ**CDM** Cold dark matter cosmological model which includes the presence of a **cosmological constant**, Λ.

Line blanketing Change in the temperature structure of stellar atmospheres caused by the opacities of a large number of emission lines, in particular at shorter wavelengths.

Look-back time Light travel time.

Luminosity distance A **standard candle** method relating an object's observed and intrinsic luminosities.

Lutz–Kelker bias Systematic bias that causes measured stellar parallaxes to be larger than their actual values.

Lyman-α forest Ensemble of absorption lines caused by the Lyman-α transition of neutral hydrogen (at a rest-frame wavelength of 1216 Å) in the spectra of high-z galaxies and quasars.

Main-sequence fitting Matching of the main defining features of an observed cluster **Hertzsprung–Russell diagram** at an unknown distance either to their counterparts in a well-calibrated reference diagram or to theoretical **isochrones**.

Malmquist bias Selection effect affecting samples of objects that are flux limited, causing an apparent increase in average luminosity with distance. Malmquist biases of the first and second kind refer to the magnitude- and distance-dependent varieties, respectively.

Masers Microwave amplification by stimulated emission of radiation: the microwave analogues of lasers operating at optical and near-infrared wavelengths.

Mass quadrupole The quadrupole moment, $Q \sim$ (source mass)(source size)^2, is in essence a measure of how far from spherically symmetric a radiating source is.

Metallicity Heavy-element content. In practice, this refers to either iron content or the abundance of elements heavier than hydrogen and helium.

Mira variables Stars in their late evolutionary phases that occupy the classical **instability strip** where it crosses the asymptotic giant branch.

Monte Carlo experiments Class of computational algorithms which rely on repeated random or pseudo-random sampling.

Moving groups method Method of distance determination to a cluster of stars with coherent proper motions, which are all moving towards a common convergence point. Identification of the latter provides sufficient detail to translate each star's apparent motion into a real space velocity. The distance to a star in the cluster can be deduced by comparing its actual speed with how fast it appears to be moving.

Novae See **cataclysmic variables**.

OH forest Region of the near-infrared spectrum (longwards of roughly 6700 Å) which is highly contaminated by **telluric lines**.

Optical depth, τ Measure of transparency (i.e. the fraction of radiation which is not scattered or absorbed along a given path length): $I(\tau) = I_0 e^{-\tau}$, where I_0 and I are the intensity at source and the observed intensity, respectively.

Ostriker–Vishniac effect Kinetic **Sunyaev–Zel'dovich effect**, a Doppler-type effect.

Pair-instability supernovae Very massive stars (with initial masses $\geq$140 $M_\odot$) will attain very high core temperatures, so that photons spontaneously form electron–positron pairs. Because of the subsequent reduction of the photon pressure supporting the star's outer layers, a collapse followed by ignition of their oxygen core is triggered which vaporizes the star.

Pan-STARRS Panoramic Survey Telescope and Rapid Response System, of which the first telescope is based at Haleakala Observatory on Maui, Hawai'i (USA).

Parsec An annual **trigonometric parallax** angle of one arcsecond corresponds to a distance of 1 pc (3.086×10^{16} m or 3.26 lightyears).

Pauli exclusion principle Quantum-mechanical principle affecting electrons, protons and neutrons which implies that no two electrons can have the same four quantum numbers (including the electron energy level or shell, n, the subshell, l, the specific orbital, m_l, and the electron spin, m_s).

P Cygni profile Feature in stellar spectra in which the presence of both absorption and emission in the spectral line profile indicates the existence of a gaseous outflow.

Peculiar velocity Random deviation (of several hundred to up to 1000 km s^{-1}) from a galaxy's general recessional velocity (the large-scale **Hubble flow**) caused by locally dominant gravitational attractive forces.

Period–luminosity(–colour) relation Representation of variable star behaviour: more luminous stars are expected to have longer pulsation periods, which is somewhat dependent on **metallicity** (hence introducing a colour effect).

Photometric parallax Equivalent to the **spectroscopic parallax**, but using colours instead of spectral features to estimate stellar luminosity classes.

Photometric redshift Redshift determination based on observations in (usually) broad-band filters, which relies on assumptions made about an object's spectral features, in particular about the presence of clearly identifiable characteristics which move across filters as a function of redshift.

Planetary nebulae luminosity function Number of planetary nebulae as a function of magnitude, usually given in a narrow passband centred on their strong [O III]λ5007 Å emission line and characterized by a universal, sharp cut-off at bright magnitudes.

Population I Originally defined by Walter Baade as the stars in the solar neighbourhood, with typical **Hertzsprung–Russell diagrams** similar to those of open star clusters and an overall blue colour index, because the brightest stars are blue and, hence, young: blue main-sequence stars and blue (super)giants are always massive stars and, therefore, short-lived. These are mostly found in spiral discs.

Population II Originally defined by Walter Baade as the stars of globular clusters and the bulge of M31, with typical **Hertzsprung–Russell diagrams** similar to those of globular clusters and an overall red colour index, because the brightest stars are red giants and, hence, old.

Pre-main-sequence stars Newly forming stars undergoing gravitational contraction (lasting less than 1% of their total lifetimes) which have not yet reached the **zero-age main sequence**, following a phase as protostars. Low-mass ($< 2\ M_\odot$) pre-main-sequence stars include **T Tauri** and FU Orionis stars, while pre-main-sequence stars with masses between 2 and 8 $M_\odot$ include Herbig Ae/Be stars.

Prior probability (prior) Probability distribution in **Bayesian** statistics which represents the uncertainty about a quantity or hypothesis before the evidence (data) is taken into account.

Proper distance Distance to an astronomical object at a given time during the cosmic expansion of the Universe.

Proton–proton chain Set of fusion reactions by which stars convert hydrogen to helium in $\lesssim 1\ M_\odot$ stars.

Pulsars Highly magnetized, rotating neutron stars that emit a beam of electromagnetic radiation.

Pulsation parallax See **Baade–Wesselink method**.

Pulse timing parallax Parallax measurement based on quantification of the curvature of an arriving **pulsar** wavefront.

Quasar parallax Proposed inverted **trigonometric parallax** approach to determine distances to high-redshift quasars using the size of their **broad-emission-line regions** as **standard ruler**.

Rayleigh–Taylor instability Instability of an interface between two fluids of different densities, which occurs when the lighter fluid pushes its heavier counterpart.

Redshift See **cosmological redshift**.

Reverberation mapping Determination of the linear sizes of quasar/active galactic nucleus **broad-emission-line regions** and the mass of the central supermassive black holes based on light travel time measurements.

Riemannian geometry Riemannian metrics describe a broad range of standard geometries, such as non-Euclidean and Euclidean, spherical and hyperbolical geometries, as well as metrics with properties which vary from point to point.

Rotational parallax Measurement of rotation curves of galaxies in the Local Group based on high-precision astrometry, thus allowing an independent, local calibration of the **Tully–Fisher relation**.

RR Lyrae Pulsating, old, low-mass stars with periods from 0.2 to 0.8–0.9 days, occupying the **instability strip** where it crosses the horizontal branch.

RV Taurus stars Long-period **Population II Cepheid** stars.

s-process Slow neutron capture stellar nucleosynthetic process that occurs at relatively low neutron density ($10^5 - 10^{11}$ neutrons cm^{-2} s^{-1}) and intermediate temperature (typical of asymptotic giant branch stars).

Schwarzschild radius Radius at which, if all the mass were compressed inside of that distance, the escape speed would equal the speed of light.

Second-parameter effect This refers to the wide variety of horizontal branch morphologies in globular clusters. While metallicity differences can account for some of the colour spread observed along the horizontal branch, a second parameter (probably age) is required for a more satisfactory description.

Secular parallax Statistical approach to measuring the **trigonometric parallax**, enhanced by employing the Sun's motion through the Milky Way to gain a longer baseline.

Sgr A* Nonthermal radio source often associated with the Galactic Centre, believed to be located within 1 pc of the central supermassive black hole.

Shapiro time-delay effect Slowing down of light rays in the presence of a strong gravitational field.

Shklovsky method Statistical distance scale to planetary nebulae. It requires knowledge of a planetary nebula's absolute Hβ flux and angular size, and is based on the assumptions that every planetary nebula has the same ionized mass, is optically thin and can be characterized by a constant electron density and filling factor.

Simple stellar population 'Simple' stellar populations are based on the idea that any **stellar population** formed in one or more 'single bursts', which implies that all stars within the population have the same age and **metallicity** (since they formed from the same gas cloud at the same time). The only variable, for any given population, is the stellar **initial mass function**.

Sloan Digital Sky Survey (sdss) Multipassband imaging and spectroscopic redshift survey using the 2.5 m telescope at Apache Point Observatory (New Mexico, USA). The third phase (sdss iii) has been running since 2008 and is expected to conclude in 2014.

Sound horizon Horizon at the time of recombination (the **epoch of recombination**), which is hence a fixed physical scale at the **surface of last scattering**.

Spacetime Mathematical model which combines space and time into a single continuum.

Spaenhauer diagram Diagnostic diagram used to detect selection biases in observational samples, showing the sample objects' logarithmic distances as a function of their absolute magnitudes. In such a diagram, objects of a given, constant apparent magnitude occupy a straight line.

Spectroscopic parallax Method which relies on spectral classification of target stars, based on measurements of their spectral line widths, combined with their position in the **Hertzsprung–Russell diagram**, which allows determination of their intrinsic luminosities and, hence, absolute magnitudes. The **distance modulus** resulting from comparison with their apparent magnitudes enables distance derivation.

Spherical harmonics Solutions to the **Laplace equation**.

Square Kilometre Array Ambitious next-generation radio telescope, consisting of up to a few hundred individual 'stations' (antennas), with a total collecting area of one million m^2.

SRa variables Semi-regular type 'a' variables are very similar to **Mira variables**, yet have amplitudes $\Delta V < 2.5$ mag.

Standard candle Object of known luminosity and, hence, absolute magnitude.

Standard ruler Object of known linear size.

Statistical parallax See **secular parallax**.

Stellar population synthesis Method to build up models of star clusters or galaxies based on the idea that any **stellar population** formed in one or more 'single bursts'.

Stellar populations Stellar samples jointly defined by their dominant age (i.e. time since the last episode of star formation), **metallicity** and kinematics (which describes whether they exhibit ordered or chaotic motions in e.g. a disc-like or spheroidal configuration).

Stretching–luminosity correlation Relationship between the width of a supernova light curve and its luminosity.

Subdwarf fitting See **main-sequence fitting**.

Sunyaev–Zel'dovich (SZ) effect Distortion of the **cosmic microwave background (CMB) radiation** – allowing detection of density perturbations in the early Universe – through **inverse Compton scattering** by high-energy electrons, where the low-energy CMB photons gain energy through collisions.

Super-Chandrasekhar white dwarf White dwarf with a mass in excess of the **Chandrasekhar mass limit**, possibly formed through **double-degenerate** evolution.

Supernova light echoes Reflection of a supernova-generated flash of light off scattering interstellar dust particles.

Supernova Type Ia Exploding, collapsing carbon–oxygen white dwarf that has accreted sufficient matter to overcome its plasma's **electron degeneracy pressure**, resulting in rapid ignition of carbon fusion in most of the stellar core.

Supernova Type II Type of **cataclysmic-variable** massive ($\gtrsim 8\ M_\odot$) star that undergoes core collapse and a subsequent violent explosion.

Surface of last scattering Horizon at redshift $z \sim 1089$ where matter and radiation decoupled shortly after the Big Bang.

Surface brightness fluctuations Technique of distance determination to galaxies with unresolved stars, taking advantage of the discreteness of the stellar composition.

SX Phoenicis stars See **CY Aquarius**.

T Tauri stars Rapidly rotating, highly variable **pre-main-sequence stars** – including the youngest visible F-, G-, K- and M-type stars – with masses of $<2\ M_\odot$.

Telluric spectral line Spectral line caused by contamination by the Earth's atmosphere.

Thermal pulsations Thermal pulses occur when asymptotic giant branch stars run out of helium fuel. They derive their energy from fusion of hydrogen in a thin shell surrounding the inert helium shell, but the latter switches back on occasionally, hence leading to thermal pulsations.

Thomson scattering Elastic scattering of electromagnetic radiation by freely moving charged particles.

Time dilation Change of apparent time due to the effects of gravity, as described by Einstein's theory of **general relativity**.

Trigonometric parallax Half the angular size of the apparent ellipse on the sky traced by (nearby) stars during the course of a year owing to the Earth's orbital motion around the Sun. See also **absolute parallax**.

Triple-α process Set of nuclear fusion reactions at temperatures of $(1-3) \times 10^8$ K by which three ^{4}He nuclei (α particles) are transformed into carbon in stars with masses $>0.5\ M_\odot$.

Tully–Fisher relation Distance indicator for spiral galaxies based on the premise that the intrinsic luminosity of a galaxy, L, should increase with the amplitude of the mean orbital velocities, v_c, as $L \propto v_c^4$, provided the product of the galaxy's **(effective) surface brightness** and its mass-to-light ratio is constant.

21 cm tomography Three-dimensional mapping of redshifted neutral hydrogen.

Unbiased plateau Region in the **Spaenhauer diagram** that is unaffected by **Malmquist bias**.

Very Long Baseline Interferometry Radio interferometric technique using the longest attainable baseline, often spanning multiple countries and sometimes even including space-based radio observatories.

Virial theorem The virial theorem implies that the average speed with which test masses (e.g. stars in a galaxy) orbit the system's centre increases with the system's mass, $M_{\rm virial}$, as $M_{\rm virial} = \eta(R_{\rm eff}\sigma_0^2)/G$, where $R_{\rm eff}$ is the system's **effective radius**, σ_0 its central velocity dispersion, G the Newtonian gravitational constant and η a proportionality constant.

W Ursae Majoris contact binary systems Type of **eclipsing binary stars**, where the surfaces of both components are in contact with each other.

W Virginis stars **Population II Cepheids**. Prototypes include **BL Herculis** and **RV Taurus** stars.

Wave strain A measure of the fractional size distortion caused by a passing **gravitational wave**.

Wesenheit relation Reddening-free **period–luminosity relation**.

White dwarf cooling sequence Path in the **Hertzsprung–Russell diagram** traced by white dwarfs as they slowly become fainter and redder when cooling down.

Wielen dip Plateau (or shallower slope) in the stellar luminosity function between approximately $M_V = 6$ and 9 mag.

Wilkinson Microwave Anisotropy Probe (WMAP) NASA (National Aeronautics and Space Administration) satellite used for measuring minuscule differences (< 0.0002 K) in the **cosmic microwave background radiation**, launched in 2001.

Wilson–Bappu effect The tight correlation between the width of the emission core in (usually) the Ca II K absorption line with absolute magnitude for late-type, cool stars.

Zero-age main sequence Locus in the **Hertzsprung–Russell diagram** where stellar luminosity is a function of the star's mass (and where stars are in **hydrostatic equilibrium** and spend most of their lifetimes), reached by **pre-main-sequence stars** at the onset of hydrogen burning (thermonuclear fusion) in the stellar core (for stellar masses >0.08 $M_\odot$, depending on **metallicity**).

Figure Credits

Chapter 1

Figure 1.1 *(Left)* Shapley H 1918 *Astrophys. J.* **48**, 154–181.
Figure 1.2 Trumpler RJ 1930 *Lick Obs. Bull.* **XIV**, 154–188.
Figure 1.3 Trumpler RJ 1930 *Lick Obs. Bull.* **XIV**, 154–188.
Figure 1.4 National Radio Astronomy Observatory/Associated Universities, Inc./National Science Foundation, Yusef-Zadeh *et al.* 2004.
Figure 1.5 Reid MJ *et al.* 2009 *Astrophys. J.* **705**, 1548–1553.
Figure 1.6 Gillessen S *et al.* 2009 *Astrophys. J.* **707**, L114–L117.
Figure 1.7 Růžička A *et al.* 2009 *Astrophys. J.* **691**, 1807–1815.
Figure 1.8 Schaefer BE 2008 *Astron. J.* **135**, 112–119.
Figure 1.9 Sandstrom KM *et al.* 2007 *Astrophys. J.* **667**, 1161–1169.
Figure 1.10 Schweizer F *et al.* 2008 *Astron. J.* **136**, 1482–1489.
Figure 1.11 Blakeslee JP *et al.* 2009 *Astrophys. J.* **694**, 556–572.
Figure 1.12 Riess AG *et al.* 2009 *Astrophys. J.* **699**, 539–563.

Chapter 2

Figure 2.1 European Space Agency.
Figure 2.2 Michael Perryman/Wikimedia Commons, licensed under the Creative Commons Attribution–Share Alike 3.0 Unported license.
Figure 2.3 Lindegren L 2010 *Proc. Int'l Astron. Union Symp.* **261**, 296–305.
Figure 2.4 European Space Agency.
Figure 2.5 Graphics: Milde Science Communication; Hubble Space Telescope image: NASA/ESA, and the Hubble Heritage Team; radio images: A. Brunthaler, Max-Planck-Institut für Radioastronomie, Germany.
Figure 2.6 Dravins D *et al.* 1997 In *Proc. ESA Symp. 'Hipparcos – Venice '97'* (ed. Perryman MAC), *ESA Spec. Publ.* **402**, 733–738.
Figure 2.7 Kraus S *et al.* 2007 *Astron. Astrophys.* **466**, 649–659.
Figure 2.8 Boden A and Quirrenbach A 2008 *Proc. Int'l Astron. Union Symp.* **248**, 36–43.
Figure 2.9 de Grijs R *et al.* 2002 *Mon. Not. R. Astron. Soc.* **331**, 245–258.
Figure 2.10 de Grijs R *et al.* 2002 *Mon. Not. R. Astron. Soc.* **331**, 245–258.
Figure 2.11 Wikimedia Commons, licensed under the Creative Commons Attribution–Share Alike 3.0 Unported license.

An Introduction to Distance Measurement in Astronomy, First Edition. Richard de Grijs.

Figure 2.12 V. Belokurov and the Sloan Digital Sky Survey.
Figure 2.13 Jurić M *et al.* 2008 *Astrophys. J.* **673**, 864–914.

Chapter 3

Figure 3.1 Commonwealth Scientific and Industrial Research Organisation, Astronomy and Space Science: http://outreach.atnf.csiro.au/education/senior/astrophysics/stellarevolution_formation.html.
Figure 3.2 Wikimedia Commons, licensed under the Creative Commons Attribution–Share Alike 3.0 Unported license.
Figure 3.3 Baade W 1944 *Astrophys. J.* **100**, 137–146.
Figure 3.4 Majewski SR 1999 Stellar populations and the formation of the Milky Way. In: *Globular clusters*, 10th Canary Islands Winter School of Astrophysics (eds Martínez Roger C, Pérez Fournón I and Sánchez F), pp. 43–107. Cambridge University Press.
Figure 3.5 Stanek KZ and Garnavich PM 1998 *Astrophys. J.* **503**, L131–L134.
Figure 3.6 Grocholski AJ and Sarajedini A 2002 *Astron. J.* **123**, 1603–1612.
Figure 3.7 Mager VA *et al.* 2008 *Astrophys. J.* **689**, 721–731.
Figure 3.8 Cho D-H and Lee S-G 2002 *Astron. J.* **124**, 977–988.
Figure 3.9 Kudritzki RP *et al.* 2003 *Astrophys. J.* **582**, L83–L86.
Figure 3.10 Renzini A *et al.* 1996 *Astrophys. J.* **465**, L23–L26.
Figure 3.12 European Southern Observatory.
Figure 3.13 Bono G *et al.* 2010 *Astrophys. J.* **715**, 277–291.
Figure 3.14 Madore BF *et al.* 2009 *Astrophys. J.* **695**, 988–995.
Figure 3.15 Wood PR *et al.* 1999 *Proc. Int'l Astron. Union Symp.* **191**, 151–158.
Figure 3.16 Gingold RA 1985 *Mem. Soc. Astron. It.* **56**, 169–191.
Figure 3.17 Clementini G *et al.* 2003 *Astron. J.* **125**, 1309–1329.
Figure 3.18 Della Valle M and Livio M 1995 *Astrophys. J.* **452**, 704–709.
Figure 3.19 Santander-García M *et al.* 2008 *Astron. Astrophys.* **485**, 117–126.
Figure 3.20 NASA, ESA and The Hubble Heritage Team (STScI/AURA).
Figure 3.21 P. Marenfeld and National Optical Astronomy Observatory/Association of Universities for Research in Astronomy/National Science Foundation.
Figure 3.22 NASA/ESA and A. Feild (STScI).
Figure 3.23 Cordes JM and Lazio TJW 2002 Unpublished (astro-ph/0207156v3).

Chapter 4

Figure 4.1 Freedman WL *et al.* 2001 *Astrophys. J.* **553**, 47–72.
Figure 4.2 Dunn LP and Jerjen H 2006 *Astron. J.* **132**, 1384–1395.
Figure 4.3 Jordán A *et al.* 2007 *Astrophys. J. Suppl. Ser.* **171**, 101–145.
Figure 4.4 Jordán A *et al.* 2006 *Astrophys. J.* **651**, L25–L28.
Figure 4.5 Gnedin OY and Ostriker JP 1997 *Astrophys. J.* **474**, 223–255.
Figure 4.6 Kundu A *et al.* 1999 *Astrophys. J.* **513**, 733–751.
Figure 4.7 Ciardullo R 2002 *Astrophys. J.* **577**, 31–50.
Figure 4.8 de Grijs R and Peletier RF 1999 *Mon. Not. R. Astron. Soc.* **310**, 157–167.

Figure 4.9 Rhee MH 2004 *J. Kor. Astron. Soc.* **37**, 15–39.
Figure 4.10 Allanson SP *et al.* 2009 *Astrophys. J.* **702**, 1275–1296.
Figure 4.11 Kelson DD *et al.* 2000 *Astrophys. J.* **529**, 768–785.
Figure 4.12 Bender R *et al.* 1992 *Astrophys. J.* **399**, 462–477.
Figure 4.13 Peletier RF and de Grijs R 1999 *Mon. Not. R. Astron. Soc.* **300**, L3–L6.
Figure 4.14 de Grijs R and Peletier RF 1999 *Mon. Not. R. Astron. Soc.* **310**, 157–167.
Figure 4.15 Hogg DW *et al.* 2004 *Astrophys. J.* **601**, L29–L32.

Chapter 5

Figure 5.1 M. Blanton and the Sloan Digital Sky Survey.
Figure 5.3 Perlmutter S *et al.* 1999 *Astrophys. J.* **517**, 565–586.
Figure 5.4 Wheeler JC and Harkness RP 1990 *Rep. Prog. Phys.* **53**, 1467–1557. doi: 10.1088/0034-4885/53/12/001.
Figure 5.5 Kasen D, University of California at Berkeley/Lawrence Berkeley National Laboratory, USA.
Figure 5.6 Goldhaber G *et al.* 2001 *Astrophys. J.* **558**, 359–368.
Figure 5.7 Hamuy M and Pinto PA 2002 *Astrophys. J.* **566**, L63–L65.
Figure 5.8 Bartel N *et al.* 2007 *Astrophys. J.* **668**, 924–940.
Figure 5.9 Schaefer BE 2007 *Astrophys. J.* **660**, 16–46.
Figure 5.11 Suyu SH *et al.* 2010 *Astrophys. J.* **711**, 201–221.
Figure 5.12 Bonamente M *et al.* 2006 *Astrophys. J.* **647**, 25–54.
Figure 5.13 NASA/WMAP Science Team.

Chapter 6

Figure 6.1 de Grijs R *et al.* 2005 *Mon. Not. R. Astron. Soc.* **359**, 874–894.
Figure 6.2 Maíz Apellániz J 2005 In *The Three-Dimensional Universe with Gaia* (eds Turon C, O'Flaherty KS and Perryman MAC), *ESA Spec. Publ.* **576**, 179–182.
Figure 6.3 Butkevich AG *et al.* 2005 *Mon. Not. R. Astron. Soc.* **362**, 321–330.

Chapter 7

Figure 7.1 van Leeuwen F 2009 *Astron. Astrophys.* **497**, 209–242.
Figure 7.3 Vianello G *et al.* 2007 *Astron. Astrophys.* **473**, 423–427.
Figure 7.4 Kokkotas KD 2002 Gravitational wave physics. In *Encyclopedia of Physical Science and Technology* (ed. Meyers RA), 3rd edn, vol. 7, pp. 67–85. Copyright 2004, reproduced with permission from Elsevier.
Figure 7.5 Mao Y *et al.* 2008 *Phys. Rev. D* **78**, 023529.
Figure 7.7 Ciardullo R 2006 In *Planetary Nebulae Beyond the Milky Way* (eds Stanghellini L, Walsh JR and Douglas N), *ESO Astrophys. Symp.*, pp. 79–90 (reproduced with permission of Springer Science+Business Media), with design elements reproduced with permission of Stuart Robbins: http://jtgnew.sjrdesign.net/extras_foundations_distanceladder.html (accessed 20 January 2011).

Index

(Page numbers in *italics* refer to figures on the relevant pages)

An Introduction to Distance Measurement in Astronomy, First Edition. Richard de Grijs.